CLIFFORD, TOM NEVILLE
AFRICAN MAGMATISM AND TECTONIC

D1327638

AFRICAN MAGMATISM AND TECTONICS

A

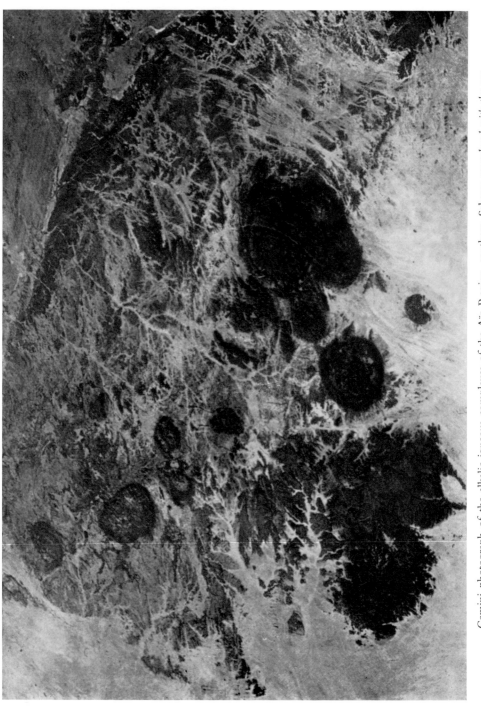

Gemini photograph of the alkalic igneous complexes of the Aïr Region, southern Sahara; *reproduced with the permission of N.A.S.A., U.S.A.* The inset area and regional setting are shown on Figs. 7 and 2 respectively of Chapter

. . . by Black and Girod

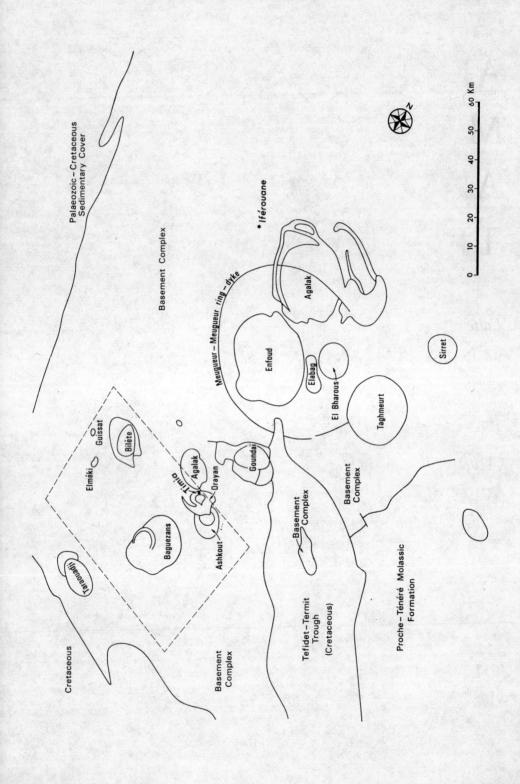

Palaeozoic – Cretaceous
Sedimentary Cover

Basement Complex

•Iférouane

Meugueur – Meugueur ring – dyke

Enfoud

Agalak

Elabag

El Bharous

Taghmeurt

Sirret

Guissat

Elméki

Bilète

Baguezans

Timia

Agalak

Orayan

Goundaï

Ashkout

Basement
Complex

Basement
Complex

Taraouadji

Cretaceous

Basement
Complex

Tefidet – Termit
Trough
(Cretaceous)

Proche – Ténéré Molassic
Formation

N

| 0 | 10 | 20 | 30 | 40 | 50 | 60 Km |

AFRICAN MAGMATISM AND TECTONICS

Edited by
T. N. CLIFFORD and I. G. GASS
University of Leeds

A volume in honour of
W. Q. KENNEDY FRS

OLIVER & BOYD · EDINBURGH

OLIVER & BOYD
Tweeddale Court Edinburgh EH1 1YL
(A Division of Longman Group Ltd)
ISBN 0 05 001709 8

First published 1970
© The Authors 1970
All rights reserved

Set in Monotype Bembo 270 and printed in
Great Britain by T. & A. Constable Ltd, Edinburgh

CONTENTS

v

PREFACE

Perhaps because of the enormous size of Africa and the still limited, though rapidly growing, coverage by detailed geological mapping there are remarkably few books summarising specific elements of continental geology. In this volume a number of studies have been carefully selected to illustrate the relationship between two important geological processes—magmatism and tectonics. Most of the contributions are presented chronologically and their time relationships are outlined in Chapter 1. However, in the final part of the volume a number of more general papers deal with the composition of African acidic igneous rocks, the structural environments of kimberlites and other irruptives, and convective processes. We feel that Africa is particularly well suited to this kind of treatment which is directed towards certain aspects of continental history rather than attempting a rigorous comprehensive coverage. Indeed, we are hopeful that this volume may influence others to consider analogous works on structure, sedimentation and mineralisation, and we shall feel gratified indeed if this becomes the first of a series of Earth Science books dealing with Africa.

This volume is dedicated to Professor W. Q. Kennedy, F.R.S., by a number of his students, friends and colleagues, as a token of their esteem and affection. By its very title, it illustrates the three fields in which his influence as a teacher and researcher has had its greatest scientific impact. Nevertheless, it was no easy task to select a limited number of authors from the wide circle of Professor Kennedy's colleagues and former students. The choice of contributors was ours alone and we were influenced by the following factors: that the contributor should have worked on problems concerning tectonics and magmatism in Africa; that works of synthesis or of regional implication would be particularly valuable; and that a wide temporal and spatial spread of interest was desirable. It is a matter of some satisfaction to us that of the twenty-two persons invited to contribute, all accepted; this is a clear and tangible reflection of the high esteem in which Professor Kennedy is held.

Throughout the preparation of this volume we have received valuable assistance from members of the Department of Earth Sciences at the University of Leeds, and we are grateful to Professor R. M. Shackleton for placing the clerical, secretarial and draughting facilities of the Department at our disposal. We are particularly indebted to Dr Dorothy H. Rayner who unstintingly placed her wide editorial experience at our disposal; Mr R. C. Boud for draughting many of the text figures; Miss Pamela C. Bennion, Mrs Vivienne P. Rex and Mrs Marjorie Ward for a prompt and efficient clerical service; Dr Joan M. Rooke and Mrs E. R. Nutt for their assistance in preparing the subject and author indexes; and Mr T. F. Johnston and numerous other colleagues who have given help and encouragement during the editing of the manuscripts. Finally we would thank the contributors for their promptness in dealing with editorial requests and for their tolerance throughout.

T. N. CLIFFORD
I. G. GASS

March, 1969

LIST OF CONTRIBUTORS

BICHAN, R., Robertson Research Company Limited, Llanddulas, Abergele, Denbighshire, U.K.

BLACK, R., Laboratoire Associé de Géochronologie (C.N.R.S.), Departement de Géologie et Minéralogie, Faculté des Sciences, 5 Rue Kessler, Clermont-Ferrand, France.

BLOOMFIELD, K., Geological Survey and Mines Department, P.O. Box 9, Entebbe, Uganda.

CAHEN, L., Musée Royal de l'Afrique Centrale, Tervuren, Belgium.

CLIFFORD, T. N., Department of Earth Sciences, University of Leeds, Leeds, U.K.

COX, K. G., Grant Institute of Geology, Kings Buildings, West Mains Road, Edinburgh, U.K.

DAWSON, J. B., Department of Geology, University of St Andrews, St Andrews, U.K.

GARSON, M. S., Institute of Geological Sciences, 5 Princes Gate, London, U.K.

GASS, I. G., Department of Earth Sciences, University of Leeds, Leeds, U.K.

GIROD, M., Laboratoire de Petrographie, Sorbonne, 1 Rue Victor Cousin, 75 Paris (Ve), France.

HARRIS, P. G., Department of Earth Sciences, University of Leeds, Leeds, U.K.

KING, B. C., Department of Geology, Bedford College, Regent's Park, London, U.K.

ROOKE, J. M., Department of Earth Sciences, University of Leeds, Leeds, U.K.

SHACKLETON, R. M., Department of Earth Sciences, University of Leeds, Leeds, U.K.

SIMPSON, E. S. W., Department of Geology, University of Cape Town, Rondebosch, South Africa.

VAIL, J. R., Department of Geology, University of Khartoum, Khartoum, Sudan.

VILJOEN, M. J., Department of Geology, University of the Witwatersrand, Johannesburg, South Africa.

VILJOEN, R. P., Department of Geology, University of the Witwatersrand, Johannesburg, South Africa.

VINCENT, P. M., Faculté des Sciences, Université Fédérale du Cameroun, B.P. 812, Yaounde, Cameroun.

VON KNORRING, O., Department of Earth Sciences, University of Leeds, Leeds, U.K.

WHITESIDE, H. C. M., Anglo-American Corporation of South Africa Limited, 45 Main Street, Johannesburg, South Africa.

WOOLLEY, A. R., British Museum (Natural History), Department of Mineralogy, Cromwell Road, London, U.K.

WILLIAM QUARRIER KENNEDY

Facing page xi

WILLIAM QUARRIER KENNEDY
— AN APPRECIATION

There are commonly two motives for publishing a volume in honour of an individual scientist. One is sentiment: those who have worked with a man, learned from him and been excited by his ideas, are glad to show their affection and gratitude by contributing to such a volume or simply by possessing it. The second is to use the unifying force of one man's scientific thought and work to draw out from a diversity of individuals a body of work on one theme which can have considerably more force than the same work published in separate contributions.

William Quarrier Kennedy was born in 1903 at Bridge of Weir in Renfrewshire. He went to school and university in Glasgow and spent many years in the Scottish Office of the Geological Survey. On retirement he went back to Elie in Fifeshire. Scotland has thus been a pervasive influence throughout much of his life, and even if one were not constantly reminded by the voice, one would be aware of the many Scottish strands in his character; he is bold and prepared to take a chance, intuitive and impatient with trivialities. He arouses devotion and affection, in those who know him best, through his enthusiasm, his willingness to share his ideas, his modesty, and his genius.

Kennedy's first degree, in 1926, was in agriculture; he graduated in geology a year later. At that time the Professor at Glasgow was J. W. Gregory, one of the greatest of geological explorers, who had made his reputation with a brilliant and daring pioneer expedition to the great Rift Valley of East Africa in 1892-3. The expedition which Kennedy himself, many years later, took to the formidable Ruwenzori Range was partly aimed at solving some of the same problems of rift tectonics to which Gregory first drew his interest. After graduating from Glasgow, Kennedy went to Zürich where, under the guidance of Professor Paul Niggli, he attended a wide range of courses and carried out research on the geology and ore deposits of Traversella in the Alps. From Niggli he absorbed the thorough knowledge and exact skills of mineralogy and petrology that have been so important in controlling his imagination, for a characteristic quality of his scientific work is that it is so strictly disciplined and checked by exact and careful observation.

On leaving Zürich, Kennedy was appointed to the Scottish Geological Survey. Here he found himself amongst colleagues of such diverse and formidable powers as J. E. Richey, E. M. Anderson, H. H. Read and E. B. Bailey; it was an era of intense geological activity and discoveries in Scotland,

and Kennedy was soon in the forefront of this advance. As was common on the Survey, part of his work was mapping in the Highlands and part in the Midland Valley. In the Western Highlands he and Richey established for the first time, from evidence of well-preserved sedimentary structures, a stratigraphical succession in the Moine Schists. Their initial attempt to interpret the complex structures of the Morar anticline was followed many years later by Kennedy's interpretation of the north-west Caledonian front, in which he demonstrated the existence of a re-folded pile of nappes composed of Lewisian and Moinian rocks. More dramatic was his recognition of a 65 mile wrench along the Great Glen Fault. But the most prescient and far-reaching of his work from that period was probably his paper with E. M. Anderson on crustal layers and the origin of magmas. In it he focussed attention on the differences in composition of melts derived from different levels at a time when most petrologists were attributing the diversity of igneous rocks to differentiation from a single parent magma.

Kennedy left the Geological Survey in 1945 on his appointment to the Chair of Geology in Leeds. He took over a department whose reputation, since the time of P. F. Kendall, had been primarily as a centre for the study of the stratigraphy, palaeontology and glaciology of Yorkshire and northern England. Kennedy transformed the Department; emphasis shifted towards more general problems of petrogenesis and tectonics while mineralogical, petrological and chemical equipment and skills were developed. In 1955 Sir Ernest Oppenheimer, Chairman of the Anglo-American Corporation, agreed to provide funds to establish a Research Institute of African Geology in Leeds. The Corporation supported such research in order 'to understand the reasons for the emplacement of metal ores, their conditions of emplacement and a search for these conditions in Africa'. The creation of the Institute gave Kennedy the opportunity for his most important and fruitful geological work. Three main lines of research have been followed; the first was concerned with igneous rocks in relation to ore concentrations. Such was the study of the Nuanetsi Igneous Province in Rhodesia. The ring complexes there appeared to be related structurally to copper mineralisation at Messina and they are similar in age and mode of emplacement to the tin-bearing younger granites in Nigeria. Other igneous rocks studied included kimberlites, carbonatites and layered basic intrusives, all of which contain economic mineral concentrations. The second line of research has been on African pegmatites, their mineral chemistry, distribution and age in relation to metamorphism and structure. These rocks contain exceptional concentrations of certain elements and they too throw light on chemical differentiation in the crust. The third, and probably the most important project, was a structural analysis of the Precambrian rocks that carry the most important African mineral concentrations. The work which Kennedy initiated has been concerned especially with the periphery of the Rhodesian Craton and with South-West Africa. The changes in the views of African structure that have emerged from this and comparable studies add fundamentally to our understanding of ore distribution.

The organisation and direction of the research in Africa absorbed a great deal of Kennedy's energy over many years. Many of the scientific results were not published under his name, for, characteristically, he gave an immense amount in ideas and stimulus to his colleagues and students. Anyone who has worked with him is aware too that Kennedy finds it hard to bring work to its final stages, so that much of his work and ideas are embedded in the publications of others. The largest single project undertaken by the Research Institute was the mapping, petrological and geochemical study of the Nuanetsi ring complexes in Rhodesia. This project was initiated and super-vised throughout by Kennedy, and the final memoir published by the Royal Society in 1965 represents 17 man-years of work; typically, Kennedy's name is not amongst the list of authors. His ideas on the structural differentiation of Africa, on the significance of the West African Craton and on the continent-wide importance of the Pan-African tectonothermal episode appeared un-obtrusively in the Annual Reports of the Research Institute of African Geology.

His total geological achievement in Africa can best be judged from the theses and publications which are listed in the Annual Reports of the Research Institute of African Geology. They constitute a massive contribution to know-ledge of many parts of Africa and the inspiration behind them is largely Kennedy's.

While organising and actively carrying out work in Africa, Kennedy also continued to think about some of the problems which had arisen from his Survey work in Scotland. His paper on the use of mineral assemblages in calc-silicate bands in the Moine Schists as indices of progressive regional metamorphism and his paper on the significance of thermal structure in the Scottish Highlands were important contributions to metamorphic geology. His paper on the tectonic evolution of the Midland Valley of Scotland con-tained ideas that had been germinating for many years.

The problems that have been of the greatest interest to Kennedy are the source of magmas, the relation of magma types to crustal structure, the structure and metamorphism of the Scottish Highlands, and the evolution of the African continent. To these problems he has contributed ideas of lasting value. Many strands of his interests were woven together through his work in the Research Institute of African Geology.

R. M. SHACKLETON

Bibliography of W. Q. Kennedy, F.R.S. to 1969

1929 (with Parker, R. L.). Beobachtungen an einem Pyritkristall von Traversella. *Schweiz. miner. petrog. Mitt.*, **9**, 200.

1931 The parent magma of the British Tertiary Province. *Summ. Progr. geol. Surv. G.B.*, Pt. 2 (for 1930), 61.
On composite lava flows. *Geol. Mag.*, **68**, 166.
The igneous rocks, pyrometasomatism and ore deposition at Traversella, Pied-mont, Italy. *Bull. Suisse Miner. Petrog.*, **11**, 76.

1932   (with MacGregor, A. G.). The Morvern-Strontian 'Granite'. *Summ. Progr. geol. Surv. G.B.*, Pt. 2 (for 1931), 105.

1933   Trends of differentiation in basaltic magmas. *Am. J. Sci.*, **25**, 239.
       Composite auto-intrusion in a Carboniferous lava flow. *Summ. Progr. geol. Surv. G.B.*, Pt. 2 (for 1932), 83.
       (with Dixon, D. E.). Optically uniaxial titanaugite from Aberdeenshire. *Z. Kristallog.*, **86**, 112.

1935   The influence of chemical factors on the crystallisation of hornblende in igneous rocks. *Miner. Mag.*, **24**, 203.
       (with Read, H. H.). The differentiated dyke at Newmains, Dumfriesshire, and its contact and contamination phenomena. Abstract. *Proc. geol. Soc. Lond.*, **1296**, 91.

1936   (with Read, H. H.). The differentiated dyke of Newmains, Dumfriesshire, and its contact and contamination phenomena. *Q. J. geol. Soc. Lond.*, **92**, 116.
       An occurrence of greenalite chert in the Ordovician rocks of the Southern Uplands of Scotland. *Miner. Mag.*, **24**, 433.
       (with Dixon, B. E.). Hydro-amphibole from south Devon. *Z. Kristallog.*, **94**, 280.
       (with Richey, J. E.). The succession of the Moine Schists of western Inverness-shire. Abstract. *Rep. Brit. Ass. Adv. Sci.*, 345.

1937   Greenalitschiefer in ordovizischen Gesteinen der südlichen Hochländer Schottlands. Abstract. *Zentralbl. Miner.*, Abt. A, No. 2, 61.
       (with Harvey, C. O.). An apatite rock from Dumfriesshire, Great Britain. *Summ. Progr. geol. Surv. G.B.*, Pt. 2 (for 1935), 53.

1938   (with Anderson, E. M.). Crustal layers and the origin of magmas. *Bull. volc.*, **3**, ser. II, 23.
       (with Richey, J. E.). The succession and structure of the Moine Schists in the Morar district, western Inverness-shire. Abstract. *Proc. geol. Soc. Lond.*, **1338**, 52.

1939   The Great Glen Fault. Abstract. *Proc. geol. Soc. Lond.*, **1354**, 42.
       (with Richey, J. E.). The Moine and Sub-Moine Series of Morar, Inverness-shire. *Bull. geol. Surv. G.B.*, **2**, 26.

1940   (with MacGregor, M. *et al.*). Discussion on the boundary between the Old Red Sandstone (Devonian) and the Carboniferous (Scotland). *Ann. Rep. Brit. Ass. Adv. Sci.* (for 1939–40), 256.

1943   The oil-shales of the Lothians—structure—Area IV, Philpstoun. *Wartime Pamph. geol. Surv. G.B.*, **27**.
       Commercial mica in Scotland. Pt. I: Characteristics of commercial mica. *Wartime Pamph. geol. Surv. G.B.*, **34**.
       Commercial mica in Scotland. Pt. II: Preliminary description of some occurrences north of the Great Glen. *Wartime Pamph. geol. Surv. G.B.*, **34**.

1945   Transcurrent movement exemplified by a fault in the West Lothian oil-shale field. *Trans. geol. Soc. Glasgow*, **20**, 287.

1946   (with Pringle, J.). On algal limestones at the base of the Burdiehouse limestone, near Burdiehouse, Midlothian. *Geol. Mag.*, **83**, 149.
       The Great Glen Fault. *Q. J. geol. Soc. Lond.*, **102**, 41.

1947   (with Richey, J. E.). Catalogue of the active volcanoes of the world. *Bull. volc.*, **7**, Suppl.

1948   On the significance of thermal structure in the Scottish Highlands. *Geol. Mag.*, **85**, 229.
       Crustal layers and the origin of ore deposits. *Schweiz. miner. petrog. Mitt.*, **28**, 222.

1949   Zones of progressive regional metamorphism in the Moine Schists of the Western Highlands of Scotland. *Geol. Mag.*, **86**, 43.

1951   Sedimentary differentiation as a factor in the Moine-Torridonian correlation. *Geol. Mag.*, **88**, 257.

1952   (with Bailey, E. B.). Moinian-Torridonian problem. Skye to Loch Carron. *Int. geol. Congr.*, **18** (**13**), 273.

1953   The tectonics of the Morar anticline and the structure of the Moine nappe complex. Abstract. *Proc. geol. Soc. Lond.*, **1503**, iii.

1954   A general introduction. In *The tectonic control of igneous activity*, p. 2. 1st Inter Univ. geol. Congr., Univ. Leeds.

1955   The tectonics of the Morar anticline and the problem of the north-west Caledonian front. *Q. J. geol. Soc. Lond.*, **110**, 357.

1958   The tectonic evolution of the Midland Valley of Scotland. *Trans. geol. Soc. Glasgow*, **23**, 106.
       (with Von Knorring, O.). The mineral paragenesis and metamorphic status of garnet-hornblende-pyroxene-scapolite gneiss from Ghana (Gold Coast). *Miner. Mag.*, **31**, 846.

1959   The formation of a diffusion reaction skarn by pure thermal metamorphism. *Miner. Mag.*, **32**, 26.

1962   The scientific approach to mineral prospecting. *Optima*, **12**, 213.
       Some theoretical factors in geomorphological analysis. *Geol. Mag.*, **99**, 304.

1964   The structural differentiation of Africa in the Pan-African (± 500 m.y.) tectonic episode. *8th Ann. Rep. Res. Inst. African Geol.*, *Univ. Leeds*, 48.

1965   The influence of basement structure on the evolution of the coastal (Mesozoic and Tertiary) basins of Africa. In *Salt basins around Africa*, p. 7. Inst. Petrol., London.

In press (with Harris, P. G. and Scarfe, C. M.) Volcanism versus plutonism—the effect of chemical composition. In *Mechanism of igneous intrusion*, (Ed. N. Rast), Liverp. geol. Soc.

In preparation   The West African Craton.

T. N. CLIFFORD

I   The structural framework
of Africa

ABSTRACT. *This volume is intended as a venue for the consideration
of the complex relationships that exist between irruptive activity and the
tectonic evolution of the African crustal segment. At least five major orogenic events punctuated the pre-Silurian history of Africa, and each contributed to the virtual complete cratonisation of the continent by Lower
Palaeozoic times. Orogenesis thereafter was peripheral and confined to
north-western Africa and the Cape Fold Belt (middle Palaeozoic—early
Mesozoic orogenesis), and the Atlas Mountains (Alpine orogenesis). In
the largely cratonic environment of Africa since Lower Palaeozoic times,
there have been extensive outpourings of volcanic rocks, together with the
emplacement of a variety of sub-volcanic complexes whose general alignment
has been controlled, at least in part, by crustal patterns established during
the pre-Silurian orogenies.*

'THAT an intimate relationship of a general character exists between igneous
activity and crustal movements has long been recognised. . . . The nature of
the relationship is neither simple nor obvious, however, and even in the case
of vulcanicity . . . it is apparent that the phenomena are not confined to any
single type of tectonic environment.' Thus wrote Professor Kennedy in the
text of a lecture on the 'Tectonic Control of Igneous Activity', given at the
first Inter-University Geological Congress held at Leeds in 1953. In this
volume in his honour, the relationships between magmatic activity and the
tectonic processes which operated during the evolution of Africa are considered. A number of examples are presented of igneous events which
punctuated the time-span of development of the continent, and this intro-
ductory chapter is intended to 'set the scene' by outlining the major features
of African geology, both as an illustration of the logic of the selection and
order of the chapters which follow, and as a guide to the reader who wishes
to relate individual chapters one to another, and to African development as
a whole.

Africa occupies a particularly critical position in any discussion of earth
structure. It is the keystone of continental drift hypotheses past and present,
the classic locality for rift-faulting, the type locality for kimberlites and, since
it has been largely continental since early Palaeozoic times (except for
northern and north-western parts), it offers an extensive record of early
crustal history. At least seven major orogenic events are recorded and are,
from oldest to youngest: (*i*) *c.* 3000 m.y. ago; (*ii*) 2500-2800 m.y. ago
(Shamvaian Orogeny); (*iii*) 1850±250 m.y. ago (includes the Eburnian

1

Orogeny and the Huabian Orogenic Episode); (*iv*) 1100 ± 200 m.y. (Kibaran Orogeny); (*v*) 550 ± 100 m.y. (Damaran-Katangan or Pan-African Orogeny); (*vi*) middle Palaeozoic–early Mesozoic orogenies (Acadian and Hercynian) of north-western Africa and the Cape Fold Belt; and (*vii*) the Alpine Orogeny of the Atlas Mountains (see Fig. 1). It need scarcely be mentioned that there

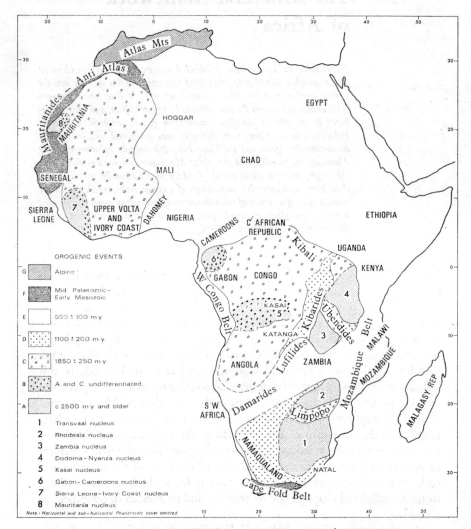

FIG. 1. Generalised map of the major orogenic structural units of Africa.

are many examples in Africa, of regions affected by older orogenies in this sequence, which were subsequently involved in more youthful ones; for example, the Upper Precambrian rocks of the southern tip of the continent were folded, and intruded by granite plutons 500-600 m.y. ago, and then involved in the orogeny of the Cape Fold Belt in late Palaeozoic–early Mesozoic times. Many of these examples will be briefly mentioned later.

However, the primary purpose of this chapter is to consider these various orogenic events taking examples from regions where the orogenic effects are still preserved in their 'primary' form, undisturbed by subsequent orogeny (Fig. 1). Of course, even in regions of this kind, the orogenic event may be recorded in at least two ways: either as deformation of the cover rocks which are generally recognisable as geosynclinal; or, as rejuvenated floor rocks (basement with respect to the geosynclinal sediments) from which the cover rocks have been removed, or on which they were never deposited. These two imprints of orogeny have been termed the 'geosynclinal facet' and the 'vestigial facet (or vestigeosyncline)' respectively (Clifford, 1968a, p. 306), and particularly good examples of these facets occur in belts of 1100 ± 200 m.y. and 550 ± 100 m.y. orogenesis (pp. 12 and 13).

It is, of course, impossible to deal with all of these varied aspects of African structural development. Indeed there are a number of texts which deal in some detail with, for instance:

(*i*) the general geology of the continent (Haughton, 1963; Furon, 1963);

(*ii*) the geology and geochronology of *specific regions* such as northern and central Africa (Cahen, 1961, 1963), southern Africa (Nicolaysen, 1962), West Africa (Rocci, 1965; Black, 1967) and equatorial Africa (Cahen and Snelling, 1966), and certain *time events* such as 550 ± 100 m.y. orogenesis (Clifford, 1963, 1967; Kennedy, 1964, 1965), and 1100 ± 200 m.y. orogenesis (Nicolaysen and Burger, 1965; Cahen *et al.*, 1967);

(*iii*) the pre-Silurian history of the continent as a whole (Clifford, 1968a).

In addition there are two comprehensive tectonic maps: the 'Esquisse structural provisoire de l'Afrique' on a scale of 1 : 10,000,000 (Furon *et al.*, 1958); and the 'International tectonic map of Africa' on a scale of 1 : 5,000,000 published by UNESCO and the Association of African Geological Surveys, Paris (1968).

In this chapter it is logical to deal with African structural development starting with the antique sequences which were affected by the most ancient orogenesis more than 2500 m.y. ago; in this way the structural evolution of the continent may be used to provide a framework, in time and space, for specific events which are dealt with in detail in subsequent chapters.

Eight important remnants, or nuclei, of ancient rocks are shown in Figs. 1 and 2: (1) Transvaal; (2) Rhodesia; (3) Zambia; (4) Dodoma-Nyanza; (5) Kasai; (6) Gabon-Cameroons; (7) Sierra Leone-Ivory Coast; and (8) Mauritania. With the exception of the Zambia nucleus, for which no age data are available, all are characterised by rock sequences affected by folding and regional plutonic events 2500 m.y. ago and earlier. Rocks of similar antiquity occur in the Hoggar, Nigeria, Malagasy Republic and elsewhere, in belts affected by later orogenies; in this chapter they are dealt with in terms of the latter.

The southernmost example is the Transvaal nucleus, which has been variously referred to as the Kaapvaal or Transvaal Craton (Pretorius, 1964; Cahen and Snelling, 1966; Anhaeusser *et al.*, 1968). A great variety of metasediments, metavolcanics and plutonic rocks of age greater than 2700 m.y. has been recognised in the exposed inliers within this nucleus (Du Toit, 1954, p. 39); and perhaps the best-studied example is the Barberton Mountain Land of the eastern Transvaal. Detailed work carried out over a period of more than 50 years has shown that the mountainland is a complex 'schist belt' of folded sediments and volcanics older than 3000 m.y., intruded by basic and ultrabasic rocks and by extensive granites around the margins (Visser *et al.*, 1956; Ramsay, 1965; Roering, 1965; Anhaeusser, 1966). In view of this extensive geological and geochronological knowledge, the Barberton Mountain Land provides a good example of the early part of African history. In this volume, recent work is summarised by Viljoen and Viljoen (Chapter 2) and particular emphasis is placed on the thick sequences of lavas (the Onverwacht) erupted during the development stage of the schist belt sequence, and plutonic acidic rocks emplaced during and after the deformation of that sequence.

To the north, in the Rhodesia nucleus, a similar general relationship between remnant schist belts and large bodies of engulfing granitic rocks has been described by Macgregor (1951a). The rocks of the schist belt sequences have been subdivided into: a lower Sebakwian System of psammitic and pelitic rocks, ironstones, limestones, and some lavas; a middle Bulawayan System composed mainly of basic and intermediate lavas with interbedded ironstones, psammitic and pelitic sediments, conglomerate, and limestones; and an upper Shamvaian System composed largely of psammitic sediments, pelites, dolomite and some ironstone. K-Ar, Rb-Sr and U-Th-Pb mineral ages place a minimum age of 2600-2700 m.y. on this succession (Nicolaysen, 1962). However, if these rocks are correlatives of the Barberton succession of the Transvaal, as was suggested by Du Toit (1954, p. 54), then they must be older than 3000 m.y. Of particular importance in the Rhodesia nucleus is the Great Dyke, an intrusive body trending north-north-east for over 330 miles, and transgressing the structure of the schist belts. This impressive igneous mass is largely composed of basic and ultrabasic rocks and is at least 2500 m.y. old, and may be as old as 2800 m.y. (Allsopp, 1965). New data on the composition of this unique feature of the African crust are given by Bichan (Chapter 3).

To the north, the geology of the oldest rocks of the Zambia nucleus (Fig. 1), and even its real antiquity, are poorly understood. In contrast, the Dodoma-Nyanza nucleus has been studied in some detail and consists broadly of two contrasted portions: (*i*) the major part of central Tanzania composed largely of granite, granodiorite, acid gneiss and migmatite associated with the folded metamorphic rocks of the Dodoman System; and (*ii*) a northern portion in northern Tanzania, western Kenya, and eastern Uganda where the Nyanzian System, composed of acid and basic volcanic rocks, quartzites, pelites and banded ironstones, is overlain by the Kavirondian System consisting essentially of arenaceous and argillaceous sediments and volcanics.

Radiometric data show that the Dodoman System is older than 2300 m.y. and may be considerably older (Cahen and Snelling, 1966). Both the Nyanzian and Kavirondian Systems have ages greater than 2550 m.y. and the former must be older than 2900 m.y., the Rb-Sr isochron age obtained for the post-Nyanzian Masaba granite of Uganda (Old, 1968).

In contrast to the Transvaal, Rhodesia and Dodoma-Nyanza nuclei, those of Kasai, Gabon-Cameroons and West Africa cannot be so precisely delimited (Fig. 1). Typical of these is the Kasai nucleus where the ancient rocks consist of gneissic granitic rocks, and a charnockitic suite of noritic gabbros, enderbites, charnockites, granulites and leptynites. On the basis of radiometric dating carried out by Ledent *et al.* (1962), a provisional age of 2700 m.y. has been assigned to the granitisation and charnockitisation. A similar relationship between granitic and charnockitic rocks has been recorded from the poorly defined Gabon-Cameroons nucleus, in which granite occurs over very wide regions; the Ebolowa-type granites have yielded biotite and zircon Pb/α ages of 2200-2500 m.y. (Lasserre, 1964) and form a part of the charnockitic complex of the southern Cameroons, Gabon and Spanish Guinea.

An analogous case occurs in the Sierra Leone-Ivory Coast nucleus (Fig. 1) where the rocks are largely a series of mica schists, quartzites, calcareous rocks, ironstones and metamorphosed basic and acid volcanics, intruded by syn- and late-kinematic granitic rocks and pegmatites (see Haughton, 1963, pp. 9-10; Black, 1967). Andrews-Jones (1968) has shown that the sedimentary-volcanic sequences in the Kenema district of Sierra Leone reach the amphibolite and granulite facies of metamorphism, and that granitic gneisses yielding a Rb-Sr whole rock isochron age of 2800 m.y. are structurally conformable with these high-grade rocks. To the east, in the Ivory Coast, granite associated with the charnockites of Man has given a similar zircon Pb/α age of 2680 m.y. and a biotite age of *c.* 2000 m.y. (Bonhomme, 1962).

Comparable age patterns have been obtained from the Mauritania nucleus (Fig. 1) to the north, where a group of metasediments and metavolcanics of the Tasiast Group is associated with migmatitic granites and pegmatites ascribed to the Sattle (or Stal) Ogmane granitic complex. The Sattle Ogmane phase of late-kinematic granitic rock emplacement is represented by a heterogeneous suite of rocks which have given mineral and whole rock ages of 2500-2600 m.y. (*ibid.*; Vachette, 1964a).

Subsequent to their orogenic deformation, some of the nuclei (Fig. 2) were the sites of sedimentation and vulcanicity unconformably laid down on the eroded roots of the folded schist belts and their associated granitic rocks. The clearest example of this ancient deposition is the Transvaal nucleus where the folding of the schist belts and the emplacement of granitic rocks was followed by the deposition of 50,000 ft of sedimentary and volcanic rocks of the Dominion Reef, Witwatersrand, Ventersdorp and Transvaal Systems (Haughton, 1963, pp. 117-126; Nicolaysen, 1962, pp. 575-576). Extrusive igneous rocks form the major part of the 10,000–12,000 ft succession of the Ventersdorp System and also occur in the underlying Dominion Reef and Witwatersrand Systems. Together, these three systems constitute the 'Witwatersrand Triad'; and Whiteside (Chapter 4) presents an up-to-date review of the volcanic rocks of this economically important and strati-graphically critical sequence. The fact that the lowest members rest on basement older than 2900 m.y., and that a quartz-porphyry high in the Ventersdorp sequence has given an age of 2300±100 m.y. (Van Niekerk and Burger, 1964), clearly demonstrates the time-span of this example of early Precambrian deposition. The deposition of the overlying Transvaal System was brought to an end with the intrusion of the great Bushveld Igneous Complex of ultrabasic, basic and acid intrusive rocks some 1950± 200 m.y. ago (Nicolaysen et al., 1958). This complex has an areal extent of 23,000 square miles and its emplacement was accompanied by reheating of older rocks (Allsopp, 1962); the Palabora carbonatite complex, one of the oldest carbonatites in Africa, also dates from about that time (Holmes and Cahen, 1957, p. 75).

The Bushveld Complex was approximately coeval with pronounced oro-genic events which affected large parts of the continent and which are reflected by ages of 1850±250 m.y. The most conspicuous effects of this orogeny are in West Africa, particularly the Ivory Coast, the Upper Volta region, Mali, Senegal, Mauritania and, perhaps, southern Morocco (Fig. 2). In the Ivory Coast, Upper Volta and surrounding regions a major part of the Precambrian consists of a strongly folded succession of sediments and vol-canics referred to the Birrimian System (Haughton, 1963, p. 12). These rocks are widely invaded by the extensive syntectonic granites for which Rb-Sr whole rock and mica ages fall within the range 1850–2050 m.y. (Bonhomme, 1962; Vachette, 1964b) and reflect the date of the Eburnian Orogenic Episode. Similar ages have been obtained from granites in Mali and Senegal (Fig. 2) which are intrusive into supposed correlatives of the Birrimian (Bonhomme, 1962; Bassot et al., 1963). Moreover, in Mauritania to the north, the Amsaga Series of folded gneisses with quartzites, mica schists and calc-silicate rocks, contains concordant and discordant granites that have yielded Rb-Sr biotite ages in the 1950–2000 m.y. range suggesting that these too were emplaced during the Eburnian Episode; the wider influence of this event is demonstrated by ages of 2000±200 m.y. given by biotites from granites in the Mauritania nucleus (Vachette, 1964a) and by pegmatites and dolerites in the Sierra Leone-Ivory Coast nucleus (Andrews-Jones, 1968; Bonhomme, 1962).

To the south, in Gabon (Figs. 1 and 2), a series of pegmatites cutting 'basement' gneisses have given muscovite Rb-Sr ages of 1900–1950 m.y. (Vachette, 1964c) that are consistent with those of the Ivory Coast and other regions to the northwest; similar ages from granites in the Cameroons and Gabon indicate an extension of this activity (Lasserre, 1964; Cahen and Snelling, 1966, p. 134). Still further south, the crystalline rocks which occupy the major part of interior Angola are largely granitic rocks and gneissses, and biotite ages from the granites fall in the 1450-1700 m.y. range (Mendes, 1964, 1966). The significance of these younger dates in the 1850±250 m.y. range is not known, but they are broadly consistent with: U-Pb zircon and Rb-Sr whole rock isochron ages for the Franzfontein granitic rocks, emplaced during the Huabian Episode, which occur in a basement inlier around Franzfontein,

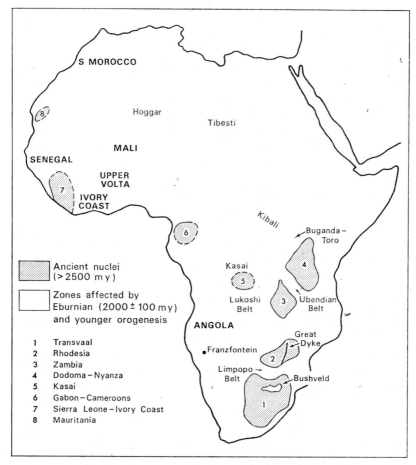

FIG. 2. The ancient nuclei of Africa, distinguished from zones affected by Eburnian and more recent orogenies.

within the late Precambrian–early Palaeozoic orogenic belt in South-West Africa (Fig. 2) (Clifford et al., 1962; Clifford et al., 1969); and with 1600-1700 m.y. ages from the 'Eburnian domaine' of Ivory Coast and Mauritania (Vachette, 1964a, b).

In the eastern part of Africa, a number of linear orogenic belts have yielded ages in the 1600-2100 m.y. range. These include the Kibali, Buganda-Toro, Ubendian, and Limpopo Belts (Fig. 2). In the Kibali Belt of the northern part of the Congo, the Kibali Group includes schists, quartzites, conglomerates, carbonate rocks, and volcanics; the sequence is cut by post-

Kibalian granites. Galena from gold-reefs in the Kibali Group has yielded an age of 1840 ± 100 m.y., and Rb-Sr ages of 1850-2050 m.y. have been obtained for pegmatite minerals (Cahen, 1961, p. 546; Cahen and Snelling, 1966, p. 59); these data suggest that the Kibalian Orogenic Episode is older than 1900 ± 100 m.y.

The region of the largely metasedimentary Buganda-Toro sequence of southern, western and south-western Uganda (McConnell, 1959) has yielded a consistent concentration of ages in the 1800 ± 100 m.y. range, which are considered to represent the late stages of orogeny (Cahen and Snelling, 1966, p. 58). To the south, a similar younger limit can be placed on the diastrophism of the Ubendian-Rusizi Belt of southern Tanzania, northern Zambia and the eastern Congo. This northwest-southeast trending belt (Fig. 2) is largely occupied by crystalline rocks including metamorphosed sediments, migmatites and biotite gneisses (Quennell and Haldemann, 1960). Exactly how much older than 1800 m.y. the Ubendian orogenesis really was, is still not known; a brief summary statement on available age data is given by Cahen (Chapter 6).

In the Kasai Province (Figs. 1 and 2) a number of crystalline rock units have a general east-north-east trend (Cahen, 1963) and include the older granitic and charnockitic basements (see p. 5) separated by an intermediate zone of younger sediments of the Luiza Series—a sequence of metasedimentary schists and gneisses that rests unconformably on the basement of the Kasai nucleus and is thus younger than 2700 m.y. Rb-Sr ages of 2000-2100 m.y. for muscovite and microcline from schists and pegmatites respectively give a general age to the metamorphism and orogenesis of the Luiza Series (Ledent et al., 1962). To the south of the Kasai region, a belt of folding affects the Lukoshi Formation (Fig. 2); muscovite from a pegmatite cutting a manganese member of the Lukoshi sediments and lavas has yielded a Rb-Sr age of 1845 m.y. (ibid.) and provides a younger limit for the Lukoshian orogenesis.

Bordering the Limpopo River in the southern part of Rhodesia and the northern part of South Africa, the east-north-east trending Limpopo Orogenic Belt stretches from Botswana eastwards almost to the Mozambique border (Fig. 2). The crystalline rocks of this belt consist of metasediments, basic and ultrabasic rocks, and granitic rocks, gneisses and pegmatites. Van Breemen et al. (1966) record Rb-Sr biotite ages of 2000 m.y. from these rocks (see also Nicolaysen, 1962) but they also consider that an event older than 2500 m.y. was responsible for the high-grade metamorphism in the belt.

In summary, then, it is clear that ages in the 1850 ± 250 m.y. range are widespread in southern and West Africa (Clifford, 1968a). In addition to the Ivory Coast and the Upper Volta, orogenic activity reflected by ages of 1800-2100 m.y. in the following regions has been tentatively ascribed to the Eburnian Orogenic Episode: southern Cameroons; the belt of the Luiza Series of Kasai; the Lukoshi Belt of the Congo; the Ubendian-Rusizi Belt of East Africa; the Kibali and Buganda-Toro Belts of Congo and Uganda; and the Limpopo Belt of Rhodesia, Botswana and South Africa. The significance, within the 1850±250 m.y. range, of 1600-1700 m.y. ages in Angola, West Africa and elsewhere is still uncertain, though in northern South-West Africa they represent the Huabian Episode of granite emplacement of batholithic dimensions (Clifford et al., 1969). It has been suggested that, in West Africa, they may reflect an orogenic phase related to, but later than, the 1800-2100 m.y. event (Kennedy, in preparation).

Since the end of 1850±250 m.y. orogenesis, a number of segments of Africa have acted as stable blocks (Fig. 3): the Rhodesia-Transvaal Craton;

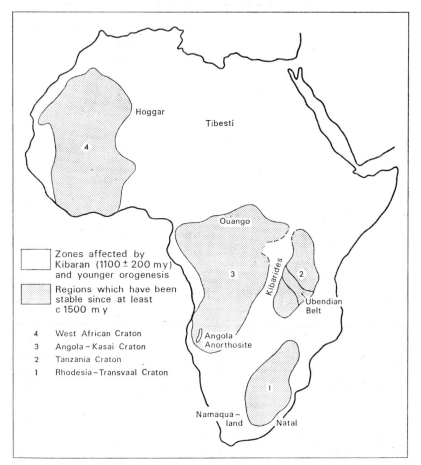

Hoggar

Tibesti

Ouango

Kibarides

Zones affected by
Kibaran (1100 ± 200 m y)
and younger orogenesis

Regions which have been
stable since at least
c 1500 m y

4   West African Craton
3   Angola – Kasai Craton
2   Tanzania Craton
1   Rhodesia – Transvaal Craton

Ubendian
Belt

Angola
Anorthosite

Namaqua –
land   Natal

FIG. 3. Regions which have remained stable since the end of 1850±250 m.y. orogenesis, distinguished from zones affected by orogenesis during Kibaran and more recent orogenies.

the Tanzania Craton; the Angola-Kasai Craton; and the West African Craton (Clifford, 1966). Within these, later sediments and volcanics laid down up to the present time have suffered largely epeirogenic movements. The long-continued stability of these regions may be illustrated from the Transvaal, where the horizontal and sub-horizontal rocks of the Waterberg-Loskop Systems of quartzites, sandstones, shales, conglomerates and some volcanics overlie the Bushveld Complex, and are therefore younger than 1950 ± 200 m.y.; these systems are older than 1400 m.y., and may even be older than 1800 m.y. (Oosthuyzen and Burger, 1964).

TABLE 1

Dated extrusives	System	Dated intrusives
		Pilanesberg (c. 1300 m.y.) Leeuwfontein (1420 m.y.) and Premier kimberlite (1100-1750 m.y.)
	WATERBERG	Granophyre (1790 m.y.)
	LOSKOP	Bushveld Complex (1950 m.y.)
	TRANSVAAL	
Andesite and quartz porphyry (2300 m.y.)	VENTERSDORP	Gaberones Granite? (c. 2350 m.y.)
	WITWATERSRAND	Great Dyke and Modipe Gabbro (c. 2600 m.y.)
Andesite, dacite and quartz porphyry (2800 m.y.)	DOMINION REEF	

It is significant that, after the emplacement of the Bushveld Complex, this general region continued as an igneous province with the emplacement of a series of dolerites and alkali complexes (including the Pilanesberg) and the Premier kimberlite pipe. Many of these bodies are discussed by Vail (Chapter 16); it suffices, therefore, to say that the alkali complexes have ages in the 1200-1450 m.y. range (Schreiner and Van Niekerk, 1958; Oosthuyzen and Burger, 1964), whereas the Premier kimberlite is certainly older than 1100 m.y. and may be as old as 1750 m.y. (Allsopp et al., 1967). A number of these irruptives and some older ones already mentioned in this chapter, are summarised in Table 1, and are of particular interest in that they represent benchmarks in the recent summary by McElhinny et al. (1968) of African palaeomagnetic data; their summary maps of pole variations during the lengthy time-interval from 2700 to 1300 m.y. ago, based largely on southern African igneous rocks, are presented in Fig. 4.

It is clear that important sedimentary deposition and igneous activity took

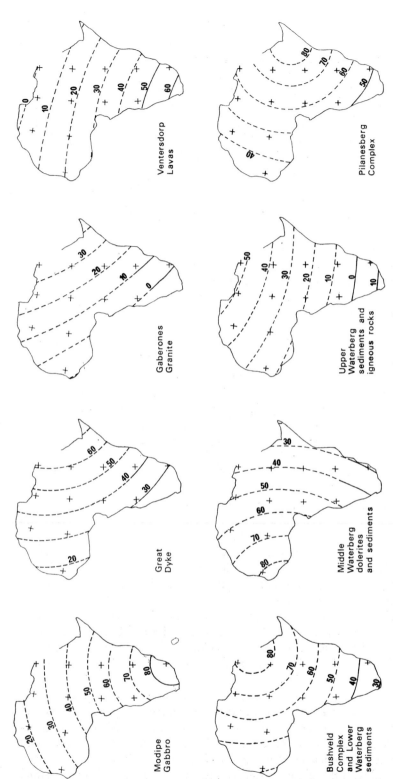

Fig. 4. Precambrian palaeolatitudes for southern Africa (modified after McElhinny et al., 1968); *reproduced with the permission of the Editor of Reviews of Geophysics.* The palaeolatitudes are based on determinations for southern African sites; their possible extensions across other parts of Africa are suggested by broken lines.

place on the site of the stable Rhodesia-Transvaal Craton. Equally significant was the emplacement of the anorthosite body of southern Angola and northern South-West Africa (Fig. 3); this mass, the largest of its kind in the world, is dealt with by Simpson (Chapter 5). Pegmatite from this complex has given an age of about 1200 m.y. which is taken to be either the actual age of the anorthosite, or at least a minimum age for its emplacement (*ibid.*). It lies within the stable environment of the Angola-Kasai Craton, which was cratonic at the time of the Kibaran Orogeny 1100±200 m.y. ago; and it thus has a similar structural position to the Bushveld Complex (1950±200 m.y. old) emplaced in the stable environment of the Transvaal nucleus during the Eburnian and Huabian Orogenic Episodes 1850±250 m.y. ago.

In contrast to the stable Rhodesia-Transvaal, Tanzania, Angola-Kasai and West African Cratons, after the Eburnian and Huabian Episodes certain zones were the sites of active accumulation of geosynclinal sediments and/or were affected by subsequent linear orogenesis during the Kibaran Orogeny 1100±200 m.y. ago. Of particular importance are the Kibaride Belt of Central Africa, and the Namaqualand (Orange River)-Natal Belt of southern Africa (see Figs. 1 and 3). The former is occupied by 30,000 ft of geosynclinal sediments of the Kibara and correlative groups which rest on the eroded roots of the older Ubendian and Lukoshian Belts and thus have a maximum age of 1800-2000 m.y.; they are intruded by pre- and syntectonic granites yielding ages of 1300 m.y. and 1250 m.y. respectively (Cahen *et al.*, 1966; Cahen *et al.*, 1967); a wide variety of K-Ar, Rb-Sr and U-Th-Pb ages of mineral and whole rock samples from pegmatites, veins and granites fall in the 850-1100 m.y. range. It is unnecessary to deal with this important segment of African geology in this paper, because the principal elements of its stratigraphic and structural development are summarised by Cahen (Chapter 6).

From the eastern Congo, the northern end of the Kibaride Belt (Figs. 1 and 3) swings westwards from south-west Uganda and disappears beneath the younger sediments of the Congo Basin. The extent of the belt in that direction is, therefore, uncertain. However, in the Central African Republic (Fig. 3), ages of biotites from granites in the Ouango Massif fall in the 900-1000 m.y. range (Roubault *et al.*, 1965) and are in accord with ages in the Kibaride Chain to the southeast.

Holmes (1951) was the first to suggest a connection between the Kibaride Belt and the Namaqualand Belt of South Africa and South-West Africa (Fig. 3) and he considered that both regions were part of a single geosyncline. Radiometric studies of samples from the Namaqualand Belt have given consistent mineral ages in the 900-1250 m.y. range (Nicolaysen and Burger, 1965), demonstrating that these two belts are isochronic. However, despite the similarity of their age patterns, no extensive geosynclinal sequences equivalent in age to the Kibara Group have been identified in the Namaqualand Belt. The country rocks are mainly granite and granite gneiss, with remnants of sediments and volcanic rocks of the Kheis System previously correlated with the antique systems (older than 2600 m.y.) in the Transvaal and Rhodesia. In the light of these data Nicolaysen and Burger (*ibid.*, p. 500) maintain that the rocks of the Kheis System were 'intensely metamorphosed and reconstructed when they moved into giant upwellings of basement "infrastructure" about 1000 m.y. ago'. On this reasoning, the Namaqualand Belt clearly represents a Kibaran vestigeosyncline (Clifford, 1968a; and p. 3 above) in which 1000 m.y. old tectonism is represented only by the rejuvena- tion of ancient basement.

To the south, these crystalline rocks disappear beneath younger rocks, largely Karroo (see Furon *et al.*, 1958). Further east, however, along the Natal Coast (Fig. 3) a group of crystalline rocks appear from beneath the younger Palaeozoic cover and include a variety of metasediments, meta- volcanics, gneisses and granitic plutons. Radiometric data have given evidence of 950–1150 m.y. old gneissification and granite emplacement; Nicolaysen and Burger (1965) have suggested a structural connection between the Namaqualand Belt and this region (see Fig. 1).

Finally, a number of youthful ages in the 850–1300 m.y. range have been recorded from the Ubendian-Rusizi Belt of the Congo, Tanzania and northern Zambia and include Rb-Sr and K-Ar determinations from granites, pegmatites and carbonatite (Aldrich *et al.*, 1958; Cahen and Snelling, 1966); the structural significance of these ages, however, is still not known.

After 1100±200 m.y. (Kibaran) orogenesis, the consolidated Kibarides were added to the Angola-Kasai and Tanzania Cratons to form a single large stable unit, the Congo Craton, and the Namaqualand-Natal Belt was added to the Rhodesia-Transvaal Craton to form the Kalahari Craton (Clifford, 1963, 1967). These two stable blocks, together with the West African Craton (Kennedy, 1964, 1965) thus represent units that were stable after Kibaran orogenesis and during the subsequent major orogeny in Africa, 550±100 m.y. ago. This is best demonstrated by the extensive regions of little-deformed Upper Precambrian rocks on the Congo Craton (Cahen, 1963; Clifford, 1968a, p. 321), and by the even older sequences (Transvaal System, Water- berg-Loskop Systems) of the Kalahari Craton, which have been unaffected by any orogeny. In addition, the West African Craton has a longstanding history of stability since the end of 1850±250 m.y. orogenesis (Kennedy, 1964; Black and Girod, this volume, p. 185).

In contrast to these three units (Fig. 5), geological and geochronological data clearly demonstrate that a major part of the continent was affected by late Precambrian–early Palaeozoic orogenesis (the Damaran-Katangan or Pan-African Orogeny) reflected by widespread ages of 550±100 m.y. The complex pattern of zones affected by this orogenesis can be subdivided into two types:

(*i*) zones of orogenically deformed Upper Precambrian geosynclinal sediments; for example, Central Africa (Katanga System), South- West Africa (the Outjo System), the Lower Congo (Western Congo System), Egypt (? Hammamat Series) and southern Sahara (? Pharusian).

(ii) zones of rejuvenated basement; for example, in Central and East Africa (the Mozambique and Zambesi Belts), the Malagasy Republic, the Cameroons, the Central African Republic, Nigeria, the Hoggar and western Sierra Leone.

Of these, the first group represents the geosynclinal facet of orogenesis, while the second group is the vestigial facet or vestigeosyncline (Clifford, 1968a,

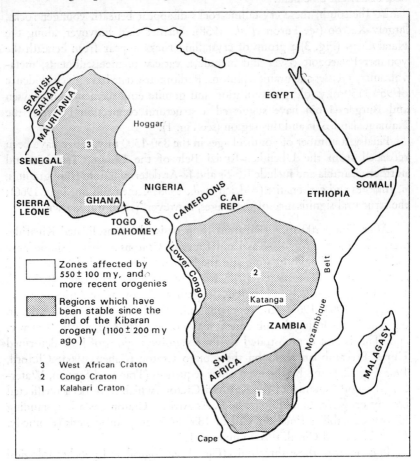

FIG. 5. Regions which have remained stable since the end of the Kibaran Orogeny, distinguished from zones affected by Damaran-Katangan (Pan-African) and more recent orogenies.

pp. 306 and 310). The similarity of the age patterns of these facets, together with their obvious geographical and structural continuity indicate that these zones represent different structural levels through a sinuous mountain chain.

Of the geosynclinal regions, the depositional and deformational history of the Katangan (≡ Katanga System), the type Upper Precambrian of Africa, is discussed in detail by Cahen (Chapter 6); it is sufficient to say here, therefore, that it consists largely of a lower (Roan) unit of arkoses, sandstones, shales, conglomerates, and dolomites, overlain by the Mwashya and the 'Grand Conglomérat' tillites, shales and sandstones, and an uppermost (Kundelungu) sequence of conglomerates, limestones, shales, sandstones, and quartzites. In the Lufilian Arc (Lufilides: Fig. 1) this sequence is strongly folded and thus contrasts markedly with the flatlying correlative sequences on the Congo Craton to the north. To the south, moreover, the sequence becomes progressively metamorphosed and intruded by granite batholiths (De Swardt and Drysdall, 1964). A mirror image of this thermal increase has been suggested for the northern part of Rhodesia where regional metamorphic grade increases broadly from south to north (Macgregor, 1951b). To the west, in the Damarides, the extensive sequence of Upper Precambrian (the Outjo System) in South-West Africa consists of two correlative facies (Damara and Otavi) both of which are strongly folded (Fig. 1). These facies differ in that the Otavi is largely an unmetamorphosed miogeosynclinal sequence, whereas the Damara is a regionally metamorphosed 'eugeosynclinal' sequence. In the latter the regional metamorphism culminates in a high-grade central zone profusely intruded by granite and pegmatite (Martin, 1961, 1965; Clifford, 1963, 1967).

In the Lower Congo, the Western Congo System, which is considered to be correlative with the Katanga System, is strongly folded on the west (Fig. 5). The sequence consists of a basal Sansikwa System of conglomerates, shales and quartzites, overlain by tillite with intercalations of lava, and then a thick overlying succession of clastic and calcareous sediments of the Haut Shiloango, Schisto-Calcaire and Schisto-Gréseux Systems; together the last three systems show strong lithological similarities with the Katanga and Outjo Systems of Central and South-West Africa respectively.

Within this system of late Precambrian–early Palaeozoic belts of orogeny a number of other sequences have been ascribed to an Upper Precambrian age: the Pharusian System of the Hoggar (Black, 1967); the Hammamat Series, and underlying sequences, of Egypt (Schürmann, 1961); the Inda Ad Series and parts of the so-called Basement Complex of Somali Republic and Ethiopia (Greenwood, 1961; Rogers et al., 1965); certain of the sequences (Turoka) of East Africa (Sanders, 1965); the Rokel River Series of Sierra Leone (Allen, 1968); and the Voltaian-Buem sequence of Ghana (Grant, 1967) (Fig. 5).

Dated events within all of these show a broad grouping of ages in the 450–680 m.y. range and include the emplacement of synorogenic and late orogenic pegmatite and granite, regional metamorphic mineral growth or modification, and uranium mineralisation (Cahen, 1961; Clifford, 1967, 1968a; Kennedy, 1964; Nicolaysen, 1962; Cahen and Snelling, 1966).

In addition to these zones of Upper Precambrian geosynclinal sedimentation, extensive coeval tectonothermal activity undoubtedly took place in regions where Upper Precambrian rocks are poorly represented or are absent. Of these, the Mozambique Belt (Fig. 5) represents one of the fundamental features of Africa. First defined and delineated by Holmes (1951), this belt consists of high-grade crystalline rocks, has an approximate north-south orientation through Rhodesia, Mozambique, Malawi, Tanzania and Kenya, and is characterised by ages of 400-700 m.y. (Cahen, 1961). Igneous and metasomatic activity associated with the Malawi segment of the Mozambique Belt are described in this volume by Bloomfield (Chapter 7). This belt, extending over much of East Africa, certainly contains rocks of many ages, including a number of rejuvenated antique sequences: the Nyanzian of Kenya (Sanders, 1965); the Bulawayan of Rhodesia (Johnson, 1968); and the Dodoman of Tanzania (Hepworth et al., 1967). Although generally oriented north-south, it becomes east-west in Rhodesia, and is locally known as the Zambesi Belt, which is structurally contiguous with the folded Upper Precambrian geosynclinal facet of Central Africa.

The incorporation of basement rocks in the zones of this orogeny is also well demonstrated in the Malagasy Republic, where the following pre-Karroo crystalline sequence has been recognised: (1) Androyan System; (2) Graphite System; (3) Vohibory System; (4) Sahatany System (including the Cipolin Series); and (5) Quartzite Series (Haughton, 1963). Of these, systems 1-3 make up the major part of the crystalline rocks of the island (see Furon et al., 1958); common lead and Pb/α ages suggest that the Vohibory System is older than 1500-2100 m.y., and galena ages indicate that the Cipolin Series is older than 1100 m.y. (Holmes and Cahen, 1957, pp. 56 and 83). On the basis of geochronological data, Delbos (1964, p. 1855) concludes that a low-grade metamorphism affected the areally restricted Quartzite Series 600 m.y. ago, and that widespread ages in the 460-550 m.y. range represent an event of major importance in which three interrelated phenomena are recognisable: (1) granitisation giving rise to the stratified granites, 550 m.y.; (2) pegmatite mineralisation, 500 m.y.; and (3) rejuvenation of biotites, 460 m.y. That these complex events reflect the imprint of an orogeny is beyond doubt (Emberger, 1958). However, if this region ever contained geosynclinal sediments related to that orogeny then they have been removed, with the possible exception of the locally preserved Quartzite Series. The Malagasy Republic thus represents the vestigial facet of activity of the 550 ± 100 m.y. orogenesis largely preserved in rocks of much greater antiquity.

On the western side of equatorial Africa, Lasserre (1964) has shown that a uniform pattern of 500-650 m.y. ages is present over a large part of the Cameroons in rocks which have long been regarded as 'Lower' and 'Middle' Precambrian (see Fig. 5). These rocks consist of a variety of metasediments, granulites and migmatites intruded by granite and synkinematic dunite and gabbro (Haughton, 1963). Similarly in the Central African Republic, a consistent pattern of 500-650 m.y. ages has been obtained from terrain believed to be very antique (ibid.; Roubault et al., 1965). In addition to this very uniform pattern, however, certain mineral samples have yielded Rb-Sr ages greater than 800 m.y.

In West Africa, two regions of late Precambrian—early Palaeozoic activity have been recognised and quantitatively defined by radiometric dating: an eastern region extending westwards and northwards from the Cameroons through Nigeria, Dahomey, Mali and into the Hoggar (Fig. 1); and a western region extending from Sierra Leone northwards through Senegal and Mauritania. The two zones, separated by the ancient stable region of the West African Craton (Kennedy, 1964), are discussed at some length by Black and Girod (Chapter 9) and only a brief résumé of data need therefore be given here.

In the southern part of the eastern region, in Dahomey, Ghana, Togo, Mali and Nigeria (Fig. 5), the most extensive rock system within the late Precambrian–early Palaeozoic belt is the Dahomeyan consisting essentially of granites, gneisses, migmatites and metasediments. Radiometric ages yielded by micas from the gneisses and concordant granites all fall in the 450–600 m.y. range (Jacobson et al., 1964), but an orthogneiss (Kouandé) believed to be intrusive into the Dahomeyan has given a whole rock Rb–Sr age of 1650± 220 m.y., suggesting a minimum age for at least part of the Dahomeyan (Bonhomme, 1962, p. 34).

To the north, the structural continuation of this belt is well exposed in the Hoggar region (Fig. 5) where the following tectonic-stratigraphic units have been recognised: (i) Suggarian ($\equiv$ Dahomeyan (?)); (ii) Pharusian; and (iii) Nigritian. It has been suggested that, during late Precambrian–early Palaeozoic orogenesis, movement took place along the north-south thrust faults which delimit the blocks of Suggarian gneisses, marbles and quartzites and Pharusian metasediments and metavolcanics, and that the life of this portion of the orogenic belt came to an end with the deposition of the Nigritian molasse and correlative sequences (see Black and Girod, Chapter 9). Radiometric ages from the Hoggar largely fall in the 500–650 m.y. range and it is believed that syntectonic granite and pegmatite emplacement and major metamorphism took place at that time. Nevertheless, geochronological studies of the western part of the Hoggar have given whole rock ages of 2800 m.y. and biotite ages of 1700–1800 m.y. clearly indicating the presence of a crystalline basement (Ouzzalian $\equiv$ Suggarian?) of great antiquity (Ferrara and Gravelle, 1966).

In contrast to the eastern region of late Precambrian–early Palaeozoic orogenesis in West Africa, the western region is less well defined by radiometric age determinations; moreover, it is characterised by a more extended Phanerozoic orogenic history. In Sierra Leone (Fig. 5), it forms a well-defined orogenic belt extending across the country from south-east to north-west (Kennedy, in preparation; Allen, 1968); this zone extends northwards into Guinea where it disappears beneath unconformably overlying horizontal sandstones ('grés horizontaux') that pass upwards into fossiliferous Silurian strata. To the north, similar sequences to those of the marginal orogenic belt of Sierra Leone and Guinea, reappear and form part of the great 'Mauritanide Zone' (Fig. 1) which extends along the western margin of the continent from Senegal and Portuguese Guinea in the south, through Mauritania and the Spanish Sahara, and into Morocco (Sougy, 1962; Bassot et al., 1963). This major zone has been ascribed to the Hercynian revolution (Sougy, 1962) and consists of two sectors: a southern sector comprising mainly metamorphic

B

rocks and folded sediments of uncertain age; and a northern sector in which fossiliferous Devonian rocks are involved.

Geochronological data from this western region of West Africa indicate an extensive late Precambrian–Phanerozoic orogenic belt in which important tectono-thermal pulses in middle and late Palaeozoic times are recorded in the Mauritanide Zone by ages in the 300-350 m.y. and 200-250

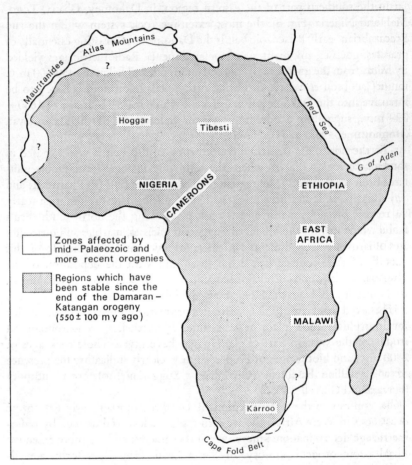

Fig. 6. Regions which have been stable since the end of Damaran-Katangan (Pan-African) orogenesis distinguished from zones affected by Hercynian and Alpine orogenesis.

m.y. ranges; these two events accord with the Acadian and Hercynian Orogenies respectively (Kennedy, in preparation). In addition, however, the ages in the 450-600 m.y. range sporadically recorded in Senegal and Mauritania suggest that the Mauritanide Zone coincides with the late Precambrian orogenic belt (*ibid.*). Such an analysis integrates, at least structurally, the Mauritanide Zone with the marginal orogenic zone identified along the coastal regions of Sierra Leone and characterised by 450-550 m.y. ages.

In summary, it is clear that important late Precambrian–early Palaeozoic orogenesis is now largely expressed in the geosynclinal facet of Upper Precambrian sedimentation in Central Africa, South-West Africa, and elsewhere; and as a vestigial facet of rejuvenated basement rocks in the southern Cameroons and the Central African Republic, the Mozambique Belt, the Malagasy Republic and perhaps the Dahomeyan of West Africa. In addition, parts of the orogenic belt were involved in later orogenesis; for example, in the Hercynian movements of the Cape Fold Belt of South Africa, and in the Acadian and Hercynian Orogenies of the Mauritanide Zone of West Africa— these represent rejuvenated facets of orogenesis.

Within these differing facets, orogenic activity is reflected by ages of 450-680 m.y., and a number of terms have been used for events in this range; for example, the Mozambiquian Orogeny in the Mozambique Belt (Holmes, 1965), the Riphean Orogeny in the Hoggar (Black, 1966), and Pan-African Orogeny (Kennedy, 1964) or Damaran-Katangan Orogeny (Clifford, 1968a), for Africa as a whole. This, however, is not the place for discussions on nomenclature, but the importance of this major event is recognised by all authorities. It affected an enormous area, perhaps half of the continent, and thereafter Africa became essentially a stable segment of the earth's crust and has remained so (Fig. 6; African Craton I of Fig. 7). Only minor zones of mobility in the southern, northwestern and northern extremities survived as sites of later folding, either in later Palaeozoic to early Mesozoic times, in the Cape Fold Belt and the Mauritanides (De Villiers, 1944; Furon et al., 1958; Sougy, 1962), or in Tertiary times in the Atlas Mountains (De Sitter, 1956).

Figs. $2 \rightarrow 3 \rightarrow 5 \rightarrow 6$ illustrate the successive changes in the structural development of Africa, and clearly show that the overall effect of African orogenesis has been the progressive cratonisation of the continent with the ultimate formation of the African 'high craton' (African Craton II) of the present day; the 'flow-sheet' of that process is illustrated in Fig. 7. There is no suggestion of lateral accretion of new crust during this process (Clifford, 1968a, p. 402). Instead, these changes have been considered in terms of variations in the size of mantle convection cells (ibid.; Clifford, 1968b) and it has been argued that the structural data for the more youthful part of African history suggest that motivating sub-crustal convection cells increased in size from late Precambrian–early Palaeozoic times to Upper Palaeozoic–Tertiary times (ibid.). If this reasoning is extrapolated back in geological time, the cells may have been very small about 2750-3000 m.y. ago (ibid.), and this is consistent with views expressed by Macgregor (1951a) and Talbot (1967, 1968) on the role of convection in the ancient gregarious granite batholiths of the Rhodesian nucleus (see p. 4).

It is clear from Figs. 6 and 7 that by early Palaeozoic times the structural scene had been set for the spectacular Mesozoic to Recent vulcanicity for which Africa is particularly well known. For this reason, a substantial part of this volume is concerned with selected examples from now classic regions. Black and Girod (Chapter 9) deal with the important Mesozoic to Recent

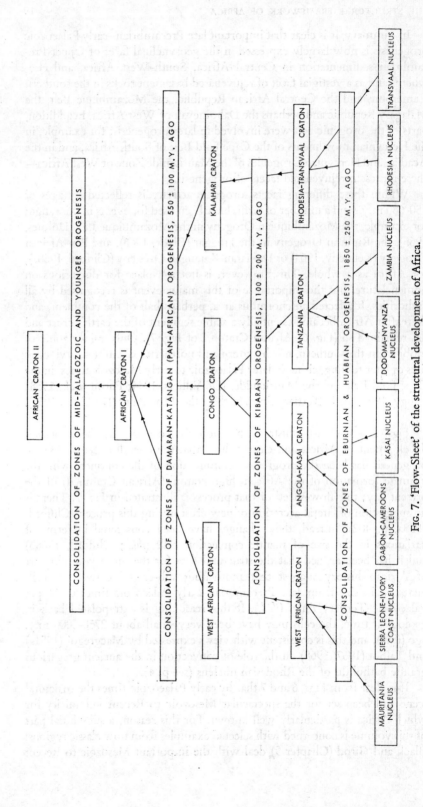

Fig. 7. 'Flow-Sheet' of the structural development of Africa.

igneous activity of West Africa and discuss the relationship between structural environment and magma-type. Cox (Chapter 10) discusses the vulcanism which extended over large parts of southern Africa during the Karroo Period, whereas Woolley and Garson (Chapter 11) mainly consider the Cretaceous igneous activity in the Chilwa Alkaline Province of Malawi; radiometric ages of 160–200 m.y. have been obtained for the Karroo magmatism, whilst dates of 100–140 m.y. have been yielded by the Chilwa irruptives (*ibid.*). Pertinent to both the study by Cox, and that by Woolley and Garson, is Vail's treatment (Chapter 16) of dykes and related irruptive rocks in eastern Africa as a whole. The Tertiary to Recent Volcanic Province of the East African Rift System, to the north, is described by King (Chapter 12); igneous activity of a similar age in the Ethiopian Rift–Red Sea–Gulf of Aden region is reviewed by Gass (Chapter 13); and the Tertiary to Recent vulcanism in the Tibesti Massif of north Africa is described in detail for the first time in English by Vincent (Chapter 14).

All these authors emphasise the close relationship between vulcanism and faulting and warping and/or doming in time and space, and it is clear that there has been an intimate interplay between crustal movements and more fundamental deep-seated mechanisms. In the latter regard, there is now a large body of opinion in support of some form of convection or turbulence in the mantle, and the relationship between temperature changes and mass movement on the one hand, and the composition of resulting vulcanism on the other, are reviewed in this volume by Harris (Chapter 18). Moreover, a number of other authors invoke the influence of convection from extreme points of the geological timescale. For example, it is considered to have played a major role in the formation of the Great Dyke of Rhodesia, 2500–2600 m.y. ago (Bichan, Chapter 3); in Karroo vulcanism and the fragmentation of Gondwanaland (Cox, Chapter 10); and in the tectonism and Cretaceous to Recent vulcanism in the Ethiopia–Red Sea–Gulf of Aden region (Gass, Chapter 13). However, it should perhaps be stressed that, in contrast to widely quoted oceanic models (Wilson, 1963), neither Gass nor Cox consider that there is any areal coincidence between major upwelling in mantle convection systems on the one hand, and vulcanicity and rift-faulting on the other; they believe that the latter more probably took place 'on the backs' of the convecting cells. In the case of the Ethiopia–Red Sea–Gulf of Aden region, Gass suggests that the development of a localised diapiric lithothermal system produced vulcanism with attendant doming and crustal attenuation.

For a number of years, it has been considered that the most cogent evidence of transfer of material from the mantle is the presence of xenoliths of peridotite and eclogite in kimberlite pipes. These interesting bodies are discussed in this volume by Dawson (Chapter 15) who reviews not only the type area of Kimberley in South Africa, but also the distribution and structural environment of kimberlites elsewhere in the continent (see also Harris, this volume, p. 434). Kennedy (1964) has pointed out that *diamondiferous* kimberlites are restricted to the West African, Congo, and Kalahari Cratons

(Fig. 5). However, it has been suggested by Clifford (1966) that economically diamondiferous kimberlites only occur in regions which have not been orogenically deformed since 1850±250 m.y. ago (see Fig. 3), namely the Rhodesia-Transvaal, Tanzania, Angola-Kasai and West African Cratons; these *older cratons* are also characterised by important deposits of Au, Cr, Fe, Pt and asbestos (*ibid.*).

The older cratons stand in marked contrast to the *younger orogens* (Fig. 3), represented by zones which suffered orogenic deformation during and since the Kibaran Orogeny, 1100±200 m.y. ago. From an economic point of view, these zones are characterised by important Cu, Pb, Zn, Co, Be, Sn, W, and Nb-Ta mineralisation (*ibid.*). In this volume, Von Knorring (Chapter 8) presents a comprehensive review of pegmatites in the equatorial and southern African portions of these orogenic belts, and notes the presence of a western 'tin-belt' (the Kibaran-Damaran Belt) and an eastern 'rare-earths belt' (the Mozambique Belt); he argues that since the former includes orogenic belts of two different ages, the geochemical character of the belt may reflect fundamental geochemical variations of the earth's crust. In this context a complementary chapter of wide geological and geographical sweep is presented by Rooke (Chapter 17) and gives the results of a statistical geochemical study of anorogenic and orogenic acid igneous rocks of a wide range of ages and from a large number of localities in eastern, southern and western Africa. Comparing average analyses she notes that, irrespective of age, Al, Ca, Mg and Sr are relatively high in acidic rocks from orogenic environments, whilst Fe^{3+}, Na, Mn, Ga, Li, Nb, Zr, and Y are concentrated to a greater extent in anorogenic acid rocks. Whilst the sampling can hardly be considered to be random, this work offers a significant 'stepping-stone' in the delineation of geochemical sialic crustal provinces in Africa. A greater knowledge of such provinces is clearly of great importance in view of the fact that a number of authors in this volume invoke crustal fusion in the genesis of significant quantities of acidic igneous rocks, particularly of Phanerozoic age (Black and Girod, p. 202; Woolley and Garson, p. 259; and Vincent, p. 317).

In this Kennedy Volume, therefore, a number of individual works are presented by Earth Scientists actively engaged in problems of African structure, magmatism, and geochemistry. That the conclusions may differ in emphasis in different chapters is a challenging reflection of what we still do not know; the over-riding point of agreement is that without diastrophism, in the broad sense of Billings (1960), there is no igneous activity. Together, the chapters define a large number of the parameters which are presently being used to identify the causal mechanisms which link crustal tectonics with igneous activity. Particularly important amongst these is the prior structural history of various regions for it has been demonstrated that magmatism in West Africa, clearly aligned parallel to the late Precambrian—early Palaeozoic zone of Nigeria and the Hoggar, differs markedly from that of the West African Craton (Black and Girod, p. 185); that the site of Karroo sedimentation and vulcanicity in southern Africa was related to the pre-existing pattern of the pre-Silurian Mozambique and Limpopo Belts (Cox, p. 213); that the faults of the major East African Rift System and the associated volcanic centres are largely localised along the linear Ubendian, Kibaran and Mozambique Orogenic Belts (King, p. 263); and that basement structure has played an important role in the development of the Tibesti Volcanic Province (Vincent, p. 303). In the light of these examples it is difficult to resist the conclusion that the late Palaeozoic to Recent epeirogenic faulting, warping and flexuring posthumously reflect the structural heterogeneity imposed during pre-Silurian times. Perhaps, then, compositional variations in petrographic subprovinces (as, for example, between the Eastern and Western Rift Valleys) may reflect, in part at least, differences in crustal composition resulting from longstanding differences in crustal history.

Acknowledgements

I am grateful to: Drs Ian G. Gass and Dorothy H. Rayner for critically reading the manuscript of this paper; Mr R. C. Boud for draughting Figs. 1, 2, 3, 5 and 6 and for his help, and that of Mr T. F. Johnston, in preparing Figs. 4 and 7. My work in Africa was initiated whilst Professor Kennedy was the Director of the Research Institute of African Geology (Leeds, U.K.); I thank him for his inspiring encouragement, and the Anglo-American Corporation Ltd for its generous financial support of the institute.

REFERENCES

ALDRICH, L. T., WETHERILL, G. W., DAVIS, G. L., and TILTON, G. R. 1958. Radioactive ages of micas from granitic rocks by Rb–Sr and K–A methods. *Trans. Am. geophys. Un.*, **39**, 1124.

ALLEN, P. M. 1968. The stratigraphy of a geosynclinal succession in western Sierra Leone, West Africa. *Geol. Mag.*, **105**, 62.

ALLSOPP, H. L. 1962. Rb–Sr age measurements on total rock and separated-mineral fractions from the Old Granite of the central Transvaal. *J. geophys. Res.*, **66**, 1499.

—— 1965. Rb–Sr and K–Ar age measurements on the Great Dyke of Southern Rhodesia. *J. geophys. Res.*, **70**, 977.

—— BURGER, A. J., and VAN ZYL, C. 1967. A minimum age for the Premier kimberlite pipe yielded by biotite Rb–Sr measurements, with related galena isotopic data. *Earth Planetary Sci. Let.*, **3**, 161.

ANDREWS-JONES, D. A. 1968. Petrogenesis and geochemistry of the rocks of the Kenema district, Sierra Leone. *12th Ann. Rep. Res. Inst. African Geol., Univ. Leeds*, 20.

ANHAEUSSER, C. R. 1966. Facets of the granitic assemblage on the northwest flank of the Barberton Mountain Land. *Inform. Circ. 32, Econ. Geol. Res. Unit, Univ. Witwatersrand, Johannesburg.*

—— MASON, R., VILJOEN, M. J., and VILJOEN, R. P. 1968. A reappraisal of some aspects of Precambrian Shield geology. *Inform. Circ. 49, Econ. Geol. Res. Unit, Univ. Witwatersrand, Johannesburg.*

BASSOT, J. P., BONHOMME, M., ROQUES, M., and VACHETTE, M. 1963. Mesures d'âges absolus sur les séries précambriennes et paléozoiques du Sénégal oriental. *Bull. Soc. géol. Fr.*, **5**, 401.

BILLINGS, M. P. 1960. Diastrophism and mountainbuilding. *Bull. geol. Soc. Am.*, **71**, 363.

BLACK, R. 1966. Sur l'existence d'une orogénie riphéenne en Afrique occidentale. *C.R. Acad. Sci. Paris*, **262**, 1046.

—— 1967. Sur l'ordonnance des chaines métamorphiques en Afrique occidentale. *Chron. Mines Rech. Min.*, **364**, 225.

BONHOMME, M. 1962. Contribution a l'étude géochronologique de la plate-forme de l'Ouest Africain. *Annls. Fac. Sci. Univ. Clermont*, **5** (5).

CAHEN, L. 1961. Review of geochronological knowledge in middle and northern Africa. *Annls. N.Y. Acad. Sci.*, **91** (2), 535.

—— 1963. Grands traits de l'agencement des éléments du soubassement de l'Afrique centrale. Esquisse tectonique au 1/5,000,000. *Annls. Soc. géol. Belg.*, **85**, B183.

—— DELHAL, J., and DEUTSCH, S. 1967. Rubidium-strontium geochronology of some granitic rocks from the Kibaran Belt (central Katanga, Rep. of the Congo). *Annls. Mus. roy. Afr. centr., Sci. géol.*, **59**.

—— DELHAL, J., and MONTEYNE-POULAERT, G. 1966. Age determinations on granites, pegmatites and veins from the Kibaran Belt of central and northern Katanga (Congo). *Nature, Lond.*, **210**, 1347.

—— and SNELLING, N. J. 1966. *The geochronology of equatorial Africa*, Amsterdam.

CLIFFORD, T. N. 1963. The Damaran episode of tectono-thermal activity in South-West Africa, and its regional significance in southern Africa. *7th Ann. Rep. Res. Inst. African Geol., Univ. Leeds*, 37.

—— 1966. Tectono-metallogenic units and metallogenic provinces of Africa. *Earth Planetary Sci. Let.*, **1**, 421.

—— 1967. The Damaran episode in the Upper Proterozoic–Lower Paleozoic structural history of southern Africa. *Spec. Pap. geol. Soc. Am.*, **92**.

—— 1968a. Radiometric dating and the pre-Silurian geology of Africa. In *Radiometric dating for geologists*, p. 299. (Eds. E. I. Hamilton and R. M. Farquhar), London.

—— 1968b. African structure and convection. *Trans. Leeds geol. Ass.*, **7**, 291.

—— NICOLAYSEN, L. O., and BURGER, A. J. 1962. Petrology and age of the pre-Otavi basement granite at Franzfontein, northern South-West Africa. *J. Petrology*, **3**, 244.

—— ROOKE, J. M., and ALLSOPP, H. L. 1969. Petrochemistry and age of the Franzfontein granitic rocks of northern South-West Africa, *Geochim. et cosmochim. Acta*, **33**, 973.

DELBOS, L. 1964. Mesures d'âges absolus sur les séries précambriennes de Madagascar. *C.R. Acad. Sci. Paris*, **258**, 1853.

DE SITTER, L. U. 1956. *Structural geology*, New York.

DE SWARDT, A. M. J., and DRYSDALL, A. R. 1964. Precambrian geology and structure in central Northern Rhodesia. *Mem. geol. Surv. N. Rhodesia*, **2**.

DE VILLIERS, J. 1944. A review of the Cape Orogeny. *Annls. Univ. Stellenbosch*, **22**, Ser. A, 183.

DU TOIT, A. L. 1954. *The geology of South Africa*, 3rd ed., Edinburgh.

EMBERGER, A. 1958. Les granites stratoïdes du Pays Betsileo (Madagascar). *Bull. Soc. géol. Fr.*, **8**, 537.

FERRARA, G., and GRAVELLE, M. 1966. Radiometric ages from western Ahaggar (Sahara) suggesting an eastern limit for the West African Craton. *Earth Planetary Sci. Let.*, **1**, 319.

FURON, R. 1963. *Geology of Africa*, Edinburgh.

—— et al. 1958. Esquisse structurale provisoire de l'Afrique au 1 : 10,000,000. *Int. geol. Congr., Ass. Serv. géol. Afr.*, Paris.

GRANT, N. K. 1967. Complete late Pre-Cambrian to early Palaeozoic orogenic cycle in Ghana, Togo and Dahomey. *Nature, Lond.*, **215**, 609.

GREENWOOD, J. E. G. W. 1961. The Inda Ad Series of the former Somali Protectorate. *Overseas Geol. Mineral Resources*, **8**, 288.

HAUGHTON, S. H. 1963. *The stratigraphic history of Africa south of the Sahara*, Edinburgh.

HEPWORTH, J. V., KENNERLEY, J. B., and SHACKLETON, R. M. 1967. Photogeological investigation of the Mozambique Front in Tanzania. *Nature, Lond.*, **216**, 146.

HOLMES, A. 1951. The sequence of Pre-Cambrian orogenic belts in south and central Africa. *Int. geol. Congr.*, **18** (14), 254.

—— 1965. *Principles of physical geology*, London.

—— and CAHEN, L. 1957. Géochronologie africaine 1956. *Acad. roy. Sci. colon., Sci. nat., Mem. in-8°., n.s.*, **5** (1).

JACOBSON, R. R. E., SNELLING, N. J., and TRUSWELL, J. F. 1964. Age determinations in the geology of Nigeria, with special reference to the older and younger granites. *Overseas Geol. Mineral Resources*, **9**, 168.

JOHNSON, R. L. 1968. Structural history of the western front of the Mozambique Belt in northeast Southern Rhodesia. *Bull. geol. Soc. Am.*, **79**, 513.

KENNEDY, W. Q. 1954. A general introduction. In *The tectonic control of igneous activity*, p. 2. 1st Inter-Univ. geol. Congr., Univ. Leeds.

—— 1964. The structural differentiation of Africa in the Pan-African (±500 m.y.) tectonic episode. *8th Ann. Rep. Res. Inst. African Geol., Univ. Leeds*, 48.

—— 1965. The influence of basement structure on the evolution of the coastal (Mesozoic and Tertiary) basins of Africa. In *Salt basins around Africa*, p. 7. Inst. Petrol., London.

—— (in preparation). The West African Craton.

LASSERRE, M. 1964. Mesures d'âges absolus sur les séries précambriennes et paléozoïques du Cameroun (Afrique équatoriale). *C.R. Acad. Sci. Paris*, **258**, 998.

LEDENT, D., LAY, C., and DELHAL, J. 1962. Premières données sur l'âge absolu des formations anciennes du 'socle' du Kasai (Congo méridional). *Bull. Soc. belge Géol.*, **71**, 223.

MACGREGOR, A. M. 1951a. Some milestones in the Precambrian of Southern Rhodesia. *Trans. geol. Soc. S. Afr.*, **54**, xxvii.

—— 1951b. A comparison of the geology of Northern and Southern Rhodesia and adjoining territories. *Int. geol. Congr.*, **18** (14), 111.

MARTIN, H. 1961. The Damara System in South-West Africa. *C.C.T.A. southern reg. Comm. Geol.*, Pretoria, 91.

—— 1965. *The Precambrian geology of South West Africa and Namaqualand*. Precambrian Res. Unit, Univ. Cape Town.

McCONNELL, R. B. 1959. The Buganda Group, Uganda, East Africa. *Int. geol. Congr.*, **20**, *Asoc. Serv. geol. Afr.*, 163.

McELHINNY, M. W., BRIDEN, J. C., JONES, D. L., and BROCK, A. 1968. Geological and geophysical implications of paleomagnetic results from Africa. *Rev. Geophys.*, **6**, 201.

MENDES, F. 1964. Ages absolus par la méthode au strontium de quelques roches d'Angola. *C.R. Acad. Sci. Paris*, **258**, 4109.

—— 1966. Ages absolus par la méthode au strontium de quelques roches d'Angola. *C.R. Acad. Sci. Paris*, **262**, 2201.

NICOLAYSEN, L. O. 1962. Stratigraphic interpretation of age measurements in southern Africa. In *Petrologic studies: a volume in honor of A. F. Buddington*, p. 569. (Eds. A. E. J. Engel, H. L. James and B. F. Leonard), Geol. Soc. Am.

—— and BURGER, A. J. 1965. Note on an extensive zone of 1000 million-year old metamorphic and igneous rocks in southern Africa. *Sci. Terre*, **10** (3–4), 497.

—— DE VILLIERS, J. W. L., BURGER, A. J., and STRELOW, F. W. E. 1958. New measurements relating to the absolute age of the Transvaal System and of the Bushveld Igneous Complex. *Trans. geol. Soc. S. Afr.*, **61**, 137.

OLD, R. A. 1968. The geology of part of south-east Uganda. Unpublished Ph.D. thesis, Univ. Leeds.

OOSTHUYZEN, E. J., and BURGER, A. J. 1964. Radiometric dating of intrusives associated with the Waterberg System. *Annls. geol. Surv. S. Afr.*, **3**, 87.

PRETORIUS, D. A. 1964. *6th Ann. Rep. Econ. Geol. Res. Unit, Univ. Witwatersrand*, Johannesburg.

QUENNELL, A. M., and HALDEMANN, E. G. 1960. On the subdivision of the Precambrian. *Int. geol. Congr.*, **21** (9), 170.

RAMSAY, J. G. 1965. Structural investigations in the Barberton Mountain Land, eastern Transvaal. *Trans. geol. Soc. S. Afr.*, **66**, 353.

ROCCI, G. 1965. Essai d'interprétation de mesures géochronologiques. La structure de l'Ouest Africain. *Sci. Terre*, **10** (3-4), 461.

ROERING, C. 1965. The tectonics of the main gold-producing area of the Barberton Mountain Land. *Inform. Circ. 23, Econ. Geol. Res. Unit, Univ. Witwatersrand, Johannesburg.*

ROGERS, A. S., MILLER, J. A., and MOHR, P. A. 1965. Age determinations on some Ethiopian basement rocks. *Nature, Lond.*, **206**, 1021.

ROUBAULT, M., DELAFOSSE, R., LEUTWEIN, F., and SONET, J. 1965. Premières données géochronologiques sur les formations granitiques et cristallophylliennes de la République Centre-Africaine. *C.R. Acad. Sci. Paris*, **260**, 4787.

SANDERS, L. D. 1965. Geology of the contact between the Nyanza Shield and the Mozambique Belt in western Kenya. *Bull. geol. Surv. Kenya*, **7**.

SCHREINER, G. D. L., and VAN NIEKERK, C. B. 1958. The age of a Pilanesberg dyke from the central Witwatersrand. *Trans. geol. Soc. S. Afr.*, **61**, 197.

SCHÜRMANN, H. M. E. 1961. The Riphean of the Red Sea area. *Geol. Fören. Stockh. Förh.*, **83**, 109.

SOUGY, J. 1962. West African fold belt. *Bull. geol. Soc. Am.*, **73**, 871.

TALBOT, C. J. 1967. Rock deformation at the eastern end of the Zambezi Orogenic Belt, Rhodesia. Unpublished Ph.D. thesis, Univ. Leeds.

—— 1968. Thermal convection in the Archaean crust? *Nature, Lond.*, **220**, 552.

VACHETTE, M. 1964a. Essai de synthèse des déterminations d'âges radiométriques de formations de l'Ouest Africain (Côte d'Ivoire, Mauritanie, Niger). *Annls. Fac. Sci. Univ. Clermont*, **25**, 7.

—— 1964b. Nouvelles mesures d'âges absolus de granites d'âge éburnéen de la Côte d'Ivoire. *C.R. Acad. Sci. Paris*, **258**, 1569.

—— 1964c. Ages radiométriques des formations cristallines d'Afrique équatoriale (Gabon, République centrafricaine, Tchad, Moyen Congo). *Annls. Fac. Sci. Univ. Clermont*, **25**, 31.

VAN BREEMEN, O., DODSON, M. H., and VAIL, J. R. 1966. Isotopic age measurements on the Limpopo Orogenic Belt, southern Africa. *Earth Planetary Sci. Let.*, **1**, 401.

VAN NIEKERK, C. B., and BURGER, A. J. 1964. The age of the Ventersdorp System. *Annls. geol. Surv. S. Afr.*, **3**, 75.

VISSER, D. J. L. *et al.* 1956. The geology of the Barberton area. *Spec. Publ. geol. Surv. S. Afr.*, **15**.

WILSON, J. T. 1963. Hypothesis of earth's behaviour. *Nature, Lond.*, **198**, 925.

M. J. VILJOEN and R. P. VILJOEN

2   Archaean vulcanicity and
      continental evolution in the
      Barberton Region, Transvaal

ABSTRACT. *The Barberton Mountain Land of the Transvaal is an extensively developed and well preserved greenstone belt. With an age of about 3400 m.y., the layered assemblage of this belt is the oldest known on earth.*

The stratigraphy commences with a thick volcanic succession termed the Onverwacht Group which attains a thickness of 50,000 ft and has been divided into six formations, each with its own distinctive assemblage and association of rock types. Ultramafic varieties, many in the form of lavas, are abundant in the lower three formations and are associated with primitive mafic lavas and felsic tuffs. Sediments are generally poorly developed in these formations. In the upper formations, tholeiitic basalt is the most abundant rock type, but felsic lavas and pyroclasts, together with cherts, become conspicuous. Early intrusive sills of ultramafic material have been emplaced into the lower formations in certain areas; they are differentiated and a variety of rock types is represented.

The granitic rocks which surround and intrude the belt are described and a new threefold classification is proposed. This is based on age, petrology, chemistry and mode of emplacement. The oldest granites (tonalitic gneisses) range in age from 3400 to 3200 m.y. and intrude the Onverwacht volcanics in a diapiric fashion giving rise to arcuate schist belts. At about 3000 m.y. ago the tonalitic gneisses were reconstituted within broad zones by a potash metasomatic event which at higher levels gave rise to the formation of a potash-rich, homogeneous, 'hood' granite. Between about 2800 and 2500 m.y. a series of young, sharply transgressive granite plutons was emplaced.

An attempt is made to assess the geological significance of the Barberton belt and surrounding granites with respect to the early development and evolution of this part of the African crust.

Introduction

EARLY Precambrian granite–greenstone terrains often constitute the most stable as well as the most ancient portions of the continents. They appear to be remarkably similar in all of the better known shield areas of the world (Anhaeusser *et al.*, 1969), the most distinctive features being their great age, together with the invariable occurrence of the so-called 'greenstone belts'. The latter all commence with a volcanic assemblage and are terminated by a sedimentary succession. These belts appear to 'float' in the form of synclinorial 'keels' or 'rafts' in a complex granitic terrain.

Younger earth processes such as metamorphism, metasomatism and cata-

clasis have, in certain areas and often within broad linear zones, resulted in
the reworking and reconstitution of the ancient granite–greenstone material
as well as varying amounts of younger cover. In Africa these younger metamorphic mobile belts, together with the ancient granite–greenstone terrains
which they encircle, form a striking pattern.

The two most southerly of the old granite–greenstone areas on the
African continent are known as the Kaapvaal and Rhodesian Cratons
(Pretorius, 1964; Anhaeusser *et al.*, 1969). The Kaapvaal Craton of South
Africa is for the most part covered by shallow-dipping, younger Precambrian

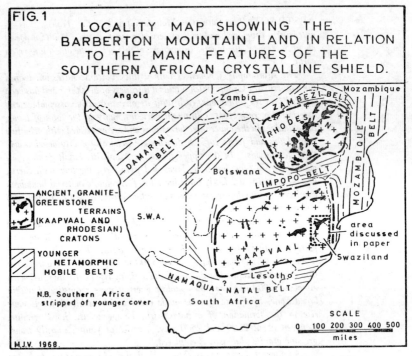

FIG. 1.

strata. In the eastern Transvaal, however, erosion has removed much of this
sedimentary cover, exposing to the south an extremely well-developed and
preserved greenstone belt surrounded by a variety of granites (Fig. 1). This
belt, commonly known as the Barberton Mountain Land (Hall, 1918;
Visser *et al.*, 1956; Anhaeusser *et al.*, 1967), has an age of about 3400 m.y.,
and probably constitutes the oldest clearly recognisable assemblage of rocks
known on earth (Fig. 2).

It is difficult, in the space available, to present a comprehensive account of
the varied geological aspects of the Barberton region. An attempt is, however, made to outline briefly most of the important geological features and
in accordance with the general theme of this volume, emphasis has been
placed on the latest findings and ideas concerning the lowermost volcanic

assemblage of the belt and the enveloping granites.[1] The significance of these rocks with respect to the early development and evolution of the earth's crust in this region of southern Africa is also discussed.

General geology

The rocks of the Barberton greenstone belt have collectively been referred to as the Swaziland System (Hall, 1918). The lower volcanic succession has been termed the Onverwacht Series whereas the overlying sedimentary assemblages have been referred to as the Fig Tree Series and the Moodies System (Visser et al., 1956). It is suggested that in view of the fact that definitive time-stratigraphic controls in the form of fossils or detailed radiometric dating are lacking, a lithostratigraphic nomenclature, based on the recommendation of the International Subcommission on Stratigraphic Terminology (1961), Truswell (1967) andNewton (1968), should be adopted. With this new classification, the Swaziland System now becomes the Swaziland Sequence, and the Onverwacht and Fig Tree Series and the Moodies System now become Groups.

The predominantly volcanic Onverwacht Group attains a thickness of over 50,000 ft and constitutes the initial magmatic phase of the Swaziland Sequence. It consists dominantly of basaltic lavas with a preponderance of ultramafics towards the base and more felsic lava types towards the top. Sediments, largely cherts, limestones and shales, as well as pyroclasts, are of minor importance.

In most instances the essentially sedimentary Fig Tree Group conformably overlies the Onverwacht Group, and is composed mainly of pelitic sediments together with siliceous chemical precipitates. The succession attains a thickness of 7,000 ft and can be divided into two main pulses of sedimentation. The first of these consists mainly of the 'greywacke suite'; higher in this sequence the greywackes become finer, and shale and banded iron formations are well represented. The second major pulse of sedimentation commenced with the accumulation of grits and coarse greywackes and ended with the deposition of crystalline trachytic tuffs and agglomerates with minor feldspathic greywackes and greywacke conglomerates. Sedimentary structures indicate that the greywackes were transported by turbidity currents.

The Moodies Group, which attains a thickness of 14,000 ft, overlies the Fig Tree Group, in places conformably, and in others unconformably. The dominant rock types are conglomerates, quartzo-feldspathic sandstones, subgreywackes, siltstones and shales. Three major pulses of sedimentation have been recognised within the group. These invariably commence with conglomerates or sandstones and are terminated by minor volcanics as well as shale and banded magnetic jaspilite horizons. Sedimentary structures in the Moodies rocks are indicative of a shallow-water environment of deposition.

[1] Detailed accounts of aspects of the geology and geochemistry of the volcanic and granitic rocks of the Barberton region are to be published in a special volume (vol. 2, 1970) of the Geological Society of South Africa dealing with South Africa's contribution to the IUM Project.

The Onverwacht, Fig Tree and Moodies Groups respectively, have some similarities with the well-established ophiolite, flysch and molasse assemblages of geosynclinal environments (Anhaeusser *et al.*, 1967). More recently, however, it has become apparent that such analogies cannot be applied rigorously to the Barberton belt which appears to have a geotectonic setting unique to the early Precambrian (Anhaeusser *et al.*, 1969). The surface distribution of the three groups of the Swaziland Sequence is shown in Fig. 2.

The rocks of the Swaziland Sequence are surrounded by a variety of intrusive granites which are responsible for deformation as well as contact metamorphism. Radiometric age determinations indicate ages within the

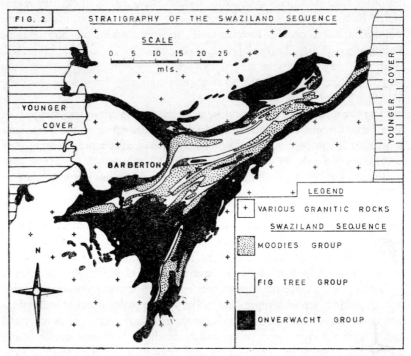

FIG. 2.

range of 3400 m.y. to about 2500 m.y. for these granites (Allsopp *et al.*, 1962, 1968; De Gasparis, 1967; R. Davies and E. J. Oosthuyzen, personal communication).

The entire belt has been subjected to a low-grade regional greenschist facies of metamorphism, which has been upgraded in various places to the amphibolite facies, due to contact metamorphism caused by granite emplacement.

The broad structure of the mountain land is one of a strongly deformed, synclinal, boat-shaped remnant with a dominant east-north-easterly trend. The main deviations from this fundamental grain are narrow, tapering, arcuate protuberances of generally schistose Onverwacht volcanics which trend in various directions (Fig. 2). Conspicuous features within the mountain land are a series of well-developed, east-north-easterly-trending, pre-dominantly tight synclinal folds with steeply dipping limbs. Locally these folds have been deformed by the intrusion of diapiric granite plutons. In general the anticlinal structures between the major synclines are poorly developed and major fault zones often separate large juxtaposed synclines. These faults are extensive, regional features which divide the belt into a series of narrow, east-north-easterly trending blocks, parallel to the regional grain (Roering, 1965).

Stratigraphy of the Onverwacht Group

THE CONCEPT OF THE JAMESTOWN IGNEOUS COMPLEX

The Onverwacht volcanics are exceptionally well developed and preserved in the southern part of the Barberton Mountain Land, and it was here that

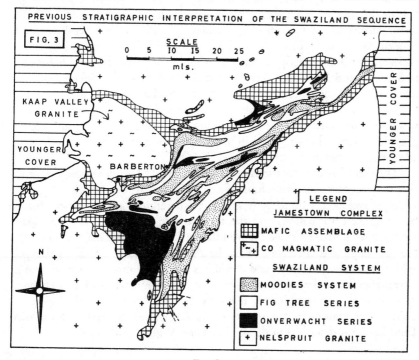

FIG. 3.

the group was first recognised and described by Hall (1918). The remainder of the mafic rocks surrounding the mountain land were considered to represent an intrusive post-Onverwacht succession which Hall (1918) termed the 'Jamestown Series'. Following Hall, Visser *et al.* (1956) retained the Onverwacht Series in the southern part of the belt, but considered the 'Jamestown Series' to constitute the mafic assemblage of a post Moodies intrusive complex. This complex, termed the 'Jamestown Igneous Complex', was thought to have the hornblende granodioritic mass of the Kaap Valley granite as a comagmatic felsic phase. This interpretation is illustrated in Fig. 3.

It has recently been shown (Viljoen and Viljoen, 1967), however, and has now been officially accepted (Ferguson, *coordinator*, 1967) that the majority of mafic, as well as many of the ultramafic rocks, such as amphibole, talc, and chlorite schists, and serpentinites, previously grouped into the Jamestown Igneous Complex, merely represent the more strongly metamorphosed equivalents of the Onverwacht Group as defined in the type area in the southern part of the belt; the presently accepted stratigraphic interpretation is shown in Fig. 2.

DETAILED STRATIGRAPHY OF THE ONVERWACHT GROUP

In an earlier paper (Viljoen and Viljoen, 1967) a threefold subdivision of the Onverwacht Group in the type area was proposed. In the light of new field evidence, however, it was found necessary to make further subdivisions so that in the present paper the Onverwacht Group has been divided into six formations. Each of these is characterised by a distinctive assemblage and association of rock types, the distribution of which is shown in Fig. 4. The rocks are generally steeply or vertically dipping and have been folded about two major anticlines, the Onverwacht and Steynsdorp anticlines, and an intervening syncline, the Kromberg syncline. Farther east and into Swaziland the sequence has been folded about another series of more closely spaced synclines and anticlines (Fig. 4). Faulting, intrusion of granite and stratigraphic variation have resulted in the drastic thinning and/or elimination of all of the formations in certain areas (Fig. 4).

The three lower formations of the Onverwacht Group are distinctive in that they contain abundant ultramafic material and are characterised by very poorly-developed sediments (Fig. 5). Representing an important break between these formations and the three upper formations, is a persistent sedimentary horizon termed the Middle Marker. From this marker upwards, the petrology and chemistry of the basaltic rocks shows a marked change, with felsic lavas, pyroclasts and sediments becoming more conspicuous and ultramafics much less abundant (Fig. 5). The significance of this change, as well as the composition of the Onverwacht Volcanics with respect to gold mineralisation in the southern portion of the Barberton Mountain Land has been described in detail elsewhere (Viljoen *et al.*, 1969).

The Sandspruit Formation

The Sandspruit Formation constitutes the lowermost recognisable unit of the Onverwacht Group. It is everywhere in contact with the surrounding granites which metamorphose, extensively intrude and fragment the rocks so that the formation is preserved for the most part as groups or stringers of closely packed xenoliths. The assemblage has an approximate thickness of 7000 ft; an unknown amount of the lower part, together with possible lower formations, has been completely eliminated by intrusive granite.

The rocks consist mainly of ultramafic bands, lenses and pods, interlayered and closely associated with minor mafic bands and lenses. Petrologically the

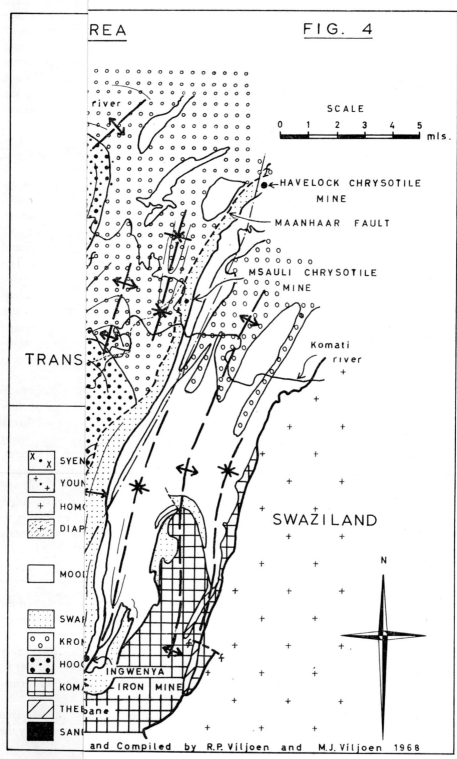

REA   FIG. 4

SCALE

0   1   2   3   4   5
                              mls.

● HAVELOCK CHRYSOTILE
         MINE

MAANHAAR FAULT

MSAULI CHRYSOTILE
        MINE

Komati
river

TRANS

SWAZILAND

N

SYEN

YOUN

HOMO

DIAP

MOOI

SWAF

KRON

HOOG

KOMA

THEB

SAN

INGWENYA
IRON MINE

and Compiled by R.P. Viljoen and M.J. Viljoen 1968

ultramafics range from almost pure serpentinites consisting largely of antigorite with minor amounts of chlorite and magnetite, through antigorite-chlorite-tremolite varieties, to tremolite-chlorite rocks. They constitute about 60-70% of the succession, and the metabasalts, in the form of dark green to black hornblende or actinolitic-hornblende amphibolites with minor amounts of soda plagioclase, make up the remainder; occasional narrow sedimentary interlayers occur.

The Theespruit Formation

The Theespruit Formation attains a thickness of 6200 ft and comprises a succession of metamorphosed mafic lavas characterised by persistent, interlayered, felsic tuffs. Serpentinised ultramafics, often as conformable and persistent bands, but also occurring as pods and lenses, are present, together with talc, chlorite talc and carbonate schists.

The main distinguishing feature is the occurrence of the felsic tuffs, which grade from white, fine-grained, siliceous, often friable rocks consisting almost exclusively of quartz with minor amounts of sericite and pyrophyllite, to bedded, reworked, fine- and coarse-grained felsic tuffs and agglomerates. The latter are often aluminous, containing, in addition to quartz and sericite, conspicuous andalusite, pyrophyllite and chloritoid. In places these rocks grade into more mafic varieties. The felsic horizons are frequently overlain by, or closely associated with, narrow, generally impersistent, bands and lenses of black, siliceous, cherty sediments. Primitive fossil forms have recently been reported in some of the latter rocks (Engel *et al.*, 1968) which make these the oldest known forms yet found.

The metabasalts consist largely of hornblende, actinolitic-hornblende and soda plagioclase (mainly oligoclase) with actinolite and chlorite becoming more conspicuous away from the granite contact. Pillow structures, spherulites and variolites, although not plentiful, are encountered in places. The ultramafics are comprised of antigorite, with minor amounts of tremolite, chlorite, and magnetite.

The Komati Formation

The Komati Formation attains a thickness of 11,500 ft, and consists of an alternating sequence of amphibolitised, pillowed and massive basalts and ultramafic bands. An important feature is the absence of interlayered sediments or felsic rocks, although intrusive bodies of feldspar and quartz porphyry are plentiful and rather typical. The ultramafics predominate in the lower half of the formation where a striking feature is the sympathetic thinning of individual ultramafic and pillow basalt horizons. The extrusive nature of most of the ultramafics has recently been conclusively demonstrated, in particular by the finding of pillow structures near the top of some of the bands (Viljoen and Viljoen, in preparation). The ultramafics are in places remarkably fresh-looking with a dark blue-black colour. Microscopic examination, however, reveals that they have been extensively serpentinised although in some cases well-preserved kernels of olivine, rimmed by antigorite and split by magnetite partings, form a mesh constituting most of the rock. This rock, originally an olivine-rich peridotite, grades into a variety consisting of antigorite with varying amounts of tremolite, chlorite and magnetite. The main minerals encountered in the metabasalts are actinolite and soda plagioclase with lesser amounts of anthophyllite, chlorite and cummingtonite.

The lavas become sheared and more talcose towards the top of the formation, with an accompanying change in chemistry. In addition, bands and lenses of carbonate, possibly in part sedimentary and associated with palagonitic tuffs, are a notable feature in many instances just below the Middle Marker (Viljoen *et al.*, 1969).

The Hooggenoeg Formation

The Komati Formation is overlain by a narrow chert and carbonate sediment which, in view of its position in the sequence and its remarkable persistence, has been referred to as the Middle Marker. This horizon constitutes the base of the Hooggenoeg Formation and is one of the most significant units within the entire Onverwacht Group. It has an extremely wide areal extent and marks an important change and probable time break within the volcanic pile as noted previously.

In the type area the Hooggenoeg Formation attains a maximum thickness of 15,900 ft (see Fig. 5). A difference in the mineralogical and chemical composition of the basalts renders them more competent and resistant to erosion, resulting in an abrupt change in the topography above the Middle Marker.

The most striking feature of the formation is the cyclic nature of the vulcanicity. Each cycle (of which there are five or more) commences with a large accumulation of basalt which passes up into a narrow zone of andesitic to dacitic or more felsic lava. The latter in turn is capped by black and white chert which terminates each individual cycle. The mafic component of the cycles becomes thinner higher up in the succession and varies from 4500 ft for the lowermost cycle to about 600 ft for some of the uppermost cycles. The top of the Hooggenoeg Formation is characterised by a distinctive, broad, felsic zone capped by a substantial chert.

The lavas are predominantly tholeiitic basalts, frequently containing well-developed pillow structures, but more generally having the form of massive sill-like flows, at times over 1000 ft in thickness. They are generally less altered than the often strongly carbonated and silicified pillow lavas into which they grade. Chlorite and actinolite are the main minerals encountered as well as partly sericitised sodic plagioclase, together with leucoxene and minor secondary quartz. Pyroxene remnants are present in some sections and carbonate is often encountered.

The felsic lavas, which in places contain well-developed pillow structures, together with spherulites, variolites and amygdales, are fine-grained, somewhat massive, light greenish-grey rocks. They are generally altered and

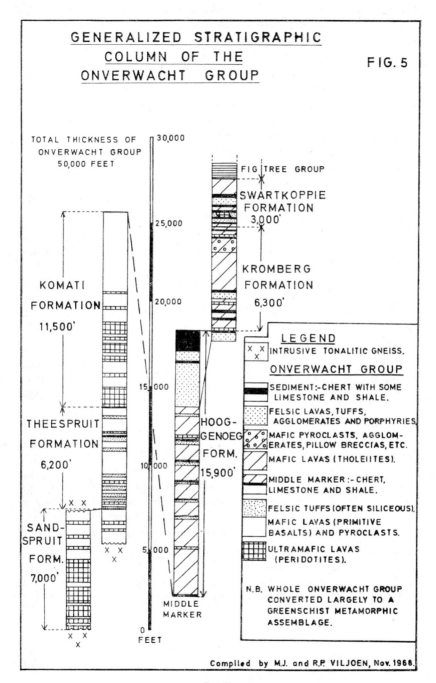

FIG. 5.

strong silicification is a common feature. There appears to be a genetic relationship between the cherts and these felsic lavas, and it is envisaged that much of the silica necessary for the formation of the cherts has been derived largely from the felsic lava just after extrusion.

In the type area, the upper portion of the Hooggenoeg Formation is composed of a thick zone of acid material in the form of felsic lavas, feldspar and quartz porphyries, and felsic pyroclasts. Further to the east in the Kromberg Syncline (Fig. 4), this zone is composed mainly of water-worked, felsic, aluminous tuffs with associated agglomeratic interlayers. Both in the type area and in the Kromberg Syncline, the upper felsic zone is overlain by a substantial black and white chert which, in the type area, contains ferruginous and shaly interlayers. Although not nearly as widespread as in the lower formations, fairly continuous ultramafic horizons, probably in the form of early sills or flows, are well developed in places, especially on the east and west limbs of the Kromberg Syncline. These often occur immediately above major chert horizons and appear to constitute the first phase of the volcanic cycles described above. Most of these ultramafics have undergone magmatic differentiation to yield a number of rock types including dunites, wehrlites, harzburgites and websterites.

A radiometric (U-Pb) age determination on zircons from the upper felsic zone of the formation has yielded an age of 3360 m.y. (Van Niekerk and Burger, 1967). This is the greatest age yet obtained from the Swaziland Sequence rocks directly, and has been taken as part of the evidence for placing the age of this assemblage at about 3400 m.y.

The Kromberg Formation

As in the case of the Hooggenoeg formation, the overlying Kromberg Formation consists dominantly of massive flows of slightly metamorphosed tholeiitic basalt (Fig. 5) with numerous, generally rather narrow, intervening zones of well-developed pillow lavas. Three zones of apparently conformable ultramafic material occur, and felsic lavas are again developed and closely associated with chert horizons, together with calc-silicate and carbonate layers. The most distinctive feature of the formation is the occurrence of a great variety of pyroclastic rocks in the form of palagonitic tuffs, cross-bedded mafic tuffs, and aquagene tuffs as well as pillow breccias similar to those described by Carlisle (1963). A variety of agglomerates also occurs. In places, black, carbonaceous and siliceous, shaly zones are developed within the chert horizons; remains of primitive life forms have recently been reported from these (Engel et al., 1968).

Where best developed and exposed, in a gorge cut by the Komati River through the Kromberg Syncline, the Kromberg Formation attains a thickness of 6300 ft (Fig. 5).

The Swartkoppie Formation

In the southern part of the Barberton Mountain Land the formation occurs to the east of the Stolzburg Syncline and in a strip along the Swaziland border where it attains a thickness of about 3000 ft (Steyn, 1965) (Fig. 5). The latter zone has been strongly sheared, and is separated from the lower formations of the Onverwacht Group by the major Maanhaar Fault (Fig. 4). The main rocks encountered are green and grey schists derived from felsic and intermediate volcanics as well as mafic schists derived from basaltic rocks. Well-developed banded black and white cherts, together with minor pyroclastic horizons and perhaps greywacke layers, are important components of the stratigraphy. A number of conformable ultramafic bands and lenses occur within the succession and, during the phase of deformation which resulted in the strong shearing of the Swartkoppie zone, appear to have been drawn out into a number of discrete pods. Two such pods contain the most important chrysotile asbestos deposits in the mountain land; for example, the Havelock body in Swaziland and the Msauli body in the Transvaal (Fig. 4).

The formation is better known and probably better developed in the northern portion of the mountain land where it has previously been grouped into the Fig Tree Series (Visser et al., 1956). Here the main rock types are banded, black, white, grey and green cherts, together with sericitic 'grey' and 'green' schists and talc-carbonate rocks. It is in these cherts that some of the initial fossil discoveries in the Barberton belt were made (Barghoorn and Schopf, 1966; Pflug, 1966).

Rocks of the Swartkoppie Formation conformably underlie the Fig Tree Group but as yet their relationship to the Kromberg Formation is unknown. In spite of this it is proposed that, because of the predominance of volcanic material as opposed to clastic sedimentary material, as well as the lack of banded ironstones, this formation be placed in the Onverwacht Group.[1]

CORRELATION OF THE ONVERWACHT GROUP

On the basis of the distinctive rock types and associations of the six formations of the Onverwacht Group in the type area, it has been possible to effect a lithostratigraphic correlation of the majority of the pre-Fig Tree rocks (Anhaeusser et al., 1966). In most areas the occurrence of aluminous felsic schists is widespread, invariably associated with metabasalts and ultramafics. As shown earlier, this assemblage characterises the Theespruit Formation as defined in the Onverwacht type area. Also present in certain areas is the association of metabasalts and ultramafics which typify the Komati Formation. Thus, of the six formations recognised in the south, it would appear that the Theespruit and Komati Formations are the only two which are extensively developed away from the type area (Fig. 6). Local variations may exist but these are generally never great enough to change the fundamental identity of the distinctive assemblages of these two formations.

[1] A recent reconnaissance investigation in the area to the east of the Stolzburg Syncline (Fig. 4) has indicated the presence of a substantial thickness of argillaceous sediments and banded ferruginous cherts interlayered with felsic volcanics and pyroclasts and lying apparently conformably above the Kromberg Formation. This assemblage, previously mapped as Fig Tree, might represent a transitional zone from the Onverwacht Group into the Fig Tree Group. Its position relative to the Swartkoppie Formation is, however, unknown so that the question of whether it should be placed in the Fig Tree or in the Onverwacht Group is as yet unresolved.

THE DIFFERENTIATED ULTRAMAFIC BODIES OF THE BARBERTON MOUNTAIN LAND

The large majority of ultramafic rocks occurring within the Barberton Mountain Land are considered by the authors to constitute an integral part of the Onverwacht Group.

Most of the ultramafics appear to take the form of subaqueous flows, but in certain instances along the northwestern region of the belt, early intrusive sills of Onverwacht age have been emplaced into the lower formations of the group. These bodies have undergone marked differentiation, on the

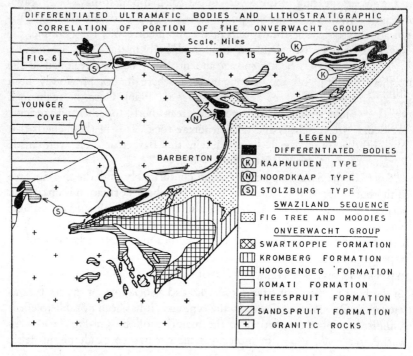

FIG. 6.

basis of which three distinct types have been recognised (Fig. 6) as follows: Kaapmuiden type; Stolzburg type; and Noordkaap type.

The Kaapmuiden type

Between Kaapmuiden and Malelane in the northeastern portion of the Barberton Mountain Land (Fig. 6), three large, nearly vertical ultramafic bodies have been emplaced. These have an average thickness of about 2000-2500 ft, and have all been subjected to the same structural disturbance and metamorphism which has affected the surrounding volcanics and sediments.

Each body is characterised by four distinctive layers or rock units. The lower zone, comprising on an average about 60% of the total, consists largely of dunite. This is followed by an orthopyroxenite zone, which constitutes on the average about 20% of the total of a particular body, and invariably forms conspicuous resistant ridges. Narrow websterite and anorthositic gabbro-norite zones comprise the remainder of the differentiated sequences. All three bodies have late-phase intrusive lenses of coarse pegmatoid consisting of plagioclase, pyroxene and amphibole. Magnesite deposits are commonly developed in the basal dunite layers of these bodies.

Details of the geology and geochemistry of the Kaapmuiden type ultramafic bodies are to be published in a special volume (vol. 1, 1970) of the Geological Society of South Africa entitled *Symposium on the Bushveld Igneous Complex and other layered mafic intrusions.*

The Stolzburg type

Somewhat similar to the bodies described above, are a group of well-layered differentiated ultramafics referred to as the Stolzburg type (Fig. 6). There are four known occurrences of this kind, typified by an alternating sequence of orthopyroxenites and serpentinised dunites. Chrysotile asbestos deposits are frequently developed in the serpentinised dunite layers of these bodies.

The Noordkaap type

In the eastern part of the Jamestown schist belt in the vicinity of Noordkaap occurs a third type of differentiated ultramafic complex referred to as the Noordkaap type. Two bodies, each with several cycles of differentiation, form prominent features in this area. The main rock types, now extensively serpentinised and metamorphosed, are considered to have been peridotites, pyroxenites and gabbros (Anhaeusser, 1969).

THE CYCLIC EVOLUTION AND GEOCHEMISTRY OF THE ONVERWACHT VOLCANICS

A fundamental aspect and one of the most striking features of the Onverwacht Volcanic Group is the cyclic nature of the vulcanicity and sedimentation. Two major cycles, each occurring at least twice in the succession and each containing numerous smaller subcycles of a similar nature, have been recognised. These cycles evolve upwards with time but remain unmistakable due to their diagnostic rock types and associations.

The Komati type cycle is characterised by a close association of alternating mafic and ultramafic lava horizons without sedimentary or felsic interlayers. This type is represented by the Sandspruit and Komati Formations which show a decrease upwards in the amount of ultramafic material from 60% in the Sandspruit to 30% in the Komati Formation.

The Hooggenoeg type cycle is distinguished by the association of mafic lava, felsic lava or tuff and siliceous, often cherty sediment. This cycle type is represented by the Theespruit, Hooggenoeg and Kromberg Formations and shows an evolutionary trend upwards; the Theespruit Formation being characterised by felsic tuffs generally capped by very narrow and impersistent chert horizons, whereas the Hooggenoeg and Kromberg Formations contain felsic lava interlayers invariably capped by well-developed cherts and carbonate sediments.

Initial geochemical results (Viljoen and Viljoen, in preparation) indicate that the volcanics belong to a calc-alkaline igneous suite. Tholeiitic basalts constitute by far the most common rock type with comparatively minor amounts of andesites, dacites and more felsic lavas. Alkali basalts or members of the alkali line of basalt descent and their metamorphic equivalents appear to be largely absent. Spilites which are common in more recent geosynclinal volcanic piles are also absent.

Metamorphism, silicification and carbonitisation have had an effect on the original chemistry of the Onverwacht volcanics in many areas. Nevertheless, the chemistry indicates a remarkable trend of igneous differentiation within the basaltic rocks. The latter are all of a 'primitive' nature, the chemistry of those above the Middle Marker closely approaching the chemistry of oceanic tholeiites. Most of the basalts of the three lower formations seem to be even more primitive and are unlike any widespread class of basalt yet described. The largely extrusive ultramafics of the lower three formations also have a distinctive chemistry which is unlike that of any well-established class of ultramafic yet described. The name 'komatiite' for these distinctive mafic and ultramafic rocks has been proposed (Viljoen and Viljoen, in preparation).

The geology of the granitic rocks surrounding the Barberton Mountain Land

A detailed investigation of the granitic terrain in Swaziland by Hunter (1961, 1965 and 1968) has laid a sound foundation on which some of the concepts developed in this paper have been based. Investigations of the granites in the Transvaal have been largely of a reconnaissance nature and include work by Visser et al. (1956), Van Eeden and Marshall (1965), Anhaeusser (1966) and Roering (1967). We have investigated the granitic terrain immediately to the south of the mountain land as well as undertaking reconnaissance investigations of the whole area. From this, a map of the granitic rocks, with the Swaziland granites modified after Hunter, has been compiled and given in Fig. 7. Three main granite episodes, each with its own distinctive tectonic style, mode of emplacement, age, petrology and chemistry, have been recognised and have resulted in the formation of four main granite types as follows: ancient tonalitic gneisses; homogeneous hood granite; Nelspruit gneisses and migmatites; and a number of 'young' granite plutons.

THE ANCIENT TONALITIC GNEISSES

The oldest granitic rocks in the area comprise a variety of biotite-tonalite gneisses, grading into hornblende varieties and including minor granodioritic gneisses. They have yielded Rb-Sr and U-Pb ages ranging from about 3400 to 3200 m.y. (Allsopp et al., 1962; R. Davies and E. J. Oosthuyzen, personal communication).

FIG.7

CL
AN
TO

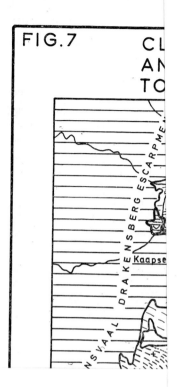

The typical features of these granites and their relationship to the Swazi-land Sequence can clearly be seen to the south and west of the mountain land where they underlie low-lying country (Fig. 7). In that area the gneisses tend to form a series of generally elliptical or circular, often discrete plutons or domes, varying in diameter from 25 miles (Kaap Valley Granite) to one mile (Fig. 7). These plutons are intrusive into the lower formations of the Onverwacht Group, and although not always strikingly apparent, are slightly transgressive in most instances. Many of them are partly separated from each other by arcuate tongues of Onverwacht volcanics, the best developed of which are known as the Jamestown and Nelshoogte schist belts (Fig. 7). These, together with other arcuate tongues, appear to have formed by downfolding and downsagging, concomitant with the slow forceful up-welling of more or less discrete, circular, largely plastic, diapiric bodies. To the south and west, away from the main development of the Onverwacht, the tongues of volcanic material within the tonalitic gneisses become frag-mental and broken into numerous xenoliths.

That these gneisses account for most of the thermal and dynamic meta-morphism of the Onverwacht rocks is shown by the orientation and nature of their contained foliations and lineations which frequently lie parallel to those in the surrounding metabasalts. The foliation in the granites is caused by the alignment of platy minerals and is accentuated by the parallel align-ment of metamorphosed mafic xenoliths.

On petrological and chemical grounds, two distinct types of tonalitic gneiss have been recognised. These are the Kaap Valley type in which the main constituents are hornblende, oligoclase, microcline and quartz, with minor amounts of biotite and chlorite; and the more widespread Nelshoogte type characterised by biotite with little or no hornblende. Few of the gneisses in this area have as yet been dated although preliminary investigations (E. J. Oosthuyzen, personal communication) indicate an age of over 3200 m.y. for the Kaap Valley Granite.

In the vicinity of Lochiel (Fig. 7) the low-lying tonalitic gneiss terrain disappears under a younger, high-lying, homogeneous, granite which forms the Lochiel plateau. Farther south in Swaziland, this plateau gives way once again to low-lying country underlain by a variety of gneisses. In that area the latter have been divided into the 'Ancient Gneiss Complex', the 'Tonalitic Gneisses' and the 'Granodiorite Suite' (Hunter, 1968). The first two, of which the 'Tonalitic Gneiss' represents the supposed granitised equivalent of the 'Ancient Gneiss Complex', are considered by Hunter to constitute a pre-Swaziland Sequence floor. The 'Granodiorite Suite', on the other hand, is considered by that author to be post-Swaziland Sequence in age. It is our opinion that the majority of these gneisses bear a marked similarity to the diapiric gneisses intrusive into the Onverwacht Group to the south of the Barberton Mountain Land (described above), and it is contended that most of them are of an identical nature and origin (Viljoen and Viljoen, in preparation). In no place in Swaziland, however, is the relation-ship of these gneisses to the Swaziland Sequence to be seen and it is conceivable, as suggested by Hunter (1968), that portions of the 'Ancient Gneiss Complex' in particular, may represent vestiges of a granitised, pre-Swaziland Sequence floor. As evidence for this Hunter cites structural features as well as the occurrence of certain rock types which are not typical of metamorphic or metasomatic equivalents of the Onverwacht Group. The abundant mafic and ultramafic lenses and pods within the gneisses, however, rather than being resisters within a granitised, pre-Swaziland succession as suggested by Hunter, are considered by the authors to represent in the main, xenoliths of Onverwacht material. If this interpretation is correct, then the provisional age of 3400 m.y. which has recently been obtained for some of these rocks (R. Davies, personal communication) would imply that the Onverwacht Group is at least this age and probably older.

THE HOMOGENEOUS HOOD GRANITE

The next major granite type is the so-called Homogeneous granite of Hunter (1968); previously termed the 'G4' and 'late-orogenic' granite and first recognised and described in Swaziland (Hunter, 1965). This is a potash-rich homogeneous granite with a Rb-Sr age of 3070 m.y. (Allsopp *et al.*, 1962) and occurs for the most part in a broad belt along the eastern flank of the mountain land in north-western Swaziland and the adjoining areas of the Transvaal. It is very largely confined to elevated areas and for this reason the term homogeneous 'hood' granite is proposed and has been used in this paper. Other smaller areas of this granite occur in the southern part of Swaziland (Fig. 7).

The granite is typically a homogeneous, grey, medium- to coarse-grained rock composed mainly of quartz and microcline, with lesser amounts of plagioclase and biotite; a notable feature of the granite is a complex stockwork of pegmatites of variable size, and the general absence of xenoliths of mafic and ultramafic material.

In many instances the granite passes gradationally downward into the underlying gneisses described previously. Where the topography is steep the gradational change takes place fairly rapidly, but where the present day erosion surface is close to the plane of the contact, complex relationships are encountered (Hunter, 1968). In some cases the gneisses grade upward, first into porphyroblastic varieties, which in turn grade, by a decrease in the content of mafic minerals, into the normal grey hood granite (*ibid.*). In some instances the gneisses approaching the hood granite grade into zones of well-banded gneiss which become strongly contorted and folded and associated with younger veins and dykes of more leucocratic granitic material. All of these features are typical of those encountered within the complex migmatitic terrain of the Nelspruit area, and as will be shown later it is our contention that they formed in the same way.

For the formation of the granite, the authors visualise a major thermal episode about 3000 m.y. ago. Intrusive dykes of granitic magma and/or granitic *ichors* rich in potash, invaded and metasomatised the older, largely tonalitic gneiss and greenstone terrain within roughly linear and vaguely east-north-easterly trending belts. This left adjoining zones of ancient tonalitic gneisses completely unaffected. At higher levels the granitic magma and/or ichors, together with the mobile leucocratic fractions derived from reconstituted tonalitic gneisses and associated greenstone remnants, merged and mixed to form the high-level potash-rich homogeneous hood granite.

THE NELSPRUIT GNEISSES AND MIGMATITES

No detailed investigation has as yet been carried out on the complex granitic terrain, commonly referred to as the Nelspruit Granite (Visser et al., 1956), which flanks the Barberton Mountain Land along its northern side (Fig. 7). As will be shown, this terrain probably represents the lower-lying part of a region which was subjected to the potash metasomatic event which at higher levels gave rise to the homogeneous hood granite discussed above. As such it does not constitute a distinctive class of granite as in the case of the other granites discussed, but has been treated separately because of its wide areal extent and distinctive field characteristics. The granite generally consists of a complex variety of migmatites, banded and contorted gneisses (frequently porphyroblastic), together with more homogeneous leucocratic granitic phases. The typical contorted and banded, often somewhat porphyroblastic gneiss consists of microcline, quartz, soda plagioclase and biotite. Field and laboratory evidence suggest that the banding has resulted from a process of metamorphic differentiation which gave rise to the formation of more mobile potash-rich felsic bands, and darker more mafic bands or 'basic behinds'. Pegmatitic and leucocratic granitic veins consisting largely of microcline and quartz are widespread in certain areas and, together with amphibolitic xenoliths and banded gneisses often containing mafic schlieren, give rise to areas of complex migmatites. A narrow mobile leucocratic intrusive phase of the Nelspruit migmatite occurs along the immediate northern margin of the mountain land (Fig. 7) (Viljoen, 1964; Anhaeusser and Viljoen, 1965). It is associated with coarse mica, potash feldspar and quartz pegmatites, particularly in the Consort Mine area, and has suffered subsequent intense cataclasis.

All of the evidence from the Nelspruit migmatitic terrain indicates an episode of widespread potash metasomatism, mobilisation, plasticisation, and granite and pegmatite intrusion. Indications are that this episode was of the same nature as that which in other areas led to the formation of the homogeneous hood granite. First the Rb-Sr age of 2990 m.y. obtained for the migmatites (De Gasparis, 1967) is, when experimental errors are considered, very close to the Rb-Sr age of 3070 m.y. obtained for the hood granite in Swaziland (Allsopp et al., 1962). It is significant that patches of a homogeneous leucocratic granite, remarkably similar in chemistry and topographic level to the widespread zones of homogeneous hood granite to the south of the Barberton Mountain Land, occur within the Nelspruit terrain (Fig. 7). Also encountered between Kaapmuiden and Noordkaap is an ovoid body of Nelspruit migmatites and gneisses, defined by stringers and xenoliths of Onverwacht material. This pattern, as shown earlier, is typically developed where early tonalitic gneiss plutons have intruded the Onverwacht Group in a diapiric fashion, and strongly suggests the earlier existence of a now reconstituted tonalitic body. A study of the alkali content (Na_2O and K_2O) of the Nelspruit gneisses supports this as well as the contention that the latter represent the root zone of a previous homogeneous hood granite cover. The abundance of the alkalies in the Nelspruit gneisses and migmatites has been found to be transitional between the abundances in the tonalitic gneisses and the abundances in the homogeneous hood granite (Viljoen and Viljoen, in preparation).

The above evidence suggests that much of the Nelspruit terrain might have represented a zone of tonalitic gneisses and incorporated xenoliths which was subjected to the thermal, granitic and metasomatic event which gave rise, in higher-lying areas, to the homogeneous hood granite of the southern area. Erosion in the Nelspruit terrain, however, has largely removed the higher-lying, homogeneous material (except for the small remnants mentioned above).

THE YOUNG GRANITE PLUTONS

Ten medium- to very coarse-grained, often conspicuously porphyritic, granite plutons, with sharply-defined contacts, occur in the area. These bodies are younger than the granitic rocks discussed above and are distinctive in their petrology, chemistry and mode of emplacement. They characteristically cause abrupt truncation of earlier formed structures and trends, usually displacing or eliminating the formations into which they intrude. Very narrow contact metamorphic aureoles and lack of dynamic metamorphism in the form of flattening or stretching suggests that emplacement took place at high crustal levels. Faults sometimes encompass individual plutons suggesting that a process akin to cauldron subsidence might have been operative during emplacement. These plutons are easily recognised by resistant outcrops of rounded, boulder strewn 'castle koppies'. They are generally porphyritic with scattered phenocrysts of zoned, potash feldspar.

These granites can be divided on mineralogical and chemical grounds into two groups which are for convenience termed the *older* and *younger* granite plutons (Viljoen and Viljoen, in preparation).

The older granite plutons

Within the older group are classed the Salisbury Kop and Dalmein Granites, as well as the so-called 'older pluton' of Swaziland (Hunter, 1968). The Salisbury Kop Granite with a provisional Rb-Sr age of 2830 m.y. (E. J. Oosthuyzen, personal communication) is the only granite of this group which has as yet been dated. These plutons are generally porphyritic with scattered phenocrysts of zoned potash feldspar. They are both medium- and coarse-grained and often pinkish in colour with characteristic light-blue opalescent quartz. The bulk chemistry, and in particular the alkali content (Viljoen and Viljoen, in preparation) suggests that they constitute a distinct group.

The younger granite plutons

Seven younger coarse-grained porphyritic plutons also occur in the area. They are homogeneous, vary in colour from pink to grey and are composed of potash feldspar (microcline, perthitic microcline and orthoclase), quartz, plagioclase (mainly oligoclase) and biotite. Minor amounts of hornblende occur and fairly common accessory minerals include allanite, zircon, apatite, sphene and magnetite (unter, 1968). Xenoliths are rare and pegmatites are entirely lacking. Also occurring occasionally are late coarse-grained granitic dykes of the same type as the pluton which they cut. Radiometric age determinations indicate that the younger plutons have Rb-Sr ages ranging from 2500 to 2650 m.y. (Allsopp *et al.*, 1962; De Gasparis, 1967).

To the south of the Barberton Mountain Land, within the ancient tonalitic gneisses, a syenite plug known as Bosmanskop occurs (Fig. 7) and appears to have been emplaced in the same way as the young granite plutons.

The geological evolution of the early continental crust in the Barberton Region

A fundamental question which arises in a discussion of the early geological evolution of the continents is that concerning the constitution and nature of the earth's primitive crust. In the Barberton region, the oldest recognisable granitic event, that which gave rise to the formation of the ancient tonalitic gneisses, has completely granitised and/or eliminated the pre-Onverwacht floor. Thus although granitised vestiges of this material may be present, evidence concerning the nature of the early crust must come largely from the rocks comprising the Swaziland Sequence. Many of the features of the Onverwacht Group, and particularly the composition of the lower formations, suggest an origin on an extremely thin crust. The abundance of ultramafic material which can be shown to have been emplaced very largely in the form of a mobile molten magma is of significance in this respect. For such a dense high-temperature magma, which could well represent a near total melt of the upper mantle, to reach the surface of the earth, implies an extremely thin, possibly negligible crust, together with a high geothermal gradient. The associated basaltic material, presumably representing a partial melt of the mantle, has characteristics which make it more primitive than oceanic tholeiites. The latter are generally considered to be amongst the most primitive basalts known, and are confined to areas where a typical thin, oceanic crust is present. The above features of the Onverwacht Group are consistent with its formation on a thin, possibly basic crust. The presence of minor cross-bedded quartzitic horizons in some of the lower formations, however, as well as the abundant felsic tuffs in the Theespruit Formation, are suggestive of the presence of early sialic material. We favour the existence of an early, widespread thin crust of a general sialic nature which wedged out in the area underlying the main depository of the Onverwacht Group and might even here, in part, have been of a basic composition. The features and composition of the overlying sedimentary groups, and particularly the Moodies Group, are suggestive of rocks derived largely from a sialic source and deposited on a somewhat thicker and more stable crust than that on which the Onverwacht was deposited. This supports the existence of a substantial sialic crust adjacent to, and probably also underlying the Barberton Mountain Land during Moodies times. If the inference of a thin, and possibly in part basic, crust below the Barberton Mountain Land in pre-Onverwacht times is correct, then it is probable, as suggested by Engel (1966), that a certain amount of thickening of this crust occurred during Fig Tree and largely before Moodies times. Whether this early thickening actually occurred, or whether a fairly substantial sialic crust (at least in the area adjoining the main depository of the mountain land) was present from pre-Onverwacht times, is difficult to determine. It is possible that some idea regarding the nature of this early sialic crust might be obtained from a detailed study of the composition of the Fig Tree and Moodies sediments.

The question now remains as to how the subsequent thickening which led to the formation of the present stable continental crust occurred. This final thickening is considered to have been due almost entirely to the emplacement of the various granites described in this paper. These granites have in most cases eliminated or else reconstituted the earliest crust.

The rapid thinning of the stratigraphy away from a thick pile along the central east-north-east trending axis of the Barberton belt suggests deposition in an elongate trough or downbuckle. It is significant that the east-north-east trend is a regional one discernible in most of the greenstone belts of the Kaapvaal and Rhodesian Cratons (Fig. 1). This implies some fundamental control over the siting and development of the Barberton and other greenstone depositories; possibly an early fold or fracture pattern formed across the primitive crust as a result of rotational forces as suggested by Anhaeusser et al. (1969). The huge pile of the Onverwacht volcanics accumulated in this trough with the development, in adjacent areas, of a much thinner veneer of lava. The weight of this dense pile of mafic lavas in the main trough probably led to downsagging and the initiation of the large fold structures. This was probably accompanied by an isostatic uplift on either side. The uplifted material, consisting at this stage of a fairly thick sialic crust with a thin lava cover in places, was then eroded to supply the sediments for the Fig Tree and the Moodies Groups. All of the sedimentation took place within the deep central trough and was accompanied by further downsagging and folding. Very little lateral compression was apparently involved in the development of the folds, the whole process probably being largely one of near vertical isostatic readjustment. This is supported by the absence of deformational features such as cleavage and flattening (except along the immediate granite contacts) that typically accompany similar tight folds formed by compression (H. Martin, personal communication). The preponderance of large well-developed synclines and the almost total lack of intervening anticlines also suggests slumping rather than compression, with the major fault zones probably representing large slide planes.

Initially the ancient gneisses appear to have been very extensive and to have represented in part granitised, and in part completely mobilised pre-Onverwacht crust. They probably intruded a rather thin accumulation of Onverwacht volcanics which must have had considerable lateral extension away from the central trough as is evidenced by the wide distribution of xenoliths in the granitic terrain. Nearer to the main keel of the mountain land, the gneisses intruded as discrete plutons resulting in the formation of arcuate protuberances of Onverwacht material. The latter represent the only major deviation of the structure of the belt from the early regional east-north-east trend. The emplacement of the tonalitic gneisses was thus the last event which had any significant effect on the overall structure of the belt.

The next major granitic event occurred at about 3000 m.y. ago with potash metasomatism and reconstitution of the existing tonalitic gneiss and greenstone terrain within well-defined, broad zones. This event gave rise to the homogeneous, high-lying hood granite. Where the latter was removed by erosion, the reconstituted material, considered by the authors to be exemplified by the Nelspruit gneiss and migmatite terrain, was exposed.

The last granitic event was the emplacement, starting at about 2800 m.y. ago and ending at 2500 m.y. ago, of the younger porphyritic plutons which caused sharp truncation of all previously formed structures.

Three distinct episodes and styles of granitic emplacement are thus involved in the evolutionary picture outlined above with each one being responsible for further thickening and stabilisation of the early crust. Except for the intrusion of an extensive north-easterly and north-westerly-trending swarm of diabasic dykes as well as some younger Karroo dolerite dykes, the young plutons were the last major magmatic episode to have affected the now stable and thickened Kaapvaal crustal fragment in this area.

One of the most important and fundamental problems, as yet not convincingly answered, is the source and origin of the tremendous amount of granitic material. These granites are totally unlike the various orogenic granites associated with more recent geosynclines and, as suggested by Roering (1965) and Engel (1966), appear to have been derived directly from some differentiation process in the mantle, and to have caused thickening of the continental crust perhaps by a process of sialic underplating.

Finally, it is the authors' contention that, because of the remarkable development, state of preservation, and exposure of what could well represent the almost complete range of rock types to be found in a typical granite-greenstone terrain, the 9000 square mile crustal fragment described in this paper might well be useful as a model for the elements and evolution of the early granite-greenstone terrains, not only of Africa, but of the shield areas of the world (Viljoen and Viljoen, in preparation).

Acknowledgements
The work was carried out as part of South Africa's Upper Mantle Project and was financed by the South African National Committee for the Upper Mantle Project. The authors would like to express their thanks to Dr J. F.

Truswell of the Geology Department, University of the Witwatersrand, and Dr H. L. Allsopp of the Bernard Price Institute for Geophysical Research, for critically reading the manuscript and making useful suggestions, and to Mrs M. S. McCarthy for typing the manuscript.

REFERENCES

ALLSOPP, H. L., ROBERTS, H. R., SCHREINER, G. D. L., and HUNTER, D. R. 1962. Age measurements on various Swaziland granites. *J. geophys. Res.*, **67**, 5307.
—— ULRYCH, T. J., and NICOLAYSEN, L. O. 1968. Dating some significant events in the history of the Swaziland System by the Rb-Sr isochron method. *Can. J. Earth Sci.*, **5**, 605.
ANHAEUSSER, C. R. 1966. Facets of the granite assemblage on the northwest flank of the Barberton Mountain Land. *Inform. Circ.* **32**, *Econ. Geol. Res. Unit, Univ. Witwatersrand, Johannesburg.*
—— 1969. The stratigraphy, structure and gold mineralization of the Jamestown and Sheba Hills areas of the Barberton Mountain Land. Unpublished Ph.D. thesis, Univ. Witwatersrand, Johannesburg.
—— MASON, R., VILJOEN, M. J., and VILJOEN, R. P. 1969. Reappraisal of some aspects of Precambrian Shield geology. *Bull. geol. Soc. Am.*, **80**, 2175.
—— ROERING, C., VILJOEN, M. J., and VILJOEN, R. P. 1967. The Barberton Mountain-Land—a model of the elements and evolution of an Archaean Fold Belt. *Inform. Circ.* **38**, *Econ. Geol. Res. Unit, Univ. Witwatersrand, Johannesburg.*
—— and VILJOEN, M. J. 1965. The base of the Swaziland System in the Barberton-Noordkaap-Louw's Creek Area, Barberton Mountain Land. *Inform. Circ.* **25**, *Econ. Geol. Res. Unit, Univ. Witwatersrand, Johannesburg.*
—— VILJOEN, M. J., and VILJOEN, R. P. 1966. A correlation of pre-Fig Tree rocks in the northern and southern part of the Barberton Mountain Land. *Inform. Circ.* **31**, *Econ. Geol. Res. Unit, Univ. Witwatersrand, Johannesburg.*
BARGHOORN, E. S., and SCHOPF, J. W. 1966. Micro-organisms three billion years old from the Precambrian of South Africa. *Science*, **152**, 758.
CARLISLE, D. 1963. Pillow breccias and their aquagene tuffs, Quadra Island, British Columbia. *J. Geol.*, **71**, 48.
DE GASPARIS, A. A. A. 1967. Rb-Sr isotopic studies relating to problems of geochronology on the Nelspruit and Mpageni Granites. Unpublished M.Sc. thesis, Univ. Witwatersrand, Johannesburg.
ENGEL, A. E. J. 1966. The Barberton Mountain Land: clues to the differentiation of the earth. *Inform. Circ.* **27**, *Econ. Geol. Res. Unit, Univ. Witwatersrand, Johannesburg.*
—— NAGY, B., NAGY, L. A., ENGEL, C. G., KREMP, G. O. W., and DREW, C. M. 1968. Alga-like forms in the Onverwacht Series, South Africa. Oldest recognised life-like forms on earth. *Science*, **161**, 1005.
FERGUSON, J., *co-ordinator*, 1967. Report on the findings of the Upper Mantle Steering Committee to the Barberton Mountain Land to investigate ultramafics in fold belts. Unpublished report to C.S.I.R. National Upper Mantle Committee, South Africa.
HALL, A. L. 1918. The geology of the Barberton Gold Mining District. *Mem. geol. Surv. S. Afr.*, **9**.
HUNTER, D. R. 1961. The geology of Swaziland. *Geol. Surv. Swaziland.*
—— 1965. The Precambrian granitic terrain in Swaziland. Abstract. *Geol. Soc. S. Afr.*, *8th Ann. Congr.*
—— 1968. The Precambrian terrain in Swaziland with particular reference to the granitic rocks. Unpublished Ph.D. thesis, Univ. Witwatersrand, Johannesburg.

INTERNATIONAL SUBCOMMISSION ON STRATIGRAPHIC TERMINOLOGY 1961. Stratigraphic classification and terminology. *Int. geol. Congr.*, **21 (25)**.

NEWTON, A. R. 1968. Correlation and nomenclature in the Precambrian. *In* Symposium on the Rhodesian Basement Complex. *Trans. geol. Soc. S. Afr.*, **71** (Annex.), 215.

PFLUG, H. D. 1966. Structured organic remains from the Fig Tree Series of the Barberton Mountain Land. *Inform. Circ.* **28**, *Econ. Geol. Res. Unit, Univ. Witwatersrand, Johannesburg.*

PRETORIUS, D. A. 1964. *6th Ann. Rep. Econ. Geol. Res. Unit, Univ. Witwatersrand, Johannesburg.*

ROERING, C. 1965. The tectonics of the main gold-producing area of the Barberton Mountain Land. *Inform. Circ.* **23**, *Econ. Geol. Res. Unit, Univ. Witwatersrand, Johannesburg.*

—— 1967. Non-orogenic granites in the Archean geosyncline of the Barberton Mountain Land. *Inform. Circ.* **35**, *Econ. Geol. Res. Unit, Univ. Witwatersrand, Johannesburg.*

STEYN, M. v. R. 1965. Basal rocks of the Swaziland System in the Steynsdorp Valley and Fairview areas of the Barberton Mountain Land. Abstract. *Geol. Soc. S. Afr., 8th Ann. Congr.*

TRUSWELL, J. F. 1967. A critical review of stratigraphic terminology as applied in South Africa. *Trans. geol. Soc. S. Afr.*, **70**—in press.

VAN EEDEN, O. R., and MARSHALL, C. G. 1965. The granitic rocks of the Barberton Mountain Land in the Transvaal. Abstract. *Geol. Soc. S. Afr., 8th Ann. Congr.*

VAN NIEKERK, C. B., and BURGER, A. J. 1967. A note on the age of acid lava of the Onverwacht Series of the Swaziland System. Abstract. *Geol. Soc. S. Afr., 10th Ann. Congr.*

VILJOEN, M. J. 1964. The geology of the Lily Syncline and portion of the Eureka Syncline between the Consort Mine and Joe's Luck Siding, Barberton Mountain Land. Unpublished M.Sc. thesis, Univ. Witwatersrand, Johannesburg.

—— and VILJOEN, R. P. 1967. A reassessment of the Onverwacht Series in the Komati River Valley. *Inform. Circ.* **36**, *Econ. Geol. Res. Unit, Univ. Witwatersrand, Johannesburg.*

VILJOEN, R. P., SAAGER, R., and VILJOEN, M. J. 1969. Metallogenesis and ore control in the Steynsdorp Goldfield, Barberton Mountain Land, South Africa. *Econ. Geol.*, **64**, 778.

VISSER, D. J. L., *et al.* 1956. The geology of the Barberton area. *Spec. Publ. geol. Surv. S. Afr.*, **15**.

C

R. BICHAN

3   The evolution and structural setting of the Great Dyke, Rhodesia

ABSTRACT. *The Great Dyke is a northerly-trending linear body some 480 km long and having an average width of 5·8 km. For most of its length it is intruded into a basement complex of Lower Precambrian batholithic granites that separate areas of older sediments and volcanics. It consists essentially of four layered lopolithic complexes, each of which has a cyclic sequence of ultramafic rocks overlain by a gabbroic capping. Recent detailed investigation of the Hartley Complex, particularly the geochemical variations of the whole rocks and separate mineral phases, suggests that each complex formed by pulsatory injection of magma derived from a parent source. Calculation of the bulk composition of one ulse impindicates that the rocks which fused to produce the Great Dyke magma were compositionally unlike any present-day upper mantle magma genetic zones. An examination of the structural setting of the Great Dyke suggests that the evolution of the ancient Rhodesian nucleus, the formation of the marginal orogenic belts and the emplacement of the Great Dyke have relied on a common convective pattern in the upper mantle.*

Introduction

The Great Dyke is a northerly-trending linear body some 480 km long and with an average width of 5·8 km. From north to south it can be divided into four layered lopolithic complexes, Musengezi, Hartley, Selukwe and Wedza (see Fig. 1, and Worst, 1960), in which the layering dips inwards at a shallower angle than the steeply-inclined dyke contacts.

Each complex consists of cyclic sequences of ultrabasic rocks, which reach a maximum exposed thickness of 2100 m in the Hartley Complex and are overlain by a gabbroic capping 900 m thick (Fig. 2). Ideally each ultrabasic cycle has a basal chromite seam, forming a sharp footwall contact with underlying pyroxenites, followed upwards by peridotites and then pyroxenites. In the Hartley Complex, 11 chromite seams are developed, although in many cycles, particularly in the lower part of the ultrabasic succession, the pyroxenite phase is absent and the chromites are contained entirely within olivine-rich rocks. The gabbroic capping has a sharp contact with the underlying ultramafic rocks and consists of a lower zone of anorthositic gabbro followed upwards by true gabbros and norites which at the very top give way to quartz gabbros.

From his observations, Worst (*ibid.*) concluded that three major events had taken place in the history of the Great Dyke: (*i*) formation of a linear

51

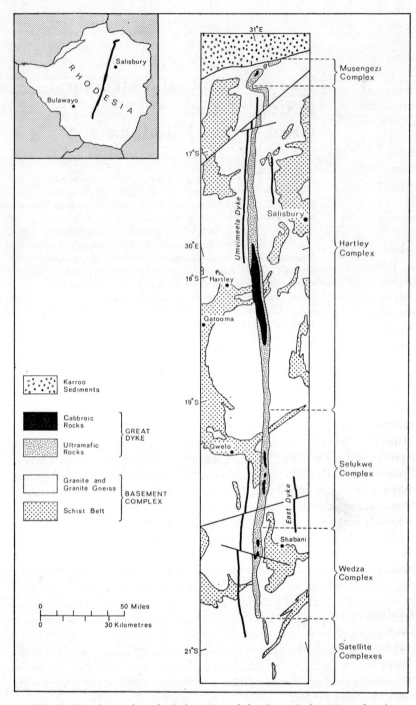

Fig. 1. Location and geological setting of the Great Dyke. Note also the position of the Umvimeela and East Dykes.

zone of weakness; successive heaves of magma were injected through fissures developed at four positions along this line until they met a horizontal plane of weakness in the earth's crust where they spread out laterally and differentiated as individual units; (*ii*) subsidence of the floor and formation of a graben with layers sagging into their present synclinal structure and concomitant shearing of the dyke contacts; (*iii*) erosion down to the present level with removal of any lateral extension of the layers beyond the present dyke margins.

For most of its length the Great Dyke intrudes the Rhodesian cratonic block which consists mainly of batholithic granites of Lower Precambrian age (*c.* 3000 m.y.) that separate older sedimentary and volcanic enclaves termed the schist belts. The available isotopic age determinations for the dyke suggest that it could be as young as 2240 ± 370 m.y. (Faure *et al.*, 1963). However, more recent work by Allsopp (1965) using Rb-Sr and K-Ar dating techniques on separate minerals indicates that the age of the dyke is at least 2530 ± 30 m.y. and is possibly as great as 2800 m.y.; Rb-Sr determinations on whole rock samples gave an age of 2550 ± 410 m.y.

Borehole information indicates that the rocks in the immediate vicinity of the contact plane have suffered shearing and desilication. Outside this zone, although thermal effects are more obvious, they never appear to extend into the country rock for more than 90 m at any point along the dyke margins. Xenoliths of basement granite and schist belt rocks are to be found in many parts of the dyke; they are generally restricted to the gabbros, and may well be roof pendants which were incorporated during the final stages of the igneous cycle.

Since Carl Mauch first indicated the position of the Great Dyke on a geological map (see Harger, 1934), a number of geologists have studied various aspects of its geology. Notable among these were the early reports of Mennell (1908), Zealley (1918), Lightfoot (1927) and Wagner (1929). More recently Lightfoot (1940) has discussed the petrology of part of the Hartley Complex, and Weiss (1940) published results of gravimetric and earth magnetic surveys. Further petrological information has been presented by Hess (1950) and in a comprehensive work by Worst (1958, 1960 and 1964) who has dealt with the petrogenesis and structure of the dyke as a whole.

The purpose of the present investigation is to examine the evolution of the Great Dyke in more detail and to suggest a possible explanation of its structural setting.

Evolution

On the basis of the scheme presented by Worst (1960), that the Great Dyke was formed from the crystallisation of successive heaves of magma, each complex can be divided into a series of major synclinal units which from the top downward are: Unit 1, gabbroic rocks; and Units 2, 3, 4, 5, the cyclic units. Ideally each cycle has a basal chromite seam followed upwards by peridotites and pyroxenites (Fig. 2).

In order to examine the validity of Worst's hypothesis of repeated

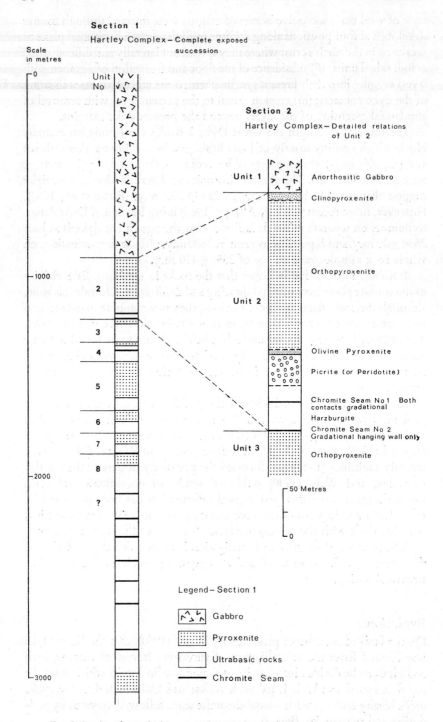

FIG. 2. Section through the exposed succession in the Hartley Complex, together with the detailed relationships of Unit 2.

magmatic injection, the internal and contact relations of Unit 2 in the Hartley Complex have been investigated in detail.

PETROGRAPHY

The rocks of Unit 2 are generally medium-grained with chromite, olivine and bronzite as the cumulus minerals and clinopyroxene, plagioclase, ortho-clase, quartz, mica and sulphides as intercumulus material. Depending on which mineral occurs as the dominant cumulus phase, the rock formed is a chromitite, peridotite (harzburgite or picrite) or pyroxenite. Modal varia-tions of the main mineral phases are illustrated in Fig. 3.

Chromite occurs both as the dominant phase in the chromite seams and as disseminated crystals in all olivine-bearing rocks. Although chromite crystals in individual seams are 1–2 mm in diameter, occasionally 7 mm, they seldom exceed 0·25 mm in the peridotites. The composition of chromites from individual seams differs considerably from those separated from olivine-rich rocks; in particular, the Cr/Fe ratio in the seams is much higher (2·37–2·73) than in the disseminated crystals (1·67–2·32). *Olivine*, Fo_{80-93}, is the most prominent phase in the picrites and harzburgites and always shows some secondary serpentinisation. In the harzburgites most crystals are between 1 and 2 mm in maximum dimension whereas in the picrites 3–4 mm crystals are common. *Orthopyroxene*, occurring as bronzite, is present in nearly every rock type; in the pyroxenites it forms the primary cumulus phase, whereas in the harzburgites and picrites it occurs only as poikilitic plates which occupy 30–50% of the volume of the harzburgites where they reach 3 cm in length and about 30% of the picrites where they are up to 5 cm long.

Clinopyroxene occurs throughout the sequence but is most prominent in the pyroxenites where it occurs as poikilitic plates except in the top 6 m where an upwards change in the mode of occurrence of the clinopyroxene from intercumulus to cumulus crystals is accompanied by a gradation from an ortho- to a clinopyroxenite. Although *plagioclase*, occurring mainly as labradorite and bytownite, has a similar distribution to clinopyroxene it only occurs as cumulus crystals in the gabbroic sequence of Unit 1 (see Fig. 2).

Minor intercumulus minerals include biotite, which is widespread, and muscovite, quartz, orthoclase and apatite which are restricted to the top of the pyroxenites. Sulphides, mainly pyrrhotite, pentlandite, chalcopyrite and pyrite, are developed as intercumulus crystallisation products in two horizons of Unit 2: a lower zone at the base of the peridotites and an upper zone in the top 40 m of the pyroxenites.

STRATIGRAPHIC VARIATIONS

In an investigation of stratigraphic variations of Unit 2, three features merit close examination: (*i*) the relation of Unit 2 to the underlying pyroxenites of Unit 3; (*ii*) the variations of individual rock types and their mutual relations within Unit 2 itself; and (*iii*) the relation of Unit 2 to the gabbroic capping of the complex.

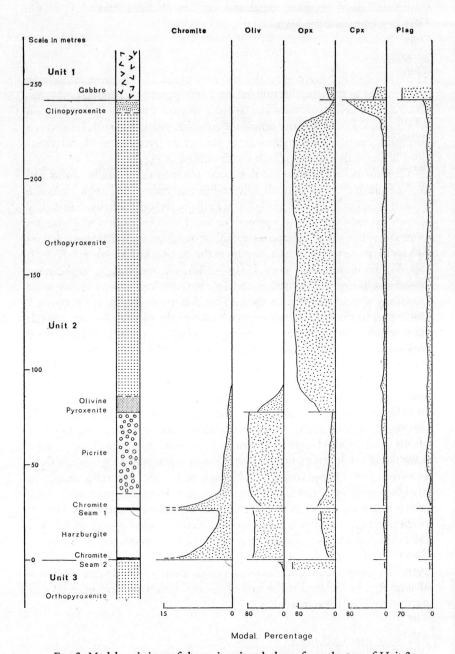

FIG. 3. Modal variations of the main mineral phases from the top of Unit 3, across Unit 2 and into the base of Unit 1 (see Fig. 2). (Oliv, Olivine; Opx, Orthopyroxene; Cpx, Clinopyroxene; Plag, Plagioclase).

Relation of Unit 2 to the underlying pyroxenites of Unit 3

Over the greater part of the outcrop, the contact between Unit 2 and the underlying rocks is a sharp junction between the basal chromite layer (Chromite Seam 2) and the pyroxenites (Fig. 3). However, in certain areas, particularly near the centre of the Hartley Complex and towards the northern and southern extremities, Chromite Seam 2 occurs as much as 6 m above the top of the pyroxenites. In these areas, harzburgites and pyroxenites are in contact, with both sharp and irregular junctions. Furthermore, modal analyses of the 3 m of pyroxenite immediately underlying the contact reveals that in many instances these otherwise almost monomineralic pyroxenites are enriched in olivine and chromite.

Variations of individual rock types within Unit 2

The most important marker horizons within Unit 2 are Chromite Seams 1 and 2 (Fig. 3). In general, Chromite Seam 2 has a sharp footwall contact against Unit 3 pyroxenites and a gradational hanging wall, whereas Chromite Seam 1 always has gradational upper and lower contacts and is contained entirely within harzburgites. Furthermore, although both chromite seams are commonly compact, they do split into several minor seams in the central part and around the northern end of the complex. Detailed descriptions of the stratigraphical and chemical variations are given in Bichan (1969).

The harzburgites in the basal part of Unit 2 are mineralogically homogeneous and normally consist of 60–70% olivine in poikilitic association with an orthopyroxene host that forms 20–30% of the rock; minor amounts of clinopyroxene, chromite, plagioclase and mica are present. These oikocrysts of orthopyroxene increase slightly in size from the base upwards and in the immediate hanging wall of Chromite Seam 1 consist entirely of a continuous growth of poikilitic orthopyroxene containing cumulus olivine.

The harzburgites grade upwards over approximately 3 m into the picrites which differ from them in that they are slightly coarser and contain more plagioclase (6–12% as against 4–8%). Although the picrites are stratigraphically higher in the sequence they contain, on the average, 10% more olivine than the harzburgites; modal analyses reveal that the olivine : orthopyroxene ratio in the harzburgites is 2·7 whereas it is about 5·4 in the picrites. This change is entirely due to a decrease in the volumetric percentage of orthopyroxene oikocrysts although their average size increases from 3 to 5 cm. The top of the picrite has a sharp contact with the overlying olivine pyroxenite and across this contact, from the picrite to the pyroxenite, the olivine content drops and the orthopyroxene content rises markedly (Fig. 3).

Upwards over a distance of 9–12 m the olivine gradually disappears and a pyroxenite composed of 80–90% orthopyroxene, up to 10% plagioclase and minor amounts of clinopyroxene and mica is typical. Almost simultaneously with the disappearance of the olivine small amounts of free quartz appear in the interstices.

Upwards in the sequence the pyroxenites remain remarkably homogeneous until, 36-42 m below the base of the overlying gabbro, small interstitial sulphide grains appear which increase in abundance upwards but never exceed 9% of the mode. With the influx of the sulphides, alkali feldspar, frequently in graphic intergrowth with the quartz, and muscovite appear and the development of these late-stage minerals was accompanied by corrosion and alteration of the cumulus bronzite crystals. The most significant change in the pyroxenites takes place about 6 m below the base of the gabbros where the amount of clinopyroxene begins to increase markedly (Fig. 3) from a value of 5-10% until, in the rocks immediately below the gabbro contact, almost 90% of clinopyroxene is present. As the amount of clinopyroxene increases, its poikilitic habit changes until it forms discrete cumulus crystals. The formation of the clinopyroxenite is accompanied by a coarsening in grain size until, in the top 30 cm of the pyroxenites, crystals reach 1·5 cm in length.

Relation of Unit 2 to the overlying gabbros of Unit 1

In contrast to the lower contact of Unit 2, the upper junction of the clinopyroxenites with the gabbroic rocks of Unit 1 is invariably sharp. However, as will be seen later (p. 62) significant compositional changes take place in the plagioclase as the contact is approached from below.

GEOCHEMISTRY

Rocks

Out of 34 analyses for major and trace elements completed on specimens from the top of Unit 3, throughout Unit 2 and into the base of the gabbroic rocks of Unit 1, representative major element analyses have been selected to show the general compositional differences between the major rock types. These analyses, together with their C.I.P.W. norms, are listed in Table 1 and graphical representation of the variation of certain elements from these and other analyses is depicted in Fig. 4. That diagram shows clearly that a sharp compositional break takes place at the contacts of Units 2 and 3; namely at the junction between harzburgite and pyroxenite, and at the contact of the gabbroic rocks of Unit 1 with the pyroxenites of Unit 2. Ignoring the break around Chromite Seam 1, within Unit 2 the overall vertical changes from the base to the top of the sequence are an increase in SiO_2, CaO and Al_2O_3 (1·78-6·40%), especially in the top 9-12 m for CaO, corresponding to a change from ortho- to clinopyroxenite, and a decrease in MgO, total iron oxides and $MgO/(FeO + Fe_2O_3)$ (Fig. 4).

In the vicinity of Chromite Seam 1, which is contained entirely within harzburgite, there is an increase from the footwall to the hanging wall rocks in Cr_2O_3 and a decrease in total iron oxides and SiO_2. For MgO and $MgO/(FeO + Fe_2O_3)$ there is a slight increase in both the footwall and hanging wall rocks.

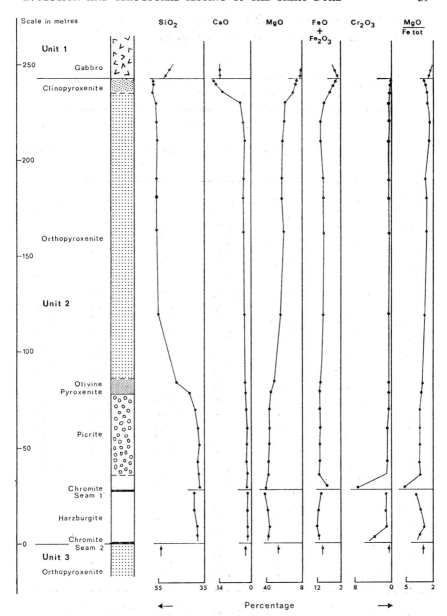

FIG. 4. Variation in the major element concentration in rocks from the top of Unit 3, across Unit 2 and into the base of Unit 1 (Fe tot = total iron as FeO).

Minerals

From X-ray fluorescence analysis of pure chromite samples separated from peridotite host rocks, together with X-ray diffraction determination of the forsterite content of olivines, using the method of Hotz and Jackson (1962), and universal stage determinations of the anorthite content of the plagioclase minerals (Chudoba, 1933), significant vertical variations have been found to exist in the composition of these minerals (Fig. 5).

TABLE 1

Chemical analyses and C.I.P.W. norms of selected rock types from the top of Unit 3, Unit 2 and the bottom of Unit 1

	UNIT 3		UNIT 2			UNIT 1
	Pyroxenite	Harzburgite	Picrite	Olivine Pyroxenite	Pyroxenite	Gabbro
SiO_2	54·26	39·49	40·23	47·29	55·20	51·65
TiO_2	0·20	0·10	0·17	0·22	0·16	0·09
Cr_2O_3	0·53	1·53	0·51	0·47	0·46	0·10
Al_2O_3	2·17	1·78	1·90	2·76	3·67	19·55
Fe_2O_3	1·11	5·38	2·71	2·42	1·04	0·94
FeO	8·41	6·08	8·37	7·81	8·58	2·93
MnO	0·16	0·14	0·14	0·17	0·19	0·07
MgO	29·83	38·02	37·41	33·13	26·21	9·60
CaO	2·19	1·41	1·33	2·47	3·41	13·41
Na_2O	0·44	0·14	0·20	0·37	0·37	1·04
K_2O	0·14	0·12	0·11	0·30	0·13	0·26
H_2O+	0·25	5·71	5·32	2·09	0·46	0·54
H_2O-	0·06	0·29	0·31	0·31	0·25	0·12
P_2O_5	0·27	0·03	0·07	0·03	0·02	0·06
CO_2	—	—	1·18	—	—	—
	100·02	100·22	99·96	99·84	100·15	100·36
q	—	—	—	—	1·50	3·44
or	0·83	0·71	0·65	1·77	0·77	1·54
ab	3·72	1·19	1·69	3·13	3·13	8·80
an	3·53	3·86	3·96	4·99	7·97	48·73
di	4·41	2·29	1·72	5·59	7·01	13·80
hy	77·84	20·36	18·80	41·86	76·53	21·88
ol	5·98	55·49	61·16	35·42	—	—
mt	1·61	7·80	3·93	3·51	1·51	1·36
cr	0·78	2·25	0·75	0·69	0·68	0·15
il	0·38	0·19	0·32	0·42	0·30	0·17
ap	0·64	0·07	0·16	0·07	0·05	0·14
H_2O	0·30	6·00	5·63	2·40	0·71	0·61
'Others'	0·24	—	1·18	—	—	—

In the case of the chromite, there is a general upwards increase in the total iron content of the chromites and a decrease in the Cr_2O_3, $Cr_2O_3/(FeO+Fe_2O_3)$ and $MgO/(FeO+Fe_2O_3)$ with important breaks occurring in each trend as Chromite Seam 1 is approached. In the case of Cr_2O_3 this break in trend is represented by an increase towards the seam in both the foot-wall and hanging wall whereas total iron shows a sharp drop and $Cr_2O_3/(FeO+Fe_2O_3)$ and $MgO/FeO+Fe_2O_3$ increase above this horizon.

Although there is a general decrease in the forsterite content of the olivines from Fo_{90} to Fo_{82} from the base of Unit 2 upwards there is an increase in the

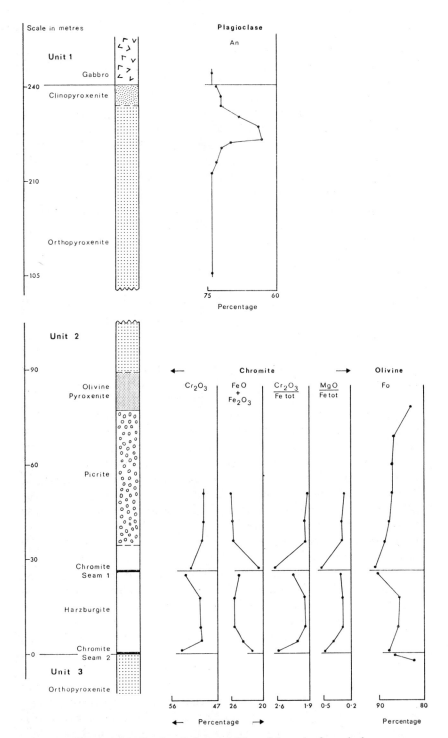

FIG. 5. Variation in the chemical composition of chromite from the lower part of Unit 2, the forsterite content (Fo) of olivine from the top of Unit 3 and the lower part of Unit 2, and the anorthite content (An) of the plagioclase in the vicinity of the Unit 1—Unit   contact (Fe tot=total iron as FeO).

Fo content in the pyroxenites in the very top of Unit 3 (Fig. 5); for example, in one instance the change is from $Fo_{82.5}$ at 91·6 cm below, to $Fo_{86.1}$ at 7·6 cm below the base of Unit 2. Within Unit 2 anomalies exist in the compositional trend of the olivines in the vicinity of Chromite Seam 1. In general, the Fo content increases towards the seam and in the footwall this increase appears to be restricted to the 1-2 m of harzburgite immediately underlying the seam.

Anorthite determinations of the intercumulus plagioclase in the pyroxenites at the top of Unit 2 and in the gabbros at the base of Unit 1 show that there is an upwards decrease in the anorthite content of the pyroxenites (Fig. 5) from An_{74} to An_{64} until the top part is reached where a reversal of this trend takes place and the anorthite content increases to An_{73} which is comparable in composition to the cumulus plagioclase crystals in the over-lying gabbros.

DIFFERENTIATION

In major layered basic and ultrabasic intrusions the first problem to be con-sidered before other irregularities can be examined is the origin of the layering. Two main schools of thought exist. The first that they are the pro-ducts of successive magmatic injections from a parent magma reservoir has been visualised by Lombaard (1934) to explain the major units of the Bush-veld Complex, by Van der Walt (1941) in a study of the western Bushveld chrome ores, by Brown (1956) and Wadsworth (1961) to explain the major layering in the ultrabasic rocks of Rhum, and by Irvine and Smith (1967) to explain the cyclic units of the Muskox Intrusion. The second concept of layering relies on the crystallisation of the complete complex from one injection of magma with the layers owing their origin to several agencies, the most important of which is convective overturning of the cooling magma. This hypothesis gained status after the classic work of Wager and Deer (1939) on the Skaergaard Intrusion and has more recently been invoked by Cameron and Emerson (1959) and Cameron (1963) in the study of the chro-mite deposits of the Bushveld Complex, and by Hess (1956, 1960) and Jackson (1961) to explain the major layers of the Stillwater Complex.

Discussing the petrogenesis of the Great Dyke, Worst (1960) concluded from mainly lithological evidence that the dyke was formed by a series of magmatic pulses which had been injected from a parent magma reservoir. The present results support this view but define the mechanism more closely. It is suggested that the layers of Unit 2 resulted from two successive liquid injections from a parent magma source. The first of these spread out over a horizontal floor which was almost, but not completely, crystallised from the preceding injection. Sharp contacts formed where consolidation of under-lying rocks was complete, and irregular contacts where it was not complete and pools of partially consolidated mush remained. In the latter areas enrich-ment in both the amount and forsterite value of the olivine took place. Sub-sequent differentiation of this first impulse resulted by gravitative settling of chromite and olivine with the more dense chromite crystals settling first to

form Chromite Seam 2, followed by harzburgite containing only disseminated chromite grains.

Before crystallisation of this impulse was completed the second injection
took place with the same crystallisation and settling mechanism operating as
before, resulting in the formation of Chromite Seam 1. The difference in
petrographic character between the harzburgites immediately above and
below Seam 1, together with the change in major element concentration in
whole rocks of similar type above and below this seam, and with sympathetic
variations in the ratios of $MgO/(FeO + Fe_2O_3)$ in olivines and whole rocks, all
support the view that Chromite Seam 1 and overlying rocks crystallised from
a separate magma inflow. However, the gradational footwall to Chromite
Seam 1 and the reversal in the trend of the forsterite content of the olivines
and the Cr_2O_3 content of the chromites immediately under this horizon
indicate that mixing of pre-existing and new liquids took place. This enrichment of the olivine and chromite could have taken place by diffusion of
material from the overlying magma into the interstices of the underlying
mush, as envisaged by Wager *et al.* (1960, pp. 77-79) to explain adcumulus
growth. The new liquid invading the interstices would be at a higher temperature so that solution would take place to retain equilibrium and form more
forsteritic olivines and chromium-enriched chromites.

Succeeding the formation of Chromite Seam 1, differentiation resulted
in the formation of the harzburgites which grade upwards into picrites
(Fig. 5). An apparently abrupt change takes place as picrites give way to
olivine pyroxenites. Closely-spaced modal analyses have shown that their
contact is represented by an upwards drop in the volume percentage of
olivine from greater than 75% to less than 50% and a change in the mode
of formation of orthopyroxene from intercumulus to cumulus crystals.
Furthermore, the presence of very thick sequences of virtually monomineralic
olivine-rich and pyroxene-rich rocks, which have a similar grain size and
specific gravity, suggest that gravity settling could not have effectively
separated these minerals had they both been present as primary crystals in the
same liquid. It seems therefore that the change from picrite to olivine pyroxenite delineates the boundary between cumulus olivine and pyroxene
precipitation with the overlap being represented by the presence of olivine
crystals in the base of the pyroxenites.

In the remainder of the pyroxenites no significant breaks occur, differentiation proceeding normally until the topmost part of the succession of Unit 2
is reached (Fig. 5) where a sharp increase in the concentration of calcium and
alumina is reflected in the formation of a clinopyroxenite. The abrupt break
at the top of the pyroxenites with the formation of the gabbroic rocks can
only be adequately explained by a further magmatic injection. It is suggested
that a separate magmatic injection succeeded the formation of the pyroxenites
after they had crystallised to such an extent as to present a solid floor for the
new liquid phase, but before they had cooled sufficiently for contact thermal
effects to take place. This concept of repeated injection is supported by the
increase in anorthite content of the interstitial plagioclase in the top of the

pyroxenites, suggesting that the intercumulus material was enriched in calcium by diffusion from the overlying magma.

CALCULATED COMPOSITION OF ONE LIQUID IMPULSE

As no marginal chill phase is present, direct conclusions concerning the composition of the Great Dyke magma are not possible. However, the ubiquitous presence of hypersthene and the absence of nepheline in the norm indicate tholeiitic affinities. On the basis of the normative tetrahedron of Yoder and Tilley (1962, p. 352) the rocks vary from olivine tholeiites, with

TABLE 2

Thickness and average composition of the major rock types of Unit 2 from Chromite Seam 1 upwards.

	Chromite	Per	Ol-Py	Or-Py	Cl-Py
Thickness in metres	0·15	55	7·6	146	6·1
SiO_2	—	41·60	46·81	55·90	56·63
TiO_2	0·61	0·14	0·19	0·14	0·10
Cr_2O_3	51·68	1·63	0·55	0·45	0·33
Al_2O_3	11·79	2·46	2·82	3·81	5·34
Fe_2O_3	9·90	5·01	3·28	1·08	0·07
FeO	13·33	6·23	7·58	8·64	5·73
MnO	—	0·14	0·16	0·19	0·14
MgO	12·48	40·54	35·49	25·77	16·48
CaO	—	1·94	2·51	3·51	14·51
Na_2O	—	0·18	0·32	0·37	0·47
K_2O	—	0·11	0·27	0·10	0·10
P_2O_5	—	0·04	0·04	0·02	0·07
Total	99·79	100·02	100·02	99·98	99·97

Designations: *Chromite*, Chromite Seam 1 (gangue free); *Per*, Peridotite; *Ol-Py*, Olivine Pyroxenite; *Or-Py*, Orthopyroxenite; *Cl-Py*, Clinopyroxenite. All analyses calculated water free; analyst, H. R. Bichan (except for chromite which is from Worst, 1964, p. 217).

the degree of undersaturation reaching 25·56% in the olivine-rich assemblages to tholeiites containing 6·54% normative quartz.

Evidence has been presented which indicates that the Great Dyke formed from pulsatory injections of liquid from a parent magma source. In particular, it was suggested that Unit 2 of the Great Dyke formed from two such impulses the second of which differentiated to form Chromite Seam 1 at its base followed upwards by harzburgite, picrite and pyroxenites. In addition, it was concluded that this impulse had almost completely consolidated before the succeeding injection took place. Accepting that these rocks fractionated from a uniform liquid in a virtually closed system, a reasonable approximation of the composition of this liquid can be deduced from the thickness and composition of the rock types present (Table 2) (for calculation, see Holmes,

1921, p. 392). In Table 3 this calculated composition is compared with the analyses of chill zones from other major layered intrusions.

The tholeiitic nature of all these intrusions is evident from the presence of normative hypersthene and furthermore the undersaturated nature of the

TABLE 3

Calculated chemical composition of the second liquid of Unit 2 of the Great Dyke (e) compared to probable parental magmas of other layered intrusions.

	(a)	(b)	(c)	(d)	(e)
	Bushveld	Stillwater	Skaergaard	Muskox	Great Dyke
SiO_2	50·58	50·95	48·38	51·11	51·91
TiO_2	0·66	0·46	1·33	1·07	0·14
Cr_2O_3	0·01	0·04	0·05	—	0·79
Al_2O_3	15·24	17·74	18·48	13·66	3·48
Fe_2O_3	1·04	0·26	1·24	1·18	2·14
FeO	10·08	9·94	8·65	9·16	7·90
MnO	0·23	0·15	0·12	0·18	0·18
MgO	8·30	7·71	8·16	9·78	29·61
CaO	11·31	10·53	10·86	11·32	3·38
Na_2O	2·24	1·88	2·44	1·80	0·32
K_2O	0·19	0·24	0·21	0·64	0·11
P_2O_5	0·12	0·09	0·08	0·10	0·03
Total	100·00	99·99	100·00	100·00	99·99
q	—	—	—	—	—
or	1·11	1·45	1·22	3·78	0·67
ab	18·91	15·88	20·65	15·20	2·73
an	30·97	39·50	38·81	27·30	7·70
di	20·62	10·01	11·79	22·91	7·01
hy	19·69	31·40	10·65	24·49	63·42
ol	5·68	0·23	12·18	2·21	13·83
mt	1·51	0·37	1·81	1·72	3·11
il	1·26	0·88	2·52	2·04	0·27
ap	0·30	0·20	0·20	0·24	0·07
cr	0·02	0·07	0·07	—	1·16

(a) *Bushveld* (Wager and Brown, 1968, p. 355); (b) *Stillwater* (Hess, 1960, p. 152); (c) *Skaergaard* (Wager, 1960, p. 378, except for Cr_2O_3 which is calculated from data by Wager and Mitchell, 1951); (d) *Muskox* (Smith and Kapp, 1963, p. 33). Analyses calculated water free; blanks indicate no available analysis.

Great Dyke liquid is emphasised by the formation of almost 14% of normative olivine. Comparison with tholeiites investigated experimentally by Yoder and Tilley (1962) illustrates that the rocks of the Bushveld, Stillwater, Skaergaard and Muskox compare in composition to this magma type. The Great Dyke however is similar only in the concentration of SiO_2, total iron and manganese (Table 3). In all other major constituents distinct differences exist: for example, the Great Dyke has *more* MgO and Cr_2O_3 but *less* Al_2O_3, CaO, alkalies, P_2O_5 and TiO_2.

The composition of this Great Dyke liquid does not resemble any of the recognised 'magma types'. It is however similar in composition to the average orthopyroxenite given by Nockolds (1954, p. 1022) although the presence of clinopyroxene and plagioclase in all Great Dyke rocks causes higher calcium and alumina values.

It must be emphasised that this derived composition is not representative of the parent magma but only of one liquid impulse. This calculation is also open to criticism as there is no absolute proof that the later differentiates of Unit 2 have been retained. However, the presence of muscovite, quartz and orthoclase in the higher regions of the pyroxenite, particularly in the clino-pyroxenites, and the absence of minor silicic apophyses in the country rocks suggest that little has been lost in the form of volatiles.

From the not unreasonable assumption that the composition given in Table 3 (anal. (e)) at least approximates to the composition of one magmatic impulse and that this was totally liquid at the time of injection, further speculations are possible. There is for example a major compositional difference between the postulated Great Dyke magma and those of Phanerozoic layered basic complexes. As most recent experimental evidence and petrological reasoning suggests that basic magmas originate by partial fusion of a peridotite or garnet peridotite upper mantle, it seems that the magma either came from a mantle that differed compositionally from that of later epochs or the magma was a product of a different magma genetic environment.

Search of the literature has not revealed any experimental evidence that liquids close to orthopyroxenite in composition can be produced by partial fusion of peridotites in any pressure-temperature environment. It is therefore tentatively concluded that the rocks that fused to produce the Great Dyke magma were compositionally unlike any present-day upper mantle magma genetic zones. This is a tentative proposal and it is in no way suggested that this is universal, as perfectly normal basaltic rocks comparable in age to the Great Dyke occur elsewhere in Africa.

Structural setting

Any student of Great Dyke geology cannot fail to be impressed by the magnitude and regularity of this remarkable body of rocks. In particular, the Great Dyke and its two accompanying intrusions, the Umvimeela and East Dykes, occur in one distinct and narrow zone transgressing almost the entire length of the Rhodesian Craton from north to south (Fig. 1). To explain its formation, therefore, a search has to be made for a mechanism which will provide such a unidirectional effect.

The Rhodesian cratonic nucleus mainly consists of granitic rocks that separate older sedimentary and volcanic enclaves termed the schist belts. This association of circular or elliptical batholiths partially or completely surrounded by belts of metamorphic rock led Macgregor (1951) to suggest that their formation was closely related. He postulated that the original crust of sedimentary or volcanic rocks rested upon a less dense granitic substratum which, when mobilised, rose through the overlying rocks which were then squeezed and metamorphosed between the ascending blocks. It was further suggested by Macgregor (*ibid.*), and later supported by Talbot (1967), that the circular shape of individual batholiths was due to localised convection currents established during the uprise of the granites.

In assessing the available radiometric dates for the pre-Silurian geology of Africa, T. N. Clifford (1968a) concludes that acceptance of MacGregor's ideas on the relationship of deformation and gregarious batholith emplacement in turn suggests that the ages of 2500-2700 m.y. 'may approximate to the last important episode of orogenesis in the Rhodesian nucleus'. Moreover, isotope studies of the Great Dyke by Allsopp (1965) using Rb-Sr and K-Ar dating techniques on separate minerals indicates that the age of the Dyke is at least 2530 ± 30 m.y. and is possibly as great as 2800 m.y.

During the last few decades the mechanism of sub-crustal convection has been widely invoked to explain major tectonic features, particularly the frequent occurrence of orogenic belts around cratonic blocks. Recently, T. N. Clifford (1968b) has illustrated the application of this concept in the development and growth of stable cratons in Africa throughout geological time.

Taking into account the isotopic age of the Great Dyke and Macgregor's gregarious batholiths and the exposure of orogenic belts on three sides of the Rhodesian cratonic block, it is now possible to construct a hypothetical model which will explain their relationship. In Fig. 6 four stages in the development of the Great Dyke and those related features are schematically envisaged:

(*a*) Establishment of a convection cell extending underneath the entire Rhodesian Craton giving rise to directions of major heat flow at the centre and margins of the cratonic block; in addition, this would cause an overall increase in the temperature of the granitic substratum of the crust which, when mobilised, would move upwards to 'intrude' and deform the cover rocks, giving rise to the gregarious batholiths and complementary schist belts (Macgregor, 1951; Talbot, 1967).

(*b*) Tension in the cratonic block immediately above the convective updraft resulting in a split into one major zone and two subsidiary fractures with movement apart due to the outward movement of the convection currents. This mechanism would also initiate compression and downwarping in the marginal orogenic zones.

(*c*) Intrusion of the Great Dyke (GD), Umvimeela (UD) and East Dykes (ED) and continuing deformation in the orogenic zones.

(*d*) Waning of the convection system and heat flow pattern resulting in slumping of the Great Dyke into the graben with concomitant shearing of the dyke margins and sagging of the layering into its present synclinal structure.

It can be argued that the orogenic belts surrounding the Rhodesian Craton are much younger than the Great Dyke: the Mozambique and Zambesi

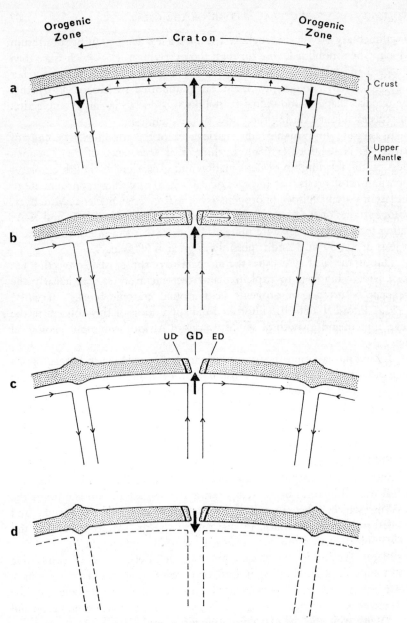

FIG. 6. Section through the earth's crust and upper mantle to illustrate the effects of a hypothetical convection system in the formation of the Rhodesian Craton, Great Dyke and orogenic belts:

(a) Establishment of convection cells in upper mantle and resulting heat flow pattern (indicated by arrows).

(b) Splitting of cratonic block with movement apart and concomitant embryonic stage of marginal orogenic zones.

(c) Intrusion of Great Dyke (GD), Umvimeela Dyke (UD) and East Dyke (ED).

(d) Waning of convection system and heat flow pattern resulting in slumping of Great Dyke into the graben.

Orogenic Belts on the east and north respectively give radiometric dates $c.\,500$ m.y.; the Limpopo Belt on the south, $c.\,2000$ m.y.; and on the west any existing orogenic belt is covered by more recent sediments. However, Van Breemen et al. (1966) consider that an event older than 2500 m.y. was responsible for some of the high-grade metamorphism of the Limpopo Belt. Furthermore, rejuvenation of ancient orogenic belts is not uncommon and indeed T. N. Clifford (1968a) has invoked this process in several instances throughout Africa. It is certainly possible therefore that orogenesis accompanied the emplacement of the Great Dyke.

In the Canadian Shield, P. M. Clifford (1968) has pointed out that regular patterns of dyke swarms, together with the flood basalt-dolerite provinces, indicate tensional zones of continental proportions for which he suggests the most reasonable explanation is subcrustal flow in the form of convection cells with the basalt-dolerite dyke provinces localised over the position of convective updraft. It is pertinent to note that the Muskox Intrusion (Irvine and Smith, 1967), which is remarkably similar in many respects to the Great Dyke, lies at the northern end of the main tensional zone (Clifford, P. M., 1968) that runs from the south-east of the Superior Province northwards for almost 3000 km through the Churchill, Slave and Bear Provinces. Although there are as yet not enough regional data on the large dyke swarms cutting across the West Australia cratonic block, the size and similar trend of many of the large dykes, such as the Binneringie Dyke and Jimberlana 'Dyke' (G. J. H. McCall, personal communication), may again be related to major mantle processes.

Conclusion

From the petrographic, stratigraphic and geochemical data presented, it appears that individual complexes of the Great Dyke formed from the crystallisation of separate magma impulses derived from a parent source. In many inflows gravitative settling of early-formed crystals gave rise to a basal chromite seam followed upwards by peridotites and pyroxenites. The absence of contact chill or thermal effects suggests that the impulses followed one another rapidly and in some instances consolidated before the preceding injection had been completed, so that the underlying liquid was enriched in certain elements.

As no marginal chill phase is present in the Great Dyke as a whole, calculation of the bulk composition of the initial liquid is not possible. However, the ubiquitous presence of normative hypersthene indicates tholeiitic affinities. Calculation of the bulk composition of *one impulse* suggests that at least part of the Great Dyke crystallised from a liquid approximating to an orthopyroxenite in composition.

By considering the magnitude and structural setting of the Great Dyke in relation to the evolution of the ancient Rhodesian nucleus and marginal orogenic belts, it seems that the intrusion of the dyke formed part of a sequence of events. Initial establishment of a major convection cell in the upper mantle under the entire Rhodesian nucleus heated and mobilised the granitic sub-stratum which moved upwards to give rise to MacGregor's (1951) gregarious batholiths. Persistence of the convection pattern eventually caused tension and intrusion of the Great Dyke over the position of convective updraft and corresponding orogenesis around the cratonic margins where downdraft and compression existed. With waning of the convection system slumping of the Great Dyke into the graben took place.

Acknowledgements

Part of this work was carried out in the Research Institute of African Geology, University of Leeds as part of a Ph.D. thesis. The author would like to express his sincere thanks to: Dr I. G. Gass who supervised the work; the Anglo-American Corporation of South Africa who provided financial support for field work; and the teaching and technical staff of the Department of Earth Sciences, University of Leeds.

REFERENCES

ALLSOPP, H. L. 1965. Rb–Sr and K–Ar age measurements on the Great Dyke of Southern Rhodesia. *J. geophys. Res.*, **70**, 977.

BICHAN, R. 1969. Origin of chromite seams in the Hartley Complex of the Great Dyke, Rhodesia. *Econ. Geol. Monog.*, **4**, 95.

BROWN, G. M. 1956. The layered ultrabasic rocks of Rhum, Inner Hebrides. *Phil. Trans. r. Soc. Lond.*, **240**, Ser. B, 1.

CAMERON, E. N. 1963. Structure and rock sequences of the Critical Zone of the eastern Bushveld Complex. *Min. Soc. Am. Spec. Pap.*, **1**, 93.

—— and EMERSON, M. E. 1959. The origin of certain chromite deposits in the eastern part of the Bushveld Complex. *Econ. Geol.*, **54**, 1151.

CHUDOBA, K. 1933. *The determination of feldspars in thin section.* (English translation), London.

CLIFFORD, P. M. 1968. Flood basalts, dike swarms and sub-crustal flow. *Can. J. Earth Sci.*, **5**, 93.

CLIFFORD, T. N. 1968a. Radiometric dating and the pre-Silurian geology of Africa. In *Radiometric dating for geologists*, p. 299. (Eds. E. I. Hamilton and R. M. Farquhar), London.

—— 1968b. African structure and convection. *Trans. Leeds geol. Ass.*, **7**, 291.

FAURE, G., HURLEY, P. M., FAIRBAIRN, H. W., and PINSON, W. H. 1963. Age of the Great Dyke of Southern Rhodesia. *Nature, Lond.*, **200**, 769.

HARGER, H. S. 1934. An early Transvaal geological map by Carl Mauch. *Trans. geol. Soc. S. Afr.*, **37**, 1.

HESS, H. H. 1950. Vertical mineral variation in the Great Dyke of Southern Rhodesia. *Trans. geol. Soc. S. Afr.*, **53**, 159.

—— 1956. The magnetic properties and differentiation of dolerite sills: discussion. *Am. J. Sci.*, **254**, 446.

—— 1960. Stillwater Igneous Complex, Montana: a quantitative mineralogical study. *Mem. geol. Soc. Am.*, **80**.

HOLMES, A. 1921. *Petrographic methods and calculations*, London.

HOTZ, P. E., and JACKSON, E. D. 1962. X-ray determinative curve for olivines of composition Fo_{80-95} from stratiform and alpine-type peridotites. *Prof. Pap. U.S. geol. Surv.*, **450-E**, 101.

IRVINE, T. N., and SMITH, C. H. 1967. The ultramafic rocks of the Muskox Intrusion, North-West Territories, Canada. In *Ultramafic and related rocks*, p. 38. (Ed. P. J. Wyllie), New York.

JACKSON, E. D. 1961. Primary textures and mineral associations in the Ultramafic Zone of the Stillwater Complex, Montana. *Prof. Pap. U.S. geol. Surv.*, **358**.

LIGHTFOOT, B., 1927. Traverses along the Great Dyke of Southern Rhodesia. *Short Rep. geol. Surv. S. Rhodesia*, **21**.

—— 1940. The Great Dyke of Southern Rhodesia. *Proc. geol. Soc. S. Afr.*, **43**, xxvii.

LOMBAARD, B. V. 1934. On the differentiation and relationships of rocks of the Bushveld Complex. *Trans. geol. Soc. S. Afr.*, **37**, 5.

MACGREGOR, A. M. 1951. Some milestones in the Precambrian of Southern Rhodesia. *Proc. geol. Soc. S. Afr.*, **54**, xxvii.

MENNELL, F. P. 1908. *The Rhodesian mines handbook.* Publication No. 4, Rhodesian Museum.

NOCKOLDS, S. R. 1954. Average chemical compositions of some igneous rocks. *Bull. geol. Soc. Am.*, **65**, 1007.

SMITH, C. H., and KAPP, H. E. 1963. The Muskox Intrusion; a recently-discovered layered intrusion in the Coppermine River area, North-West Territories, Canada. *Min. Soc. Am. Spec. Pap.*, **1**, 30.

TALBOT, C. J. 1967. Rock deformation at the eastern end of the Zambezi Orogenic Belt, Rhodesia. Unpublished Ph.D. thesis, Univ. Leeds.

VAN BREEMEN, O., DODSON, M. H., and VAIL, J. R. 1966. Isotopic age measurements on the Limpopo Orogenic Belt, southern Africa. *Earth Planetary Sci. Let.*, **1**, 401.

VAN DER WALT, C. F. J. 1941. Chrome ores of the western Bushveld Complex. *Trans. geol. Soc. S. Afr.*, **44**, 79.

WADSWORTH, W. J. 1961. The ultrabasic rocks of southwest Rhum. *Phil. Trans. r. Soc. Lond.*, **244**, Ser. B, 21.

WAGER, L. R. 1960. The major element variation of the layered series of the Skaergaard Intrusion and a re-estimation of the average composition of the hidden laycred series and of the successive residual magmas. *J. Petrology*, **1**, 364.

—— and BROWN, G. M. 1968. *Layered igneous rocks*, Edinburgh.

—— BROWN, G. M., and WADSWORTH, W. J. 1960. Types of igneous cumulates. *J. Petrology*, **1**, 73.

—— and DEER, W. A. 1939 (re-issued in 1962). Geological investigations in East Greenland. Pt. III. The petrology of the Skaergaard Intrusion, Kangerdlugssuak, East Greenland. *Medd. Grønland*, **105** (4).

—— and MITCHELL, R. L. 1951. The distribution of trace elements during strong fractionation of basic magma—a further study of the Skaergaard Intrusion, East Greenland. *Geochim. et cosmochim. Acta*, **1**, 129.

WAGNER, P. A. 1929. *The platinum deposits and mines of South Africa*, Edinburgh.

WEISS, O. 1940. Gravimetric and earth magnetic measurements of the Great Dyke of Southern Rhodesia. *Trans. geol. Soc. S. Afr.*, **43**, 143.

WORST, B. G. 1958. The differentiation and structure of the Great Dyke of Southern Rhodesia. *Trans. geol. Soc. S. Afr.*, **61**, 283.

—— 1960. The Great Dyke of Southern Rhodesia. *Bull. geol. Surv. S. Rhodesia*, **47**.

—— 1964. Chromite in the Great Dyke of Southern Rhodesia. In *Geology of some ore deposits in southern Africa*, p. 209. (Ed. S. H. Haughton), Vol. 2, Geol. Soc. S. Afr.

YODER, H. S., and TILLEY, C. E. 1962. Origin of basalt magmas: an experimental study of natural and synthetic rock systems. *J. Petrology*, **3**, 342.

ZEALLEY, A. E. V. 1918. The occurrence of platinum in Southern Rhodesia. *Short Rep. geol. Surv. S. Rhodesia*, **3**.

H. C. M. WHITESIDE

4 Volcanic rocks of the Witwatersrand Triad

ABSTRACT. *The Witwatersrand Triad comprises the three lowest members of the Proterozoic in South Africa, namely the Ventersdorp, Witwatersrand and Dominion Reef Systems. The extrusive rocks of these systems are described and recent information, some of it unpublished until now, is reviewed. Stratigraphic columns of the Ventersdorp and Dominion Reef Systems have been compiled and it has been suggested that the Ventersdorp stratigraphy should be modified in the light of recent information from diamond drilling.*

Introduction

The Witwatersrand Triad is the name given by Hamilton and Cooke (1960) to the three lowest members of the Proterozoic in South Africa:

<blockquote>
Ventersdorp System

Witwatersrand System $\begin{cases} \text{Upper Division} \\ \text{Lower Division} \end{cases}$

Dominion Reef System
</blockquote>

These three systems are closely related both areally and structurally and, as Brock and Pretorius (1964a) state, 'their relationships suggest that the three systems collectively constitute a major geological cycle' (Fig. 1).

Because the Witwatersrand System, and more especially the Upper Division, contains the greatest concentration of gold known in the world, it has been studied in detail and much information is available particularly with regard to structural, sedimentational and economic aspects. It is only comparatively recently that attention has been paid to the other two members of the triad—the Ventersdorp because it overlies the Upper Witwatersrand and knowledge about it is essential to prospecting, and the Dominion Reef because interest is being directed towards the uranium-bearing conglomerates at its base.

Rocks of the triad, excluding such possible correlatives as the Wolkberg (Witwatersrand) and Insuzi (Dominion Reef), have been found over some 80,000 square miles in the southern Transvaal, northern Orange Free State and northern Cape Province. To describe such a vast area in detail is impossible within the compass of this paper. For this reason, emphasis is given to generalisation and to new information which is not generally available.

Dominion Reef System

Molengraaff (1905) named the sediments which were locally exposed around the Dominion Reef Mine, the Dominion Reef Formation and considered that they might be of Witwatersrand age, equivalent to the Orange Grove quartzites. The overlying lavas he assigned to the Vaal River System (Ventersdorp). Nel (1934) included the overlying lavas in the Dominion Reef but still allocated it to the Lower Witwatersrand System. Truter (1949) considered that the Dominion Reef lavas and sediments and their correlatives should be assigned to a separate system, principally because the correlatives (the Insuzi, Godwan, Uitkyk, Zoetlief and Koras Formations) are found widely distributed and in some cases lying below beds thought to be Witwatersrand age (for example: Insuzi below Mozaan in the south-eastern Transvaal and Swaziland; Godwan below Wolkberg in the north-eastern Transvaal). This view was upheld by Von Backström (1952, 1962), Nel and Verster (1962) and Van Eeden et al. (1963).

Recently, Pretorius (1966) has suggested that the Dominion Reef beds should again be relegated to the lowest series of the Witwatersrand System because it has been shown that many of the lithologically similar correlatives are not the chronological equivalents of the Dominion Reef beds. He considers that the Dominion Reef Series proper is confined to the north-western portion of the Witwatersrand Basin.

The succession in the type locality, near the village of Hartbeesfontein which is 18 miles west of Klerksdorp in the western Transvaal, has been established as follows (Fig. 2):

(a) Upper or Acid Volcanic Formation   6700 ft
(b) Lower or Basic Volcanic Formation   2100 ft
(c) Basal Sedimentary Formation   120 ft

The Sedimentary Formation is absent in places but, where fully developed, comprises a basal conglomerate overlain by sericitic quartzite, pebbly quartzite and lenticular bodies of quartz-pebble conglomerate. These in turn are overlain by a persistent conglomerate which is in part auriferous, followed by cross-bedded quartzites up to the base of the lavas.

The greater part of the Basic Volcanic Formation consists of slightly schistose, grey-green to pale grey, fine-grained andesite. Amygdales are common, more especially near the top and bottom of flows, and are composed of quartz, chlorite, calcite and epidote. The lavas have a microfelsitic texture made up of microlites of feldspar and fibrous flakes of amphibole set in a turbid, cryptocrystalline groundmass of chlorite, quartz, iron ores, sericite, zoisite and epidote, which are largely alteration products. Tuffs and tuffaceous breccias occur throughout the formation, generally marking the tops of major flows. Three distinctive horizons have been noted within the andesitic lavas: a bed of quartzite up to 100 ft thick near the base; a quartz-feldspar porphyry some 150 ft thick; and a contorted, somewhat tuffaceous flow breccia, 50 ft thick, which occurs about 500 ft up in the sequence.

The Acid Volcanic Formation overlies the Basic Volcanic Formation with no disconformity and consists predominantly of rhyolites with subordinate flows of andesitic lava and beds of tuff. Textural variations in the rhyolites have enabled this formation to be subdivided into three members. The Lower

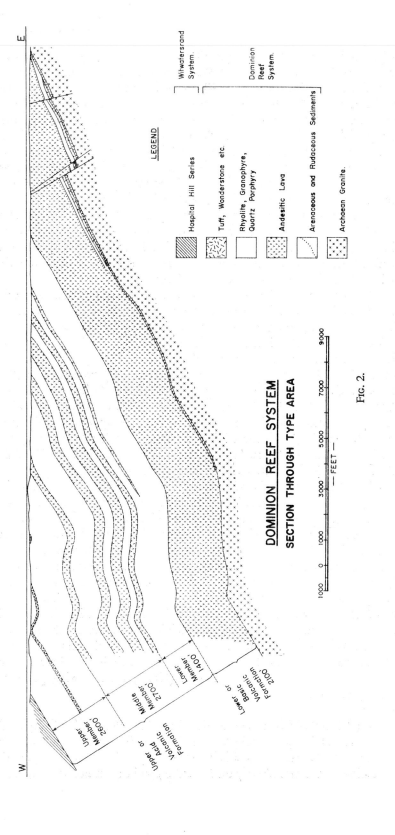

DOMINION REEF SYSTEM

SECTION THROUGH TYPE AREA

LEGEND

	Hospital Hill Series	Witwatersrand System.
	Tuff, Wonderstone etc.	
	Rhyolite, Granophyre, Quartz Porphyry	
	Andesitic Lava	Dominion Reef System.
	Arenaceous and Rudaceous Sediments	
	Archaean Granite.	

1000   0   1000   3000   5000   7000   9000
— FEET —

FIG. 2.

Upper Member 2600'

Middle Member 2700'

Lower Member 1400'

Upper or Acidic Volcanic Formation

Lower or Basic Volcanic Formation 2100'

Member is predominantly porphyritic rhyolite, blue-grey to dark grey in colour, with plagioclase phenocrysts up to 6 mm in length, in a micro-crystalline groundmass of feldspar, quartz and chlorite. The Middle Member, which is about 2700 ft thick, consists of alternating flows of rhyolite and andesite. Five flows of andesitic lava have been recognised within the main body of the rhyolites, which are similar to those of the Upper and Lower Members. The Upper Member, up to 2600 ft thick, is characterised by a thick, fine-grained, 'cherty' lava at the base, composed of altered feldspar and quartz phenocrysts in a siliceous and feldspathic cryptocrystalline ground-mass. Above this are banded lavas, pink and dark grey tuffs and 'wonder-stone', a metamorphosed volcanic ash composed essentially of pyrophyllite.

Occasional thin beds of red, black and olive-green tuffs are intercalated in the lavas. All the lavas of the Basic and Acid Volcanic Formations have been silicified to some extent and have been subjected to metamorphism with the development of cordierite in places.

Rocks of the Dominion Reef System are also exposed at several other localities (Fig. 3). North-eastwards from the type area towards the town of Ventersdorp, outcrops are lithologically similar to the type section but the thickness of the volcanic members decreases to a few feet in the neighbour-hood of Ventersdorp. To the west and north-west of the type area, the rocks of the Dominion Reef System are largely covered by younger formations but near Ottosdal they crop out prominently. The thickness of the Basic Volcanic Formation in this region appears to diminish westwards so that, locally, the Acid Volcanic Formation may rest directly on the Sedimentary Formation which, with several intercalations of andesitic lava, may here attain a thickness of 400 ft. In addition to these two regions the Dominion Reef System is found in outcrop around the Archaean granite dome at Vredefort. The lavas, basic to intermediate in composition, are amygdaloidal and generally very altered. Together with subordinate lenticular beds of metamorphosed sediments, the maximum thickness attained is 800 ft.

In earlier works, several authors have noted both lithological and strati-graphic similarity between the Dominion Reef System and the Insuzi, Uitkyk, Godwan and Zoetlief Formations. The latter are widespread, extend-ing from the northern Cape into Botswana, through to the northern Trans-vaal (where they are well exposed in the Zoutpansberg near Louis Trichardt), and into the eastern Transvaal, Swaziland and northern Natal. The main reason for equating them with the Dominion Reef System is the similarity of the lithology and also, in part, the difficulty of placing them anywhere else in the geological column. In regard to the similarities between that system and these presumed correlatives (Fig. 1) the following points are worthy of note:

(a) Zoetlief (the type area lies about 30 miles north of Vryburg): quartz and feldspar porphyries overlie coarse feldspathic sediments with intercalated flows of andesitic lavas; the porphyries locally transgress the sediments and rest directly on Archaean granite (Van Eeden et al., 1963).

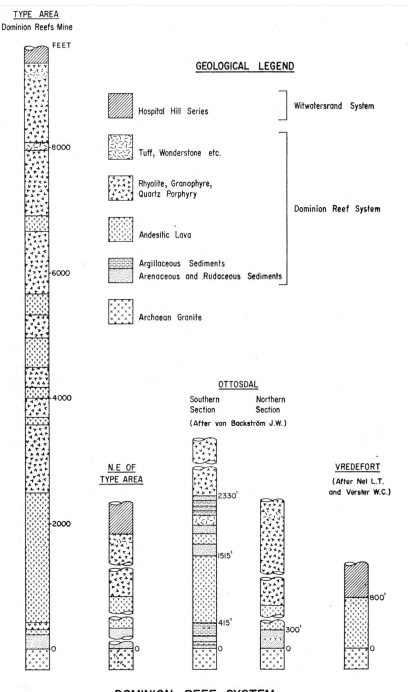

TYPE AREA
Dominion Reefs Mine

GEOLOGICAL LEGEND

Hospital Hill Series   ] Witwatersrand System

Tuff, Wonderstone etc.

Rhyolite, Granophyre,
Quartz Porphyry

Andesitic Lava   Dominion Reef System

Argillaceous Sediments
Arenaceous and Rudaceous Sediments

Archaean Granite

OTTOSDAL

Southern   Northern
Section   Section

(After von Backström J.W.)

N.E OF
TYPE AREA

VREDEFORT

(After Nel L.T.
and Verster W.C.)

DOMINION REEF SYSTEM
COMPARATIVE GEOLOGICAL COLUMNS

FIG. 3.

Recent Pb-isotope dating of acid lavas indicates that the Zoetlief volcanic sequence was extruded 2500-2700 m.y. ago (Van Niekerk and Burger, 1968). Since the Dominion Reef has yielded a somewhat older age of 2820 ± 100 (Allsopp, 1964), Van Niekerk and Burger (1968) suggest that the Zoetlief succession can most appropriately be accommodated as a separate stratigraphic system.

(b) Mafeking: acid lavas and tuffs with thin flows of basic lavas at their base lie on Archaean granite. Similar acid lavas and tuffs occur in the south-eastern portion of Botswana and have been correlated by Boocock and Van Straten (1962) with the Dominion Reef System.

(c) Gaberones-Thabazimbi: here there are two stages, a lower sedimentary stage which locally reaches a thickness of 1000 ft but which thins rapidly westward, and an upper volcanic stage composed of felsitic to porphyritic rhyolites at the top and amygdaloidal to porphyritic andesites at the bottom.

(d) In the Zoutpansberg mountains there are great thicknesses of andesitic lavas intercalated with arenaceous sediments. These were previously assigned to the Waterberg System.

(e) The Uitkyk and Godwan beds, which crop out at intervals along the base of the eastern escarpment from Potgietersrus to Kaapsehoop, in the northern and eastern Transvaal (Fig. 1), exhibit a lithology similar, in certain aspects, to that of the Dominion Reef System; more especially the Godwan beds which are composed of a lower division of shales, quartzites and conglomerates and an upper division of andesitic to acid lavas, agglomerates and tuffs.

(f) Amsterdam–Vryheid–Nkandhla: in several localities from Amsterdam in the south-eastern Transvaal, through Swaziland to Nkandhla in Zululand, thick successions of volcanics and sediments were classified as the Pongola System, equivalent to Upper and Lower Witwatersrand. Later these beds were divided into: the Mozaan (equivalent to Lower Witwatersrand); and the Insuzi (equivalent to Dominion Reef) (Du Toit, 1939; Matthews, 1967). The Insuzi was subdivided into an Upper Volcanic Stage, comprising 2800-16,000 ft of amygdaloidal andesites passing upwards, in the north, into amygdaloidal and non-amygdaloidal porphyries, and a Lower Sedimentary Stage with subordinate lavas. However, a granite intruded into the Mozaan near the Swaziland border has been dated at 3070 ± 70 m.y. (Allsopp et al., 1962), whereas the Dominion Reef System has been dated at 2820 ± 100 m.y. (Allsopp, 1964), indicating that both the Insuzi and the Mozaan are older than the Dominion Reef and emphasising the necessity for further radiometric data before correlation can be satisfactorily established.

Witwatersrand System

In contrast to the Dominion Reef System, extrusive rocks are comparatively rare in the Witwatersrand System, which principally consists of quartzites, shales and conglomerates.

The name 'Witwatersrand Beds' was first given by W. H. Penning in 1891 to the quartzites and conglomerates which crop out in the vicinity of Johannesburg but the term was restricted to beds of the Upper Division. The Lower Division, at that time, was called the Hospital Hill Series and correlated with formations in the Barberton Goldfields. The nomenclature at present accepted is:

Witwatersrand System
- Upper Division
  - Kimberley-Elsburg Series
  - Main-Bird Series
- Lower Division
  - Jeppestown Series
  - Government Reef Series
  - Hospital Hill Series

The Bird Amygdaloidal Diabase and the Jeppestown Amygdaloid are the only two lavas known within the system (Fig. 4) apart from extrusives in the Stinkfontein and Wolkerg Formations which may be correlatives of the Witwatersrand System.

The Bird Amygdaloidal Diabase is absent on the Central Rand but has been identified in the eastern portion of the Witwatersrand Basin from East Rand Proprietary Mines (12 miles east of Johannesburg) through the East Rand to Greylingstad, a village 55 miles south-east of Johannesburg. It also occurs in the Evander Goldfields, an outlier of Witwatersrand beds 30 miles east of the East Rand Basin, but is not developed in the western Transvaal or in the Orange Free State (O.F.S.) Goldfields. Often associated with the diabase is a light green to khaki arenaceous shale with abundant chloritoid which overlies the diabase and is separated from it by 20–40 ft of quartzite (Antrobus and Whiteside, 1964).

The Bird Amygdaloidal Diabase is a green, fine-grained altered basaltic lava with small amygdales of calcite, chlorite and chalcedony throughout. On the East Rand it is intercalated with one or two beds of quartzite and varies in thickness from 125 to 330 ft increasing in thickness eastwards and southwards and attaining a maximum of approximately 1000 ft at Greylingstad. In the Evander Goldfield the average thickness is 350 ft.

The Jeppestown Amygdaloid, near the middle of the Jeppestown Series, is more widespread than the Bird Amygdaloidal Diabase and has been exposed in boreholes and mine workings over the whole of the Witwatersrand Basin with the exception of the Central Rand and the Evander Goldfields. Nevertheless it is little known, probably because, occurring as it does in the Lower Division of the Witwatersrand, it is seldom closely associated with auriferous conglomerates. The lavas have, however, been seen in a number of boreholes. They are generally dense andesites, pale grey to dark green in colour, which occasionally contain small, dark green, stubby phenocrysts; the amygdales are filled with white quartz or dark green chlorite with a white rim, and often merge into each other, giving rise to dumb-bell shapes. Agglomerates and tuffs are sometimes present and these also have the same grey to dark green colour, often with a brownish tinge.

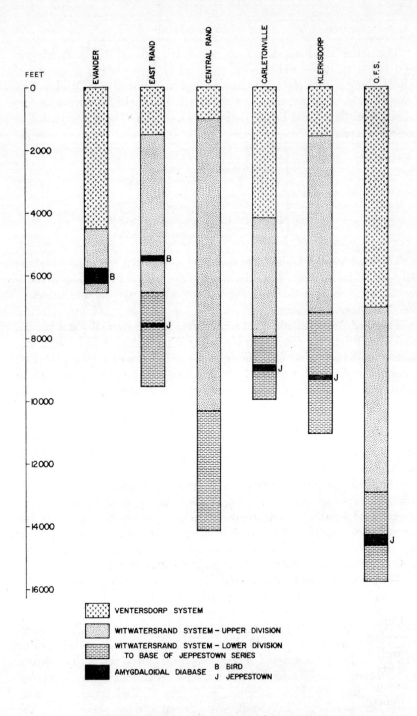

FEET

VENTERSDORP SYSTEM

WITWATERSRAND SYSTEM – UPPER DIVISION

WITWATERSRAND SYSTEM – LOWER DIVISION
TO BASE OF JEPPESTOWN SERIES

AMYGDALOIDAL DIABASE   B BIRD
                        J JEPPESTOWN

VENTERSDORP AND WITWATERSRAND SYSTEMS

COMPARATIVE GEOLOGICAL COLUMNS

FIG. 4.

These characteristics are sufficiently diagnostic for the Jeppestown Amygdaloid to be identified with certainty. It varies in thickness from 100 to 300 ft over most of the Witwatersrand Basin and, with several intercalations of quartzites, reaches a maximum known thickness of 1400 ft near Heilbron, a town in the Orange Free State, 40 miles south-east of Vredefort.

Ventersdorp System

The first description of lavas of this system was given by Stow (1874), who called them 'The Amygdaloidal Rocks of Pniel', named after a mission station about 20 miles north-west of Kimberley. Similar lavas were found in

TABLE 1

	Formation	Representative rocks
UPPER	Allanridge	Dark green amygdaloidal lavas
	Bothaville	Quartzites, conglomerates, subordinate dark shales
	———————————— unconformity ————————	
MIDDLE	Rietgat	Green-grey amygdaloidal and porphyritic lavas; minor intercalated tuffaceous, calcareous and cherty shales; impure limestones; quartzites; conglomerates
	Makwassie	Quartz and feldspar porphyries; green-grey amygdaloidal lavas; minor intercalated sediments
	New Kameeldoorns	Conglomerates; quartzites; tuffaceous, calcareous and cherty shales; impure calcareous rocks
	———————————— unconformity ————————	
LOWER	Langgeleven, or Lower Volcanic	Grey-green pyroclastics and amygdaloidal lavas

the Klipriviersberg hills south of the Central Rand and locally termed the 'Klipriviersberg Amygdaloid'. These volcanic rocks were followed, almost continuously, along the Vaal River and found to link up with the lavas at Pniel, so Molengraaff suggested that the name 'Vaal River System' be given to them. However, similar lavas and associated sediments had also been found near Ventersdorp, 75 miles west of Johannesburg, and Hatch (1903) proposed the title of 'Ventersdorp System' which was adopted and is now standard. Later, Du Toit (1907) described a group of predominantly acid volcanics in the Vryburg–Mafeking area, the Zoetlief Series, and considered them to be the lower member of the Ventersdorp System, the upper member being the Pniel Series.

Numerous authors contributed descriptions of rocks of this system from

D

various localities but it was not until the late 1930s that Nel (1935), Beetz (1936) and Jacobsen (1943) attempted to unravel the stratigraphy. Truter and Strauss (1943) and Van Eeden (1946) suggested that the Zoetlief Series, around Taungs, Zoetlief and Schweizer-Reneke (villages along the border between the Transvaal and northern Cape Province), should be correlated with the Dominion Reef System. Later, Borchers (1950) and Matthysen (1953) contributed greatly to our knowledge of the Ventersdorp System when they recognised that the succession was not continuous but interrupted by disconformities.

TABLE 2

	Formation	Representative rocks
		————————————————————unconformity————————
UPPER	Allanridge	Dark green amygdaloidal lavas
	Bothaville	Quartzites, conglomerates, subordinate dark shales
		———————————————————————— disconformity————————
MIDDLE	Rietgat	Green-grey amygdaloidal and porphyritic lavas; minor intercalated tuffaceous, calcareous and cherty shales; impure limestones; quartzites, conglomerates
	Makwassie	Quartz-porphyries with minor intercalations of andesitic lavas, tuffs and, rarely, sediments
LOWER	Langgeleven	Andesitic lavas and tuffs, including the New Kameeldoorn sub-stage of sediments and agglomerates
	Vaal Bend	Shales; tuffs; quartzites; conglomerates; andesitic lavas; pyroclastics; feldspar- and quartz-feldspar porphyries
		————————————————————— unconformity————————

The most comprehensive recent contribution has been made by Winter (1965a) whose classification of the system, shown in Table 1, was based on the type locality near Bothaville, where deep diamond drilling had traversed a great thickness of Ventersdorp rocks. On this classification Winter suggested that the 'primary subdivisions can be correlated with the Klipriviersberg Series, the Zoetlief Series, and the Pniel Series'. This is a straightforward cross correlation but Winter continues: 'A more stratigraphically correct subdivision would be to add the Klipriviersberg Series to the Upper Division of the Witwatersrand System, to elevate the Zoetlief Series to the rank of a System and to add the Pniel Series to the Transvaal System'. This would be a radical change which has not been generally accepted so far.

Strydom (1968) and others who have had access to diamond drill cores which were not available to Winter, have proposed a modification and simplification of Winter's scheme, shown in Table 2. There are four principal innovations in this scheme. First, there are major unconformities above the Allanridge Formation and below the Vaal Bend Formation. In the second place, Winter's unconformity below the Bothaville Formation has been modified to a disconformity because 'although the Bothaville overlies the older Ventersdorp Formations disconformably and overlaps on to pre-Ventersdorp rocks, there appears to be no great time interval between the Bothaville and the underlying Rietgat Formations' (Strydom, 1968). Thirdly, the New Kameeldoorns Formation is relegated to a sub-stage of the Langgeleven Formation. Strydom (1968) points out that 'where faulting was not accompanied by fissures ejecting pyroclasts, the troughs were infilled by lava more or less uninterruptedly. The New Kameeldoorns Stage can only be distinguished where these pyroclasts and sediments are developed.' Finally, the Vaal Bend Formation has been added, and is named for the prominent bend in the Vaal River north-west of Bothaville where considerable exploratory diamond drilling has recently been completed. This formation has been recognised in boreholes over a wide area, especially to the west and south-west of Klerksdorp.

Strydom's classification has much to recommend it and fits in well with views which he and I hold, as yet unpublished, on the tectonic history of the Greater Witwatersrand Basin.

The Ventersdorp System reaches its greatest development in the northern Orange Free State and western Transvaal and both Winter and Strydom have used this region as their type area. The approximate maximum thickness of Strydom's divisions (1968) are:

Allanridge Formation	1500 ft	Makwassie Formation	2000 ft
Bothaville Formation	1000 ft	Langgeleven Formation	7000 ft
Rietgat Formation	2000 ft	Vaal Bend Formation	3000 ft

A comparison of sections in the Bothaville–Klerksdorp area shows that thicknesses are extremely variable. In a borehole on the Farm Syferfontein, 18 miles west of Klerksdorp, the six formations amount to 4000 ft (Fig. 5, Section V); at the other extreme, a borehole on the Farm Hart van Boomtuin, about 20 miles south-west of Klerksdorp, traversed 10,000 ft of Ventersdorp rocks representing only the Allanridge, Bothaville, Makwassie and Langgeleven Formations, the Rietgat and Vaal Bend being absent (Fig. 5, Section I).

Since little detail is as yet available on the Ventersdorp System it is worth adding some recent discoveries. Winter (1963) described algal structures in cores from a diamond drill hole on the Farm Ellenboogleegte, about 40 miles south-west of Bothaville. These stromatolites occur in impure dolomitic limestones in the Rietgat Formation and were found at depths of approximately 3500 and 3900 ft below surface. Winter considers the environment in which these algae grew 'to have been shallow lakes which occasionally dried up.'

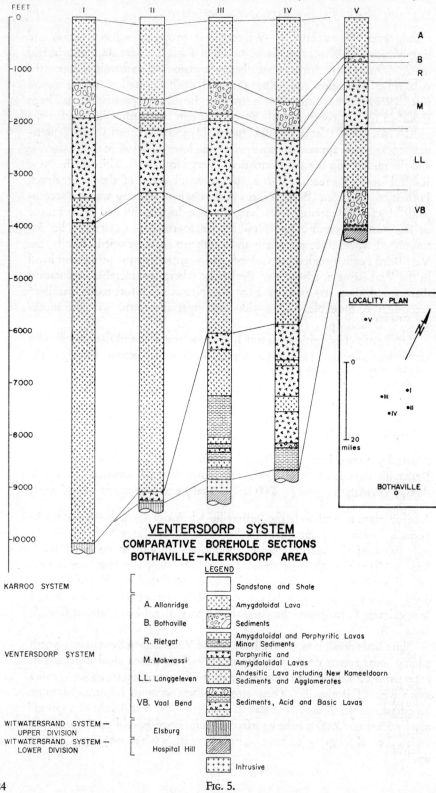

VENTERSDORP SYSTEM
COMPARATIVE BOREHOLE SECTIONS
BOTHAVILLE—KLERKSDORP AREA

LEGEND

KARROO SYSTEM — Sandstone and Shale

VENTERSDORP SYSTEM
A. Allanridge — Amygdaloidal Lava
B. Bothaville — Sediments
R. Rietgat — Amygdaloidal and Porphyritic Lavas Minor Sediments
M. Makwassi — Porphyritic and Amygdaloidal Lavas
LL. Langgeleven — Andesitic Lava including New Kameeldoorn Sediments and Agglomerates
VB. Vaal Bend — Sediments, Acid and Basic Lavas

WITWATERSRAND SYSTEM — UPPER DIVISION — Elsburg
WITWATERSRAND SYSTEM — LOWER DIVISION — Hospital Hill

+ + + Intrusive

84

FIG. 5.

Pillow structures were first recorded in 1961 by the geologists on Western Deep Levels Gold Mine near Carletonville in the western Transvaal; the pillows immediately overlie the Ventersdorp Contact Reef, indicating 'that certain areas of the V.C.R. were under water where the lava flowed over the area' (Knowles, 1966). Since then other zones of pillows have been seen, notably one about 250-300 ft above the Ventersdorp Contact Reef, suggesting an advancing and retreating shore line during early Ventersdorp times.

Finally, a remarkable feature of the Langgeleven Formation is the persistence of 'markers', such as a zone of coarsely porphyritic lava with feldspar phenocrysts up to three inches in length, near the base of the formation. This lava has been recorded near Balfour, in the South Rand Goldfield, in the Evander Goldfield, on the East, Central and West Rand, on the Far West Rand and in the Klerksdorp area. Other less extensive markers are the 'purple zones' of the Orange Free State and Klerksdorp Goldfields, which lie some distance above the porphyritic lava.

Conclusions

When considering the time scale of the Witwatersrand Triad, it should be noted that the Dominion Reef System, the oldest member of the triad, has been dated at 2800 m.y. whereas zircon from a quartz porphyry lava considered to occur near the top of the Ventersdorp System, the youngest of the triad, has given a U-Pb age of 2300 m.y. (Allsopp, 1964; Van Niekerk and Burger, 1964, 1968). Thus, the time span of the triad is comparable to that of virtually the whole of the Phanerozoic.

The maximum total thickness of the triad is probably of the order of 50,000 ft, of which nearly half is lavas. Not very much is known about the feeders which supplied this vast amount of volcanic material. In the East and Central Rand, numerous dykes have been encountered in underground workings, and these are considered to have been feeder channels from which the lavas were extruded. In the western Transvaal, especially in the vicinity of Klerksdorp, breccia-filled pipes and fissures have been noted which probably represent the vents of explosive eruptions. In this context, it should be noted that agglomerates and tuffs are more common in the Ventersdorp sequence in the western Transvaal and Orange Free State than on the eastern side of the Witwatersrand Basin.

Knowledge of the volcanic rocks of the triad is essential to mining and exploration geologists in South Africa as the eruptives are intimately associated with the gold- and uranium-bearing conglomerates of all three systems. One example of the necessity of knowing these rocks is found in the Dominion Reef area where lavas of both the Dominion Reef System and the Ventersdorp System may crop out in close proximity to each other. In this instance it is very important to be able to differentiate between them as this will assist the elucidation of the structure and also give a fair approximation of the depth to underlying gold- and uranium-bearing conglomerates.

Perhaps even more necessary is correct dating and recognition of some of the so-called correlatives of the Dominion Reef and Witwatersrand Systems. In this respect there is great scope for more radiometric dating.

Acknowledgements

I am grateful to the Anglo-American Corporation of South Africa, Limited, for permission to publish this paper and to use information from company reports.

To the many geologists who have worked on the Dominion Reef System and the Ventersdorp System I wish to extend my thanks, as it would not have been possible to have written this paper without the vast amount of information contributed by them. My especial thanks go to Messrs P. M. Strydom and A. J. Jager for their help in supplying information and for their discerning criticism.

REFERENCES

ALLSOPP, H. L. 1964. Rubidium-strontium ages from the western Transvaal. *Nature, Lond.*, **204**, 361.
—— ROBERTS, H. R., SCHREINER, G. D. L., and HUNTER, D. R. 1962. Rb-Sr age measurements on various Swaziland granites. *J. geophys. Res.*, **67**, 5307.
ANTROBUS, E. S. A., and WHITESIDE, H. C. M. 1964. The geology of certain mines in the East Rand. In *The geology of some ore deposits of southern Africa*, p. 125. (Ed. S. H. Haughton), Vol. 1, Geol. Soc. S. Afr.
BEETZ, P. F. W. 1936. Contribution to the geology of the Klerksdorp district from the results of drilling activities by the Western Reefs Exploration and Development Company Limited. *Trans. geol. Soc. S. Afr.*, **39**, 223.
BOOCOCK, C., and VAN STRATEN, O. J. 1962. Notes on the geology and hydrogeology of the Central Kalahari region, Bechuanaland Protectorate. *Trans. geol. Soc. S. Afr.*, **65**, 125.
BORCHERS, R. 1950. The Odendaalsrus-Virginia goldfield and its relation to the Witwatersrand. Unpublished D.Sc. thesis, Univ. South Africa.
BROCK, B. B., and PRETORIUS, D. A. 1964a. An introduction to the stratigraphy and structure of the Rand Goldfield. In *The geology of some ore deposits of southern Africa*, p. 25. (Ed. S. H. Haughton), Vol. 1, Geol. Soc. S. Afr.
—— and PRETORIUS, D. A. 1964b. Rand basin sedimentation and tectonics. In *The geology of some ore deposits of southern Africa*, p. 549. (Ed. S. H. Haughton), Vol. 1, Geol. Soc. S. Afr.
DU TOIT, A. L. 1907. Geological survey of portions of the divisions of Vryburg and Mafeking. *Geol. Comm.*, *Cape of Good Hope*.
—— 1939. *Geology of South Africa*, Edinburgh.
HAMILTON, G. N. G., and COOKE, H. B. S. 1960. *Geology for South African students*, 4th ed. Johannesburg.
HATCH, F. H. 1903. The boulder beds of Ventersdorp, Transvaal. *Trans. geol. Soc. S. Afr.*, **6**, 95.
JACOBSEN, W. 1943. Ausbildung und Petrographie der südafrikanischen Ventersdorp-Formation im südwestlichen Transvaal und nördlichen Oranje-Freistaat. *Neues Jb. Miner. Geol. Abh. Bd.*, **78**, 217.
KNOWLES, A. G. 1966. Palaeocurrent study of the Ventersdorp Contact Reef at Western Deep Levels, Limited, on the Far West Rand. Unpublished M.Sc. thesis, Univ. Witwatersrand.
MATTHEWS, P. E. 1967. The pre-Karroo formations of the White Umfolozi inlier, northern Natal. *Trans. geol. Soc. S. Afr.*, **70**—in press.

MATTHYSEN, J. L. 1953. 'n Nuwe stratigrafise indeling van die Ventersdorp Sistum. Unpublished M.Sc. thesis, Univ. Pretoria.

MOLENGRAAFF, G. A. F. 1905. Note on the geology of a portion of the Klerksdorp district with special reference to the development of the Lower Witwatersrand beds. *Trans. geol. Soc. S. Afr.*, **8**, 16.

NEL, L. T. 1934. The Witwatersrand System in the Klerksdorp-Ventersdorp area. *Bull. geol. Surv. S. Afr.*, **1**.

—— 1935. Geology of the Klerksdorp-Ventersdorp area. *Spec. Publ. geol. Surv. S. Afr.*, **9**.

—— and VERSTER, W. C. 1962. Die geologie van die gebied tussen Bothaville en Vredefort. Explanation of Sheets 2726B and 2727A. *Geol. Surv. S. Afr.*

PRETORIUS, D. A. 1966. Recorrelation of alleged Witwatersrand correlatives. *8th Ann. Rep. Econ. Geol. Res. Unit, Univ. Witwatersrand, Johannesburg*, 17.

STOW, G. W. 1874. Geological notes on Griqualand West. *Q. J. geol. Soc. Lond.*, **30**, 581.

STRYDOM, P. M. 1968. Anglo-American Corporation of South Africa Limited, Company Reports 1966, 1967 and 1968.

TRUTER, F. C., and STRAUSS, C. A. 1941. The pre-Transvaal rocks at Taungs, Cape Province. *Trans. geol. Soc. S. Afr.*, **44**, 161.

—— 1949. A review of volcanism in the geological history of South Africa. *Proc. geol. Soc. S. Afr.*, **52**, xxix.

VAN EEDEN, O. R. 1946. Korrelasie van sekere voor-Transvaal-gesteentes in die distrik Schweizer-Reneke. *Trans. geol. Soc. S. Afr.*, **49**, 277.

—— DE WET, N. P., and STRAUSS, C. A. 1963. The geology of the area around Schweizer-Reneke. Explanation of Sheets 2724B and 2725A. *Geol. Surv. S. Afr.*

VAN NIEKERK, C. B., and BURGER, A. J. 1964. The age of the Ventersdorp System. *Annls geol. Surv. S. Afr.*, **3**, 75.

—— and BURGER, A. J. 1968. Pb-isotope dating of the Zoetlief System, South Africa. *Earth Planetary Sci. Let.*, **4**, 211.

VON BACKSTRÖM, J. W. 1952. The Dominion Reef and Witwatersrand Systems between Wolmaransstad and Ottosdal, Transvaal. *Trans. geol. Soc. S. Afr.*, **55**, 53.

—— 1962. Die geologie van die gebied om Ottosdal, Transvaal. Explanation of Sheets 2625D and 2626C. *Geol. Surv. S. Afr.*

WINTER, H. DE LA R. 1963. Algal structures in the sediments of the Ventersdorp System. *Trans. geol. Soc. S. Afr.*, **66**, 115.

—— 1965a. The Ventersdorp System in the Western Transvaal. *7th Ann. Rep. Econ. Geol. Res. Unit, Univ. Witwatersrand, Johannesburg*, 19.

—— 1965b. The stratigraphy of the Ventersdorp System in the Bothaville district and adjoining area. Unpublished Ph.D. thesis, Univ. Witwatersrand.

E. S. W. SIMPSON

5   # The anorthosite of southern Angola : a review of present data

ABSTRACT. *The great complex anorthosite body of southern Angola is the largest known mass of its type, having an exposed outcrop area of 12,500 km² plus an estimated additional 4000 km² concealed beneath younger sediments. Two contrasted rock types are represented: (i) a pale massive anorthosite of Adirondack type, for which a metasomatic origin by calcium metasomatism of the original banded amphibolitic and granitic gneisses has been proposed but not adequately substantiated; and (ii) a dark troctolitic anorthosite which shows frequent igneous lamination and weak cryptic layering, comprising the whole of the northern part of the massif but occurring in the south as multiple and banded intrusions into the pale massive anorthosite. The development of mafic and ultramafic rocks is confined to relatively narrow marginal zones of the massif in some areas.*

The age of the pale massive anorthosite is c. 1260 m.y. or more, the minimum age of the troctolitic type being late Precambrian.

Introduction

LOCATED in the extreme south of Angola, and extending southward across the Cunene River into South-West Africa, the great complex anorthosite mass of southern Angola is the largest known body of its type. It is exposed as a longitudinally elongated massif over a distance of 300 km, varying in width between 30 and 100 km: the total exposed outcrop area is 12,500 km², to which may be added an estimated 4000 km² concealed beneath the cover of younger sediments (Bembe/Otavi and Kalahari). Geological and tectonic maps by Mouta (1954, 1955) show the regional setting of the anorthosite complex.

Basic rocks were apparently first recognised in this area by Rego Lima in 1898 and they were described by Pereira de Souza in a series of publications, notably those of 1906 and 1916. Later, in 1921, Pereira de Souza published analyses of norite and gabbro and these were reproduced by Torre de Assuncao *et al.* (1954). In an early summary account, Mouta and O'Donnell (1933, pp. 22-25) noted that the complex is largely made up of coarse-grained anorthosite becoming finer grained and more melanocratic towards the margins. Much later, Montenegro de Andrade (1950) gave a comprehensive account of published descriptions of the mass and this was followed, in 1958, by a paper by Stone and Brown in which they gave a detailed description of a series of specimens and noted the existence of cryptic layering in the basic rocks of the complex.

Owing mainly to its remote location, the anorthosite complex has received very little systematic or detailed study. Since the last published summary of information (Simpson and Otto, 1960), a brief description of dolerite dykes carrying abundant anorthosite inclusions has been published by Alves et al. (1966), and Köstlin (1967) has made a fairly detailed field and laboratory study of the extreme south-western area, south of the Cunene River, west of 13° 30′ E.

Through the courtesy of the Director of the Instituto de Investigacão Cientifica de Angola, I was able to undertake a 12-day reconnaissance of the anorthosite exposures north of the Cunene River, in 1964, in the company of Dr C. A. de Matos Alves and F. E. de V. Lapido-Loureiro and their assistants. The accompanying map (Fig. 1) has been compiled from observations made during that brief study (which was greatly aided by the availability of excellent 1 : 100,000 topographic maps and full aerial photograph coverage) combined with data collected in northern South-West Africa by both Köstlin (1967) and myself on a previous visit to the southern area in 1958.

With few exceptions the anorthosite area is easily accessible to four-wheel drive vehicles. The northern half is relatively flat with low ridges and isolated tors diminishing in both frequency and elevation northwards, although exposures are rarely completely absent except in the extreme north between Olivenca a Nova and Vila Paiva Couceiro. The southern region (south of 16° 30′ S.) is deeply dissected by the Cunene River and its tributaries, providing excellent exposures in rugged terrain having a relief of 500 m and more.

Lithology and structure

COUNTRY ROCKS

The anorthosite complex is bordered by a heterogeneous assemblage of granitic, basic and gneissic rocks which have only been studied in some detail by Köstlin (1967) over a restricted area in the south-west. Granitic and granodioritic gneisses predominate, accompanied by extensive zones of banded gneisses, amphibolites, metasediments and metavolcanics of the Epupa and other formations, and unfoliated granites around Oncocua, Otchinjau and near the northern margin. These rocks, of mixed origin, represent the products of a variety of metamorphic, metasomatic and magmatic processes and most are assigned to the albite-epidote-amphibolite and amphibolite facies.

As indicated on Fig. 1, long tongues and scattered patches of metasediment, quartzo-feldspathic gneiss or coarse pegmatite are intimately associated with the pale massive anorthosite south of Epupa and in the Cacuio area. A large irregular body of metasediment and gneiss extends eastward from Epupa almost reaching the deeply embayed eastern margin of the anorthosite mass to the north of Chitado, and numerous elongated bodies and scattered patches of granite and gneiss extend north-eastward across the exposed width of the dark troctolitic anorthosite north of Pocolo. With few exceptions the granitic rocks which occur in close association with the anorthosite characteristically develop a brown or maroon coloration due to hematite staining of feldspar, accompanied by the presence of much epidote.

Apophyses of red aplite, granite and syenite clearly show intrusive relationships with the anorthosite in areas of good exposure and are provisionally
ascribed to the effects of rheomorphism during emplacement of the anorthosite.

Large-scale faulting is associated with the large granite-gneiss embayment
east of Epupa and north of Enjandi. Emplacement of anorthosite in this
area appears to have been structurally controlled, and this example serves to
emphasise the necessity for a proper understanding of the regional tectonic
pattern and history before the mode of emplacement and structure of the
anorthosite mass can be adequately assessed. A similar approach has been
successfully applied by Michot (1960) to the anorthosite problem in south-
western Norway, but the area surrounding the Angola anorthosite has never
been adequately mapped or studied more deeply than reconnaissance level.

THE ANORTHOSITE COMPLEX

The three main lithological types of the anorthosite mass comprise the dark
troctolitic and pale massive anorthosite, and marginal mafic and ultramafic
rocks. Titaniferous magnetite bodies are sporadically developed.

Pale massive anorthosite

This is the predominant rock type exposed in the southern half of the complex and it is not characteristically developed in the area north of Nihiquilo.
It is similar in many respects to the Adirondack and southern Norwegian
anorthosites, being massive, predominantly leucocratic with plagioclase of
intermediate composition (An_{50}-An_{66}) and occasional minor amounts of
orthopyroxene and/or clinopyroxene, pale grey or brown when fresh, and
generally medium- to coarse-grained, some patches containing plagioclase
crystals up to a metre in length. The occasional presence of minor olivine in
addition to pyroxene (notably around Enjandi and extending towards
Chitado) suggests that types transitional to the troctolitic anorthosite are
developed in some areas.

Being easily weathered, the massive anorthosite underlies areas of low-
lying terrain where it is not well exposed, but such areas are conspicuous on
aerial photographs owing to their white or pale tones. Saussuritisation of
plagioclase and the development of epidote are common, particularly
adjacent to the margins and remnant xenoliths and patches of reddish
quartzo-feldspathic country rock where foliated anorthositic gneisses with
streaks and rosettes of tremolite-actinolite are characteristically developed
(for example south and east of Epupa, south of Oncocua and in the Cacuio
area). In the extreme south-west the anorthositic gneisses frequently contain
more than 20% of amphibole (hornblende); they are then indistinguishable
from the amphibolite lenses in the adjacent mixed gneisses with which the
contacts are gradational along strike over several hundreds of metres. A
study of this area has led Köstlin (1967) to the tentative conclusion that the
massive anorthosite is the product of calcium metasomatism of the regional

mixed and banded gneisses: the amphibolitic components would require little energy for transformation while the granitic rocks would tend to persist as discrete bodies, segregation pegmatites and remnant patches within the anorthosite, as indeed they do in many areas (see Fig. 1). Although Köstlin's data and argument are not entirely convincing, his suggestion offers a sufficiently attractive explanation of the observed field relations to warrant more detailed mineralogical and geochemical investigation. It must however be borne in mind that some phases of the massive anorthosite (for instance near Nihiquilo) appear to be more closely related genetically to the troctolitic type; that occurrences of ultramafic rocks are marginal to both types; and that veins or segregations of irregular pink pegmatite and graphic granite are also found in the troctolitic anorthosite which is undoubtedly magmatic and intrusive.

Dark troctolitic anorthosite

This rock type occupies the northern part of the complex north of 16° S. and is well represented in the better exposures of the southern areas. These rocks are easily distinguished from the pale massive type by their positive relief up to 700 m; dark grey, almost black colour; tabular habit of plagioclase crystals; average grain size usually 10-15 mm, varying up to 10 cm; common development of igneous lamination and occasional faintly discernible rhythmic layering in the coarsely banded rocks which are developed over several areas south of 15° 45′ S. (see Fig. 1); almost universal lack of alteration; and consistent mineral constitution. Hololeucocratic rocks are not common and the normal mineral variation is plagioclase (An_{54}-An_{80}) 70-95%, olivine (Fa_{13}-Fa_{42}) 1-30%, orthopyroxene (Of_{15}-Of_{29}) 0-12%, clinopyroxene 0-5%, with accessory titaniferous magnetite and biotite; the total mafic mineral content is commonly 10-15%, olivine being predominant.

The fabric of the troctolitic anorthosite appears to be uniformly massive in the northern area where good outcrops are sparse. In the region between Pocolo, Chibemba and Nihiquilo, areas of conspicuously banded and laminated anorthosite with high angles of dip are clearly recognisable on aerial photographs and are shown diagrammatically in Fig. 1. Lamination and banding are recognisable in the intervening outcrops but such structures are near-horizontal. In the Cacuio area the regionally developed pale massive anorthosite is invaded by an arcuate system of discontinuous, outward-dipping dykes of massive troctolitic anorthosite up to several hundreds of metres thick. In the Zebra Mountains (between Otjijanjasemo and Enjandi) the structures shown in Fig. 1 are interpreted as an asymmetrical dome-shaped arrangement of thick intrusive sheets of laminated troctolitic anorthosite separated by screens of pale massive anorthosite, which are conspicuous where the attitude exceeds 30° but are largely hidden beneath the troctolitic anorthosite sheets near the centre of this structure where the attitude is near-horizontal. In the Zebra Mountain structure and the Cacuio arcs, the troctolitic anorthosite is clearly intrusive into the pale massive anorthosite. Multiple intrusion and/or differentiation may be responsible for the banded structures in troctolitic anorthosite north of Nihiquilo. The presence of cryptic layering in the massive troctolitic anorthosite between Pocolo and Chibemba (Gambos) was reported by Stone and Brown (1958) but was not found by the author in a series of specimens collected across the complex in the vicinity of Quihita. Cryptic layering is also reported from the Zebra Mountain structure (Simpson and Otto, 1960; Köstlin, 1967).

The enigmatic relationship between the pale massive and troctolitic anorthosite types has been summarised above. Table 1 shows Köstlin's (1967) average chemical compositions, determined spectrochemically, for the two types.

TABLE 1

Average chemical compositions of pale massive (*a*) and troctolitic (*b*) anorthosites.

	(*a*)	(*b*)
SiO_2	52·23	49·26
TiO_2	0·13	0·25
Al_2O_3	26·39	25·10
Fe_2O_3	1·12	1·15
FeO	0·56	3·58
MgO	0·66	5·28
MnO	0·04	0·06
CaO	11·97	11·28
Na_2O	4·91	3·50
K_2O	0·42	0·43
P_2O_5	1·33	0·01
Loss on ign.	1·45	0·82

Mafic and ultramafic rocks

These types appear to be confined to the border zones of the anorthosite complex and comprise norites, gabbros, troctolite, pyroxenite and serpentinite; their known distribution is shown in Fig. 1. These rocks have a very evident banded arrangement over a width of some 600-1000 m along the near-vertical margin of the complex north-west of Pocolo. They are provisionally interpreted as accumulates forming the floor-zone of the anorthosite mass.

Köstlin (1967) regards the body of norite and serpentinite near Ombuku north of Otjijanjasemo as an intrusive plug, and has described three satellitic basic and ultrabasic intrusions in the gneiss terrain: the Ejau anorthosite gabbro west of Epupa, the Otjijanjasemo troctolite, and the Etengua layered dunite west of Otjijanjasemo. I have located yet another, the Lundu gabbro, some 16 km south of Vila João de Almeida in the northern region.

Titaniferous magnetite

Titaniferous magnetite is a constant accessory mineral in the troctolitic anorthosite but in some localities (for example, north of Chitado) it is a major constituent, making up some 20% of the rock which then resembles the ilmeno-norite of the Tellnes orebody north of Jøsingfjord in southern Norway. Massive bodies of titaniferous magnetite are frequently found in both varieties of anorthosite near inclusions and remnant patches of red granitic and syenitic rock and coarse pegmatite, as for example, the prospected deposits near Almoster and the abundance of titaniferous magnetite rubble in the Cacuio and Chitado areas.

Several random samples of the massive ore were analysed by the Anglo-American Research Laboratory in Johannesburg and serve to indicate the following compositional ranges: Fe, 49·4-52·3%; TiO_2, 13·4-19·7%; V_2O_5, 0·4-0·7%.

BEMBE DOLERITES

The Bembe System (Chella Formation of Beetz, 1933) comprises a more or less horizontally disposed succession of grits, sandstones, conglomerates and limestones which cover part of the anorthosite complex near Oncocua, and extend further to the west, north and south to form prominent tabular outliers. Representatives of these rocks south of the Cunene River are assigned to the Nosib Formation and Otavi Series of the cratonic/miogeosynclinal Outjo Facies (formerly Otavi System) of the late Precambrian Damara System (Martin, 1965). The sediments in the anorthosite area are continental in character, deposited in a stable cratonic environment and reach a thickness of several hundred metres.

The large outlier of Bembe sediments north of Oncocua is invaded by dolerite sheets which increase in thickness eastwards and predominate over sediments in that part of the succession which covers the anorthosite complex west and south-west of Cacuio. In the latter area it is most difficult to distinguish between dolerite and troctolitic anorthosite both on aerial photographs and in the field. Dolerites are completely absent from the Nosib/ Otavi sedimentary succession south of the Cunene River and are not conspicuous in the Chella Escarpment west of Sá da Bandeira in the north (Fig. 1). It is therefore remarkable that the area of maximum dolerite intrusion is so closely connected with the site of an earlier, more extensive phase (or phases) of emplacement of basic igneous rocks represented by the anorthosite complex.

The southern part of the anorthosite complex is traversed by a north-westerly-trending swarm of dolerite dykes (see Fig. 1) some of which are clearly intrusive into at least the lower Bembe dolerites. The dykes are therefore regarded as feeders to the sills. The age of these basic intrusives is unknown, but is probably either late Precambrian–early Palaeozoic or Stormberg (Jurassic).

Age relations and tectonic setting

Radiometric analysis of muscovite from a pegmatite in massive anorthosite near Enjandi gave a Rb-Sr apparent age of 1260±90 m.y. (Simpson and

Otto, 1960), which has been recalculated to 1190 m.y. by Cahen and Snelling (1966). This either indicates a minimum age for the massive anorthosite, or the actual date of emplacement if the pegmatite is of rheomorphic origin. Other ages published by Mendes (1966) and Cahen and Snelling (1966) from the basement gneisses to the north and north-west of the anorthosite show a spread between 871 and 1734 m.y. with a possibly significant grouping of Rb-Sr biotite apparent ages at approximately 1000 m.y. from rocks west of Sá da Bandeira. Present geological data are too sparse to attempt a correlation between the age of emplacement of the massive anorthosite and that of any single event of regional importance; however, it corresponds approximately to the Kibaran and related orogenic episodes (Clifford, 1968). The intrusive phase of the troctolitic anorthosite is younger than much of what has been termed the pale massive type, but no data are available to indicate the age difference. The spatially related intrusion of Bembe dolerite dykes and sills can be more confidently assigned a maximum age limit of late Precambrian or early Palaeozoic.

Fig. 2 shows that whereas the Great Dyke of Rhodesia (2500 m.y.) and the Bushveld Complex (1950 m.y.) are centrally situated within the primitive Rhodesia-Transvaal Craton (Clifford, 1968) of even greater antiquity, the Angola anorthosite (1200 m.y.?) is marginally located with respect to the Congo Craton (Kennedy, 1965) of which the Angola-Kasai Craton (Clifford, 1968) forms a more ancient part. The anorthosite is therefore an integral unit in one of the primitive African cratonic nuclei, whereas the spatially coincident Bembe dolerites are possibly related in time with the great Pan-African ($\pm$ 500 m.y.) thermo-tectonic (orogenic) event discussed by Professor W. Q. Kennedy in the course of his outstanding contributions to African geology.

REFERENCES

ALVES, C. A. DE M., LAPIDO-LOUREIRO, F. E. DE V., and SILVA, L. C. DE SOUSA E. 1966. Sobre uma rocha dolerítica con encraves do Sudoeste de Angola. Bol. Inst. Invest. cient. Angola, **3**, 257.

BEETZ, P. F. W. 1933. Geology of south-west Angola, between Cunene and Lunda axis. Trans. geol. Soc. S. Afr., **36**, 137.

CAHEN, L., and SNELLING, N. J. 1966. The geochronology of equatorial Africa, Amsterdam.

CLIFFORD, T. N. 1968. Radiometric dating and the pre-Silurian geology of Africa. In Radiometric dating for geologists, p. 299. (Eds. E. I. Hamilton and R. M. Farquhar), London.

DE VILLIERS, J., and SIMPSON, E. S. W. (in press). Late-Precambrian tectonic patterns in south-western Africa. Symp. Continental Drift, Montevideo, 1967.

KENNEDY, W. Q. 1965. The influence of basement structure on the evolution of the coastal (Mesozoic and Tertiary) basins of Africa. In Salt basins around Africa, p. 7. Inst. Petrol., London.,

KÖSTLIN, E. C. 1967. The geology of part of the Kunene Basic Complex, Kaokoveld, South-West Africa. Unpublished thesis, Univ. Cape Town.

MARTIN, H. 1965. The Precambrian geology of South West Africa and Namaqualand. Precambrian Res. Unit, Univ. Cape Town.

MENDES, F. 1966. Ages absolus par la méthode au strontium de quelques roches d'Angola. C.R. Acad. Sci. Paris, **262**, 2201.

MICHOT, P. 1960. La géologie de la catazone: le problème des anorthosites, la palingénèse basique et la tectonique catazonale dans le Rogaland méridional (Norvège méridionale). *Int. geol. Congr.*, **21**, Guide to Excursion A.9.

MONTENEGRO DE ANDRADE, M. 1950. Estado actual dos conhecimentos sobre as rochas igneas de Angola. *Publ. Mus. Min. Geol.*, **27**, Univ. Coimbra, 35.

MOUTA, F. 1954. Noticia explicativa do esboço geológico de Angola (1:2,000,000). *Junta Invest. Ultramar, Lisbon.*

—— 1955. Esboço tectónico de Angola. Noticia explicativa. *Anais*, **10** (5), *Junta Invest. Ultramar, Lisbon.*

—— and O'DONNELL, F. 1933. Carte géologique de l'Angola. Notice explicative. *Minist. Colón. Lisbon.*

PEREIRA DE SOUZA, F. L. 1906–11. Alguns trechos do relatorio do Eng.° Rego Lima sobre a sua missao as minas de Cassinga em 1898. *Rev. Eng. militar.*, **16**.

—— 1916. Contributions a l'étude petrographique du sud-ouest de l'Angola. *C.R. Acad. Sci. Paris*, **162**, 692.

—— 1921. Sur quelques roches remarquable d'Angola. *C.R. Acad. Sci. Paris*, **173**, 177.

SIMPSON, E. S. W., and OTTO, J. D. T. 1960. On the Precambrian anorthosite mass of southern Angola. *Int. geol. Congr.*, **21** (13), 216.

STONE, P., and BROWN, G. M. 1958. The Quihita-Cunene layered gabbroic intrusion of south-west Angola. *Geol. Mag.*, **95**, 195.

TORRE DE ASSUNCAO, C., MOUTA, F., and BRAK-LAMY, J. 1954. Contribution a l'étude de la petrographie du sud de l'Angola. *Int. geol. Congr.*, **19** (20), 253.

L. CAHEN

6   Igneous activity and mineralisation episodes in the evolution of the Kibaride and Katangide Orogenic Belts of Central Africa

ABSTRACT. *Field relations and radiometric ages of igneous and mineralisation events form the basis of this study of the relations in space and time between the Kibaride and Katangide Orogenic Belts of Central Africa.*

The first sections of this paper summarise the data concerning the basement of the Kibaride Belt, the Kibaran stratigraphical column, and the syn-tectonic and post-tectonic events of the Kibaran cycle. The connection between the Kibaran cycle and the subsequent deposition of the Lower Katangan is then emphasised, and followed by a consideration of the Katangan succession above the 'Grand Conglomérat', and the syn-tectonic and post-tectonic events of the Katangan cycle. A discussion of molasse of the Katangan orogeny is used to retrace the history of the Katangide Belt.

Three conclusions are reached: (i) the vertical movements of the Kibaride Belt controlled to a great extent the sedimentation of the Katangan —initially these movements were 'active' and represented the late stages of the Kibaran cycle, but 'passive' movements followed, and represented the responses of the basement of the Katangan to the tectonic phases of the Katangan orogenic cycle; (ii) molasse-like deposits of each of the belts were deposited within the time range defined by radiometric dating of post-tectonic events of each cycle; and (iii) the major tectono-stratigraphic subdivisions of Precambrian rocks in Africa often comprise sediments belonging to two successive orogenic cycles—namely the late stage(s) of a particular cycle, and the early and median stages of the succeeding cycle.

Introduction

Although far from complete, knowledge of the geology of the Katanga Province of the Democratic Republic of the Congo has reached a point where it is possible to discuss the relationship between successive orogenic belts from both a spatial and a chronological point of view. This essay deals essentially with the relationship between the Kibaride Belt and the subsequent Katangide Belt. However, the present state of our knowledge of the relationships between the former and even earlier belt(s) is also summarised, as is the later history of the Katangide Belt. The age data used in this regional analysis are mainly derived from magmatic events and this paper thus comprises a review of the magmatism of three successive orogenic belts.[1]

Custom in the Congo has favoured use of the same word to name a major assemblage of beds and the orogeny that deformed them. In this paper, the suffix 'ide' will be used for spacial-tectonic units (e.g. Katangide Belt). The suffix 'an' or 'ian' will be used to designate the orogenies (e.g. Katangan orogeny) and the depositional sequences involved (e.g. Katangan) (Cahen and Lepersonne, 1967, p. 150). The adjectives 'pre-tectonic', 'syn-tectonic' and 'post-tectonic' are used to designate phenomena which occur respectively before, during or after the main phase of folding of a given belt at a given locality.

The basement of the Kibaride Belt

The Kibaride Belt involves, from south-west to north and north-east: the Kibaran (proper), Burundian and Karagwe-Ankolean successions which rest unconformably on: (a) the *Lukoshian* and the *Luizian*; (b) the *Ubendian* or *Rusizian*; and (c) the *Buganda-Toro-Kibalian* (see Fig. 1 and Cahen and Snelling, 1966, p. 68; Cahen and Lepersonne, 1967, p. 152). Geochronological data show that the pre-Kibaran belts (given in italics) are approximately the same age, and dated syn-tectonic events include metamorphism of the: (i) Luizian, 2150 m.y. (Cahen and Snelling, 1966, p. 120; Ladmirant, unpublished work); and (ii) Rusizian gneiss, c. 2100 m.y. (Cahen and Snelling, 1966, p. 64).

Post-tectonic events connected with the Luizian and Lukoshian (Fig. 1) have yielded ages of 1950 m.y. for granites and granodiorites, and about 1850 m.y. for pegmatites (*ibid.*, p. 118; and unpublished work). In the Ubendian-Rusizian, post-tectonic granites have given ages of about 1800 m.y. or somewhat older, but pegmatite and vein minerals are in the 1800-1650 m.y. range (*ibid.*, p. 66). The Buganda-Toro-Kibalian appears to be of the same general age with syn-tectonic events older than about 2075 m.y. and post-tectonic dates as young as 1725 m.y. (*ibid.*, p. 59).

Despite these apparent similarities, there appears to be a significant difference between the western and eastern (and north-eastern) margins of the Kibaride Belt. On the west, the Lukoshide and Luizide Belts terminated their history, including recordable post-tectonic uplift, about 1850 m.y. ago, whereas on the eastern and north-eastern margins, the Ubendide-Rusizide and Buganda-Toro-Kibalide Belts (hereafter referred to as the Ubendide Belt and the Kibalide Belt respectively) had a longer post-tectonic recorded history, extending to about 1700 m.y. ago.

Although the base of the Kibaran succession is not well known on the western margin of the Kibaride Belt proper, it is younger than 2150 m.y.,

---

[1] When the ages are quoted without reservation, they have been obtained by the most reliable methods (Rb-Sr isochron or concordant U-Pb). Other ages, qualified by an adverb such as 'about' or 'c.' (*circa*) are usually less reliable apparent ages. Most Rb-Sr ages are calculated using $\lambda^{87}Rb = 1.47 \times 10^{-11} yr.^{-1}$. However, when Rb-Sr ages have to be compared to U-Pb ages, two figures are given: the lower one corresponds to the above-mentioned decay constant, the higher one is obtained using $\lambda^{87}Rb = 1.39 \times 10^{-11} yr.^{-1}$.

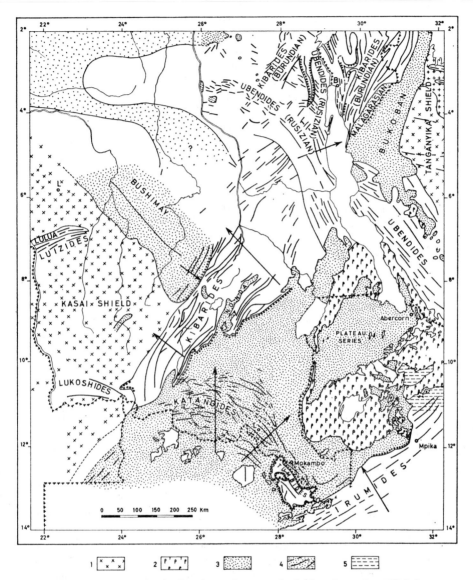

FIG. 1. Structural sketch of south-east Congo and neighbouring regions (slightly modified after Cahen, 1963a). 1, pre-Kibaran shields; 2, porphyry and alaskite; 3, Katangan (widely spaced stipples indicate Katangan buried under a thick Palaeozoic and/or Mesozoic cover); 4, folded Katangan of the Katanga geosyncline; and 5, flat-lying formations corresponding to material involved in the Irumide Belt.

The arrows indicate vergence of folds. Although the Plateau Series is represented as being entirely Katangan, this is not the case (see p. 103). L=Luina.

the age of the metamorphism in the Luizide Belt. On the eastern and north-eastern margin (Fig. 1) there is evidence of terrigenous, sometimes molasse-like, sediments connected with the Ubendide Belt and Kibalide Belt respectively. For example, the Lower Plateau Series of the Tanzania–Zambia border (Fig. 1), unconformably overlies the Ubendian and is intruded by the post-tectonic Kate granite and accompanying veins dated at *c.* 1800 m.y.; this terrigenous deposit was derived from the Ubendide Belt (Page, 1960; Cahen and Snelling, 1966, p. 67). Further north, in Burundi, the base of the Burundian portion of the Kibaran succession locally consists of thick con-glomeratic beds denoting deposition from strongly eroded uplands (Waleffe, 1966a and b). Still further north, the Singo Series of Uganda (0° 45′ N., 31° 35′ E.) is a practically undisturbed terrigenous deposit derived from the Buganda–Toro portion of the Kibalide Belt; that series is cut by the Mubende Granite, 1800 m.y. old or somewhat older, which is post-tectonic with respect to the latter belt (*ibid.*, p. 60). The Lower Plateau and Singo Series represent terrigenous deposits which were laid down at the end of the Ubendian and Kibalian orogenic cycles respectively but which were not involved in the subsequent Kibaran cycle; in contrast, the basal beds of the Burundian in Burundi (Fig. 1) represent a similar type of sediment which was folded during the Kibaran Orogeny. In that eastern section, the deposition of the Kibaran and Burundian successions began before *c.* 1800 m.y. ago, but after the syn-tectonic events of the Ubendide cycle, provisionally dated at *c.* 2100 m.y. In the south-western section, all that may be said is that sedimentation of the base of the Kibaran started after 2150 m.y.

The Kibaran stratigraphic column and syntectonic events of the Kibaran cycle

At present there are no geochronological milestones to date the deposition of the Kibaran. At least three (possibly four) successive sequences make up the Kibaran succession some 10,000 m thick or more (Table 1). The lowest sequence (Mt Kiaora) ends locally with a rhyolite and is separated by a rather pronounced disconformity from the overlying sequences which appear to form a more or less continuous succession. In the Burundian part of the Kibaride Belt (Table 1) there are three successive sequences totalling over 15,000 m. The upper sequence is disconformable or slightly unconformable on the middle sequence (Cahen and Snelling, 1966, p. 70; Cahen and Lepersonne, 1967, p. 199).

The entire succession (with the possible exception of the upper sequence of the Burundian) has been involved in the main Kibaran tectonic phase dated at 1300 ± 40 m.y. (Cahen *et al.*, 1967). This is the age of two distinct types of rock: (*i*) an intrusive porphyritic biotite granite which suffered cata-clasis and was recrystallised during the main tectonic phase; and (*ii*) biotite gneisses resulting from syntectonic granitisation of Kibaran sediments directly after the intrusion of the porphyritic granite. These two types of rock form the major granitic masses of the Kibaride Belt (Fig. 1), in both its

TABLE 1

Simplified succession and subdivision of the Kibaran of central
Katanga and the Burundian of Rwanda and Burundi[1]

Kibaran (Katanga)	Burundian (Rwanda and Burundi)
Lubudi sequence, 1500–1850 m	Miyove sequence, over 1450 m
Limestone and dolomite, often silicified, black graphitic phyllites, some sandstone, dark arkoses and conglomeratic lenses	Shales and phyllites Quartzites and quartz phyllites Quartzites and conglomerates
Mt Hakansson sequence, 1500–4000 m	Byumba sequence, 2000–2500 m
Dark-coloured phyllites and quartzites, light-coloured phyllites and quartz phyllites, quartzite, or conglomerate	Phyllites and quartzites Locally conglomerates at the base
Lufira sequence, 1300–5500 m	Lower sequence, 10,000–11,500 m
Quartzites, locally quartz phyllites and phyllites, conglomerate at base	Phyllites, often dark and zoned, quartz phyllites, quartzites, locally mafic lavas.
Mt Kiaora sequence, 1700–4300 m or more	In Burundi: coarse-grained quartzites and conglomerates at the base
Phyllites, quartz phyllites and schists, quartzite horizons, locally rhyolite at the top, locally crystalline limestone, conglomerates where the base is known	
Note. The two upper sequences may have to be merged into one sequence.	

[1] After Cahen and Lepersonne (1967, pp. 200 and 209); Cahen and Snelling (1966, p. 70); and Waleffe (1966a and b).

southern and central portions (*ibid.*). Although granites of this same age occur in northern Rwanda and Uganda (*ibid.*, p. 53; Cahen and Snelling, 1966, p. 86), the relationship of the discordant upper sequence of the Burundian to granites of this age is as yet unknown; that sequence is, however, certainly older than *c.* 1000 m.y.

The post-tectonic events of the Kibaran cycle

In the Karagwe-Ankolean portion of the Kibaride Belt (at the northern extremity of Fig. 1), a late-tectonic granitic body is 1190 m.y. old; in the Kibaran and Burundian portions of the belt (Fig. 1), the earliest dated post-tectonic intrusives are respectively: quartz veins with uranium minerals, 1055–1100 m.y. old; and quartz-microcline graphic pegmatites with some muscovite, 1100 m.y. old. Equigranular muscovite tin granites carrying some biotite have given ages in the 900–950 m.y. range. These granites are unde-formed and intrude both the Kibaran metasediments and the 1300 m.y. old syntectonic gneissic massifs. Their emplacement was followed by that of tin-bearing pegmatites, the potassic phase of which is only slightly younger, if at all, than the granites, whereas the sodic phase is some millions of years younger (Cahen et al., 1967, p. 33).

It is noteworthy that 1300 m.y. old gneisses also yield muscovite Rb-Sr ages similar to those of the neighbouring tin-granites and to the potassic phases of the tin-pegmatites, whereas muscovites from the tin-granites and pegmatites retain their true age; the biotites of all granites yield apparent ages about the same age as the sodic phase of the tin pegmatites or just a little younger (850-875 m.y.) in the central portion of the Kibaride Belt of Central Katanga (Fig. 1). It is probable that the temperature at which the older muscovites became 'closed' to strontium isotope migration approximates that at which tin-granites were formed or, at least, that at which the potassic phase of the tin-pegmatites was intruded. The temperature at which the biotites started to become 'closed systems' probably approximates that at the end of the albitic phase of the pegmatites (*ibid.*, p. 45).

The important fact that has become clear from recent studies is that post-tectonic igneous activity specific to the Kibaride Belt spreads over a period of 250-350 m.y. The successive igneous events which have been dated occurred during the post-tectonic uplift of the Kibaride Belt and this is by no means an isolated case (*ibid.*, p. 59).

'Molasse' of the Kibaran cycle and early stages in the deposition of the Katangan

CENTRAL KATANGA

The existence of molasse belonging to the Kibaran cycle has been established by Cahen and Lepersonne (1967, p. 206) and Dumont (1965). At the close of Kibaran times, in that area (9° S., 25-28° E.), three troughs formed within, behind and in front of the Kibaride Belt (see Figs. 1 and 2). They were filled with thick (500-2500 m) beds of terrigenous sediments containing thick layers of boulder conglomerates, interstratified with arkoses, often coarse-grained, and quartzitic and feldspathic grits.

In the foreland (western) trough (Fig. 2) the molasse-type beds of the lower sequence of the Bushimay (Fig. 1), are overlain by the higher sequences of the same unit which have a characteristic platform facies and which contain detrital elements derived from the Kibaran Belt (Cahen and Mortelmans, 1947; Raucq, 1957). In the southern portion of the median trough (Fig. 2), the molasse-type beds are overlain by a polymictic conglomerate which is itself overlain with slight unconformity by the 'Grand Conglomérat' and the rest of the Katangan succession (Dumont, 1965). In the eastern or hinterland trough (Fig. 2), representatives of the Upper Roan and Mwashya sequences rest on the molasse-type beds.

In general, the beds which overlie these early Katangan molasse-type sediments are disconformable or slightly unconformable and begin with polymictic conglomerates which contrast with the boulder conglomerates forming beds and lenses in the underlying beds. The boulders are essentially quartz and quartzites from the Kibaran beds. The rest of the succession is generally either conformable or pseudo-conformable on these polymictic conglomerates. In the south of the median trough, both the molasse-type beds and the pseudo-conformable polymictic conglomerates were tilted before deposition of the less tilted 'Grand Conglomérat'.

MPIKA REGION OF ZAMBIA

Near Mpika in Zambia (Fig. 1), Ackermann and Forster (1960) have described a terrigenous deposit (Bemba Beds) that originated through uplift and erosion of the Irumide Belt, which is the same age as the Kibaride Belt (Vail *et al.*, 1968; Cahen and Snelling, 1966, p. 86). This deposit is supposed to merge into the Plateau Series (Fig. 1) which has been subdivided by Page (1960) into three lithostratigraphic units which are, from oldest to youngest: the Lower Plateau Series, separated by an unconformity from the Upper Plateau Series, itself separated by a 'major' unconformity from the overlying Abercorn Sandstone. It has been noted that the Lower Plateau Series is older than 1800 m.y. and is a terrigenous deposit derived from the Ubendide Belt (Fig. 1). On the basis of an apparent age on dolerite cutting the Abercorn Sandstone of the Ufipa District of Tanzania, this formation is now dated as terminating at about 940 m.y. ago (Vail *et al.*, 1968). Furthermore, the Abercorn Sandstone has been correlated with the Bukoban (Quennell *et al.*, 1956) which continues into Burundi under the name of Malagarasian (Fig. 1) most of which is also older than about 950 m.y. (see p. 105). The Upper Plateau Series, wedged between the Lower Plateau Series and the Abercorn Sandstone, is clearly distinct from both these units but more so from the latter. It can only be a foreland equivalent to some of the beds deposited in the Irumide geosyncline (Philipps, 1955). This suggestion is in agreement with the fact that there is only a minor break between Lower and Upper Plateau Series, both involved in attenuated Irumide movements, whereas the break is less important than that below the Abercorn Sandstone. The correlation of the terrigenous beds near Mpika, which are distinctly unconformable on the Irumide Belt (Ackermann and Forster, 1960, Fig. 5) and do not belong to the Irumide geosyncline, thus cannot be with either the Lower or the Upper Plateau Series but with the Abercorn Sandstone. Therefore, if one follows the views of Ackermann and Forster (*ibid.*) on the origin of the Bemba Beds and their correlation, both the Bemba Beds and the Abercorn Sandstone belong to the base and lower part of the Katangan stratigraphic column. If the supposed correlation is incorrect, the Bemba Beds may be younger, but the Abercorn Series, by virtue of its correlation with the Bukoban and Malagarasian, definitely belongs to the lower portion of the Katangan.

COPPERBELT OF ZAMBIA AND SOUTH-EAST KATANGA

The Lower Roan of the Copperbelt (see Table 2) was not derived from uplift of an immediately preceding orogenic belt and therefore does not appear to be a molasse. However, it contains boulder and other conglomerates and many coarse sediments which indicate derivation from nearby uplands (Gysin, 1937, p. 250), which are now interpreted as hills, mainly granitic, in the pre-Katangan landscape (Mendelsohn, 1961, p. 48).

   In this context it is believed that after intrusion into the ancient Lufubu Schists, the granites, like all other rocks, were subjected to a period of erosion, 'the more resistant granites being left as hills separated by valleys carved in the

TABLE 2

Simplified succession and subdivision of the Katangan in Katanga
and the Copperbelt of Zambia[1]

Katanga	Copperbelt of Zambia and south-east Katanga
Kundelungu, see Table 3 (up to 7000 m) { Upper Kundelungu: Middle Kundelungu: Lower Kundelungu:	shale, quartzite shale, tillite shale, Kakontwe dolomite and shale
'Grand Conglomérat', up to 300 m Tillite and periglacial beds	*Tillite*
Mwashya, up to 800 m Black and zoned shales; quartzitic feldspathic sandstone; dolomitic shales; locally, pebbly mudstone	*Mwashya* Carbonaceous shale, argillite, (dolomite and quartzite)
Roan, over 1500 m	*Upper Roan*
Mofya Group or Groups: Magnesian limestones alternating with mica and quartz-bearing shales; at the top, silicified black oolite with *Girvanella*	Dolomite and argillite (quartzite, breccia)
Dipeta Group or Groups: chlorite and talc schist, feldspathic sandstone, siliceous dolomites	
'Mine Series' Group: dolomites, sandstones, shales and phyllites	Argillite and quartzite
Red chloritic sandstones with dolomitic matrix	*Lower Roan*
base not observed in Copper Zone of Katanga	Hanging wall quartzite, argillite and feldspathic quartzite (dolomite)
	Ore-argillite, impure dolomite, micaceous quartzite (graywacke, arkose)
Note. The correlation between the Roan of the copper zone of Katanga and that of the Copperbelt of Zambia and south-east Katanga is as yet uncertain	Footwall (Footwall conglomerate) argillaceous quartzite, feldspathic quartzite, 'aeolian' quartzite, conglomerates

[1] After Cahen and Lepersonne (1967, p. 232) and Mendelsohn (1961, p. 41).

softer schists' (Jordaan, in Mendelsohn, 1961, p. 325). It is now known that the earliest possible age for the commencement of Katangan sedimentation is 1300 ± 40 m.y. ago, more than 600 m.y. later than the intrusion of the post-Lufubu granites, 1975 ± 20 m.y. ago (Cahen *et al.*, 1970a). It is thus very unlikely that these hills would have resisted erosion for such a long interval. Also the presence of valleys suggests that the landscape was not one of granite inselbergs jutting out of an erosion surface. The pre-Roan basement is wedged between the Irumide and Kibaride Belts which were uplifted at the same time; it is thus highly probable that the uplift of the area between these belts, which is at present mainly made up of rocks involved in the Tumbide Belt (Fig. 1) (Ackermann and Forster, 1960; Cahen and Snelling, 1966, p. 115), also occurred at the same time. The Lower Roan sediments (Table 2) although not derived from uplifted Kibaran rocks, nevertheless are sediments derived from uplift and erosion of rocks which were elevated at the same time as the Kibaride and Irumide Belts.

We may now consider the igneous activity and still sparse geochronological data pertaining to the lower portion of the Katangan succession comprising all the beds of the type area, older than the 'Grand Conglomérat' and their equivalents elsewhere. To the north-west of the Kibaride Belt, the Bushimay (Figs. 1 and 2) comprises three sequences designated: B_0, up to 2000 m; B_1, 1200–1500 m; and B_2, over 1000 m (Cahen and Lepersonne, 1967, p. 253). It has been shown (Cahen and Snelling, 1966, p. 105) that syngenetic lead deposited near the top of sequence B_1 has a special isotopic composition indicating a less radiogenic lead than normal epigenetic lead from post-tectonic veins of the Kibaride Belt. That composition is compatible with deposition during the time-range 950–1300 m.y. Another lead from an epigenetic galena within the B_2 sequence has an istopic composition which is more radiogenic and indistinguishable from that of the epigenetic lead in the Kibarides to the west, which occurs in veins associated with the post-tectonic pegmatites and granites some 900–950 m.y. old.[1]

The Bushimay is intruded by necks that can be considered as representing the vents through which amygdaloidal pigeonite-albite lava (Delhal and Dumont, unpublished work) was poured out at the end of the known Bushimay times. Recent K-Ar datings (Snelling et al., unpublished work) yield c. 940 m.y. for these lavas.

It has already been mentioned that the Abercorn Sandstone, corresponding to the base of the Katangan, is cut by dolerite which yielded an age of c. 940 m.y. (Vail et al., 1968) corresponding to late 'Kibaran' ages of the Irumide Belt.

Another important indication of the age of the lower sequences of the Katangan is yielded by the Malagarasian succession of south Burundi, in which a gabbro sill has yielded a K-Ar apparent age of 980 m.y. (Snelling, unpublished work). Pending further results, this is taken to mean that c. 950 m.y. is a younger limit to the Bukoba or Nkoma Sandstone which, although not the youngest sequence in the Malagarasian, is unconformable on the folded Burundian beds of the Kibaride Belt (Waleffe, 1965, 1966c).

Thus it can be shown that the groups or sequences classified as the base of the Katangan succession are, in some instances, terrigenous, occasionally molasse-like deposits linked to the Kibaride Belt. The age of these lower sequences may therefore be expected to fall in the time range of the post-tectonic uplift of the Kibaride Belt (875–1300 m.y. or younger). That this is true is borne out by the available, admittedly still scanty, age data indicating that the entire Bushimay succession, a substantial portion of Malagarasian

---

[1] The 'ages' of these leads, using the Holmes-Houtermann model, are 1065–1145 m.y. and 940–1020 m.y. respectively, according to the numerical assumptions adopted.

(Bukoban)—which is directly linked to the Katangan of the type area—and the Abercorn Sandstone were deposited before *c.* 950 m.y. ago. Indeed, 950 m.y. is a younger limit not only to the molasse-like beds at the base of the Katangan, but to the platform sedimentation which followed upon it in the cratonic areas outside the Katanga geosyncline.

From all of these data, two points are worthy of note:

(*i*) at about the time the post-tectonic tin-granites of the Kibaride Belt were intruded into the Kibaran metasediments (950 m.y. ago), the basins of Katangan deposition were the site of mafic intrusive or extrusive activity; and

(*ii*) the Kibaride Belt, both in its foreland and in its hinterland areas, played an active role in controlling the sedimentation of the lower portion of the Katangan succession. In this regard it is possible to distinguish a first period of active influence during which an important and possibly rapid uplift followed by erosion produced the terrigenous sediments of the basal sequence of the Bushimay. In the Katanga geosyncline, the conglomerates which border the massifs formed by the Kibaran rocks, between latitudes 10° 30′ and 8° 30′ N., were deposited at that time. Subsequently this first period was followed by a second period of active influence during which the vertical movements of the Kibaride Belt controlled the sedimentation of the platform sequences (B_1 and B_2) of the Bushimay (Raucq, 1957). On the south-eastern border of the Kibaride Belt, in Katanga proper, the Upper Roan was deposited in a not very rapidly subsiding basin whose shore-line lay to the north-west[1], along the Kibaride Belt (Oosterbosch, 1959).

The Katangan succession above the 'Grand Conglomérat' and syntectonic events of the Katangan cycle

Of the various 'tillites' in Africa south of the Sahara, the 'Grand Conglomérat' (Table 2) is the one most likely to be of truly glacial nature on account of its thickness, continuity and abundant glacial characteristics. In Katanga, the 'Grand Conglomérat' probably represents ground moraine in the north, blending into submarine moraine towards the south, and reworked moraine still further south (Cahen, 1963b; Bellière, 1966; Cahen and Lepersonne, 1967, p. 241). It is an excellent marker between two rather contrasting portions of the Katanga succession. The beds below (Table 2) are in a large measure represented outside the geosynclinal basin whilst those above are, for the most part, limited to the geosynclinal basin or at least much more developed within it (see Fig. 1). The entire succession above the 'Grand Conglomérat' is of geosynclinal character (Bellière, 1966), in contrast to that below it, which is only partially so.

Table 3 gives the succession and present-day subdivision of the part of the

---

[1] This has been erroneously translated into north-east in Cahen and Lepersonne (1967, p. 247).

Katangan stratigraphic column which post-dates the 'Grand Conglomérat'. The Lower Kundelungu is well developed and differentiated in the southern part of Katanga where it clearly represents geosyncline deposition. Gradual uplift of areas in the south partially supplied the infilling of the basin; nevertheless, in view of the thinning out of the Lower Kundelungu towards the north, it is probable that a substantial portion of the supply of clastics was of northern provenance. The Lower Kundelungu sediments have a rather

TABLE 3

Succession and subdivision of the Kundelungu[1]

Upper Kundelungu (observed maximum: 1900 m)
  Kundelungu Plateau Group
    Tshiuswe Shale Formation: 1000 m
    Kilungu-Lupili Arkose Formation: 900 m

Middle Kundelungu (maximum: 2100 m)
  Kiaka Group (maximum: 1500 m)
    Kapenga Carbonaceous Sandstone Formation: 600 m
    Sonta Sandstone Formation: 100 m
    Sampwe Shale Formation: 500 m
    Kyafwamakemba Sandstone Formation: 10-30 m
    Kalulu Nord Shale Formation with chert horizon: 200-250 m
    Kyubo Sandstone Formation (upper portion, markedly feldspathic): 30 m
  Kalule Group (maximum: 800 m)
    Kyubo Sandstone Formation (lower portion, non-feldspathic): 30 m
    Mongwe Shale Formation: 90-100 m
    Lubudi Cement Works Oolitic Limestone Formation: 10-80 m
    Kanianga Calcareous Sandstone Formation: 40-375 m
    Lusele Pink Dolomite Formation: 20-45 m
    Kiandamu Conglomerate Formation: 10 m

'Petit Conglomérat' Tillitic Facies (at least 50 m)

Lower Kundelungu (over 2000 m, probably thicker southwards)
    Monwezi Formation: sandstones and quartzites (sub-greywackes) (*c.* 1000 m)
    Pelitic Formation: shales and slates, calcareous locally (*c.* 500 m)
    Kakontwe Formation: calcareous shales and slates, limestones and dolomites with
      stromatolites (*c.* 400 m)
    Conglomerate or Arkose Formation (local)

---

[1] After Cahen and Lepersonne (1967, p. 232) and Dumont (1967).

typical 'flysch' facies (Cahen and Lepersonne, 1967, p. 248). With the arching of the southern hinterland the 'Petit Conglomérat' (Tables 2 and 3) was deposited, and in the south represents a littoral facies incorporating abundant pebbles of the normally deeply buried Kakontwe calcitic-dolomitic limestone (Grosemans, 1935; Cahen and Lepersonne, 1967, p. 239); these pebbles disappear northwards.

During the Middle Kundelungu[2] the axis of subsidence was situated in a

[2] The Middle Kundelungu is not identical with the Middle Kundelungu of Cahen and Mortelmans (1948) which consisted of the Kalule Group only. This new subdivision is based on recent work by Dumont (1967) and supersedes the old one.

more northerly region and most of the sedimentation was of a terrigenous nature (*ibid.*, p. 248; Bellière, 1966). Bellière considers that the Middle Kundelungu is probably a flysch. However it is clear that the source of the clastic material was to the north (north-west to north-east) and that the influx of material from that direction completely overshadowed the pelitic contribution of southern origin.

After a tectonic phase, which occurred between the Kundelungu Plateau Group and the Kiaka Group (Table 3), renewed subsidence again occurred in the north (Dumont, 1967). To this portion of the Kundelungu, François (in Demesmaeker *et al.*, 1963) ascribed a 'molasse' nature. However this sequence has neither the obvious lithological characters and provenance of sediments nor the palaeogeographical distribution of a molasse; it is an infilling of an outer basin under conditions identical to those which prevailed throughout the geosynclinal history of the area.

It is clear from the above data that the subdivisions of Lower, Middle and Upper Kundelungu are not only based upon lithological criteria but correspond to well-defined episodes of the development of the Katanga geosyncline. Within this sequence, three phases of movement have been recognised: (*i*) the *post-Lower, pre-Middle Kundelungu phase*, which in Katanga produced uplifting and arching of the southern part of the country, followed by erosion and by overlap of the Middle Kundelungu (Cahen *et al.*, 1961; Cahen and Lepersonne, 1967, p. 239); (*ii*) the *Kolwezian phase* which occurred between Middle and Upper Kundelungu, and is named after Kolwezi, the main centre of mining activities in Katanga; and (*iii*) the *Kundelunguan phase*, named after the Kundelungu Plateau, where the Upper Kundelungu is well developed and its deformation clearly visible.

These three phases all belong to the Katangan Orogeny (Cahen *et al.*, 1961, p. 13). They will now be reviewed together with the relevant igneous events and geochronological data.

THE POST-LOWER, PRE-MIDDLE KUNDELUNGU PHASE

The evidence for the existence of this phase has been given above. Gabbro intrusions which are generally amphibolitised and scapolitised occur just above the top of the Upper Roan dolomites of the Zambian Copperbelt and also in the Mwashya and in the Kakontwe limestone of the Lower Kundelungu (see Tables 2 and 3) (Mendelsohn, 1961, p. 51; Jordaan, in Mendelsohn, 1961, p. 307). Similar rocks also occur in the Lower Kundelungu of south Katanga (Jamotte, 1933, and in Cahen, 1954, p. 110; Grosemans, 1934) but are not known for certain in more recent beds. These rocks are thus probably younger than amygdaloidal lavas and the dolerites of central Katanga that are interstratified within the 'Grand Conglomérat' (Dumont, unpublished work).[1] The relative time of their intrusion is still uncertain. The sills in the Copperbelt of Zambia appear to have been folded (Garlick, *in litt.*) and, in that case, might be linked to the post-Lower, pre-Middle Kundelungu

---

[1] Not as mentioned by Cahen and Lepersonne (1967, p. 245) between the 'Grand Conglomérat' and the 'Petit Conglomérat'.

tectonic phase. However, petrographically these rocks appear to be unde-formed. They are not yet dated.

A 710 m.y. or older uranium mineralisation (Cahen, 1964; Cahen and Snelling, 1966, p. 101) may also be linked to this tectonic phase. This is a minimum age and, even if not linked to this phase, represents a younger age limit to it. Pegmatite veins are found in the Lower Roan of the Copperbelt of Zambia and south-east Katanga; those that are microcline-bearing are about 840 m.y. old (Cahen et al., 1970b). It is not known if they are linked to the post-Lower, pre-Middle Kundelungu tectonic phase; however, their approximate age may be considered as an older limit for this tectonic phase.

It is possible that the post-Lower, pre-Middle Kundelungu phase is the Lufilian orogenic phase of De Swardt and Drysdall (1964) (see Cahen and Snelling, 1966, pp. 97 and 101), although the coincidence of the 710 m.y. age of uranium minerals with the apparent age of the Lusaka Granite (Zambia) must provisionally be considered as without significance; the age of this syntectonic granite is still uncertain (Snelling et al., unpublished work).

THE KOLWEZIAN PHASE

François (in Demesmaeker et al., 1963) deduced the existence of this phase from the spatial relationship between the Kolwezi nappes and the Middle Kundelungu of Katanga (Table 3) upon which they rest by means of thrust faults. This important deduction is now confirmed by Dumont (1967) who has proved the existence of an unconformity between Upper and Middle Kundelungu (Table 3). On the north-eastern margin of the Zambian Copper-belt, thrusts in the Luina Massif (Oosterbosch, 1959) and in the Mokambo Massif (Fig. 1) (Lecompte, 1933) are also ascribed to the Kolwezian phase as is the main folding in the Copperbelt itself.

A well-dated epigenetic uranium mineralisation yielding ages of 670 m.y. (Cahen et al., 1961, p. 36; Cahen and Snelling, 1966, p. 101) can be connected with this phase. This mineralisation, which occurs at Shinkolobwe and Swambo, was reworked during and after the Kundelunguan tectonic phase to give rise to the better known 620 m.y. mineralisation (Cahen et al., 1961; Derriks and Oosterbosch, 1958; Cahen, 1951). At Shinkolobwe, the two ages of mineralisation occur together and both Kundelunguan and Kolwezian phases are known at that locality. Since there is no definite uranium minerali-sation between 670 m.y. and 620 m.y., it appears that the former dates a discrete event immediately preceding that dated at 620 m.y. A still older uranium mineralisation, giving an age of 710 m.y. or older, has been men-tioned above. The coexistence at Shinkolobwe of all these mineralisations suggests that the 620 m.y. mineralisation occurred through reworking of the 670 m.y. mineralisation and that the latter is an epigenetic reworking of the 710 m.y. or older mineralisation. The older of these reworkings must have occurred during or after the Kolwezian phase but not before, since if 670 m.y. represented a pre-Kolwezian phase event, one might expect to find definite indications of reworking corresponding to the Kolwezian phase; on this reasoning the Kolwezian phase is older than 670 m.y. It is unlikely to be older than *c.* 720 m.y. (Cahen, 1970.)

THE KUNDELUNGUAN PHASE

In the northern part of Katanga (Kundelungu Plateau) this tectonic phase is evidenced by the distinct but not very intense folding of the Upper Kundelungu (Table 3); no igneous rocks linked to this phase are known.   Further south in the region of the copper mines of Katanga, the Kundelunguan phase is responsible for the refolding of nappes formed during the previous Kolwezian phase. After this refolding, transcurrent faulting with diapiric 'extrusions' occurred (Demesmaeker *et al.*, 1963); the uranium epigenetic mineralisation of Shinkolobwe, Kalongwe and other places was injected into these faults. Since these mineralisations are 620 m.y. old (Cahen *et al.*, 1961), it is obvious that the Kundelunguan phase is older than 620 m.y. as is the entire rock succession which is affected by that phase.[1] Still further south, in the Copperbelt, the rising of the Nchanga and Mokambo granitic domes (Table 4) appears to correspond to the Kundelunguan phase or to have followed shortly after it: its age is *c.* 600 m.y. (Snelling *et al.*, 1964; De Swardt and Drysdall, 1964, p. 35; Cahen and Snelling, 1966, p. 161; Cahen *et al.*, 1970c).

Post-tectonic events of the Katangan Orogeny

The last fold phase, the Kundelunguan phase, is dated at between 620 m.y. and 670 m.y. The 620 m.y. uranium mineralisation and the *c.* 600 m.y. Nchanga Red Granite are both essentially undeformed and thus 'post-folding'. In view of the fact that some slight tectonic adjustments seem to have followed the uranium mineralisation (Cahen *et al.*, 1961), the age of 620 m.y. could perhaps be taken as the frontier between syn-tectonic and post-tectonic events; if so, the Nchanga Red Granite, which may be slightly younger than the uranium mineralisation, would be classed as a post-tectonic event.

A subsequent hydrothermal event, represented by fissures and veins bearing calcite, quartz and copper minerals in the Mokambo and Kinsenda Granites, is distinctly post-folding. This hydrothermal event has been dated at 550-580 m.y. and is somewhat younger than the Nchanga Red Granite (Cahen *et al.*, 1970c). A well-dated event at 520 m.y. is a uranium mineralisation occurring in numerous localities both in Zambia and Katanga (Darnley *et al.*, 1961; Cahen *et al.*, 1961); this mineralisation comprises veins and disseminated material (*ibid.*). It should perhaps also be noted that some uranium mineralisations are younger and appear to represent reworking of the previous mineralisations, in response to slight local causes. Finally the apparent ages of biotite in the Copperbelt of south-east

---

[1] In 1961, Cahen *et al.* stated that, since the top portion of the Upper Kundelungu is not represented in the folded area where the 620±20 m.y. mineralisation occurs, its relationship to this mineralisation was unknown; this statement was repeated later (Cahen and Lepersonne, 1967, p. 244). The present state of knowledge is set out above.

Katanga and Zambia are 475 ± 20 m.y. old (see Cahen and Snelling, 1966, p. 109; Cahen, 1970).

The sequence of igneous events and of certain other dated events is recorded in Table 4.

TABLE 4

Igneous and dated mineralisation events in the Katangan of
Katanga and the Copperbelt of Zambia[1]

No.	Igneous and mineralisation events	Age in m.y.
9	Apparent age of biotite in the Copperbelt of Zambia and south-east Katanga; sulphide-bearing veins, Katanga and Zambia	475 ± 20–500 ± 20
8	Epigenetic uranium mineralisation, in veins or disseminated: Nkana, Kansanshi (Zambia); Musoshi, Kolwezi, Kamoto (Katanga)	520 ± 20
7	Hydrothermal event—fissures and veins bearing calcite, quartz and copper minerals: Mokambo and Kinsenda Copperbelt of south-east Katanga	550–580
6	Rising of Nchanga Red Granite and Mokambo domes (Copperbelt of Zambia and south-east Katanga)	c. 600
5	Epigenetic uranium mineralisation, following on the Kundelunguan tectonic phase, including transcurrent faulting: Shinkolobwe, Kalongwe and Luishya (southern Katanga)	620 ± 20
4	Epigenetic uranium mineralisation, following on the Kolwezian tectonic phase: Shinkolobwe and Swambo (south Katanga)	670 ± 20
3	Uranium mineralisation at 710 m.y. or older: Shinkolobwe (south Katanga)	≥ 710
2 or 1	Kibambale dolerites and amygdaloidal lavas, interstratified within the 'Grand Conglomérat' of central Katanga	
1 or 2	Pegmatite veins in the Lower Roan of the Copperbelt of Zambia and south-east Katanga	840 ± 40

[1] Amphibolitised and scapolitised gabbro intrusions of uncertain age (see p. 108), if pre-folding, are probably to be placed with event No. 3 or between events No. 2 and 3; if post-folding, are likely to be between events No. 9 and 8 or with event No. 7.

Late-stage terrigenous deposits ('molasse') of the Katangan Orogeny?

We have noted (p. 108) that nearly all of the Middle and Upper Kundelungu rocks are terrigenous sediments derived from extra-geosynclinal areas. Thus, despite their likeness to 'flysch', noted by Bellière (1966), they do not conform to one of the essential requisites of flysch. François has suggested that the Upper Kundelungu is a molasse (Demesmaeker et al., 1963); however it is sufficiently similar to the Middle Kundelungu to suggest a renewal, in a slightly northern area, of conditions which prevailed during sedimentation of the Middle Kundelungu. Indeed it is a conspicuous trait of the

Katangide Belt and of most of its counterparts around or in the Congo Basin that no obvious 'molasse' of this belt is known; indeed, no sediments deposited later than 620 m.y. and earlier than Upper Carboniferous (or end of Lower Carboniferous), c. 325 m.y. ago, appear to exist.

It is therefore of importance to stress that in West Africa belts of the same age as the Katangide Belt have recently been shown to have produced terrigenous deposits of molasse type. In Ghana, Togo and Dahomey, Grant (1967) has shown the existence of a complete orogenic cycle, the latest sedimentary unit of which, the molasse-like Obusum beds, is likely to be Palaeozoic. Further west, Allen (1968) has produced evidence that the Rokelide Belt in Sierra Leone was a geosyncline bordering the West African Craton and that the Taban Formation is a molasse of that belt deposited and deformed before the Silurian (ibid.; Allen, et al., 1967). Four K-Ar ages on micas from the Rokelide Belt range from 530 to 575 m.y. ($\pm$ 20 m.y.) (ibid.) and 'probably date the waning stages of the orogeny'. These ages give a younger limit for the date of the metamorphism and it is likely that the main metamorphism is significantly older (Armstrong, 1966; Cahen et al., 1967; Jaeger et al., 1961, 1966; Krümmenacher, 1961; Michot and Pasteels, 1968). In this regard, late syntectonic events in West Africa which include the older granites of Nigeria were dated at 540 m.y. or older by Jacobson et al. (1964). For these an age of 605-640 m.y. is now preferred (Snelling, 1967). This latter age is practically identical to that of the uranium mineralisation (620$\pm$ 20 m.y.) which has been taken (p. 110) as the frontier between the syn- and post-tectonic events in Katanga. It is therefore not surprising that Allen et al. (1967) conclude that in Sierra Leone, the orogeny responsible for the Rokelide Belt is 'essentially Precambrian in age and in general terms can be correlated with the Katangan cycle of Central Africa'. One may thus conclude that the molassic Taban Formation was deposited during the interval between 620 m.y. ago and the Silurian (440 m.y.).

Conclusions

This account of the succession of events in a portion of Central Africa comprising the Katanga Province of Congo, and neighbouring regions, leads to the following conclusions.

ACTIVE AND PASSIVE INFLUENCE OF THE KIBARIDE BELT ON KATANGAN SEDIMENTATION

The Katangan succession both in the type area and in the Bushimay to the north-west of the Kibaride Belt (Fig. 1) was to a great extent controlled by the vertical movements of the latter belt. During the first period, these vertical movements were 'active' and formed a part of the Kibaran orogenic cycle of which they represent the last stages; this period influenced all or most of the Katangan sedimentation below the 'Grand Conglomérat'. A second period followed and affected all the Katangan deposition above this important marker. During that period, the vertical movements of the Kibara Mountain

Range continued to influence the sedimentation of the Kundelungu formations but the movements were 'passive' responses to the tectonic phases of the
Katangan orogenic cycle. Furthermore, at that time, the Kibara Range was
not the only source of supply for the clastics in the Kundelungu succession;
other 'basement' areas, to the north-east and east of Katanga, also played a
part.

TERRIGENOUS (MOLASSE-LIKE) DEPOSITS AND THEIR GEOCHRONOLOGICAL TIME RANGE

Terrigenous deposits, eventually molasse-like, derived from the uplift of
the Ubendide Belt were deposited in the time-range 1800-2100 m.y. ago
which is also the time-range of post-tectonic events and of mica apparent
ages of the Ubendide Belt. In some areas these deposits were involved in the
Kibaran orogenic cycle. Similarly, terrigenous deposits, derived from the
uplift of the Kibaride Belt, were deposited in the 875-1300 (1200) m.y.
time-range which also represents the time-range of post-tectonic events and
of mica apparent ages of the Kibaride Belt. These deposits were also, in some
cases, involved in the succeeding Katangan orogenic cycle. The absence of
similar deposits related to the general uplift of the Katangide Belt in Central
Africa is noteworthy. However, in West Africa, molasse-like formations
were deposited in the time-range (440-620 m.y.) of post-tectonic events of
the Katangan cycle.

THE MAJOR 'TECTONO-STRATIGRAPHICAL' UNITS AND THE OROGENIC CYCLES

In Africa, the major subdivision of Precambrian rocks has traditionally been
'a series of fairly concordant beds separated from older and younger beds by
well marked unconformities. In other words, the sedimentary sequences
belonging to such a subdivision are marked by the orogeny which caused the
discordance taken as the upper limit.' The major units are of an essentially
'tectono-stratigraphical nature' (Cahen and Lepersonne, 1967, p. 150); for
example, the Katangan (formerly Groupe du Katanga or Katanga System),
the Kibaran or Karagwe-Ankolean, etc. It is obvious from Fig. 2 that the
major subdivisions do not entirely coincide with the orogenic cycles and that
the former often comprise sediments belonging to two successive orogenic
cycles: (i) formations belonging to the late stage (or stages) of a particular
cycle; and (ii) formations belonging to the early and median stages of the
succeeding cycle.

Despite the fact that no system of geosynclinal nomenclature can be
applied with perfect aptness to areas outside the one that gave rise to it, two
of the most recent systems are used below to try and characterise the Kibaran
and the Katangan. The Kibaran is made up of sediments produced during the
geosynclinal period (Aubouin, 1961, 1965), or the pre-orogenic and tectonic
stages (Hermes, 1968), of the Kibaran cycle. However, the Kibaran also
includes locally added sediments belonging to the late and post-geosynclinal
periods (Aubouin), or morphogenic and post-orogenic stages (Hermes), of

E

pre-Kibaran cycles; but excludes sediments belonging to the late or post-geosynclinal periods of the Kibaran cycle itself which have, in fact, been classified as Katangan. These latter sediments, which are thus late- or post-

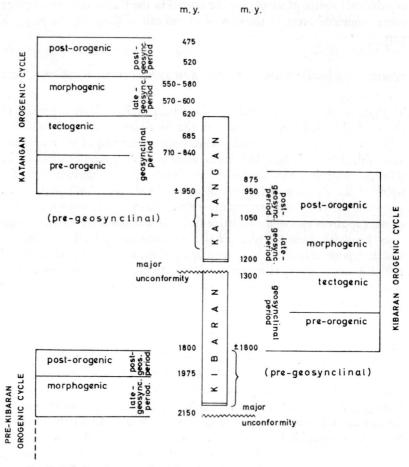

FIG. 3. Relations between the Kibaran and Katangan orogenic cycles.

Note: pre–orogenic, tectogenic, morphogenic and post–orogenic are stage names (Hermes, 1968); geosynclinal, late geosynclinal, post–geosynclinal are period names (Aubouin, 1961, 1965).

geosynclinal in respect of one cycle, might thus be called 'pre-geosynclinal' in respect of the succeeding cycle (Fig. 3).[1]

The Katangan orogenic cycle is essentially a representative of what Kennedy (1964) has called the Pan-African Cycle which is a late Precambrian to early Palaeozoic orogenic cycle. It is however incomplete as it lacks terrigenous deposits characteristic of the late- or post-geosynclinal stages of its

---

[1] This contribution was completed when an article on a somewhat similar subject came to the author's notice: P. M. Allen and A. I. Reedman, 1968. Stratigraphic classification in Precambrian rocks. *Geol. Mag.*, **105**, pp. 290-297.

development. The Katangan is a major tectono-stratigraphical unit which was built up during the late or post-geosynclinal periods of the Kibaran orogenic cycle and the geosynclinal period[1] of the Katangan (or Pan-African) cycle. It is older than 620 m.y. and entirely Precambrian.

Acknowledgements

I wish to thank my colleagues at the Musée royal de l'Afrique centrale, Tervuren, and at the University of Brussels, for permission to use some unpublished data and for helpful discussions. I am also indebted to Dr N. J. Snelling for permission to mention unpublished K-Ar age determinations.

REFERENCES

ACKERMANN, E., and FORSTER, A. 1960. Grundzüge der Stratigraphie und Struktur des Irumiden-Orogens. *Int. geol. Congr.*, 21 (18), 182.

ALLEN, P. M. 1968. The stratigraphy of a geosynclinal succession in western Sierra Leone, West Africa. *Geol. Mag.*, 105, 62.

—— SNELLING, N. J., and REX, D. C. 1967. Age determination from Sierra Leone. *15th Ann. Rep. Dept. Geol. Geophys.*, M.I.T. (for 1967), 23.

ARMSTRONG, R. L. 1966. K-Ar dating of plutonic and volcanic rocks in orogenic belts. In *Potassium argon dating*, p. 117. (Eds. O. A. Schaeffer and J. Lähringer), Berlin.

AUBOUIN, J. 1961. Propos sur les géosynclinaux. *Bull. Soc. géol. Fr.*, 3 (7), 629.

—— 1965. *Geosynclines*, Amsterdam.

BELLIÉRE, J. 1966. Les sédiments kundelungiens dans l'arc Mwashia-Bunkeya. *Annls. Soc. géol. Belg.*, 89, B357.

CAHEN, L. 1951. L'âge de la pechblende de Shinkolobwe et la limite Cambrien-Précambrien. *Bull. Soc. belge Géol.*, 60, 89.

—— 1954. *Géologie du Congo belge*, Liège.

—— 1963a. Grands traits de l'agencement des éléments du soubassement de l'Afrique centrale. Esquisse tectonique au 1/5,000,000. *Annls. Soc. géol. Belg.*, 85, B183.

—— 1963b. Glaciations anciennes et dérive des continents. *Annls. Soc. géol. Belg.*, 86, B19.

—— 1964. Eléments géochronologiques relatifs à la corrélation des terrains précambriens terminaux du Bas-Congo et du Katanga (Congo). *Ann. Rep. Géol. Min. Palaeont.*, *Mus. roy. Afr. centr.* (for 1963), 96.

—— 1970. État actuel de la géochronologie du Katanga. *Annls. Mus. roy. Afr. centr., Sci. géol.*, 65, 7.

—— DELHAL, J., and DEUTSCH, S. 1967. Rubidium-strontium geochronology of some granitic rocks from the Kibaran Belt (Central Katanga, Rep. of the Congo). *Annls. Mus. roy. Afr. centr., Sci. géol.*, 59.

—— DELHAL, J., DEUTSCH, S., GRÖGLER, N., and PASTEELS, P. 1970a. The age of the Roan Antelope and Mufulira granites (Copperbelt of Zambia). *Annls. Mus. roy. Afr. centr., Sci. géol.*, 65, 15.

—— DELHAL, J., and LEDENT, D. 1970b. On the age and petrogenesis of the microcline-bearing pegmatite veins at Roan Antelope and at Musoshi (Copperbelt of Zambia and S.E. Katanga). *Annls. Mus. roy. Afr. centr., Sci. géol.*, 65, 43.

—— DELHAL, J., LEDENT, D., and PASTEELS, P. 1970c. Isotopic data relative to the age and petrogenesis of dome-forming granites in the Copperbelt of Zambia and S.E. Katanga. *Annls. Mus. roy. Afr. centr., Sci. géol.*, 65, 69.

---

[1] Or sedimentation phase: in view of the extra-geosynclinal provenance of most of the sediments, this name used by Dietz (1963) for his model, might be more pertinent.

CAHEN, L., and LEPERSONNE, J. 1967. The Precambrian of the Congo, Rwanda and Burundi. In *The Precambrian*, p. 143. (Ed. K. Rankama), Vol. 3, New York.

—— and MORTELMANS, G. 1947. Le système de la Bushimaie au Katanga. *Bull. Soc. belge Géol.*, **56**, 217.

—— and MORTELMANS, G. 1948. Le groupe du Katanga. Evolution des idées et essai de subdivision. *Bull. Soc. belge Géol.*, **57**, 459.

—— PASTEELS, P., LEDENT, D., BOURGUILLOT, R., VAN WAMBEKE, L., and EBERHARDT, P. 1961. Recherches sur l'âge absolu des minéralisations uranifères du Katanga et de Rhodésie du Nord. *Annls. Mus. roy. Afr. centr., Sci. géol.*, **41**.

—— and SNELLING, N. J. 1966. *The geochronology of equatorial Africa*, Amsterdam.

DARNLEY, A. G., HORNE, J. E. T., SMITH, G. H., CHANDLER, T. R. D., DANCE, D. F., and PREECE, E. R. 1961. Ages of some uranium and thorium minerals from East and Central Africa. *Miner. Mag.*, **32**, 716.

DEMESMAEKER, G., FRANÇOIS, A., and OOSTERBOSCH, R. 1963. La tectonique des gisements cuprifères stratiformes du Katanga. Gisements stratiformes de cuivre en Afrique. Symposium, Part 2. *Ass. Serv. géol. Afr., Lusaka Meeting* (1962), 47.

DERRIKS, J. J., and OOSTERBOSCH, R. 1958. The Swambo and Kalongwe deposits compared to Shinkolobwe. Contribution to the study of Katanga uranium. *Proc. 2nd U.N. Int. Conf. on the peaceful uses of atomic energy*, **2**, 663.

DE SWARDT, A. M. J., and DRYSDALL, A. R. 1964. Precambrian geology and structure in central Northern Rhodesia. *Mem. geol. Surv. N. Rhodesia*, **2**.

DIETZ, R. S. 1963. Collapsing continental rises, an actualistic concept of geosynclines and mountain building. *J. Geol.*, **71**, 314.

DUMONT, P. 1965. Les formations du soubassement katangais rapportables à l'ancien 'Système du Djipidi'. *Ann. Rep. Géol. Min. Palaeont., Mus. roy. Afr. centr.* (for 1964), 56.

—— 1967. Essai de subdivision lithostratigraphique du Kundelungu supérieur. *Ann. Rep. Géol. Min. Palaeont., Mus. roy. Afr. centr.* (for 1966), 43.

GRANT, N. K. 1967. Complete late pre-Cambrian to early-Palaeozoic orogenic cycle in Ghana, Togo and Dahomey. *Nature, Lond.*, **215**, 609.

GROSEMANS, P. 1934. Roches basiques de la région de Tenke. *Annls. Serv. Mines Comité Spécial Katanga*, **5**, 8.

—— 1935. Contribution à l'étude du conglomérat de base (Petit conglomérat) du Kundelungu supérieur. *Annls. Serv. Mines Comité Spécial Katanga*, **5**, 38.

GYSIN, M. 1937. Les minerais de cuivre du Sud-Katanga. *Annls. Serv. Mines Comité Spécial Katanga*, **7**, 1.

—— 1960. L'existence de granites 'jeunes' à la frontière du district cuprifère nord-rhodésien. *Arch. Sci.* (Geneva), **13** (1), 103.

HERMES, J. J. 1968. The Papuan geosyncline and the concept of geosynclines. *Geol. Mijnb.*, **47**, 81.

JACOBSON, R. R. E., SNELLING, N. J., and TRUSWELL, J. F. 1964. Age determinations in the geology of Nigeria with special reference to the older and younger granites. *Overseas Geol. Mineral Resources*, **9**, 168.

JAEGER, E., GEISS, J., NIGGLI, E., STRECKEISEN, A., WENK, E., and WUTHRICH, H. 1961. Rb-Sr—Alter an Gesteins-glimmern der Schweizer Alpen. *Schweiz. miner. petrog. Mitt.*, **41**, 255.

—— ARMSTRONG, R. L., and EBERHARDT, P. 1966. A comparison of K-Ar and Rb-Sr ages on Alpine biotites. *Earth Planetary Sci. Let.*, **1**, 13.

JAMOTTE, A. 1933. Roches éruptives et roches métamorphiques connexes de la région comprise entre la Lufunfu et le Mualaba. Leurs relations avec les gisements de fer de la région. *Annls. Serv. Mines Comité Spécial Katanga*, **4**, 22.

JORDAAN, J. 1961. Nkana. In *The geology of the Northern Rhodesian Copperbelt*, p. 297. (Ed. F. Mendelsohn), London.

KENNEDY, W. Q. 1964. The structural differentiation of Africa in the Pan-African (±500 m.y.) tectonic episode. *8th Ann. Rept. Res. Inst. African Geol., Univ. Leeds*, 48.

KRÜMMENACHER, D. 1961. Déterminations d'âges isotopiques faites sur quelques roches de l'Himalaya du Nepal par la méthode potassium-argon. *Schweiz. miner. petrog. Mitt.*, **41**, 273.

LECOMPTE, M. 1933. Le batholite de Mokambo (Katanga) et ses alentours. *Mém. Inst. Géol. Univ. Louvain*, **7**, 129.

MENDELSOHN, F. (Ed.). 1961. *The geology of the Northern Rhodesian Copperbelt*, London.

MICHOT, J., and PASTEELS, P. 1968. Etude géochronologique du domaine métamorphique du Sud-Ouest de la Norvège (note préliminaire). *Annls. Soc. géol. Belg.*, **91**, 93.

OOSTERBOSCH, R. 1959. La série des mines du Katanga. *Bull. géol. Congo belge et Ruanda-Urundi*, **1**, 3.

PAGE, B. G. N. 1960. The stratigraphic and structural relationships of the Abercorn Sandstones, the Plateau Series and basement rocks of the Kawimbe area, Abercorn District, Northern Rhodesia. Unpublished Ph.D. thesis, Univ. Leeds.

PHILIPPS, K. A. 1955. Relationship between Muva System and Plateau Series in Northern Rhodesia. *Ass. Serv. géol. Afr.*, Nairobi Meeting (1954), 153.

QUENNELL, A. M., McKINLAY, A. C. M., and AITKEN, W. G. 1956. Summary of the geology of Tanganyika. Part 1. Introduction and stratigraphy. *Mem. geol. Surv. Tanganyika*, **1**.

RAUCQ, P. 1957. Contribution à la connaissance du Système de la Bushimay (Congo-belge). *Annls. Mus. roy. Congo belge, Sci. géol.*, **18**.

SNELLING, N. J. 1967. Age determination unit. In *Ann. Rep. Inst. geol. Sci.* (for 1966), p. 142; see also *14th Ann. Rep. Dept. Geol. Geophys., M.I.T.* (for 1966), 7.

—— HAMILTON, E. I., DRYSDALL, A. R., and STILLMAN, C. J. 1964. A review of age determinations from Northern Rhodesia. *Econ. Geol.*, **59**, 962.

VAIL, J. R., SNELLING, N. J., and REX, D. C. 1968. Pre-Katangan geochronology of Zambia and adjacent parts of Central Africa. *Can. J. Earth Sci.*, **5**, 621.

WALEFFE, A. 1965. Etude géologique du Sud-Est du Burundi (Région du Mosso et du Nkoma). *Annls. Mus. roy. Afr. centr., Sci. géol.*, **48**.

—— 1966a. Etude géologique de l'Est du Burundi et stratigraphie du Burundien. *Ann. Rep. Géol. Min. Palaeont., Mus. roy. Afr. centr.* (for 1965), 69.

—— 1966b. Observations dans la partie Sud-Ouest du Burundi (région de Makamba–Nyanza Lac). *Ann. Rep. Géol. Min. Palaeont., Mus. roy. Afr. centr.* (for 1965), 74.

—— 1966c. Quelques précisions sur la position stratigraphique du 'Nkoma' dans le Malagarasien du Burundien. *Ann. Rep. Géol. Min. Palaeont., Mus. roy. Afr. centr.* (for 1965), 82.

K. BLOOMFIELD

7   **Orogenic and post-orogenic plutonism in Malawi**

ABSTRACT. *Igneous and metasomatic activity associated with a segment of the polyphase Mozambique orogenic belt is described. Pre-kinematic rocks include granite plutons, meta-anorthosites and minor ultramafites, and there is an early kinematic suite of nepheline-syenites: the North Nyasa Alkaline Province. In southern Malawi a distinctive province of late syn-kinematic perthitic syenites and gneisses is associated with several infracrustal ring-complexes of syenitic and ultramafic rocks. Syenitisation was structurally controlled, the alkaline influx being tectono-genetic rather than a differentiation product. Post-orogenic, structurally high-level, calc-alkaline rocks of the Lake Malawi Granitic Province lie within a proto-rift zone and mark a final intrusive phase in the complex tectono-thermal history of the Mozambique Belt in this region. Finally, orogenic and epeirogenic plutonism, almost coincident in space, although separated by a wide time-gap, are briefly compared and contrasted.*

Introduction

Malawi, eastern Central Africa, lies entirely within the Mozambique Belt (Fig. 1), a complex orogenic zone that stretches southwards from Egypt to the Zambesi River and beyond, and is one of the regions affected by the widespread Pan-African thermo-tectonic episode of late Precambrian to early Palaeozoic age (Kennedy, 1965; see Clifford, this volume, p. 13).

Mapping of the 'Mozambiquian'[1] rocks of Malawi by the Geological Survey during the past 20 years has resulted in the recognition of a wide variety of orthogneisses, metasomatic and igneous rocks and in the delimitation of several contrasting geochemical provinces (Bloomfield, 1968). Many of these rocks, such as the minor basic intrusions, are of the kind normally associated with ancient orogenic environments in other crystalline terrains, but a number of plutons are considered to be sufficiently unusual and interesting to merit attention. Accordingly this review summarises present knowledge of the development of the plutons associated with the Mozambique Belt in this part of Africa and discusses possible wider structural implications. As far as possible a petro-tectonic classification is adopted.

GENERAL GEOLOGY AND STRUCTURE

Basement Complex gneisses, schists and granulites make up much of Malawi and have the general north-south 'Mozambiquian' trend. No depositional

---

[1] The term 'Mozambiquian' is used to cover the polyphase 'orogeny' that produced the metamorphic rocks of the Mozambique Belt; the rocks within this belt yield ages of 450-700 m.y. and extend northwards from the Zambesi to Uganda and beyond.

119

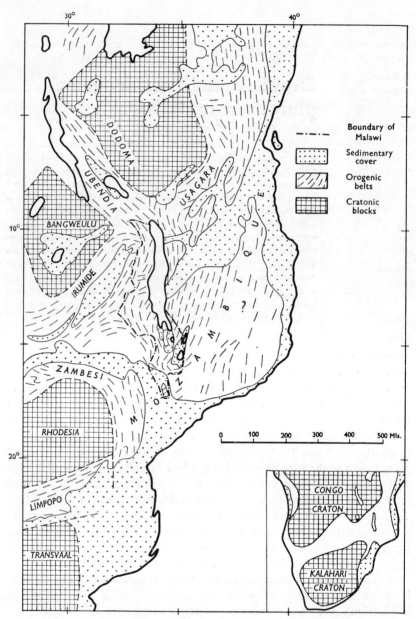

FIG. 1. The structural setting of Malawi (after Harpum, 1955; Kennedy, 1965; and Vail, 1965a and b).

features have been preserved in these rocks and any successions established are structural rather than stratigraphic. Almost all belong either to the amphibolite or to the granulite metamorphic facies; the latter is characterised by uniform linear trends and steeply-inclined isoclinal folds and the former by more complex open folding. A number of transverse structural belts show refoliation and cataclasis, and there are also several interference zones, the result of superimposed folding, which are cored by granulites and suggest thermal zoning.

The granulites comprise both banded charnockitic types and massive enderbitic rocks. There are also khondalitic units and associated quartz-feldspar granulites, marbles and high-grade pelites. The more widespread amphibolite facies rocks include semipelitic biotite-gneisses, cafemic gneisses, psammites, pelites and abundant marbles and calc-silicate granulites. Graphitic types are common, and there are local ferruginous and unusual aegirine- and nepheline-gneisses.

In southern Malawi the Basement Complex gneisses are characterised by relative richness in Mg and Ca and by the dominance of Na over K, but there is a gradual change northwards to types richer in K and Si, which may reflect an original lateral change in sediment composition (cf. Kennedy, 1951). The depositional environment was that of a stable shelf area in the north, with local deltaic conditions, and an intercratonic basin or marginal geosyncline in the south. However zones of re-activated rocks derived from older sedimentary groups may also be present.

IGNEOUS AND METASOMATIC ACTIVITY

Table 1 summarises, in a very generalised manner, the igneous and meta-somatic events associated with the Mozambique Belt in Malawi. The classification adopted is essentially petro-tectonic but the term 'kinematic' is used in preference to 'orogenic' since local movements are more readily discernible than broader orogenies and, in any event, the Mozambiquian Orogeny was almost certainly polyphase (Clifford, 1967, p. 54).

Pre-kinematic rocks

The pre-kinematic rocks represent volcanic effusives and plutonic intrusives emplaced before the main Mozambiquian tectonism. They are of two main types: meta-anorthosites and amphibolites, of which the former occur both as narrow concordant bands in the gneisses and as a larger, pod-shaped body of Adirondack-type mantled by charnockitic granulites within the core of a closed overturned antiform. The amphibolites were probably originally doleritic dykes or sills (Bloomfield and Garson, 1965, p. 81).

Minor basic and ultrabasic intrusions

Small masses of basic and ultrabasic rocks are very common within the Basement Complex gneisses and comprise numerous small metapyroxenites and metagabbros and rare pods of alpine-type serpentinised peridotite, some of which were probably composite and zoned (Bloomfield, 1958; Kirkpatrick, 1965). The metagabbros appear to be of two different ages: an older amphibolitic variety and somewhat younger hyperites (Peters, 1966).

Local hydrothermal activity and diaphthoresis converted many of the basic and ultrabasic masses to actinolite rocks and soapstones; and a late-

T<small>ABLE</small> 1

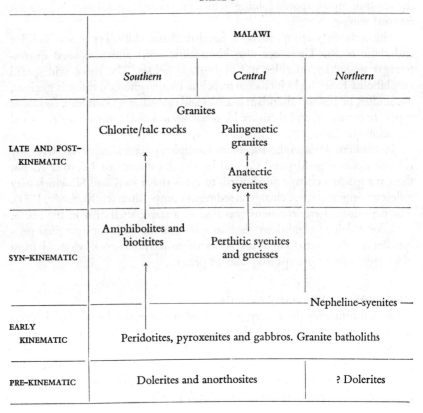

	Southern	Central	Northern
	MALAWI		
LATE AND POST-KINEMATIC	Granites Chlorite/talc rocks ↑	Palingenetic granites ↑ Anatectic syenites ↑	
SYN-KINEMATIC	Amphibolites and biotitites ↑	Perthitic syenites and gneisses	
EARLY KINEMATIC		Peridotites, pyroxenites and gabbros. Granite batholiths	—— Nepheline-syenites ——
PRE-KINEMATIC	Dolerites and anorthosites		? Dolerites

kinematic influx of potash, alumina and water produced an unusual swarm of lenticular biotitite bodies (Morel, 1955).

Early kinematic granites

Major granite batholiths occur in northern and central Malawi. Of these the Nyika Granite (Fig. 2) was intruded some 1100 m.y.[1] ago but later Mozambiquian tectonism produced extensive shearing and epidotisation (Bloomfield, 1965a, p. 45). The rather smaller calc-alkaline Dzalanyama Granite (Fig. 2) is strongly sheared and foliated along north-west to south-east lines and may be broadly coeval with the Nyika rocks (Thatcher and Walter, 1968; Thatcher and Wilderspin, in press).

Nepheline-syenites

A number of small bodies of metaluminous nepheline-syenite and a single sub-aluminous one occur in northern Malawi and constitute a distinctive sodic province (Bloomfield, 1965a). They were probably intruded in the

----

[1] A subsequent re-calculation gives an apparent age of 1300 ± 400 m.y. (Cahen and Snelling, 1966).

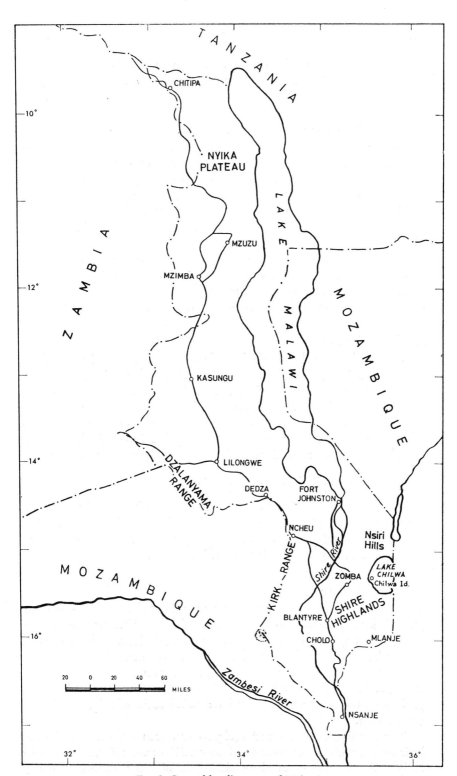

FIG. 2. General locality map of Malawi.

very late Precambrian and underwent tectonism and partial recrystallisation during the Mozambiquian Orogeny; a derivation from syenitic rocks is probable.

Syn-kinematic perthitic syenites and gneisses

Perthitic syenites and gneisses form a very distinctive and widespread geochemical province within the Mozambique Belt of southern Malawi and, as such, form the main subject of this review. Two large perthitic complexes occur within charnockitic gneisses in interference zones of structural superposition (Bloomfield, 1968) but there are also many bands and lenses within amphibolite facies rocks; an intimate association between perthitisation and tectonism is evident.

Associated with the perthitic complexes and of the same general age are a number of infracrustal ring-complexes (Bloomfield, 1965c). These have cores of metamorphosed basic and ultrabasic rock associated with peripheral calc-alkaline syenites and granites, and most take the form of synforms.

Post-kinematic granites

Structurally high-level, calc-alkaline plutons make up the Lake Malawi Granitic Province (Bloomfield, 1968). There are both syenitic and granitic phases, and older, minor basic intrusions result in strong local hybridisation. Some of the granites have a ring-form, slightly elongated north-north-west, and they appear to lie in a narrow proto-rift zone parallel to the general trend of the paragneisses. They probably mark a final intrusive phase in the complex tectono-thermal history of the Mozambique Belt in this region.

Pre- and early-kinematic plutonic activity

The numerous small bodies of basic and ultrabasic rock scattered throughout the Basement Complex gneisses of the Mozambique Belt in Malawi will not be considered in this account. Instead it is proposed to confine attention to the anorthosites and to the syenitic and granitic rocks (see Table 1).

ANORTHOSITIC GNEISSES

The anorthositic gneisses represent a very minor phase of pre-kinematic plutonic activity but they occur at a number of places within the Mozambique Belt. A band of this gneiss, up to 3000 ft wide and with a total strike length of 32 miles, occurs in the Senzani area of southern Malawi, some 50 miles north-north-west of Blantyre (Bloomfield and Garson, 1965, p. 72; Walshaw, 1965, p. 32). It is bordered by paragneisses on one side and perthite-augen gneisses on the other and contains interbanded rocks of original gabbroic composition.

The Linthipe meta-anorthosite (Fig. 3a—grouped under basic and ultrabasic rocks), with an outcrop area of almost 100 square miles, is a lenticular body with an envelope of charnockitic granulite, the foliation planes of which dip steeply westwards parallel to the contacts (Bloomfield, 1968;

Thatcher, in press). The marginal gabbroic facies has a well-developed lineation. Overall the body resembles Adirondack-type anorthosite and it is similar to the Uluguru meta-anorthosite Complex in Tanzania (Sampson and Wright, 1964) and, more particularly, to the lenticular Madagascar bodies such as the Ankafotia massif which occupies the core of a closed overturned antiform (Boulanger, 1959).

The meta-anorthosites represent pre-kinematic sills or laccoliths. The Linthipe body may have been intruded along the hinge-zone of an antiform and metamorphosed during subsequent cross-folding (Thatcher, in press).

GRANITE BATHOLITHS

Nyika Granite

This body is about 1000 square miles in area and roughly elongated north-south parallel to the trend of the surrounding gneisses; part of the contact is fault-controlled. It is a fairly uniform porphyritic biotite-granite, locally intensely sheared and epidotised. Age determinations on biotite from the granite and on the whole rock (Snelling, 1965, p. 35) suggest that the body was intruded some 1300 m.y. ago but underwent later Mozambiquian tectonism (515 m.y.). It is therefore essentially pre-kinematic in origin.

Dzalanyama Granite

This narrow, north-west-trending body is intrusive into both Basement Complex gneisses and associated Proterozoic psammo-pelites and is flanked by a wide zone of contact migmatites (Thatcher and Walter, 1968, Thatcher and Wilderspin, in press). It bears strong north-westerly foliation and local intense shearing in the same direction.

Both porphyritic and leucocratic granites have been recognised; the former are typically calc-alkaline but the latter have granodiorite/adamellite affinities.

The Dzalanyama Granite has not been dated radiometrically but, like the Nyika Granite, it suffered tectonism, although somewhat more intense, during the Mozambiquian Orogeny.

NEPHELINE-SYENITES

Small bodies of nepheline-syenite occur at a number of places in central and northern Malawi. Age determinations suggest the period of intrusion to be about 650 m.y. ago (Bloomfield, 1968) with younger biotite ages corresponding to a late Mozambiquian phase of post-metamorphic cataclasis in the country gneisses. Most of the rocks are fairly typical subsolvus biotite-nepheline-syenites some of which show crystalloblastic textures reflecting para-tectonic crystallisation. A distinctly unusual biotite-amphibole-melafoyaite makes up Kasungu Mountain (Fig. 2); it lies between a nepheline-monzonite and a malignite in chemical composition and was probably formed by a process of nephelinisation. At Kasungu the following sequence of formation is suggested: [Continued on p. 128

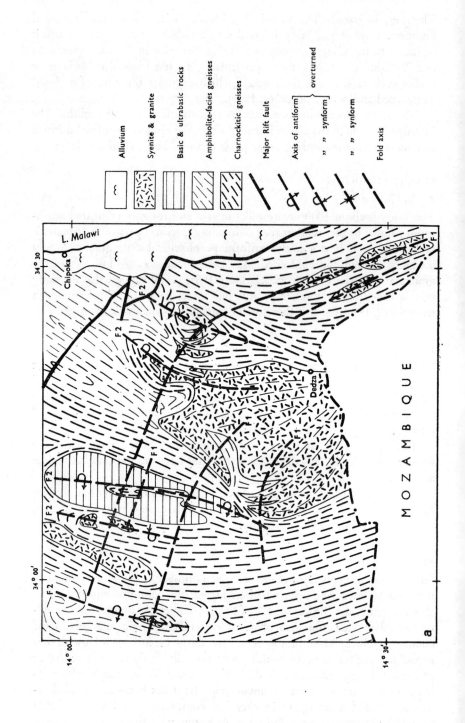

Alluvium

Syenite & granite

Basic & ultrabasic rocks

Amphibolite-facies gneisses

Charnockitic gneisses

Major Rift fault

Axis of antiform ⟩ overturned

" " synform

" " synform

Fold axis

L. Malawi

Chipoka

F2

F2

F1

F2

F2

F2

Dedza

MOZAMBIQUE

34° 30

34° 00'

14° 00'

14° 30'

F1

a

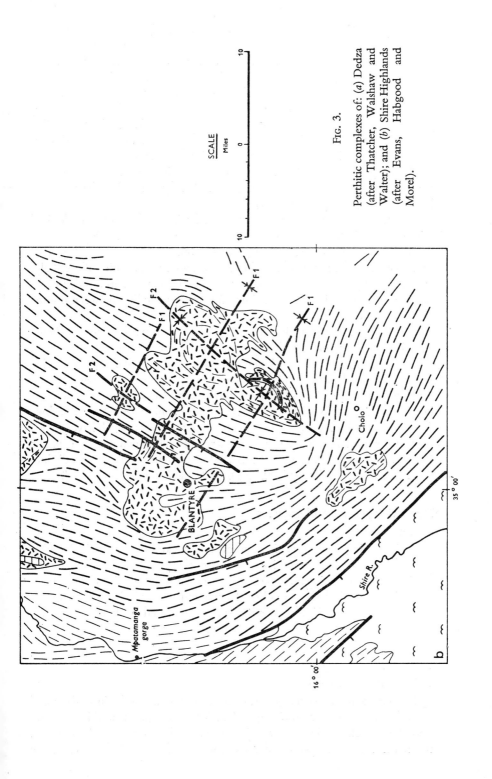

SCALE

Miles

Fig. 3.

Perthitic complexes of: (a) Dedza (after Thatcher, Walshaw and Walter); and (b) Shire Highlands (after Evans, Habgood and Morel).

(*i*) intrusion of sodic syenite or monzonite; (*ii*) intrusion of narrow appinitic dykes; and finally (*iii*) local permeation by Na and K accompanying shearing along narrow bands resulting in the formation of nepheline and some late potash feldspar.

The Ilomba Alkaline Complex in the extreme north-western corner of Malawi lies within a north-west-trending belt of gneisses, schists and granitoids in which zones of cataclasis are common. It consists in outline of a central mass of microsyenite surrounded by an incomplete ring of foyaites (locally rich in pyrochlore) which is itself partly enclosed by coarse perthosite (Bloomfield, 1959). Just to the north, in Tanzania, is the related Nachendezwaya Complex of foyaites intruded by pyrochlore-bearing carbonatite and ijolite (Horne, 1961).

At Ilomba all stages exist between perthosites and 'syenitic mylonites' with shearing in localised zones. It has been suggested that the foyaites were derived from the perthosites by a process of unmixing but that at least part of the nephelinisation was post-kinematic and metasomatic (Bloomfield, 1968). Late high-temperature hydrothermal activity introduced sodalite and carbonates into the foyaites, providing a petro-genetic link between Ilomba and Nachendezwaya.

The Ilomba complex does not fit easily into a petro-tectonic scheme since it is post-kinematic with respect to the Ubendian Orogeny (see Cahen, this volume, p. 98) and syn-kinematic with respect to the Mozambiquian Orogeny. It, and all the small nepheline-syenite bodies of north and central Malawi, form part of an extensive late Precambrian alkaline province that includes the Mbozi Syenite-Gabbro Complex in Tanzania and the Nkumbwa Carbonatite and Mivula Syenite in north-eastern Zambia which have yielded apparent ages of 745, 679 and 550 m.y. respectively (Snelling, 1962, p. 34; 1963, p. 37). It might be useful if this alkaline province were given a distinctive name and accordingly *North Nyasa Alkaline Province* is suggested.

Perthitic syenites and gneisses

INTRODUCTION

Late-kinematic perthitic syenites and gneisses make up a distinctive syenitic province within the Mozambique Belt of southern Malawi. The two largest bodies form 'perthitic complexes' within charnockitic gneisses (Fig. 3) but most of the remainder occur as broad lenses and concordant bands in amphibolite facies rocks. Similar syenitic rocks, associated with unusual infracrustal ring-complexes, are discussed on p. 133.

PERTHITIC COMPLEXES

Dedza Perthitic Complex

This complex has an outcrop area of some 240 square miles in Malawi and extends into adjacent Mozambique (Fig. 3a). It comprises perthitic syenites with local monzonitic and granitic phases, grading into perthite-gneisses: the massive or weakly-foliated rocks form the hills, and the gneisses the lower intervening ground (Bloomfield, 1965a, p. 44; Thatcher, in press).

The major fold pattern of the Dedza Complex is very characteristic of superposition (Ramsay, 1962). V-shaped trend lines converge and diverge, and there are arcuate and luniform folds, the axial traces of which swing round from north-west to north-east: isolated massive syenitic bodies occupy the cores of closed 'eyed' folds (Fig. 3a). The complex occupies the mutual interference zone of two fold systems, the first (F1) trending north-west and the second (F2) trending north-east. F1 is the regional trend of the paragneisses in this area and is characterised by uniform isoclinal folding whereas F2 has produced broad open folds.

It is significant that the Dedza Complex lies within a zone of charnockitic granulites flanked by outward-dipping biotite-gneisses and thus probably occupies the core of a major antiform. The isolated syenitic bodies to the north-east probably lie within local culminations and depressions, and the overall picture resembles type 3 interference structures of Ramsay (*ibid.*).

Normally, the margin is transitional and takes the form of permeation by alkali feldspar, but in places the Dedza syenites show cross-cutting relationships in the form of intrusion breccias and net veining. Thatcher (in press) has noted that the Linthipe meta-anorthosite is partly affected by this perthitisation process.

Shire Highlands Perthitic Complex

This is similar in many respects to the Dedza Complex and occupies an area of about 225 square miles (Evans, 1965; Morel, 1958) (Fig. 3b). It includes a completely gradational series of rocks ranging from perthitised granulites and gneisses through perthite-gneisses to anatectic perthitic syenites and perthosites. The country rocks are charnockitic granulites, and the marginal phases of the complex include many biotite-rich mafic bands and schlieren, regarded as resistors left after perthitisation. Local intrusive contacts are, however, common.

Superimposed folding is suggested by luniform and prong-shaped outcrop patterns (Fig. 3b) with a time sequence apparently the reverse of that at Dedza. An earlier north-west trend was characterised by broad arching but a younger north-east-trending phase of tighter isoclinal folding is now dominant.

The two most significant features of both the Dedza and Shire Perthitic Complexes are that they occupy structural interference zones and are contained within charnockitic granulites.

OTHER PERTHITIC SYENITES AND GNEISSES

The Cholo Mountain perthitic syenite (Fig. 3b) to the south of the main Shire Highlands Complex is a large oval body oriented north-westwards parallel to the surrounding granulites and apparently outside the zone of superimposed north-east folding (Habgood, in press). However, apart from some rare narrow concordant bands of perthite-augen gneiss, most of the other syenitic bodies lie within amphibolite facies gneisses.

Ncheu syenites

Lenticular bodies and narrow bands of perthitic syenite occupy a 50-mile-long zone trending north-west parallel to the surrounding paragneisses at the northern end of the Kirk Range in the Ncheu area (Fig. 2). The northern extension of this zone impinges upon the Dedza Perthitic Complex (Fig 3a) and has been involved in superimposed folding.

The Ncheu bodies range in composition from biotite-perthite-gneiss to perthitic syenite or, rarely, granite and form elongate concordant masses which usually show gradational contacts (Walshaw, 1965; Dawson and Kirkpatrick, in press) although there does not appear to be a regular increase in perthite across the strike but rather rapid alternation indicative of selective feldspathisation. Local mobilisation and assimilation have been noted.

In places the narrower syenite bands have been tightly folded into sharp antiforms and synforms some of which have nuclei of ultrabasic rock.

Southwards along the strike the Ncheu syenites are represented by a broad zone of perthite-gneiss containing narrow bands of perthitic syenite and there are both gradational and intrusive contacts (Bloomfield and Garson, 1965, p. 121).

South-east Lake Malawi

Holt (1961, p. 31) has described perthitic syenites and associated gneisses in the country to the south-east of Lake Malawi (Fig. 2) occurring within the cores of major folds, and thickest in the apical zones; contacts are usually gradational. Permeation and influx of alkaline material has gradually destroyed the banding in the paragneisses and consequent dilation has resulted in considerable distortion.

The regional north-south trend is followed by elongate bodies of syenite and by the axial trace of a major synform. However there is a broad belt that reflects a sudden east-west swing, probably the result of superposition, and, within this, the syenite bodies have been deformed into broad arcuate and sigmoidal folds.

Other perthite-gneisses between the south-eastern and south-western arms of Lake Malawi are interbanded with paragneisses and strongly folded about north-easterly axes (Dawson and Kirkpatrick, in press).

Shire Highlands area

Perthite-gneisses, perthitic syenites and local granitoid types occupy structural depressions and form the cores of synforms in paragneisses near Mlanje Mountain where they appear to have a sill-like form (Garson, 1965, pp. 9-10; Garson and Walshaw, in press); local feldspathisation is common. In addition narrow concordant bands of perthite-augen gneiss in the Shire Highlands illustrate the selective replacement of felsic bands by K-feldspar since thin mafic resistors can be traced without interruption for hundreds of feet (Bloomfield, 1965d; Evans, 1965). Local mobilisation is shown by angular xenoliths of gneiss and granulite. There are also rare zones of feldspathised mylonite in this area, indicative of penecontemporaneous shearing and feldspathisation.

One band of perthite-gneiss some 15 miles in length and averaging 100 ft in width coincides with the north-east-trending Shire Highlands Fault, the main Rift Valley fracture in this particular area (*ibid.*, p. 30). The conutry rocks are charnockitic gneisses although the band is, for the most part, structurally underlain by diopside-amphibolite.

Other occurrences

Perthitic syenites and gneisses are rare in central and northern Malawi but may be represented by broad zones of perthite-augen gneiss within semi-pelites near Lilongwe (Fig. 2) (Thatcher and Walter, 1968). Elsewhere there are a number of occurrences of alkali syenite, intrusive into Basement Complex paragneisses, with doubtful affinities (Bloomfield, 1968). They may either be associated with the types described above or with the nepheline-syenites discussed on p. 125.

PETROGRAPHY AND CHEMISTRY

Petrography

The following mineral assemblage, with only slight variations, is very characteristic: hornblende-biotite-(diopsidic pyroxene)-microperthite-oligo-clase-quartz. With increase in quartz there are local gradations to perthite-granites and rocks containing a significant amount of plagioclase grade into monzonites. The accessory minerals are titanomagnetite, pyrite, sphene and two distinct generations of apatite and zircon.

The perthite-gneiss—perthitic syenite series is characterised by the piece-meal replacement of plagioclase by microperthite, accompanied by the alteration of pyroxenes to hornblende and biotite. The various stages in the formation of perthitic syenite from semipelites and granulites have been described by Bloomfield and Garson (1965, p. 129). Microperthite augen have grown along and across the gneissic foliation, coalesced to form felsic bands, and eventually the mafic banding was partly destroyed with the dark minerals in discontinuous oriented clusters. Mobilisation resulted in anatectic perthitic syenites.

The mafic minerals are usually those of the associated gneisses, and syenite bodies in granulite terrain contain metastable hypersthene and diopside. The microperthites are fine vein-types or mesoperthites in which K feldspar$\gg$albite; marginal exsolution of Na-feldspar and myrmekitisation are common. Fine-grained aggregates of clear Na-feldspar surrounding individual perthite crystals may either represent original unaltered feldspar derived from the paragneisses or crushed material separated from the ex-solution rims.

Chemistry

Analyses of perthite-gneisses and anatectic syenites are given by Bloomfield (1968, p. 144). As might be expected, the former are high in total alkalies although the proportions of Na and K differ, and they are somewhat closer to the average alkali syenite in composition than to the calc-alkali type; the rocks are approximately silica-saturated. In contrast anatectic syenites from the Dedza Complex and from an appendage of the Shire Highlands Complex are almost exactly saturated. K is greater than Na in each case, and the rocks lie between calc-alkali syenites and monzonites in composition: a new term 'perthitic monzonite' is thus suggested (*ibid.*). The rocks are enriched in Ba and Sr and have fairly high V, probably replacing Fe and Mg in the mafic minerals. Finally, of two analysed syenites from the south-east of Lake Malawi, one proved to be potassic and the other sodi-potassic. These rocks are probably closer to the average alkali syenite than to the calc-alkali type.

PETROGENESIS

The possible petrogenesis of these syenitic rocks is discussed at length in Bloomfield (1965c, p. 72) and only the main conclusions, amplified by more recent work, will be given here.

Strong evidence of a replacement origin for the perthitic rocks is provided by field and petrographic characteristics. Plagioclase feldspar of the supra-crustal gneisses and granulites, initially that in the more felsic bands, has been replaced piecemeal by sodi-potassic feldspar to produce perthite-gneisses. Perthitic syenites resulted from subsequent anatexis and mobilisation. The microperthite is characteristically the high temperature form and so complete remelting of the felsic component must be invoked. Thus the associated oligoclase is more likely to represent exsolved than relic material.

It was suggested by Morel (1961) that the alkaline material responsible for the formation of the perthitic syenites and gneisses was magmatic in origin, derived from the same basaltic parent as the various mafic and ultra-mafic bodies within the Basement Complex. These latter bodies were, however, emplaced at an earlier stage in the evolution of the Mozambique Belt than the syenites, and the wide time-gap is difficult to explain. Moreover, a detailed analysis of the evidence (Bloomfield, 1965c) suggests that there are no genetic links between the two groups of rocks. The evidence for this will be considered more fully later (p. 144) but it leads to the conclusion that the association is purely tectonic, the alkaline material having risen within zones of structural weakness around the ultrabasic bodies. Analyses also show that some of the syenites are essentially alkaline, others calc-alkaline, the differences being due solely to the composition of the replaced gneisses.

It is thought that first, perthite-augen gneisses and then perthite-gneisses were developed from supracrustal rocks by the addition of Al, K and Na. Addition of further K (plus Ba) and, locally, Si accompanied the formation of perthitic syenites by anatexis. The necessary alkalies were derived directly from crustal material by partial melting during deep-seated regional meta-morphism and syenites were formed, rather than the commoner granites of other orogenic belts, because of the relative paucity of silica in those gneisses.

Termier and Termier (1956, p. 63) and others consider perthite to be a 'stress mineral' and an association between perthitisation and tectonism in metamorphic terrains has long been evident. In Malawi, the structural control of perthitisation is apparent in several places. The zones of anatectic syenite around ultrabasic plugs that form unusual infracrustal ring-complexes are considered in the next section (see below). Elsewhere an association between perthitisation and open folding has been noted (Holt, 1961, p. 31), although apart from thickening and local mobilisation within the hinge zones, somewhat in the manner of saddle reefs, there does not appear to be any association between *individual* broad folds and perthitisation. Evans (1965, p. 32) infers additional lithological control, the alkaline material ascending along so-called 'privileged paths'. The narrow band of perthite-gneiss coincident with the line of the much younger Shire Highlands Rift Fault is of considerable interest and suggests that this has been a zone of weakness since Lower Palaeozoic times (*vide* Dixey, 1956).

A closer correlation between folding and perthitisation can be made following the more recent work of Thatcher (in press) in the Dedza area (Fig. 3a). It is now thought that the location of both perthitic complexes within the interference zones of two-fold systems, and also within charnockitic granulites, is of great significance and enables an eclectic petro-tectonic hypothesis to be put forward.

Thus it is concluded that selective perthitisation along relatively permeable or felsic bands in the paragneisses and granulites took place soon after the acme of regional metamorphism coincident, in the Dedza and Shire Highlands areas, with F1 folding (Fig. 3). Incomplete expulsion of replaced Ca and Mg during the perthitisation process caused local increases in volume, and this resulted in dilation to form the concordant lenticular syenite bodies. Superimposed F2 folding about north-west axes followed and, in the interference zones, anatectic syenites were produced which, locally mobilised, were largely caught up in the folding. Thus the perthitisation process may be regarded as post-metamorphic but both syn- and late-kinematic.

At Dedza, the F1 folding was mainly isoclinal whereas the F2 phase produced a broad antiform which also exposes the granulites. A similar sequence may also be inferred in the Shire Highlands Complex where the F2 phase seems to have been more complex.

The time of formation of the perthitic syenites and gneisses in relation to the Mozambiquian Orogeny is uncertain. It is generally believed to have taken place after the main syn-kinematic phase of regional migmatisation, probably between 400 m.y. and 500 m.y. ago, i.e. during the early Palaeozoic.

Infracrustal ring-complexes

INTRODUCTION

Several unusual ring-complexes occur within the supracrustal gneisses of the Mozambique Belt in southern Malawi and are characterised by central cores of metamorphosed basic and ultrabasic rock associated with peripheral syenites and granites. Four of the complexes have been described in some

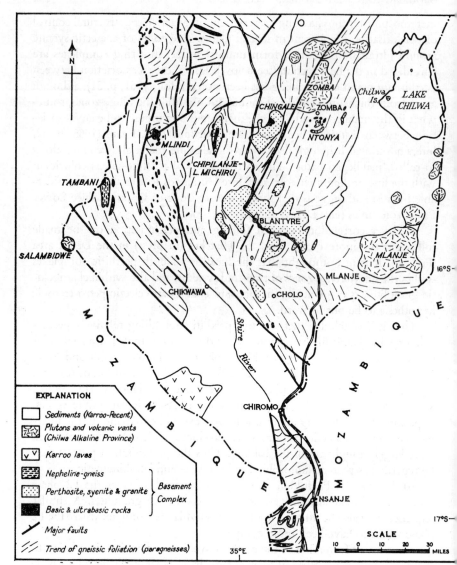

FIG. 4. Infracrustal ring-complexes of southern Malawi.

detail (Bloomfield, 1965c). They lie at about the 15° 30′ S. parallel on a line which trends N. 80° E. and, from west to east, comprise the Mlindi, Chipilanje–Little Michiru, Chingale and Ntonya Complexes: the distance from Mlindi to Ntonya is about 50 miles (Fig. 4). Other isolated ring-complexes of similar character occur at Bilila, near Ncheu (Walshaw, 1965, p. 47) and make up the Nsiri Hills body just north of Lake Chiuta (Dawson, 1966) (see Fig. 2). The Chilwa Island carbonatite/agglomerate vent of Mesozoic age penetrated a pre-existing domal structure of porphyritic anatectic syenite

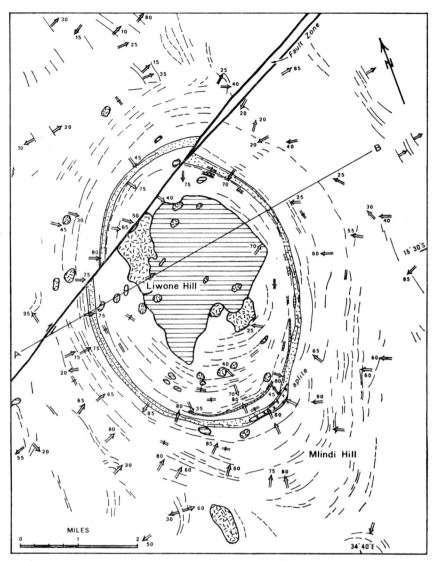

Section A–B *(Natural Scale)*

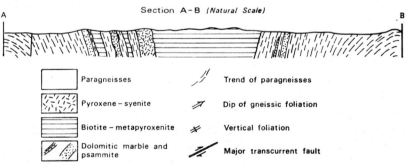

Paragneisses	Trend of paragneisses
Pyroxene – syenite	Dip of gneissic foliation
Biotite – metapyroxenite	Vertical foliation
Dolomitic marble and psammite	Major transcurrent fault

Mesozoic dyke swarm omitted

FIG. 5. Mlindi Ring-Complex

which may also have been comparable with the other ring-complexes (Garson, 1960, p. 24).

The structure and field relationships at each of the complexes will be described briefly in turn, followed by an account of the petrography and chemistry of the constituent rocks. Finally the possible mode of origin and petrogenesis of the complexes will be considered.

THE MLINDI RING–COMPLEX

A mass of biotite-metapyroxenite with local gabbroic phases some four square miles in area is surrounded by several smaller peripheral bodies of pyroxene-syenite (Figs. 4 and 5). All lie within the central core of a major closed structure in paragneisses which, in plan, is an almost perfect oval and the gneissic foliation is either vertical or dips steeply inwards (Fig. 5). A narrow discontinuous band of dolomitic marble, flow-folded and contorted, parallels a psammite band outlining the 'ring'; these rocks are typical Basement Complex types and, in any event, the ratio $^{87}Sr/^{86}Sr$ for the marble is higher than the mean value for carbonatites (Hamilton and Deans, 1963). A swarm of alkaline microsyenites trending N. 60° E. across the ring structure is of Lower Cretaceous age and unrelated to the plutonic rocks (Bloomfield and Garson, 1965, p. 173).

The ultrabasic rocks are intrusive into the paragneisses, but contacts with the syenites are usually gradational and are marked by wide zones of hybrid 'syenogabbro'. Cross-cutting dykes of somewhat finer-grained syenogabbro represent rheomorphic material. Pods and irregular masses of biotite-pyroxene-apatite rock, resembling the metapyroxenite in all essentials, occur within the syenite, which is either massive or foliated.

The sequence of events at Mlindi comprised: (*i*) intrusion of pyroxenite; (*ii*) intrusion of peripheral syenites with associated hybridisation; (*iii*) cross-cutting syenogabbro dykes; and (*iv*) pegmatite veining.

THE CHIPILANJE–LITTLE MICHIRU COMPLEX

This complex (see Fig. 4) has been subjected to strong east-west compressive forces but, apart from different proportions of the main rock types exposed, the general picture resembles that at Mlindi. A narrow mass of syenite forms the core of a closed synform or 'canoe-fold' the axial trace of which is marked by a body of metapyroxenite (Fig. 6). The metapyroxenite is completely surrounded by syenite and is elongated parallel to the regional trend of the local paragneisses. The outer junction between syenite and paragneisses is marked by a zone of feldspathisation, and there is strong hybridisation at syenite/metapyroxenite contacts. The margins of the syenite body are strongly lineated parallel to the outer contact and there are many mafic schlieren, but much of the central part is massive and contains xenoliths of gneiss and metapyroxenite. A sequence of formation similar to that at Mlindi is inferred.

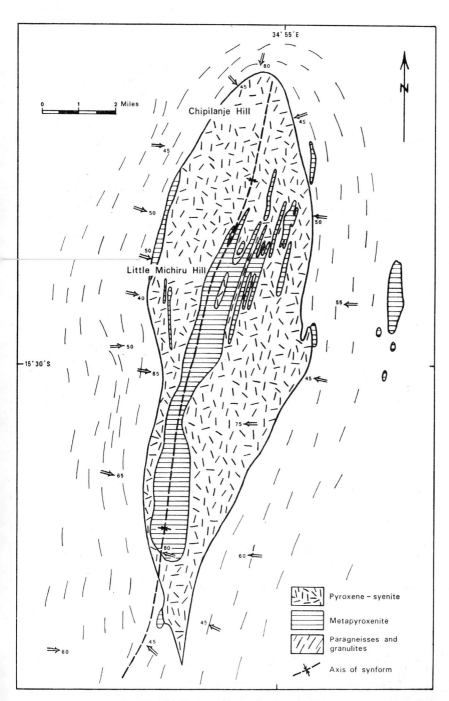

FIG. 6. Chipilanje–Little Michiru Complex.

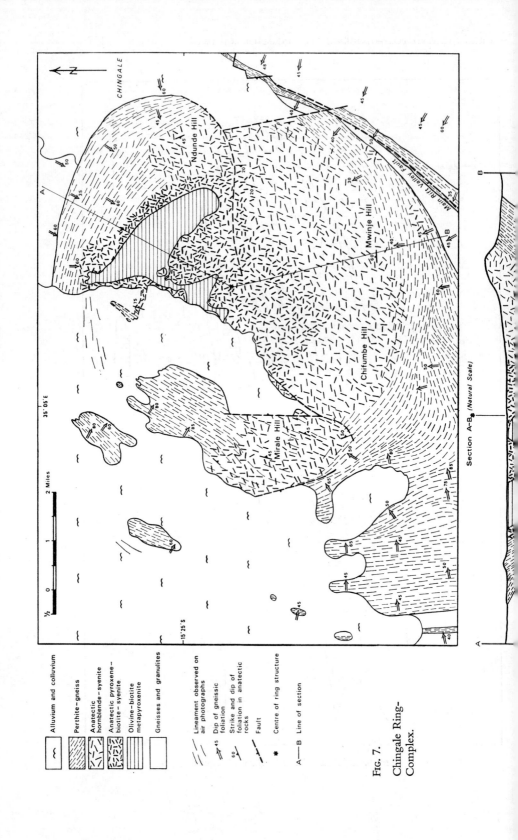

Alluvium and colluvium

Perthite–gneiss

Anatectic
hornblende–syenite

Anatectic pyroxene–
biotite–syenite

Olivine–biotite
metapyroxenite

Gneisses and granulites

Lineament observed on
air photographs

Dip of gneissic
foliation

Strike and dip of
foliation in anatectic
rocks

Fault

Centre of ring structure

A——B   Line of section

CHINGALE

N

2 Miles

½   0   1   2 Miles

35° 05′ E

–15° 25′ S

Ndunde Hill

Mwinje Hill

Chifumbe Hill

Mirale Hill

Main Rift Valley Fault

A

B

A'

Section A–B₂ (Natural Scale)

Fig. 7.

Chingale Ring-
Complex.

THE CHINGALE RING–COMPLEX

The Chingale Complex (Fig. 4) has been studied in some detail and is the largest and best-known of the four. There, an arcuate mass of perthitic syenite and perthite-gneiss, forming a discontinuous ring of hills some eight miles across, dips inwards at between 35° and 65° towards a flat arena made up of biotite-metapyroxenite surrounded by an irregular zone of hybridised syenite (Fig. 7). Both syenitic rocks and metapyroxenites are intrusive into granulites and gneisses of the Basement Complex and outer contacts appear to be sharp.

To the south the syenitic gneisses of the Chingale Complex gradually conform to the regional trend and occupy the axial zone of a northward-plunging synform. The general structure is thus similar to the Chipilanje–Little Michiru Complex.

The attitude of the metapyroxenite/syenite contact is not known but the central ultrabasic body may either have the form of a lopolith or an irregular stock. The general sequence of formation of the main rock types was as follows: (*i*) biotite-metapyroxenite; (*ii*) syenogabbro hybrids; (*iii*) anatectic perthitic syenite; (*iv*) syenogabbro dykes; and (*v*) veins of feldspar and perthosite. Stages (*iii*), (*iv*) and (*v*) occur in reverse order in some localities and this probably indicates renewed mobilisation.

THE NTONYA RING–COMPLEX

At Ntonya (Fig. 4) no central ultrabasic mass is exposed but this is thought to be due essentially to differences in present exposed level. The complex is made up of a central irregular mass of anatectic quartz-syenite surrounded by a discontinuous ring of granite, the two being separated by a screen of charnockitic granulite (Fig. 8). The granite possesses a strong arcuate lineation and a weakly-developed planar foliation which dips inwards at about 35°.

Xenoliths and roof pendants of gneiss and granulite occur within the central massive syenite, and the outer granite/gneiss junction is sharp, without any signs of contact metasomatism.

A short distance to the north-east and probably associated with the Ntonya Complex is a small body of metapyroxenite (Fig. 8), partly enveloped by lineated granite, which lies within a tight synform in the charnockitic granulites.

THE BILILA RING–COMPLEX

This complex is a narrow oval-shaped body of basic and ultrabasic rocks markedly elongated parallel to the north-east trend of the surrounding paragneisses. It is surrounded by a discontinuous envelope of granitoid and perthite-gneisses together with some perthitic syenite (Fig. 9). A well-developed planar foliation dips uniformly to the north-east and the complex is probably inclined in this direction.

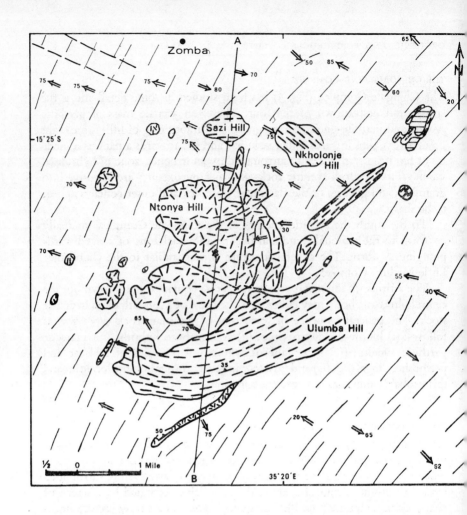

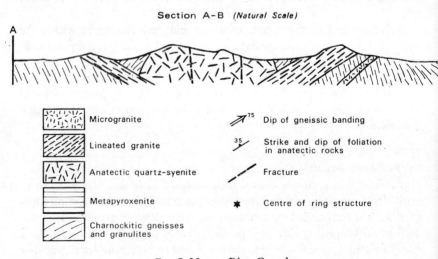

Section A–B *(Natural Scale)*

Microgranite	Dip of gneissic banding
Lineated granite	Strike and dip of foliation in anatectic rocks
Anatectic quartz-syenite	Fracture
Metapyroxenite	
Charnockitic gneisses and granulites	Centre of ring structure

FIG. 8. Ntonya Ring-Complex.

Along the margins of the complex the core rocks are in close contact with perthite-gneisses but in places they have been intruded into banded migmatitic gneisses. Metadiorites and metagabbros, with local diffuse ultrabasic pods and xenoliths of banded granulites, predominate but in places there are marginal biotite-rich phases. Unlike the other complexes, the basic rocks are little

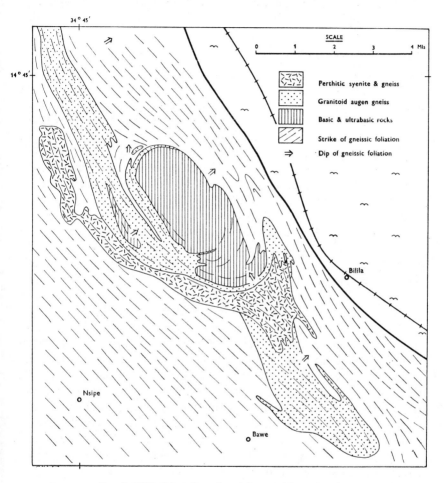

FIG. 9. Bilila Ring-Complex (after Walshaw, 1965).

affected by hybridisation although the marginal granitoid gneisses contain mafic and ultramafic schlieren.

The two small ultrabasic bodies associated with perthitic rocks to the south-east of Ncheu (Fig. 2; see Walshaw, 1965, p. 46) are probably similar to the Bilila Complex but on a smaller scale. These bodies are elongated north-west parallel to the regional trend and are partly surrounded by, and inter-folded with, perthite-gneisses. Locally, hybrid contact zones have been developed.

THE NSIRI HILLS COMPLEX

This complex is roughly circular, about 20 miles in diameter and consists mainly of anatectic perthitic syenite grading into a broad marginal zone of granites; at the outer contact granite/gneiss agmatites are locally developed. The Nsiri Complex is comparable to the structurally relatively high-level Ntonya Complex.

PETROGRAPHY AND CHEMISTRY

Ultrabasic and basic rocks

At Mlindi and Chingale (Figs. 5 and 7) biotite-metapyroxenites consisting essentially of diopside and magnesian biotite predominate; other mafic minerals developed in some rocks include olivine, hypersthene and hornblende. There are, in addition, small amounts of quartz, potash feldspar and plagioclase (oligoclase/andesine). Abundant apatite and titanomagnetite with traces of pyrite, chalcopyrite and pyrrhotite are the accessory minerals.

Chemically the biotite-metapyroxenites differ from their igneous counterparts in several respects (Bloomfield, 1965c, p. 47). They are considerably richer in potash, and the Mlindi occurrence in particular is also higher in MgO and BaO (0·89%) and lower in lime than average. This is, of course, a direct reflection of the high mica content, and a marked but possibly fortuitous chemical resemblance between these rocks and the potassic sub-volcanic types of south-west Uganda (Holmes and Harwood, 1937; Combe and Holmes, 1945) has been noted. Spectrographic analyses show a constant Co/Mg ratio and an enrichment in Ba, Sr and (in some rocks) Rb.

The metapyroxenite of the Chipilanje–Little Michiru Complex (Fig. 6), and to the north-east of the Ntonya Complex (Fig. 8) is a hypersthene-metadiallagite similar in its chemistry to the average igneous two pyroxene-pyroxenite. The metagabbros of these and the Mlindi, Chingale and Bilila Complexes were originally more felsic facies of the pyroxenites containing larger amounts of andesine and quartz. They are rather more mafic and correspondingly poorer in alumina and the alkalies than Nockolds' average pyroxene-gabbro (1954, p. 1020). The Bilila metadiorites (Fig. 9) are essentially similar to the metagabbros, but andesine appears to represent an original stable phase unassociated with any exsolved CaO minerals.

Perthitic syenites and gneisses

The perthitic syenites of Mlindi, Chipilanje and Chingale contain phenocrysts and tabular crystals of microperthite, a little interstitial oligoclase/andesine and quartz, and small mafic clots which clearly reflect the mineralogy of the neighbouring metapyroxenites. Green hornblende, biotite and magnetite have developed from the pyroxenes. With increase in mafic minerals there is a transition into syenogabbro hybrids.

Chemically the syenites from the three complexes are similar and resemble Nockolds' average calc-alkali type (*ibid.*, p. 1016) although they appear to become progressively richer in silica and alumina from west to east (see Fig. 4). Potash exceeds soda in each case, and the K_2O/Na_2O ratios of a number of rocks have been found to be constant around 1·35, reflecting approximately equal molecular proportions of Ab and Or. Spectrographic analyses show a relative richness in Ba and Sr and relatively constant Rb/K and Sr/Ca ratios.

The perthite-gneisses are normally well-foliated and consist of narrow laminae and lenticles made up of biotite and hornblende separated by microperthite porphyroblasts associated with some interstitial quartz and plagioclase. One analysed perthite-gneiss resembles an igneous calc-alkali syenite.

Quartz-syenite and granite

The quartz syenites contain distinctive red-brown biotite, both ortho- and clinopyroxenes and up to 10% quartz. The felsic minerals comprise approximately equal amounts of sodic plagioclase and microperthite. The mineralogy is a direct reflection of that of the charnockitic granulites of the basement, and the mafic minerals in particular appear to have been incorporated without recrystallisation. The rocks are of monzonitic chemical composition, similar to igneous calc-alkali syenites.

The mica in the granites is a chocolate-brown biotite; pyroxene is absent and microcline is the predominant feldspar with only small amounts of microperthite and sodic plagioclase. Abundant accessory zircon in two generations and allanite are characteristic. The granites resemble the average calc-alkali type (ibid., p. 1012) and K_2O is especially abundant. Ba and Sr are equally enriched in these rocks and in the syenites.

The trace element contents of the quartz-free and quartz-bearing rocks are not significantly different, but the latter are slightly enriched in Be, Sc, Mo and Li and impoverished in Cr, Ni, Co and Y relative to the former.

Hybrids

Two types of hybrids have been distinguished: felsic segregations in mafic and ultramafic rocks; and cross-cutting syenogabbro dykes. The former show a wide range of colour indices and their mineralogy reflects that of the host rock; there is a little plagioclase and quartz but microperthite is the predominant feldspar. The dykes have a fairly uniform composition with colour indices of between 40 and 60. Diopside and biotite are common, but hypersthene is absent. Strained cryptoperthite, antiperthite and plagioclase are the felsic minerals, and apatite is abundant.

PETROGENESIS

A detailed discussion of the petrogenesis and mode of formation of the infracrustal ring-complexes has already been published (Bloomfield, 1965c, p. 72) in which field, petrographic and chemical data were considered. The links between the complexes are sufficiently strong to support a general hypothesis which can also be applied to the Bilila and Nsiri bodies. In essence, this is but another instance of the tectonic control of perthitisation described in the previous section.

The present configurations of the ring-complexes are thought to be due essentially to differences in structural level: those in which large amounts of ultramafic rock are exposed representing a lower level than those composed of anatectic syenites and granites only. Thus there is a general rise from west to east from the Mlindi Complex to Ntonya (Fig. 4) and it may be worth speculating on the fact that the estimated difference of about 11,000 ft in structural level between Ntonya and Chingale is approximately the same as the difference between the depths of formation of syn- and post-kinematic granites according to Termier and Termier (1956, p. 431).

It is concluded that a number of composite basic and ultrabasic plutons were emplaced during an early syn-kinematic phase of the Mozambiquian Orogeny as pyroxenites, peridotites and gabbros, the line of complexes following a possible megashear in the cratonic basement. Later, during the closing stages of regional metamorphism but preceding a final phase of east-west compressive folding, dominantly potassic hydrous metasomatism, accompanied by partial desilication, was locally active to produce biotite-metapyroxenites that were at least partly rheomorphic. The influx of potash continued, and was accompanied by silica, alumina and soda—all derived from the partial melting of the supracrustal gneisses as suggested in the previous section—and this tectono-genetic alkaline material permeated the basic and ultrabasic bodies and neighbouring paragneisses to form marginal syenogabbro hybrids and, subsequently, perthitic syenites. The wave of alkali metasomatism rose upwards, around and over the favourable pipe-like plutons, but elsewhere the general migmatite level was deeper in the crust and could only rise through zones of structural weakness or selectively permeable bands. At Bilila the absence of hybrids suggests that the basic body (? having already cooled) acted as an impermeable resistant nucleus during folding.

The perthitic syenites were locally sufficiently hot and fluid to intrude the surrounding rocks as anatectic melts, in places along cone-sheet type fractures in the roof zones, themselves the result of halokinetic upward movement of rheomorphic biotite-metapyroxenite.

A further stage of granitisation proper links Chingale with the Ntonya and Nsiri Complexes. In the last of these complexes, the plagioclase of the country rocks was replaced piecemeal by K-feldspar and silica, and the whole mass mobilised and intruded as an anatectic quartz-syenite. The third and final stage of complete granitisation produced the microcline-granites from the quartz-syenites by further addition of K and Si and expulsion of bases. Unlike the syenites, the granites underwent complete fusion and were intruded as palingenetic melts along cone-sheet type fractures. The associated pegmatites and aplites were the final products of alkali metasomatism. Thus, the whole of the granite series postulated by Read (1955) is represented, albeit somewhat silica-deficient, and this ranges in space and time from autochthonous syenites and migmatites through parautochthonous types to intrusive granites.

Finally, it is of interest to note that a pre-existing ring of Basement

Complex quartz-syenite underlies the much younger Chilwa Island carbonatite vent (Fig. 4) (Garson, 1960, p. 51). Thus this is the fifth in a line of infracrustal ring-complexes trending N. 80° E. for 70 miles across the full width
of southern Malawi and possibly extrapolated westwards to the Zambesi
Orogenic Belt. This line may be the site of an old zone of weakness in the
crust and is parallel to a major transcurrent fault (Bloomfield, 1966; De
Swardt et al., 1965)—a belt of cataclasis oblique to the general 'Mozambiquian' trend (Bloomfield, 1968)—and to a line of peralkaline plutons of
Mesozoic age (Bloomfield, 1965d, p. 107).

The linearity of the zone of ring-complexes, all of which are thought to
have ultrabasic roots, may be explained by assuming that in the early stages
of the orogeny the cratonic basement yielded to compressional stress by
rupture rather than by plastic flow. Under these conditions the rigid basement fractured to form tension-joints parallel to the stress and shear-joints
making an angle of between 15° and 45° with it (De Sitter, 1956, p. 130).
If it is assumed that the principal stress direction was normal to the present
fold axes in the supracrustal gneisses, then, since these make an angle of about
30° with the line of ring-complexes, it is possible that the ultrabasic bodies
were originally intruded as pipe-like plutons along a megashear or series of
shear-joints in close echelon.

The age of the infracrustal ring-complexes is uncertain. All that can be
inferred at present is that they are late syn-kinematic and post-kinematic,
formed during the last local phase of the 'Mozambiquian' Orogeny.

The Lake Malawi Granitic Province

INTRODUCTION

A number of late- and post-orogenic granites occur near the shores of Lake
Malawi (Fig. 10), in some instances forming small islands, and they are
sufficiently distinctive to make up a petrographic province. There appears to
be a definite parallel between the zone of granitic plutons and major rift
faults.

Only one age determination is available, suggesting that the Cape Maclear
Granite (Fig. 10) was emplaced some 450 m.y. ago (Snelling, 1963, p. 38) but
it is thought that most of the other bodies are of this Palaeozoic age. Similar
granitic rocks in neighbouring Mozambique have apparent ages ranging from
380 m.y. to 593 m.y. (Mendes, in Vail and Pinto, 1966).

STRUCTURE AND FIELD RELATIONSHIPS OF INDIVIDUAL PLUTONS

Likoma Island

Likoma is quite unlike any of the other granitoids in this province, and
consists largely of graphic granite containing pelite xenoliths, strongly
sheared along east-north-east lines (Bloomfield, 1966). The following
sequence of events is suggested: (*i*) regional metamorphism of the pelites;
(*ii*) intrusion of graphic granite; (*iii*) east-north-east shearing associated with
quartz veining; (*iv*) intrusion of microtonalite dykes.

F

Sani Point, Nkhotakota

At Sani the granite body is at least $2\frac{1}{2}$ miles in width and lies on the upthrown eastern side of a prominent rift fault which is roughly parallel to the lake shore (Bloomfield, 1965a, p. 48). A weakly-developed east-west flow foliation is developed in places, and the mass is intersected by a number of north-east-trending joints and shear zones.

Senga Bay, Salima

Granites with syenitic phases crop out sporadically along the lake shore and are separated from Basement Complex gneisses to the east by a wide expanse of alluvium. However, as at Sani to the north, the western limit of the pluton is thought to be a hidden major fault with a downthrow to the west.

The northern Senga Bay granites have a faint north-westerly lineation but in the south the lineation trends east-west. Local hybridised syenite is probably due to the almost complete assimilation of mafic material and in places the granites have been greisenised. The intrusive sequence is: (i) mafic (?) dykes, later brecciated; (ii) hybrid syenite; and (iii) leucogranite dykes.

Cape Maclear Granitic Complex

This complex has been described in some detail by Bloomfield (1965b) and Dawson and Kirkpatrick (in press). It consists essentially of an oval-shape central core of biotite-granite surrounded by a discontinuous outer ring of quartz-syenite with a horseshoe-shaped outcrop pattern; the complex as a whole has an overall north-westerly elongation.

There is a strong lineation and, locally, a sub-vertical planar foliation, defined by mineral grains and mafic schlieren, in the outer syenite parallel to the margins of the body. The outer contact is not exposed and may be represented by a zone of assimilation.

The marginal phase of the granite is, in places, fine-grained and possibly chilled but there does not appear to be a sharp junction with the central granite. Hybridised mafic inclusions and xenoliths are common in the granite and one offshore island is made up of strongly contaminated adamellite.

The sequence of events appears to be: (i) ? basic intrusions; (ii) intrusion of a ring-dyke of hybridised syenite; (iii) intrusion of central granite; (iv) intrusion of aplites and pegmatites; and finally, (v) intrusion of microdiorite dykes locally accompanied by net-veining and rheomorphism.

Nkudzi Bay Complex

Some 20 miles south-east of Cape Maclear and on the lake shore, is the Nkudzi Bay Complex (ibid.) consisting of a central mass of quartz-syenite separated from a broad ring of granite by a discontinuous screen of gneiss. The outer contact of the complex is marked by a wide zone of igneous breccia and there are many xenoliths of gneiss in the granite. The central syenite contains fewer inclusions and is less contaminated than the Cape

Maclear rocks. Field relationships suggest that the first intrusions were thick ring-dykes followed in one case by a central plug and in the other by a second ring-dyke.

A north-west-trending swarm of microgranites preceded the emplacement of both the Cape Maclear and Nkudzi Bay plutons but there is also a series of net-veined microdiorites of possible early Karroo age.

It has been noted that both the Cape Maclear and Nkudzi Bay Complexes lie within a zone which, projected north-westwards, includes the Senga Bay granites and that all these plutons lie along the line of major rift faults (*ibid.*). South-eastwards this zone also includes the minor granitic bodies near Fort Johnston (see below).

Other granites

Two small bodies lie on the floor of the rift valley near Fort Johnston and one of these has a marked ring-form (Holt, 1961, p. 33). Another large, oval-shaped pluton occupies the core of a major closed synform in paragneisses trending north-north-west and makes up Pirilongwe Hill (see Fig. 10) (Walshaw, 1965, p. 72).

At Pirilongwe there are local intrusive contacts but in most places the junction is transitional with a wide zone of migmatites. Much of the body is massive with a weak mineral lineation developed near the margins. Inclusions comprise mafic xenoliths (? original dolerites), granitised gneisses and dark skialiths.

A small body of intrusive granite associated with microgranite and microtonalite dykes occurs at Domasi, near Zomba (see Fig. 4) (Bloomfield, 1965d, p. 88). This is clearly a post-kinematic pluton and in view of the association of similar rock types, it may be a member of the Lake Malawi Province, although the single age determination (*ibid.*) is ambiguous.

The massive, unfoliated Domasi Granite contains xenoliths of foliated microgranite and is intersected by dykes of lamprophyre and sölvsbergite of Mesozoic age. The associated microgranites form both dykes and larger intrusions in which strong flow foliation and shearing are common. A number of related biotite-microtonalites are net-veined and brecciated by microgranite. The following age relationships are indicated: (*i*) microtonalite; (*ii*) microgranite; (*iii*) Domasi Granite; (*iv*) Mesozoic dykes.

PETROGRAPHY AND CHEMISTRY

Granites and quartz-syenites

Most of the granites of the Lake Malawi Province are typical sub-solvus types in which biotite is the usual mafic mineral, partly replaced by muscovite and accompanied by hornblende. Microcline, some perthitic, is the dominant feldspar together with some albite/oligoclase; microclinisation of plagioclase is common. Accessories include sphene, zircon, titanomagnetite and epidote.

The Senga Bay quartz-syenites are simply silica-deficient phases of the granites but elsewhere they form distinctive units. Clinopyroxene, horn-blende and biotite occur in clusters and there are small phenocrysts of microperthite. Oligoclase is an additional feldspar.

The granites and syenites are chemically normal sodi-potassic varieties (Harpum, 1963) although some specimens at Pirilongwe and Cape Maclear are close to adamellites. In terms of Nockolds' averages (1954, p. 1012) all could be classified as slightly more calc-alkaline than alkaline.

Inclusions in the granitic rocks

Melanocratic inclusions in the Pirilongwe Granite are rich in hornblende and biotite; hypersthene-bearing xenoliths at Nkudzi Bay are probably mafic charnockitic granulites. The xenoliths of porphyritic hornblende-biotite-microtonalite and quartz-microdiorite in the Cape Maclear Granite, in places further microclinised to produce microgranites, may represent original dolerites. Similar biotite-hornblende-melasyenites at Senga Bay probably also have a doleritic parentage.

Minor intrusions

Appinites intersect the Senga Bay Granite, and dykes of biotite-microcline-microgranite with pegmatites and aplites are common at Cape Maclear and Nkudzi Bay. In addition, microtonalites are associated with the Domasi Granite and have an andesitic chemical composition ranging from relatively unaltered rocks containing bronzite and clinopyroxene to altered types with biotite and hornblende. Associated hornblende-biotite-microgranites are calc-alkaline although they are rich in small microcline phenocrysts; sphene, titanomagnetite, zircon, apatite and allanite are abundant and distinctive accessory minerals.

The very strong chemical resemblance between the microcline-granite minor intrusions and the granites forming the outer ring of the Ntonya Ring-Complex has been noted (see p. 139).

PETROGENESIS

The isolated Likoma graphic granite (Fig. 10) has a texture indicating crystallisation from a melt but its regional significance is not apparent. The Pirilongwe Granite was probably formed by local, post-metamorphic anatexis and mobilisation within the core of a synform but is probably late- rather than post-kinematic. It may well be of somewhat different origin from the other granites in the same general area.

The lakeshore granites are typical fairly high-level, post-orogenic bodies using the usual petro-tectonic classification (Harpum, 1961). The general north-north-west alignment of the plutons along the western shoreline of Lake Malawi may be due to emplacement within a zone of weakness parallel to the general 'Mozambiquian' trend, marking a final intrusive phase in the complex history of the orogenic belt in this area. Subsequent rift movements thus followed an already well-marked tectonic line.

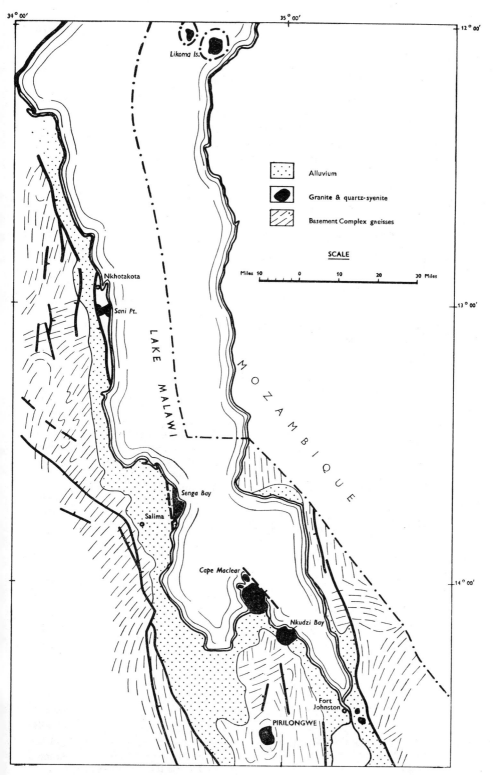

FIG. 10. Lake Malawi Granitic Province. Strong lines are faults of the rift system; tick indicates downthrow side.

TABLE 2

A comparison between orogenic and epeirogenic plutonism

Southern Malawi is the type area of the Chilwa Alkaline Province, an epeiro-genic group of plutons, volcanic vents and minor intrusions of late Jurassic to Cretaceous age (see Fig. 4); characteristic rock types are foyaites, syenites, alkaline granites and carbonatites (Bloomfield, 1965d; Garson, 1966). Thus, since this area is also the site of much of the earlier orogenic plutonism

Orogenic Plutons	*Epeirogenic Plutons*
A. *Field characteristics*	
1. Gradational or intrusive contacts	1. Sharp intrusive contacts
2. Structural association with large bodies of mafic and ultramafic rock	2. Locally associated with small amounts of thermally metamorphosed basic volcanics
3. No associated carbonatites	3. Commonly associated with peripheral carbonatite and agglomerate vents
4. Formed under orogenic conditions, partly by diapiric movement; strong structural control	4. Emplaced by cauldron subsidence under epeirogenic conditions; structural control not always apparent but related to rift faulting
B. *Petrographic characteristics*	
1. Feldspars: principally mesoperthite, some oligoclase	1. Feldspars: film-, vein- and patch-microperthite and albite
2. Nepheline absent	2. Nepheline-syenites common
3. Mafic minerals: diopside, hypersthene, hornblende, biotite	3. Mafic minerals: olivine, titanaugite, sodic pyroxenes, sodic amphiboles, lepidomelane
4. Allanite predominant accessory in granitic rocks	4. Pyrochlore predominant accessory in granitic rocks
5. Pegmatites either barren or contain recrystallised mafic minerals from country rocks	5. Pegmatites contain U and Th minerals
C. *Chemical characteristics*	
1. Usually calc-alkaline	1. Alkaline and peralkaline
2. $K > Na$	2. $Na > K$
3. Ba (>2000 ppm) and Sr (av. 1700 ppm) unusually abundant	3. Ba (av. 300 ppm) and Sr (av. 150 ppm) less than is normal for granitic rocks
4. Average contents	4. Average contents
Cr 28 ppm	Cr 12 ppm
Ni 42 ppm	Ni $<$ 10 ppm
Co 15 ppm	Co $<$ 10 ppm
V 74 ppm	V 22 ppm
Be 8 ppm	Be 22 ppm
Y 60 ppm	Y 190 ppm
5. Accessory zircons contain rather more Ni than those of epeirogenic rocks	5. Accessory zircons contain more Ca, Y, Mn, Fe and a little more Be and Th than those of orogenic plutons

described in this paper, it affords an excellent opportunity for a comparative study of the end-products of the two tectonic processes. The field, petrographic and chemical characteristics of the plutonic members of the two groups have been discussed elsewhere (Bloomfield, 1965d, p. 105) and Table 2 summarises the results.

Apart from the differences in structural setting there is a very marked geochemical contrast between the two types of plutonism in this area. Orogenic activity is calc-alkaline in character whilst epeirogenic activity is alkaline and predominantly sodic. Ba and Sr are somewhat enriched in the older rocks but the younger carry larger amounts of Be, Y, U and Th.

Summary and conclusions

That segment of the Mozambique Belt making up the greater part of Malawi shows evidence of several types of orogenic plutonism and these have been classified on a petro-tectonic basis. Pre-, syn- and post-kinematic activity are recognised but, since the 'Mozambiquian' is demonstrably polyphase it is difficult to relate them to the 'orogeny' as a whole. There could, for example, be syn-kinematic plutonism of two different ages within the wide time-span during which the orogenic belt was formed.

Pre-kinematic plutonic activity is represented by rare meta-anorthosites and small ultramafic bodies, emplaced in a pre-metamorphic sedimentary series and now isofacial with the surrounding supracrustal gneisses and granulites. The North Nyasa Alkaline Province is represented in Malawi by an early kinematic suite of predominantly metaluminous magmatic nepheline-syenites affected by Mozambiquian tectonism, principally cataclasis. Nephelinisation, partly by unmixing of perthite, is invoked.

Two major calc-alkaline granite batholiths (Nyika and Dzalanyama) were probably emplaced some 1300 m.y. ago but they too have been affected by later Mozambiquian tectonism which resulted in extensive shearing, foliation and epidotisation. These granites had a marked effect on the regional tectonic pattern.

A broadly syn-kinematic group of perthitic syenites and gneisses forms a very distinctive province within the supracrustal rocks of southern Malawi. Two large 'perthitic complexes' occur within charnockitic granulites in structural interference zones and there are also numerous broad lenses and concordant bands within amphibolite facies rocks. The syenitic rocks were formed after the main phase of syn-kinematic migmatisation by an influx of replacive tectono-genetic alkaline material (K, Na, Ba, Al) into the supracrustal gneisses and are characterised by sodi-potassic mesoperthites. The sequence of formation was: perthite-augen gneiss, perthite-gneiss, anatectic syenite. The rocks are either alkaline or calc-alkaline, depending upon the composition of the country gneisses and there are monzonitic and granitic phases. An association between perthitisation and tectonism is evident, and in one example a narrow band of perthite-gneiss is coincident with a proto-rift zone. Selective feldspathisation of relatively permeable bands in the supracrustal gneisses took place after the acme of regional metamorphism, penecontemporaneous with north-west-trending (? isoclinal) folding. Incomplete expulsion of replaced bases produced local volume increases which caused dilation, giving rise to lenticular masses of perthitic syenite and gneiss elongated along the strike. North-east cross-folding resulted in the formation of large masses of anatectic syenite, locally mobilised, in the interference zones. The perthitisation process was post-metamorphic but syn- and late-kinematic.

The perthitic syenites and gneisses were contemporaneous with a number of unusual infracrustal ring-complexes that have cores of metamorphosed ultramafic rock and peripheral calc-alkaline syenites and granites. They take the form of closed synforms. The ultrabasic bodies were emplaced as early syn-kinematic pipe-like plutons, one line following an east-west megashear in the cratonic basement, whereas others occur as isolated masses infilling *en echelon* tension joints. During the closing stages of regional metamorphism and before the final phases of folding, dominantly potassic hydrous metasomatism, accompanied by partial desilication, was locally active to produce biotite-metapyroxenites which were partly rheomorphic. This activity was followed by a more general influx of K with Si, Al and Na, all derived by partial melting of the supracrustal rocks and not from a magmatic source. Alkaline material permeated the ultramafic rocks and adjacent gneisses, rising around and over the plutons, to produce marginal syenogabbro hybrids and, subsequently, perthitic syenites. The syenites were locally sufficiently fluid to intrude the surrounding rocks as anatectic melts, in places along cone-sheet type fractures in the roof-zones caused by upward regenerative movement of rheomorphic pyroxenite. The final stage was the formation of palingenetic granite at two of the complexes and this was either intruded as a thick cone-sheet around a central mass of quartz-syenite or made up a silica-rich marginal phase. The present differences between the complexes are due to differing structural levels.

The alkaline material that gave rise to the perthitic syenites and granites of the ring-complexes was derived from the same tectono-genetic source as that which produced the widespread late-kinematic perthitic rocks of southern Malawi. This is another instance of the structural control of perthitisation: there is no consanguinity with the ultramafic rocks.

A final phase of post-kinematic plutonic activity within the Mozambique Belt is represented by the Lake Malawi Granitic Province. These structurally high-level, calc-alkaline plutons were preceded by basic minor intrusions and are locally extensively hybridised. Granitic, adamellitic and syenitic phases are found with associated minor intrusions of microgranite and microtonalite, and a number of the complexes have a distinctive ring-form slightly elongated north-north-west, parallel to the western shoreline of Lake Malawi. They appear to follow a proto-rift zone developed during the closing stages of the Mozambiquian Orogeny. It may well be that the Lake Malawi Province is derived from the rocks of structurally the highest level in the infracrustal ring-complexes.

Following Clifford (1967), I have suggested that the North Nyasa Alkaline Province was emplaced during the Katangan Episode of the polyphase Mozambiquian Orogeny and that the syenitic rocks and infracrustal ring-complexes were formed during the succeeding Damaran Episode (Bloomfield, 1968). The Lake Malawi Granitic Province is essentially immediately post-orogenic although related to the Mozambiquian tectonism.

Several geochemical provinces of different ages are represented in Malawi and may well reflect the chemical composition of the supracrustal rocks with which they are associated and from which they were probably derived. The formation of perthitic syenites depleted the gneisses in alkalies, particularly in potash, and this could be the reason for the present overall cafemic composition of the crystalline basement in this area. The later epeirogenic Mesozoic alkaline plutons are notably dissimilar and are characterised by soda.

The reason why syenites should predominate over granites in both the older and younger geochemical provinces may be a direct reflection of the silica-poor composition of the supracrustal rocks, itself probably the result of original primary sedimentary differentiation.

Acknowledgements

This review is based on research carried out by the writer and former colleagues of the Malawi Geological Survey. Grateful thanks are due to: Mr F. Habgood, Director, for permission to utilise the results of this work; Dr T. N. Clifford and Dr J. R. Vail of Leeds University for encouragement and friendly advice at all stages; and Mr J. H. Bateson and Dr E. A. Stephens who kindly read through the manuscript and made a number of constructive suggestions. Finally, sincere tribute is paid to my former teacher, Emeritus Professor W. Q. Kennedy, F.R.S., whose broad geological knowledge and flair for the synoptic view have been a constant source of inspiration. Figs. 1, 2, 3, 8 and 9 are reproduced by kind permission of the Director of Geological Survey, Zomba, Malawi.

REFERENCES

BLOOMFIELD, K. 1958. The Chimwadzulu Hill ultrabasic body, southern Nyasaland. *Trans. geol. Soc. S. Afr.*, **57**, 173.

—— 1959. The geology of Ilomba Hill, Karonga District. *Ann. Rep. geol. Surv. Nyasaland* (for 1958), 30.

—— 1965a. A reconnaissance survey of alkaline rocks in the northern and central regions. *Rec. geol. Surv. Malawi*, **5**, 17.

—— 1965b. The Cape Maclear Granitic Complex. *Rec. geol. Surv. Malawi*, **5**, 64.

—— 1965c. Infracrustal ring-complexes of southern Malawi. *Mem. geol. Surv. Malawi*, **4**.

—— 1965d. The geology of the Zomba area. *Bull. geol. Surv. Malawi*, **16**.

—— 1966. A major east-north-east dislocation zone in central Malawi. *Nature, Lond.*, **211**, 612.

—— 1968. The pre-Karroo geology of Malawi. *Mem. geol. Surv. Malawi*, **5**.

—— and GARSON, M. S. 1965. The geology of the Kirk Range-Lisungwe Valley area. *Bull. geol. Surv. Malawi*, **17**.

Boulanger, J. 1959. Les anorthosites de Madagascar. *Annls. géol. Madagascar*, **26**.

Cahen, L., and Snelling, N. J. 1966. *The geochronology of equatorial Africa*, Amsterdam.

Clifford, T. N. 1967. The Damaran episode in the Upper Proterozoic–Lower Paleozoic structural history of southern Africa. *Spec. Pap. geol. Soc. Am.*, **92**.

Combe, A. D., and Holmes, A. 1945. The kalsilite-bearing lavas of Kabirenge and Lyakuli, south-west Uganda. *Trans. r. Soc. Edinb.*, **61**, 359.

Dawson, A. L. 1966. The geology of the area west of Lake Chiuta. *Rec. geol. Surv. Malawi*, **6**, 15.

—— and Kirkpatrick, I. M. (in press). The geology of the Cape Maclear peninsula–lower Bwanje Valley area. *Bull. geol. Surv. Malawi*.

De Sitter, L. U. 1956. *Structural geology*, London.

De Swardt, A. M. J., Garrard, P., and Simpson, J. G. 1965. Major zones of transcurrent dislocation and superposition of orogenic belts in parts of central Africa. *Bull. geol. Soc. Am.*, **76**, 89.

Dixey, F. 1956. The East African Rift System. *Bull. colon. Geol. Mineral Resources*, Suppl., **1**.

Evans, R. K. 1965. The geology of the Shire Highlands. *Bull. geol. Surv. Malawi*, **18**.

Garson, M. S. 1960. The geology of the Lake Chilwa area. *Bull. geol. Surv. Nyasaland*, **12**.

—— 1965. The geology of the south Mlanje area. *Rec. geol. Surv. Malawi*, **5**, 5.

—— 1966. Carbonatites in Malawi. In *Carbonatites*, p. 33. (Eds. O. F. Tuttle and J. Gittins), London.

—— and Walshaw, R. D. (in press). The geology of the Mlanje area. *Bull. geol. Surv. Malawi*.

Habgood, F. (in press). The geology of the Cholo area. *Bull. geol. Surv. Malawi*.

Hamilton, E. I., and Deans, T. 1963. Isotopic composition of strontium in some African carbonatites and limestones and in some strontium minerals. *Nature, Lond.*, **198**, 776.

Harpum, J. R. 1955. Recent investigations in pre-Karroo geology in Tanganyika. *C.R. Ass. Serv. géol. Afr.*, Nairobi Meeting (1954), 165.

—— 1961. Granitic and metamorphic associations in Tanganyika. *Int. geol. Congr.*, **21** (26), 42.

—— 1963. Petrographic classification of granitic rocks in Tanganyika by partial chemical analysis. *Rec. geol. Surv. Tanganyika*, **10**, 80.

Holmes, A., and Harwood, H. F. 1937. The petrology of the volcanic field of Bufumbira, south-west Uganda. *Mem. geol. Surv. Uganda*, **3** (2).

Holt, D. N. 1961. The geology of part of Fort Johnston District, east of Lake Nyasa. *Rec. geol. Surv. Nyasaland*, **1**, 23.

Horne, R. G. 1961. Nachendezwaya carbonatite. *Rec. geol. Surv. Tanganyika*, **9**, 37.

Kennedy, W. Q. 1951. Sedimentary differentiation as a factor in Moine-Torridonian correlation. *Geol. Mag.*, **88**, 257.

—— 1965. The influence of basement structure on the evolution of the coastal (Mesozoic and Tertiary) basins of Africa. In *Salt basins around Africa*, p. 7. Inst. Petrol., London.

Kirkpatrick, I. M. 1965. Asbestos deposits of the Likudzi River area, Ncheu District. *Rec. geol. Surv. Malawi*, **5**, 121.

Morel, S. W. 1955. Biotitite in the Basement Complex of southern Nyasaland. *Geol. Mag.*, **92**, 241.

—— 1958. The geology of the middle Shire area. *Bull. geol. Surv. Nyasaland*, **10**.

—— 1961. Pre-Cambrian perthosites in Nyasaland. *Geol. Mag.*, **98**, 235.

Nockolds, S. R. 1954. Average chemical compositions of some igneous rocks. *Bull. geol. Soc. Am.*, **65**, 1007.

Peters, E. R. 1966. The geology of part of the Kasungu area. *Rec. geol. Surv. Malawi*, **6**, 9.

Ramsay, J. G. 1962. Interference patterns produced by the superposition of folds of similar type. *J. Geol.*, **70**, 466.

Read, H. H. 1955. Granite series in mobile belts. *Spec. Pap. geol. Soc. Am.*, **62**, 409.

Sampson, D. N., and Wright, A. E. 1964. The geology of the Uluguru Mountains. *Bull. geol. Surv. Tanzania*, **37**.

SNELLING, N. J. 1962. Age determination unit. In *Ann. Rep. Overseas geol. Survs.* (for 1960–61), p. 27.

—— 1963. Age determination unit. In *Ann. Rep. Overseas geol. Survs.* (for 1962), p. 30.

—— 1965. Age determination unit. In *Ann. Rep. Overseas geol. Survs.* (for 1964), p. 28.

TERMIER, H., and TERMIER, G. 1956. *L'evolution de la lithosphère: I. Pétrogénèse*, Paris.

THATCHER, E. C. (in press). The geology of the Dedza area. *Bull. geol. Surv. Malawi*, **29**.

—— and WALTER, M. J. 1968. The geology of the south Lilongwe Plain and Dzalanyama Range. *Bull. geol. Surv. Malawi*, **23**.

—— and WILDERSPIN, K. E. (in press). The geology of the Mchinji–Upper Bua area. *Bull. geol. Surv. Malawi*, **24**.

VAIL, J. R. 1965a. Distribution of radiogenic ages in eastern central Africa. *9th Ann. Rep. Res. Inst. African Geol., Univ. Leeds*, 27.

—— 1965b. An outline of the geochronology of the late-Precambrian formations of eastern central Africa. *Proc. r. Soc. Lond.*, **284**, Ser. A, 354.

—— and PINTO, M. S. 1966. A Série do Fingoè: contribuiçaõ para o estudo das rochas da area do Fingoè (Tete, Mocambique). *Bol. Serv. Geol. Min. Mozambique*, **33**, 33.

WALSHAW, R. D. 1965. The geology of the Ncheu–Balaka area. *Bull. geol. Surv. Malawi*, **19**.

O. VON KNORRING

8   Mineralogical and geochemical
    aspects of pegmatites from
    orogenic belts of equatorial
    and southern Africa

ABSTRACT. *The mineralogy and geochemistry of pegmatites from the
Kibaran (including Karagwe-Ankolean–Burundian) and Damaran-
Mozambiquian Orogenic Belts of equatorial and southern Africa are described. Although pegmatites of economic importance occur in the older
cratonic areas in this part of Africa, those of the orogenic zones show a
more unusual and varied composition and many constitute unique deposits
of rare-element minerals.*

Introduction

In the introduction to the Ninth Annual Report on Scientific Results of the
Research Institute of African Geology at Leeds, Professor W. Q. Kennedy
outlined some of the geological research problems connected with the
Institute's activity as follows:

1. An investigation of the basement structure of Africa aimed towards
   the recognition of the major structural elements, referring more
   particularly, in the first place at least, to a study of the nature, distribution and tectonics of the late Precambrian 'Katanga System' and its
   correlates elsewhere within the continent.
2. A study of the geochemical provinces of Africa based initially on the
   evidence of the Younger Granites together with pegmatites and
   eventually the basement granites and hydrothermal mineral deposits
   of all ages.

Although the major pegmatite regions of Africa have long been known,
large-scale exploitation of the deposits only began during the Second World
War. They are important because they constitute major world resources of
many rare-metal elements such as lithium, rubidium, caesium, beryllium,
scandium, rare earths, niobium and tantalum in addition to industrial
minerals such as mica and many precious and semi-precious gemstones.

Since 1957 I have been associated with the study of pegmatites in Uganda,
Rwanda, Rhodesia, Mozambique, Madagascar, and South-West Africa. In
this paper an attempt is made to describe and correlate the mineralogy and
geochemistry of pegmatite provinces which fall within the two main orogenic zones of equatorial and southern Africa: the Karagwe-Ankolean–
Burundian–Kibaran Orogenic Belt affected by orogenesis *c.* 1100 ± 200 m.y.
ago; and the Damaran and Mozambique Belts which suffered orogenesis
550 ± 100 m.y. ago. A vast number of new age determinations have been

made on African rocks and minerals during the past 10 years and the results have been admirably presented by Cahen and Snelling (1966). A surprisingly large number of ages have been determined for pegmatite minerals from various parts of Africa and they indicate the time of mineralisation during

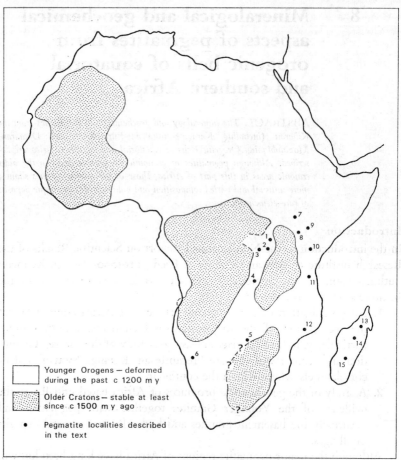

FIG. 1: 1, South-west Uganda; 2, Rwanda; 3, Eastern Congo; 4, Manono, Katanga; 5, Kamativi, Rhodesia; 6, Karibib–Usakos area, South-West Africa; 7, North-east Uganda; 8, West Suk, Kenya; 9, Baragoi, Kenya; 10, Machakos, Kenya; 11, Uluguru Mountains, Tanzania; 12, Alto Ligonha, Mozambique; 13, Tsaratanana region, Madagascar; 14, Sahatany region, Madagascar; 15, Ampandramaika–Malakialina region, Madagascar. Structural units from Clifford (1966).

the various episodes of the orogenic belts. The pegmatite minerals almost always show a younger apparent age than that of the actual orogeny within these belts, and the mineralisation appears to be connected with a distinct group of post-orogenic granites which are frequently pegmatitic in character.

In his impressive treatise on *The Mineral Resources of Africa*, De Kun (1965) has paid particular attention to the mineralogenetic provinces and their relation to the structural pattern of the African continent; he established ten mineral or orogenic-metallogenetic belts. Clifford (1966) in dealing with major structural events of the continent, has introduced the term 'tectono-metallogenic units' for regions of structural kinship characterised by major concentrations of specific economic mineral deposits. In Africa he recognises two such units: (i) *Younger Orogens*, including zones affected by orogenesis in Alpine, Hercynian, late Precambrian–early Palaeozoic and Kibaran times; and (ii) *Older Cratons* with a record of older orogenesis but which remained stable throughout those younger periods of orogenesis (see Fig. 1). Under group (i) Clifford (1966) includes beryllium, tin, tungsten and niobium-tantalum mineralisation—in other words, the major elements of the pegmatites.

Kennedy attached great importance to the distribution of pegmatites in Africa and maintained that by investigating pegmatite areas in great detail it should be possible to delineate geochemical provinces for certain rare elements. Therefore, in studying African pegmatites particular attention has been paid to their mineralogy and geochemistry. In the course of this work many rare minerals, in some cases even new mineral species or varieties, have been identified and studied in great detail. The chemistry of the columbite-tantalite group of minerals has been investigated in particular detail in order to establish the general distribution and possible content of rarer elements camouflaged in this widespread mineral group.

Pegmatites in 1100 ± 200 m.y. old orogenic belts

SOUTH-WEST UGANDA

Introduction

In connection with the pegmatite research programme at Leeds, most of the economically important pegmatites in Uganda (Fig. 1, loc. 1) have been examined since 1951.

Large-scale pegmatite mining first expanded in Uganda in the early 1940s and since that time has been concentrated in the south-western part of the country. Previously, cassiterite was mined from quartz-muscovite veins and some associated pegmatites, but rare-metal minerals, like beryl and columbite-tantalite, were only sporadically recovered. Prospecting was greatly stimulated in 1950 by a marked increase in the price of columbite, cassiterite and wolframite and production reached its peak in 1954–55. During that period old pegmatites were re-examined and many new mineralised deposits were discovered. After 1955 the demand for beryl increased and since that time south-west Uganda has been one of the world's major beryl producers.

Geology and structural control

The Karagwe-Ankolean rocks (see Cahen, this volume, p. 98) of south-west Uganda and adjoining parts of Rwanda and Tanzania are represented by an extensive group of sediments, mainly argillites and quartzites (Combe and

Groves, 1932). The sediments were laid down in water of shallow to moderate depths and subsequently deformed and metamorphosed during the
Kibaran Orogeny some 1300 m.y. ago. Two main phases of deformation
have been recognised. During the first, according to Reedman (1967), broad
zones of synclinorial or anticlinorial structures were developed, followed by
the emplacement of syn-kinematic granites at the centres of elongated domes
in the anticlinoria. During the second phase, the granitoid rocks rose into
their cover accentuating the domal structures by refolding of the mantling
sediments. Later post-tectonic granites were intruded into these domes and/or
the mantling sediments; most of the pegmatites were emplaced during this
episode. It is not, of course, surprising that mineralised pegmatites should be
found in connection with gneiss/granite structures of this type in orogenic
zones where repeated granitisation and subsequent enrichment of minor rare-
element constituents has taken place over a prolonged period of time. The
resulting pegmatites of the post-tectonic phase may occur in the form of
distinct units within the domes, but more commonly they tend to be located
in the mantling schists, amphibolites, quartzites and calc-silicate rocks.
Frequently, the post-tectonic pegmatites have invaded the country rock
to such an extent that banded gneisses and migmatites have been formed; in
this case, large areas may be mineralised, carrying the typical pegmatite
minerals as accessories. Economically, pegmatite mineralisation of this kind
can be very important but it is generally less known and commonly overlooked. Barnes and Meal (1959) have indicated that in south-west Uganda,
pegmatites are found around granitic domes which are well foliated (gneissose
in character) and have well-domed roofs. Pegmatites do not occur around
stock-like unfoliated granites, nor are they found surrounding gently domed
bodies such as the Masha granite 'arena' in Ankole.

In south-west Uganda and especially in the Ankole district, pegmatites
are mainly found within the aureole of mica schists surrounding granite-
gneiss bodies, thus corresponding to the exterior pegmatites of Gevers
(1936). Some pegmatites also occur within the granites themselves, but always
close to their margins. Of particular interest in this regard is the marked
concentration of mineralised pegmatites in an area bounded by the granitic
domes of Ntungamo, Rushenyi, Lugalama and Karenge, south-west of the
township of Ntungamo (30° 15′ E., 0° 52′ S.). Here the important rare-
element pegmatites have yielded the bulk of the beryl and columbite-tantalite
mined in south-west Uganda.

The pegmatites are frequently kaolinised to such a degree that the characteristic zoning is not always apparent; this is particularly the case with
many pegmatites in mica schists. In contrast, pegmatites emplaced in granites,
gneisses, amphibolites, and in a few cases in diopsidic rocks, are often remarkably fresh.

As the rare-element mineralisation constitutes the most important economic aspect of pegmatites, their classification may conveniently be based on
certain type minerals. For south-west Uganda the following classification is
proposed:

1. Beryl + columbite ± cassiterite
2. Beryl + columbite ± cassiterite + amblygonite ± manganese-iron phosphates
3. Beryl + columbite ± cassiterite ± lithium-phosphates + lithium silicates (spodumene, lepidolite, etc.)
4. Cassiterite + quartz + muscovite ± beryl

The majority of pegmatites in the area may be termed beryllium pegmatites, beryl being the ubiquitous rare-metal mineral. In addition varying amounts of columbite-tantalite and other niobium-tantalum minerals occur, for example, microlite, tapiolite and wodginite. Cassiterite is present in smaller amounts; and larger concentrations of this mineral are found more commonly in specific cassiterite-bearing quartz-muscovite veins (association no. 4, above).

Mineralogy and geochemistry

Quartz is one of the major minerals in most pegmatites and forms a prominent core; moreover, in tropical countries, where weathering has been extensive, these resistant quartz cores are excellent indicators of possible pegmatite deposits. Quartz crystals are, on the whole, rare in the pegmatites; at Kabira (30° 1′ E., 0° 42′ S.), however, outlines of giant, partly rose-coloured crystals several metres across do occur. This is of particular interest since crystals of rose quartz are exceptionally rare. Amethystine quartz has been observed in fissure veins at Rwemeriro (30° 11′ E., 1° 0′ S.), and semi-precious bluish opaline quartz is sometimes associated with quartz cores. There is a tendency for economic minerals to occur within or close to the core or to vein quartz; for example amblygonite, lepidolite, beryl, columbite-tantalite and tapiolite.

In most pegmatite mines the *feldspar* is partially or totally decomposed to kaolinite, pseudomorphs of which can be seen projecting into the surrounding muscovite and quartz. Such decomposition cannot be attributed to superficial weathering alone, for in some cases both fresh and altered feldspar have been seen enclosed in glassy quartz, indicating alteration prior to, or simultaneous with, the crystallisation of the quartz. It seems that the primary alteration may have been initiated during the hydrothermal stage and continued during succeeding low temperature events and finally completed by weathering. At Bulema mine (29° 44′ E., 0° 50′ S.) in Kigezi, apart from pinkish microcline, several types of albite have been noted and a conspicuous red albite with curved and undulating cleavage planes appears to be connected with the main mineralisation. In general most beryl, cassiterite, columbite, tapiolite and microlite are associated with the albite phase.

Muscovite is the main *mica* in all the mineralised pegmatites examined. Large books have been recovered from many pegmatites in the central area south-west of Ntungamo. Most commonly medium- to coarse-grained, it occurs in large masses or vein-like bodies following the quartz core. This type of mica may be mineralised with cassiterite and a variety of niobium-tantalum minerals.

Biotite is rather common as a marginal constituent in the pegmatites of Kigezi province. Lithian muscovite and a purplish lithium-poor mica have been found at Rwemeriro in Ankole (30° 11′ E., 1° 0′ S.) (see Table 3, analysis 1).

Lithium is one of the more characteristic type elements of granite pegmatites and is particularly enriched in the differentiated pegmatites which often exhibit one or more specific lithium phases of mineralisation. The only lithium mineral of economic importance in south-west Uganda is amblygonite, which occurs in a number of pegmatites in the central area south-west of Ntungamo (30° 15′ E., 0° 52′ S.) and also in the northern area in the Kabira (30° 1′ E., 0° 42′ S.) region. Amblygonite is remarkably resistant to tropical weathering and boulders of it are useful indicators in tracing mineralised pegmatite deposits. During late-stage hydrothermal processes it has been altered in places to crandallite, apatite and wavellite.

In the Kabira region, large amounts of various manganese-iron phosphates have been identified; locally gigantic nodules, up to 10 tons or more, have been observed. Both amblygonite and these phosphates are generally confined to the margin of the quartz core or the core itself. At Kabira the following phosphates have been identified: triphylite-lithiophilite, heterosite, graftonite, alluaudite, arrojadite, fillowite, amblygonite, tavorite, lazulite, barbosalite, lipscombite, frondelite, hureaulite, strengite, clinostrengite (Table 1, analyses 1, 2, 4, 5 and 7). Most of these minerals appear to be centred around, and secondary after, original triphylite-lithiophilite, but independent nodules of alluaudite, arrojadite and graftonite are also present. Similar suites of phosphates have been recorded from the Sapucaia pegmatite in Minas Gerais, Brazil (Lindberg and Pecora, 1955), the Marivolanitra pegmatite in Madagascar (Behier, 1957) and various pegmatites in the Gatumba area of Rwanda (Von Knorring, 1965b).

Although the element lithium is clearly in evidence in the phosphate minerals, apart from spodumene, lithium silicates are rare. Even the spodumene laths which are in great abundance in many pegmatites are completely altered to kaolinite and montmorillonite or are, at times, replaced by muscovite and albite. The initial concentration of lithium in the spodumene pegmatites may have been very high, but due to later hydrothermal activity and albitisation, coincident with the formation of the spodumene, lithium was released; part of it was incorporated in the coloured beryls, tourmalines and micas, and the rest migrated outwards to be absorbed by the country rocks, occasionally forming a selvage of lithium-bearing biotite or a zone of holmquistite-bearing rock. In this regard, the coloured beryl at Rwemeriro mine (30° 11′ E., 1° 0′ S.) contains up to 1% lithium oxide, the green tourmaline over 1%, and the purple micas less than 1% (see Table 3, analyses 1 and 2). It should be noted that rose to purple coloured micas are often assumed to be lepidolites; but the lithium-poor purple micas in many pegmatites can be considered lithian muscovites, the purple colouration being due to a rather high content of manganese (Table 3, analysis 1).

During 1961 the *beryl* production of Uganda exceeded 1000 tons; a high percentage of this tonnage came from the pegmatites of south-west Uganda in which this mineral is widespread. It has been observed in almost all mineralised deposits and also as an accessory mineral in some post-kinematic granites. In addition, bertrandite, euclase and the beryllian margarite have been

TABLE 1

Chemical analyses of pegmatitic phosphates

	1	2	3	4	5	6	7
Al_2O_3	—	—	1·96	tr.	—	33·42	—
Fe_2O_3	1·31	—	1·34	1·98	11·93	—	0·82
FeO	17·61	28·37	19·36	19·41	28·30	0·98	17·77
MnO	25·72	24·61	19·45	26·50	13·87	0·76	29·62
MgO	—	—	2·80	tr.	—	tr.	—
CaO	—	5·50	2·28	4·80	—	8·36	—
BaO	—	—	5·36	—	—	rt.	—
Na_2O	0·05	—	4·97	5·65	—	0·34	—
K_2O	0·04	—	0·63	0·20	—	tr.	0·04
Li_2O	8·54	—	0·20	0·11	—	4·21	1·42
H_2O+	1·62	0·52	1·45	0·43	7·34	5·36	10·32
H_2O-	0·10	—	0·10	0·13	0·30	0·07	0·10
P_2O_5	43·11	40·80	39·34	40·70	37·10	45·34	39·38
F	—	—	0·47	—	—	1·68	—
Insol.	—	—	—	0·37	0·60	0·27	—
	98·10	99·80	99·71	100·28	99·44	100·79	99·47
−O for F			0·20			0·71	
			99·51			100·08	

1. Lithiophilite from Kabira pegmatite, Ankole, S.W. Uganda. Analyst, E. Padget.
2. Graftonite from Rwanza pegmatite, Kabira, Ankole, S.W. Uganda. Analyst. O. Von Knorring.
3. Barium-arrojadite, from Buranga pegmatite, Gatumba, Rwanda. Analyst, O. Von Knorring.
4. Fillowite from Kabira pegmatite, Ankole, S.W. Uganda. Analyst, O. Von Knorring.
5. Manganoan lipscombite from Rwanza mine, Kabira, Ankole, S.W. Uganda. Analyst, O. Von Knorring.
6. Bertossaite from Buranga mine, Gatumba, Rwanda. Analyst, O. Von Knorring.
7. Hureaulite from Kabira mine, Ankole, S.W. Uganda. Analyst, E. Padget.

recorded (Gallagher and Hawkes, 1966). Beryl often forms well-developed centimetre to metre sized crystals. Occasional large irregular masses, up to 50 tons, have been recognised as at Kabira. Coloured varieties have also been identified at Rwemeriro mine; these are generally lithium-, sodium- and caesium-bearing. Bertrandite, a product of late hydrothermal alteration of beryl, has been identified in the Bulema and Kihanda mines (29° 44′ E.,

0° 50′ S.) where the altered beryl is in the form of characteristic, bipyramidal tapering crystals, which are replaced by bertrandite and muscovite and albite.

Black *tourmaline* is ubiquitously associated with most pegmatites. In places strong enrichment of this mineral is observed in the exomorphic envelope whereas the tourmaline content within the pegmatites is moderate. In some mines however, large crystal aggregates are so abundant that the mineral becomes a dominant constituent. Apart from greenish tourmaline, polychrome varieties characteristic of many lithium pegmatites have not been observed. At Rwemeriro (30° 11′ E., 1° 0′ S.) a pistachio-green tourmaline, rather like epidote, is a common constituent of the cleavelandite replacement phase; this tourmaline has rather high Mn-, Li- and F-content (Table 3, analysis 2).

Rare-earth minerals are not very common in the Karagwe-Ankolean pegmatites. Apart from small amounts of xenotime and monazite and some euxenite associated with columbites, no other rare-earth minerals have been recorded.

Tin is a characteristic element of the Karagwe-Ankolean province of Uganda. Cassiterite was the main economic mineral mined in the past and south-west Uganda is commonly referred to as the 'tin province'. An indication that tin is a ubiquitous trace element in this region is shown by its presence not only in the form of cassiterite in many pegmatites but also as a minor constituent in the niobium and tantalum minerals. All columbites contain up to 0·5% SnO_2, tapiolites up to 1%, and a new variety ferro-wodginite 11% SnO_2.

Cassiterite occurrences may be divided into three main types: cassiterite-bearing quartz veins; cassiterite-bearing quartz-muscovite veins; and cassiterite-bearing beryl pegmatites. As already pointed out by Combe and Groves (1932), all the vein deposits of cassiterite are genetically connected with the granitic domes of the region. The deposits are generally situated along the margins of the domes, in the areas adjacent to them, or in the belts of metasediment separating them. The vein deposits occur in the schists, phyllites and quartzites almost all of which are highly tourmalinised in the vicinity of the veins. Numerous cassiterite prospects are found in Kigezi district in the Lake Mutanda area centred around the Mutanda dome (Reedman, 1967). These veins are infillings of open fractures and contain quartz and cassiterite with occasional crystals of ferberite or wolframite and rare sulphides. The cassiterite-bearing quartz-muscovite veins consist largely of quartz with muscovite of pegmatitic coarseness along the walls, in joints and pockets. Cassiterite is almost invariably associated with the mica-phases. All the major tin deposits of south-west Uganda belong to this type. In the third group, the cassiterite is an accessory mineral associated with columbite and beryl. It is found in some muscovite phases within the pegmatite and associated with albite replacement units. Although further south in Rwanda these deposits are of primary importance, in Uganda they are seldom mined for cassiterite alone.

Zircon is a constant associate of niobium-tantalum minerals. It is usually intergrown with, or has crystallised close to, aggregates of niobium-tantalum minerals. The colour of these zircons is generally brown and the crystals are often curved. Hafnium-rich varieties have been observed only at Bulema (29° 44′ E., 0° 50′ S.) in small miarolitic cavities associated with tapiolite and xenotime.

With beryl, the minerals of the *columbite-tantalite* series constitute the most widespread of rare-metal minerals in pegmatites. Special attention has been paid to the chemical composition of niobium-tantalum minerals. The results presented in Table 2 show that the columbites (analyses 1-14) are predominantly niobium-rich; and it would appear that tantalites (analysis 15) are rare. However, when tantalum is present in excess, other tantalum minerals are preferentially formed, for example manganotantalite, tapiolite and microlite. In the Karagwe-Ankolean region of south-west Uganda the columbites are variable in composition and mainly niobian, the Nb/Ta ratio varying from 12 : 1 to 1 : 1. In the majority of cases the iron content exceeds that of manganese but end-members, as far as these elements are concerned, are rare, the only exception being the columbite from the Kazumu mine (30° 20′ E., 0° 55′ S.) in Ankole, where the manganese content is unusually high (Table 2, analysis 9). There is a tendency for columbites from pegmatites with spodumene-rich quartz cores to be richer in manganese.

Manganotantalite is a characteristic mineral of pegmatites with a well-developed lithium phase and in south-west Uganda the mineral has been recorded from a little known pegmatite at Jemubi River (Table 2, analysis 21) in northern Ankole, and from a recently discovered occurrence at Bugonji, north of Ntungamo (30° 15′ E., 0° 52′ S.).

Tantalum-rich tapiolite is known from the Bulema mine (29° 44′ E., 0° 50′ S.) and from several pegmatites around Nyanga mine (30° 8′ E., 0° 58′ S.). An interesting association of tapiolite and microlite has been studied in some detail at Bulema where large, irregular pockets of brown microlite up to half a ton in weight contained remnants, and also perfect crystals, of tapiolite. In some instances the tapiolite-microlite pseudomorphs consist of a core of fresh tapiolite surrounded by a shell of radiating microlite. In addition, small cavities in microlite contain aggregates of tapiolite crystals encrusted with minute, white xenotime, pink hafnian zircon and quartz crystals. Tapiolite from the Nyanga area is intimately associated with a muscovite phase in quartz-muscovite pegmatites. At Nyanga no. 2 mine, tapiolite is found together with columbite in a muscovite-rich greisen; here tapiolite is confined to the quartz aggregates, and columbite is embedded in the muscovite. Chemically, tapiolites differ from columbite-tantalites in their high tantalum and iron contents (Table 2, analyses 30-32).

The tin-bearing tantalites, ixiolite and wodginite, have been commonly identified in the past with columbite and manganotantalite respectively. A new variety of wodginite, ferro-wodginite, has been identified at Nyanga pegmatite. It occurs in well-developed black shiny crystals of distinct sphenoidal habit and is intergrown with minute tapiolite crystals (Table 2, analysis 26) (Von Knorring *et al.*, 1969a).

TABLE 2

Chemical analyses of columbite-tantalites, manganotantalites, wodginites,
ixiolites and tapiolites from African pegmatites

	1	2	3	4	5	6	7	8
Nb_2O_5	70·3	68·47	68·0	67·5	66·78	64·30	63·0	58·00
Ta_2O_5	4·8	9·17	8·1	11·5	11·34	12·19	17·0	21·89
TiO_2	1·9	1·36	2·3	—	0·85	2·36	—	—
SnO_2	0·5	0·65	1·0	0·8	0·30	—	0·1	—
MnO	14·4	8·00	10·5	9·4	5·46	7·65	9·4	13·00
FeO	6·4	12·57	10·2	10·7	14·54	12·01	10·5	6·25
	98·3	100·22	100·1	99·9	99·27	98·51	100·0	99·14

	9	10	11	12	13	14	15	16
Nb_2O_5	57·0	51·42	50·3	50·22	48·7	39·7	30·3	22·68
Ta_2O_5	23·6	29·01	28·9	30·64	32·4	39·9	51·7	61·04
TiO_2	—	—	1·2	—	—	2·4	0·9	—
SnO_2	0·3	0·24	0·5	0·23	0·3	0·4	0·4	0·25
MnO	17·2	15·40	9·0	5·32	7·3	5·4	6·9	15·27
FeO	1·8	2·69	9·7	13·12	11·2	11·9	9·7	0·67
	99·9	98·76	99·6	99·53	99·9	99·7	98·9	99·91

	17	18	19	20	21	22	23	24
Nb_2O_5	18·33	37·44	29·43	21·00	10·60	3·81	3·24	0·98
Ta_2O_5	64·12	44·51	55·27	62·22	74·05	81·44	81·88	84·39
TiO_2	1·24	—	—	0·60	—	0·07	—	—
SnO_2	0·40	tr.	1·08	—	—	—	—	0·99
MnO	12·91	15·82	12·33	13·50	14·27	12·61	13·10	13·13
FeO	2·60	1·68	4·43	1·43	0·53	1·45	1·09	0·66
	99·60	99·45	99·54	98·75	99·45	99·38	99·31	100·15

	25	26	27	28	29	30	31	32
Nb_2O_5	0·70	7·72	4·76	25·83	21·77	7·29	5·90	3·38
Ta_2O_5	82·64	65·91	67·66	36·97	45·02	76·79	78·24	80·06
TiO_2	0·10	2·44	—	9·25	8·23	—	—	—
SnO_2	—	11·17	15·52	6·30	3·08	1·13	1·30	0·54
MnO	13·89	5·60	9·95	1·86	1·51	0·46	0·83	1·91
FeO	0·40	7·05	1·79	6·46	5·77	14·44	13·25	12·92
Fe_2O_3	—	—	—	5·76	8·29	—	—	—
Sc_2O_3	—	—	—	6·00	5·10	—	—	—
	97·73	99·89	99·68	98·43	98·77	100·11	99·52	98·81

Table 2. Localities of niobium-tantalum minerals:

1. Columbite, Rwemeriro pegmatite, Ankole, S.W. Uganda.
2. Columbite, Rwanza pegmatite, Kabira, Ankole, S.W. Uganda.
3. Columbite, Kabira pegmatite, Ankole, S.W. Uganda.

4. Columbite, Kabira pegmatite, Ankole, S.W. Uganda.
5. Columbite, Nyabushenyi pegmatite, Ankole, S.W. Uganda.
6. Columbite, Neu Schwaben pegmatite, Karibib, S.W. Africa.
7. Columbite, Nyabakweri pegmatite, Ankole, S.W. Uganda.
8. Columbite, Bukangari pegmatite, Kigezi, S.W. Uganda.
9. Columbite, Kazumu pegmatite, Ankole, S.W. Uganda.
10. Columbite, Nyabushoro pegmatite, Kigezi, S.W. Uganda.
11. Columbite, Kashozo pegmatite, Ankole, S.W. Uganda.
12. Columbite, Bulema pegmatite, Kigezi, S.W. Uganda.
13. Columbite, Kayonza pegmatite, Kigezi, S.W. Uganda.
14. Columbite, Nyanga pegmatite, Ankole, S.W. Uganda.
15. Tantalite, Bulema pegmatite, Kigezi, S.W. Uganda.
16. Tantalite, Muiane pegmatite, Alto Ligonha, Mozambique.
17. Tantalite, Morrua pegmatite, Alto Lingonha, Mozambique.
18. Manganotantalite, Buranga pegmatite, Gatumba, Rwanda.
19. Manganotantalite, Namirrapa pegmatite, Alto Ligonha, Mozambique.
20. Manganotantalite, Dernburg pegmatite, Karibib, S.W. Africa.
21. Manganotantalite, Jemubi River, Ankole, S.W. Uganda.
22. Manganotantalite, Monrepos pegmatite, Karibib, S.W. Africa.
23. Manganotantalite, Okangava Ost pegmatite, Karibib, S.W. Africa.
24. Manganotantalite, Morrua pegmatite, Alto Ligonha, Mozambique.
25. Manganotantalite, Morrua pegmatite, Alto Ligonha, Mozambique.
26. Ferro-wodginite, Nyanga pegmatite, Ankole, S.W. Uganda.
27. Wodginite, Meyer Kamp pegmatite, Karibib, S.W. Africa.
28. Scandian ixiolite, Muiane pegmatite, Alto Ligonha, Mozambique.
29. Scandian ixiolite, Betanimena pegmatite, Madagascar.
30. Tapiolite, Nyanga pegmatite, Ankole, S.W. Uganda.
31. Tapiolite, Muiane pegmatite, Alto Ligonha, Mozambique.
32. Tapiolite, Namarella pegmatite, Alto Ligonha, Mozambique.

Analysts, J. R. Baldwin and O. Von Knorring.

RWANDA

Introduction

The Burundian System of Rwanda is geologically very similar to the Karagwe-
Ankolean of neighbouring south-west Uganda (Fig. 1, loc. 2). The schists,
phyllites and quartzites have a northerly trend and are continuous with the
Karagwe-Ankolean rocks on the Uganda side of the border to the north.
Generally, the same types of pegmatites and cassiterite-bearing quartz veins
are found in Rwanda as in south-west Uganda, but cassiterite is much more
prominent in the Rwanda pegmatites and constitutes the main mineral
mined. In addition to tin, tungsten is characteristic of the northern and western
parts of Rwanda and in some quartz veins and pegmatites these elements
coexist. A most striking example of mineralised zoned pegmatites is centred
around Gatumba (29° 39′ E., 1° 55′ S.), some 50 km south of Ruhengeri;
here, a variety of large pegmatites are being mined for cassiterite and other
rare-metal minerals.

Mineralogy and geochemistry

Lithium is a characteristic type element in many of the Rwanda pegmatites.
However, due to later hydrothermal activity, most of the lithium in the

pegmatite bodies migrated outwards metasomatising the country rock and forming an exomorphic halo around the deposits. As a result of this migration, the lithium content is often higher outside than inside the pegmatite

TABLE 3

Chemical analyses of pegmatite minerals

	1	2	3	4	5	6	7	8	9	10	11	12
SiO_2	48·09	36·69	60·99	59·52	78·36	77·30	52·65	50·02	45·41	43·56	62·97	41·20
TiO_2	0·00	0·21	0·15	0·15	—	—	—	0·04	—	tr.	0·00	—
Al_2O_3	30·40	39·04	11·95	17·12	17·06	17·68	22·30	34·25	38·07	35·77	17·75	35·00
Fe_2O_3	0·45	—	2·85	0·77	0·04	0·02	0·02	0·25	0·18	0·39	tr.	—
FeO	1·97	3·47	5·10	0·70	—	—	—	0·05	—	—	—	—
MnO	1·71	1·66	0·27	0·05	—	0·03	0·38	0·07	0·19	0·45	0·01	—
MgO	0·31	0·46	12·43	11·37	—	tr.	2·76	0·45	0·30	1·82	0·00	—
CaO	0·00	0·00	0·17	0·38	—	tr.	2·19	0·18	0·51	0·38	0·12	—
Na_2O	0·06	2·58	0·15	0·07	0·07	0·09	0·25	0·03	0·15	0·05	0·86	—
K_2O	10·64	0·25	tr.	0·01	0·06	0·18	0·12	0·16	0·12	0·73	0·16	—
H_2O+	2·60	3·38	2·30	4·81	0·05	0·34	9·80	11·14	11·34	11·20	2·22	6·40
H_2O-	0·00	0·00	0·10	2·24	0·05	0·26	9·72	2·49	2·07	4·84	0·02	0·09
P_2O_5	0·06	—	—	—	—	—	—	0·01	0·06	—	—	—
F	5·14	1·46	—	0·39	—	—	—	—	—	—	—	—
B_2O_3	—	10·21	—	—	—	—	—	—	—	—	—	—
BeO	—	—	—	—	—	—	—	—	—	—	11·56	17·25
Li_2O	0·26	1·25	3·73	2·58	4·56	4·40	tr.	1·23	1·50	1·08	1·18	—
Cs_2O	—	—	—	—	—	—	—	—	—	—	3·01	—
	101·69	100·66	100·19	100·16	100·25	100·30	100·19	100·37	99·90	100·27	99·86	99·94
—O for F	2·16	0·61		0·16								
	99·53	100·05		100·00								

1. Globular purplish-brown mica from Rwemeriro pegmatite, Ankole, S.W. Uganda. Analyst, M. H. Kerr.

2. Pistachio green tourmaline from Rwemeriro pegmatite, Ankole, S.W. Uganda. Analyst, M. H. Kerr.

3. Blue holmquistite from Kirengo pegmatite, Rwanda. Analyst, O. Von Knorring.

4. White holmquistite from Kirengo pegmatite, Rwanda. Analyst, O. Von Knorring.

5. White, clear petalite from Rubicon mine, Karibib, S.W. Africa. Analyst, E. Padget.

6. Pink petalite from Rubicon mine, Karibib, S.W. Africa. Analyst, E. Padget.

7. Montmorillonite after petalite, Rubicon mine, Karibib, S.W. Africa. Analyst, E. Padget.

8. Spodumene pseudomorph, Muiane mine, Mozambique. Analyst, J. R. Baldwin.

9. Spodumene pseudomorph, Namivo mine, Mozambique. Analyst, O. Von Knorring.

10. Spodumene pseudomorph, Maridge mine, Mozambique. Analyst, O. Von Knorring.

11. Beryl (morganite), Marropino mine, Mozambique. Analyst, O. Von Knorring (Vorma et al., 1965).

12. Euclase, Muiane mine, Mozambique. Analyst, O. Von Knorring.

bodies. At Kirengo mine (29° 39′ E., 1° 55′ S.), for example, lithium meta-somatism has transformed the contact zone bordering the schists into a fibrous, white holmquistite and the zone bordering the amphibolite to a holmquistite amphibolite.

Amblygonite is the main lithium mineral of economic importance and large amounts have been observed at several mines. It is frequently replaced by triplite, eosphorite and other phosphates. Buranga mine at Gatumba is noted for its rich variety of phosphate minerals which have been described by Thoreau and Bastien (1954), Altmann (1961), Gallagher and Gerards (1963) and others. As a result of field and laboratory examination of the Buranga phosphates a new lithium-calcium-aluminium phosphate, bertos-saite (Von Knorring and Mrose, 1966) has been identified (Table 1, analysis 6). It is faintly pink in colour and can be readily mistaken for amblygonite. Corroded fragments of amblygonite have been seen in bertossaite, indicating its probable formation from amblygonite by replacement during a late hydrothermal phase of crystallisation.

The phosphate minerals in the Buranga pegmatite may be divided into three major assemblages as follows:

(i) Manganese-iron nodules consisting of lithiophilite, heterosite-purpurite, ferrisicklerite, alluaudite, barium-arrojadite (a new variety), tavorite, barbosalite, lipscombite, frondelite, strengite and clino-strengite.

(ii) Amblygonite with minor lazulite, apatite and eosphorite.

(iii) Aggregates containing variable amounts of lazulite, amblygonite, brazilianite, augelite, bertossaite with some apatite, crandallite, berlinite, wavellite and a new Ba-Mn-Fe-Al phosphate. Muscovite (lepidolite) and quartz may also be present in these associations.

Spodumene is known from many pegmatites in Rwanda. At Rongi mine, about 2 km east of Gatumba (29° 39′ E., 1° 55′ S.), a large development of this mineral occurs adjacent to the quartz core. Most of the spodumene is however altered to kaolinite and in some sections even the kaolinite has been leached away and moulds of former spodumene crystals remain in the quartz. Elsewhere, especially at lower levels in the mine, some unaltered or only partly altered spodumene is found. At the upper Rongi mine nodular spodumenes, partly fresh and partly consisting of a grey, fine-grained almost porcellanous substance have been noticed; the latter consists of an intergrowth of albite and eucryptite, already described by Brush and Dana in the 1880s (see Dana, 1892, p. 426). Similar nodules have been identified in a number of mines; in ultra-violet light they show an intense red fluorescence. In viewing the large amount of altered spodumene in various pegmatites and the observed extensive exomorphic halo of holmquistite-bearing rock around these deposits, there appears to be a relationship between these two minerals, as observed by Nickel et al. (1960).

Lepidolite is of minor importance in the Rwanda pegmatites; it occurs in several deposits but never in very large quantities. On the other hand, the lithium amphibole holmquistite occurs in very large amounts especially at Kirengo mine near Gatumba. This holmquistite has been the subject of a closer study (Von Knorring and Hornung, 1965) and analyses 3 and 4 of Table 3 show the chemical composition of this mineral at Kirengo mine, where it extends into the amphibolites over 10 m from the contact with the pegmatite, and appears in different habits ranging from fibrous blue kyanite-like crystals to asbestiform varieties.

Because of the predominance of tin and tungsten mining in Rwanda the search for *beryl* has been of less importance. Consequently, very little is known about its distribution in the pegmatites. From my own observations, the Rwanda pegmatites appear to be poorer in beryllium than corresponding pegmatites of Ankole in Uganda. Nevertheless, deposits with distinct lithium and sodium phases always contain a certain amount of beryl and coloured varieties are often present. Euclase, in colourless, well-developed crystals, has been identified from the large tin pegmatite at Nyamissa near Gatumba.

Black *tourmaline* is common in various pegmatites and cassiterite-quartz veins. Coloured varieties have been identified in lithium pegmatites such as Buranga. The tourmaline is generally confined to the exocontacts of the pegmatites. Deposits bordering on amphibolites often have a contact rim of biotite some 5 cm wide, followed outwards by a zone of tourmaline.

Rare-earth minerals have been recorded in small amounts only. In Burundi a non-pegmatitic bastnäsite deposit is known from Karonge associated with barite, galena and pyrite.

Rwanda is one of the principal *cassiterite* producing countries in equatorial Africa. The tin deposits are either cassiterite-bearing pegmatites of various kinds or cassiterite-bearing quartz veins with variable amounts of muscovite. Varlamoff (1957) has studied these deposits and those of the neighbouring Congo in great detail and distinguishes several specific types in various parts of Rwanda. In the Gatumba pegmatites, I have observed cassiterite in the following major associations: (*i*) pegmatitic veins of up to 5 m wide consisting of albite, muscovite and quartz with large amounts of coarser aggregates of cassiterite; (*ii*) extensive pegmatitic sills, partly fine-grained, composed of albite, muscovite and some apatite carrying disseminated cassiterite; and (*iii*) pegmatitic schlieren and boudinaged lenses in migmatitic schist and amphibolite terrain. The cassiterite is partly disseminated throughout the pegmatite and partly concentrated in narrow, saccharoidal albitic rims along the contacts.

Smaller amounts of *niobium-tantalum* minerals are commonly associated with the cassiterite deposits of Rwanda. Varamoff (*ibid.*) estimates the columbite-tantalite content to be about 10-15% of the cassiterite mined and relates the mineralisation to the following pegmatites: muscovite pegmatites with microcline and quartz; beryl pegmatites with microcline, quartz and muscovite; and albitised pegmatites with lithium phases. In the Gatumba area columbite-tantalite, manganotantalite (Table 2, analysis 18) and microlite have been observed.

EASTERN CONGO

Important mineralised pegmatites are common west and south of Lake Kivu in the Congo (Fig. 1, loc. 3) and, as in south-west Uganda and Rwanda, they are closely associated with post-tectonic granites. In the Mumba–Numbi area west of Lake Kivu, Agassiz (in Varlamoff, *ibid.*) distinguishes the following main pegmatites: beryl pegmatites with black tourmaline, muscovite and microcline; and albitised pegmatites containing microcline, quartz, muscovite, spodumene, amblygonite and lepidolite, with cassiterite and niobium-tantalum mineralisation. Safiannikoff and Wambeke (1961) have described a rare lead-microlite and I have identified a new alkali-tantalite, rankamaite, from these deposits (Von Knorring *et al.*, 1969b). The latter mineral is found as waterworn pebbles, up to tens of grammes in weight, in heavy mineral concentrates from alluvial deposits; and it occurs as a white, fibrous matrix in which corroded grains of simpsonite, minute crystals of cassiterite, and some manganotantalite and muscovite occur.

The principal beryl pegmatites of the Congo are situated at Kobokobo (28° 7' E., 3° 6' S.) near Kamituga some 90 miles south-west of Lake Kivu (De Kun, 1965). In addition to beryl, columbite-tantalite, amblygonite and various manganese-iron phosphates are of common occurrence.

KATANGA

Within the Kibaran rocks of Katanga (see Cahen, this volume, p. 100), there are some characteristic post-tectonic, leucocratic muscovite granites that are tin-bearing. Associated with these granites, a variety of mineralised quartz veins and pegmatites constitute the extensive Katanga tin province. Perhaps one of the most spectacular tin and niobium-tantalum pegmatites in the world is Manono (27° 23' E., 7° 20' S.) in northern Katanga (Fig. 1, loc. 4). The combined length of two neighbouring deposits, Manono and Kitololo, is some 14 km and the width is in places up to 700 m. The two pegmatites are *en echelon* and are conformable with the foliation of the enclosing schists and dolerite-amphibolites. Although the pegmatites are zoned, the internal structure is complicated by numerous xenoliths of the country rocks. Spodumene is a common mineral associated with quartz and the albitic units, and it is often altered and at times replaced by greenish, finely divided micaceous pseudomorphs known as killinite. In spite of the plentiful spodumene, no holmquistite has been recorded in the contact zones.

There is a distinct enrichment of cassiterite in parts of the albite zones and in pockets of greisen along the contacts; otherwise cassiterite appears to be disseminated in minute crystals throughout the pegmatites. In the north-western part of the Manono deposit, there is a particular quartz vein carrying tantalite, some wolframite and thoreaulite. Manono is the type locality of the last of these minerals, which has since been found in other pegmatites from the Congo and Rwanda. Among additional accessories, coloured tourmaline, apatite, beryl and some lepidolite and fluorite have been recorded. It is perhaps significant that beryl and amblygonite are absent or rare in many of the cassiterite pegmatites.

KAMATIVI, RHODESIA

A pegmatitic tin belt of great economic importance is situated in the Wankie district of Rhodesia (Fig. 1, loc. 5). The mineralised area in question is some 40 km long and 2-5 km wide, and consists of schists, granites, gneisses and pegmatites. Two major groups of the latter can be distinguished: an older group of concordant tourmaline-bearing pegmatites which are generally barren; and a younger group, commonly cross-cutting, with distinct sodium and lithium phases, which are cassiterite-bearing. There are several types of pegmatite in the younger group and mining operations are mainly confined to the so-called flat-dipping pegmatites which are mineralised. These are emplaced in the metamorphic aureole close to gneiss-granites and are roughly lenticular bodies with a maximum width of some 30 m (Bellasis and Van der Heyde, 1963). Zoning in the pegmatites is not particularly well developed; quartz cores, for instance, are practically unknown. The persistent mica selvage is typical and there is some enrichment of cassiterite in this zone. In the pegmatites there are also large irregular units of fine-grained albite; these replacement bodies are characteristic and they contain corroded phenocrysts of microcline. In addition to cassiterite, columbite-tantalite is the main by-product recovered. Large bodies of spodumene, partly altered to a pinkish montmorillonite, amblygonite and some beryl are also present.

Pegmatites in the 550 ± 100 m.y. old orogenic belts

KARIBIB–USAKOS REGION OF SOUTH-WEST AFRICA

Introduction

The area consists of high-grade regionally metamorphosed sediments of the Upper Precambrian Damara Facies, trending north-east–south-west and characterised by syn-kinematic granites and post-orogenic quartz veins and pegmatites. In the Karibib–Usakos region, in particular, numerous mineral-ised pegmatites have been known for many years, mainly as important deposits of lithium, beryllium, tin, niobium-tantalum minerals and for spec-tacular finds of gem tourmaline. The pegmatites, apparently connected with late-kinematic granites, have intruded the metasediments of the Damara Facies. The mineral associations, observed within marbles, amphibolites, mica schists and quartzites, are almost identical with those in granitic rocks, thus indicating the same major source of pegmatite material and a similar mecha-nism of emplacement. The pegmatites vary to some extent in internal structure and composition, but most have a characteristic geochemical pattern. Lithium and sodium phases are well developed and among the major mineral constituents lepidolite, petalite, amblygonite, beryl and tourmaline are typical. Although the majority of the occurrences may be considered as lithium-beryllium pegmatites, there are some typical cassiterite pegmatites and also some more complex ones with copper and uranium mineralisation.

The present account is mainly concerned with the mineralogy and geo-chemistry of the pegmatites in the Karibib–Usakos area (Fig. 1, loc. 6). As regards their internal structure, zoning is well developed in most occurrences. Prominent quartz cores have been noted in many deposits, and these cores are mineralised in places and carry beryl, amblygonite, pollucite, manganotantalite, tourmaline, complex manganese-iron phosphates and occasionally bismutite.

Mineralogy and geochemistry

Numerous pegmatites in the Karibib-Usakos area contain a variety of *lithium* minerals; for example, amblygonite, lithiophilite, lepidolite, petalite, cookeite and lithium-bearing tourmaline. Geochemically the area may be considered as a typical lithium province. Only spodumene is rare; but even this mineral has been observed in the form of kaolin pseudomorphs in some quartz cores. Amblygonite (and other phosphates) are common constituents of many pegmatites, often found at the core margin or in the intermediate zone with cleavelandite, lepidolite and quartz. In some pegmatites ambly-gonite is replaced by turquoise, eosphorite and strengite. In the manganese-iron phosphate nodules the following minerals were identified: triphylite-lithiophilite, heterosite-purpurite, hureaulite, tavorite, barbosalite, lips-combite, frondelite, mitridatite, clinostrengite; in addition large masses of chalcosiderite and hematite occur.

The pegmatites in the Karibib area are particularly known for their abundance of lepidolite. In places it occurs in large units, associated with sugary albite or cleavelandite, some quartz and occasionally topaz; locally the lepidolite or other lithium-bearing micas are highly mineralised, carrying cassiterite, columbite, manganotantalite, wodginite and microlite. The lepidolite zone is often close to the quartz core of the pegmatites. Petalite is a type mineral of the Karibib pegmatites and is present in many occurrences. Its position is also close to the core and it may be associated with lepidolite, cleavelandite and amblygonite. Petalite ranges in appearance from clear, glassy white and fibrous white, to grey and pink varieties; it is seen frequently in various stages of alteration, pink kaolinised petalite being largely composed of montmorillonite (Table 3, analyses 5-7).

The rare *caesium* mineral pollucite has been recovered from the Helicon mine near Karibib where it is associated with a massive quartz core. It occurs in irregular masses, is white to creamy white in colour and is characteristically intergrown with narrow veinlets of lepidolite.

Beryl is a common mineral in practically all the pegmatites, and it is present as an abundant accessory in the late pegmatitic granites of the area. In the lithium pegmatites, it occurs as a glassy, coloured variety, frequently of gem quality aquamarine, heliodor or morganite. The structural position of beryl is variable but most commonly it is located at the core margin together with microcline or amblygonite, muscovite, lepidolite and albite. Occasionally there are specific beryl zones in the intermediate units of the larger pegmatites.

Large amounts of chrysoberyl have been observed in a microcline-muscovite-quartz pegmatite. Pistachio green, this mineral is found in irregular masses associated with bluish beryl, muscovite, quartz and some apatite. In some pegmatites hydrothermal alteration of beryl is common, and the beryl pseudomorphs consist of a lattice of bertrandite where the cavities are filled with corroded beryl fragments and a clay mineral. Phenacite and milarite have also been recorded.

The South-West African pegmatites are universally known for their coloured *tourmalines*, often of an excellent gem quality. In the lithium pegmatites the tourmalines exhibit a zonal arrangement in relation to the quartz core, which contains mainly rubellite or the polychrome pink and green variety of tourmaline; in the intermediate zones, blue indigolite, polychrome pink and blue tourmaline, or the green verdelite are present; and the black tourmaline occurs in the graphic or the border zones. Occasionally gigantic rosettes of the polychrome pink and green tourmaline are observed, and crystals up to 2 m in length have been seen projecting from an intermediate albite-lepidolite zone towards the quartz core.

Rare-earth minerals are seldom found in the pegmatites proper. A large monazite occurrence has, however, been recorded from the Namib Desert near Usakos (Von Knorring and Clifford, 1960). The monazite is the main constituent in a strontium-bearing dolomite and the composition of the monazite and the dolomite suggest that the occurrence may have carbonatitic affinities. A radiometric study of the monazite (Burger *et al.*, 1965) has yielded an age of 500 m.y. which is comparable with the general pegmatite mineralisation in the area.

Tin is a dominant type element in the pegmatites of South-West Africa and two tin zones with a north-easterly trend have been delineated. A typical cassiterite deposit is the Sandamap occurrence to the north-west of Usakos. This occurrence consists of a series of pegmatites which, in addition to cassiterite, carry columbite-tantalite, beryl, amblygonite, petalite, lepidolite and various manganese-iron phosphates. Early workings were mainly concentrated around clearly exposed quartz cores where coarse-grained cassiterite is enclosed in crumbly muscovite greisen and, in this mode of occurrence, Sandamap tin pegmatites can be compared to those of Kamativi in Rhodesia.

A unique *hafnian* zircon containing 31% hafnium oxide has been observed in a lithium pegmatite near Karibib (Von Knorring and Hornung, 1961). It is intimately associated with some tapiolite and manganotantalite and forms irregular masses in lepidolite-albite-quartz matrix.

Columbite-tantalite and manganotantalite are of common occurrence in the pegmatites of the Karibib area; in particular, it appears that tantalum is a dominant element in many lithium pegmatites (Table 2, analyses 20, 22, 23 and 27) (Von Knorring *et al.*, 1966). At the Dernburg pegmatite, close to Karibib, columbite is found abundantly in a muscovite greisen, whereas manganotantalite is confined to the quartz core. Microlite has been observed in many pegmatites, usually dispersed in lepidolite in the form of minute yellow to brown grains. Occasionally larger, yellow octahedra up to 3 cm in diameter are seen intergrown with manganotantalite; a dark-brown manganiferous microlite has also been identified. Finally, the tin-bearing wodginite has been identified in large amounts in the cleavelandite-mica intermediate zone of one of the Karibib pegmatites.

THE MOZAMBIQUE BELT IN EAST AFRICA

North-east Uganda

In the northern Karamoja district of north-east Uganda (Fig. 1, loc. 7) several radioactive anomalies connected with pegmatites have been investigated; for example, at Apeykale, Kamacharkol and Kalere. Pegmatite veins are numerous in this area and they form part of an injection gneiss complex. Near Kalere River, for example, a radioactive anomaly is caused by biotite-rich pegmatitic schlieren carrying a large amount of allanite, monazite, zircon and some rutile. On examination the following minerals were identified: ilmenite, greenish-black, sulphur-bearing monazite, a yellow monazite, zircon and rutile. These mineralised pegmatites represent an excellent example of rare-metal deposits connected with migmatitic gneisses, a paragenesis which is commonly overlooked.

Kenya

The pegmatites of Kenya (Fig. 1, locs. 8, 9 and 10) are very little known and still less exploited. Mica has been sporadically mined and on this basis three major pegmatite areas can be established; namely, West Suk, Baragoi and Machakos. In the West Suk area, muscovite has been the principal product mined but in addition some radioactive minerals have been also recovered including allanite, monazite, zircon, samarskite and columbite-tantalite. North of Kitale at Sebit, an extensive lithium pegmatite has been investigated and is, to date, the only pegmatite known in Kenya with well-developed lithium and sodium phases. Coloured beryl, partly of gem quality, microlite and columbite-tantalite have also been recorded (Du Bois, personal communication). Pegmatites are numerous in the Baragoi area (Baker, 1963) occurring in well-marked zones as concordant and cross-cutting sheets and veins. Well-developed internal zoning has been recognised in many of the larger pegmatites and books of muscovite, sometimes ruby mica of economic quality, are generally found near the outer margins of the quartz cores; some beryl has been also recovered.

Tanzania

The best known pegmatites associated with the Mozambique Belt in Tanzania are the mica occurrences of the Uluguru Mountains some 200 km west of Dar es Salaam (Fig. 1, loc. 11). These pegmatites have been described by McKie (1957) and Sampson (1962). Other pegmatites of economic importance are those of the Mikese–Handeni area, Ukaguru Mountains and Usambara–Pare Mountains (McKie, 1957). Rare-metal minerals are rare in all the pegmatite fields.

The Uluguru pegmatites form a broad zone of northerly-trending, discordant dykes, each averaging some 10-20 ft across and up to 1000 ft long, in a swarm associated with a regular system of faults and fractures in a meta-anorthositic complex. Mineralogically, the Uluguru pegmatites are also well known for the exceptional uraninite crystals which have been obtained from time to time. They are contemporaneous with the muscovite and may exhibit late-stage hydrothermal alteration to a variety of secondary uranium minerals such as kasolite, uranophane, rutherfordine (type locality), carnotite, autunite and torbernite.

THE ALTO LIGONHA AREA OF THE MOZAMBIQUE BELT IN MOZAMBIQUE

Introduction

The pegmatite province of Alto Ligonha in Mozambique (Fig. 1, loc. 12) is one of the most interesting in the world, but still comparatively little known. Some of the pegmatites have been described by Hutchinson and Claus (1956), Bettencourt Dias (1961) and others. The major pegmatites of this area are situated within a north-easterly-trending zone, some 200 km long and 50 km wide, and they exhibit a unique variety of minerals. From a geochemical point of view, the rarer elements lithium, caesium, beryllium, rare-earths, scandium, yttrium, thorium, niobium and tantalum predominate. According to Bettencourt Dias (1961), the economically important pegmatites are genetically related to a group of later equigranular granites and they are found within a zone ranging up to a few kilometres in width around the granite intrusions.

Pegmatites of all sizes are found in the Alto Ligonha area, the largest and the most impressive being Mina Muiane (38° 15′ E., 15° 44′ S.); internal structures are well developed in most pegmatites. Extensive weathering has kaolinised the feldspars with the result that pegmatites with low rare-metal content can be economically mined.

Mineralogy and geochemistry

Quartz is the major mineral in many pegmatites and often forms prominent cores. At Muiane, for example, the massive quartz core contains giant laths of altered spodumene associated with cleavelandite and muscovite. Extensive mineralisation has been noted in all pegmatites containing quartz cores of this kind. Rose quartz is found in some mines and also remarkable smoky-quartz crystals up to a ton in weight have been recovered from many pegmatites. In Muiane mine, intricate intergrowth of polychrome tourmaline with quartz crystals has been observed.

The feldspars are generally kaolinised to various degrees although it appears that albitic varieties are more resistant to alteration by weathering than microcline. At Muiane, giant crystals of microcline are rimmed with muscovite, and are found in the intermediate zones of most zoned pegmatites where beryl and columbite are also commonly concentrated; some amazonite has also been observed.

Muscovite is the principal mica in all pegmatites. Some large occurrences are entirely enveloped by a muscovite shell in contact with the schists. Elsewhere the quartz core is surrounded by mica, often lepidolite; muscovite-albite or lepidolite-albite assemblages are common. For instance, at Namivo near Morrua, a globular lithium-bearing brown mica with cleavelandite forms one of the major zones of mineralisation carrying abundant beryl, columbite, monazite, xenotime and zircon. Lepidolite greisens at Muiane and other deposits contain appreciable amounts of microlite.

Amblygonite, spodumene and various lithian micas are the major *lithium* minerals. In addition, lithium is present in the coloured tourmalines, beryls and the lithian chlorite, cookeite (Sahama *et al.*, 1968). Although triphylite-lithiophilite is so common in many pegmatites elsewhere I have not observed this mineral in the localities examined so far. Indeed, although many pegmatites in the Alto Ligonha area have well-developed lithium phases, lithium minerals are not present on the same prolific scale as in other lithium-rich pegmatitic provinces such as South-West Africa. Amblygonite is comparatively rare, but this may reflect a smaller concentration of phosphorus in the Mozambique pegmatites rather than a deficiency in lithium. Large amounts of amblygonite were observed only at Maridge near Muiane and at Morrua. Spodumene is found in great abundance in pegmatites with distinct quartz cores (Muiane); the spodumene laths are almost always altered to kaolinite, montmorillonite or cookeite (Table 3, analyses 8–10). Recrystallised, corroded spodumene and even violet kunzite are sometimes found in the kaolinised, pink pseudomorphs. Lepidolite is generally concentrated in the innermost intermediate zone in contact with the quartz core of the pegmatites.

The rare mineral pollucite occurs in Morrua and Namacotche pegmatites and it may be present in other deposits. At Morrua, large nodules of pollucite more than 2 m across were seen in 1961; this mineral has the characteristic lepidolite veinlets. Both pollucites are sodium-bearing and contain trace amounts of rubidium and thallium.

Alto Ligonha pegmatites are noted for their gem varieties of coloured *beryl*. Blue-green aquamarine is perhaps the dominant variety followed by those of greenish-yellow colour and pink morganite; colourless and pearly-black varieties are also common. At Muiane and Morrua mines, excellent tabular crystals of morganite have been recovered (Table 3, analysis 11). Gem beryl is generally found in cavities in minor amounts; at Namivo, however, a large mass consisting of intergrown aquamarine and morganite, some 6 m across, has yielded 200 tons of beryl.

The position of beryl varies to some extent in the pegmatites but it usually occurs close to the quartz core and also in the intermediate zones, especially in cleavelandite-mica zones or in connection with microcline-perthite. Bertrandite has been observed as an alteration product of beryl and excellent crystals of rare euclase have been recovered from Muiane mine (Von Knorring *et al.*, 1964) (Table 3, analysis 12).

Muiane and the neighbouring Naipa mines are the main producers of *rubellite* and *verdelite* tourmaline, in addition to polychrome varieties. These

G

tourmalines are commonly found in small cavities, together with cleave-landite, quartz and lepidolite. At times giant rubellites have been recovered; some crystals measure 70 cm in length and 25 cm across.

Among *rare-earth* minerals, monazite has been found at Nacuissupa near Muiane. A brown and an olive-green variety have been observed carrying 11 and 18% of thorium oxide respectively. Moreover, in Namivo pegmatite yellowish monazite and greenish xenotime are rather common in the cleave-landite-lithian mica zone. The monazite is associated with columbite, whereas the xenotime is intergrown with partly metamict zircon. Urani-ferous rare-earth niobotantalites are widespread and euxenite is one of the more common rare-metal minerals.

Cassiterite is a rare mineral in these pegmatites. Small amounts of alluvial cassiterite have been recovered in the Muiane area and the various niobium-tantalum minerals are generally tin-bearing, especially the scandian ixiolite, niobian wolframite, struverite and tapiolite.

Zircon is a common accessory in most pegmatites and a rare hafnian zircon containing 34% hafnium oxide has been recently recorded from the Namacotche lithium pegmatite (Quadrado and Lima de Faria, 1966).

Columbite, tantalite, manganotantalite, stibiotantalite, tapiolite, microlite, ilmenorutile and two new varieties of scandian ixiolite and a niobian wolfra-mite have been observed in various pegmatites. Morrua mine has been the main tantalum producer for many years and at one time an extensive vein of high-grade tantalite was mined. The manganotantalite from Morrua is a unique end-member; its tantalum content is one of the highest known (Table 2, analyses 24 and 25). This mineral forms radiating spheroidal aggregates closely associated with cleavelandite, lepidolite, blue zoned tourmaline, pollucite and a quartz core rich in spodumene. Tapiolite has been identified from Nacuissupa and other pegmatites (Table 2, analyses 31 and 32). Microlite is of common occurrence often intergrown with mangano-tantalite and bismutite, replacing tapiolite or disseminated in lepidolite. Stibiotantalite has also been noted from several deposits and occasionally translucent, yellowish-grey dark-rimmed crystals have been obtained.

The scandian ixiolite is of special interest being the only scandium mineral apart from thortveitite found in significant amounts. It was first noticed in 1959, in the concentrates from Muiane mine, but has since been recovered from Nacuissupa mine. The form of this unique mineral is semi-spheroidal and it is frequently intergrown with monazite, zircon, muscovite or lepido-lite. The chemical composition is distinctive and differs from ordinary columbite-tantalite in having up to 7% scandium oxide as well as significant amounts of tin, titanium and ferric iron (Table 2, analysis 28; Von Knorring *et al.*, 1969c).

MADAGASCAR

Introduction

The Precambrian geology of the island of Madagascar resembles that of adjacent parts of the Africanmainland. Recent age determinations on gneisses, granites and pegmatites show that a major orogenic event some 550 m.y. ago affected a large part of the country. The similarities in mineralisation, geochemistry and age of the pegmatites of Madagascar and Mozambique give a further indication that these regions were affected by the same orogenic events, and Cahen and Snelling (1966) regard Madagascar as the southernmost province of the Mozambique Belt.

Madagascar is known for its numerous pegmatite deposits found practically along the whole length of the island within a belt some 1200 km long, from the Daraina pegmatitic field in the north-east to the Fort Dauphin area in the south. In the present account only some of the major mineralised regions of Tsaratanana, Andriamena, Sahatany and Ampandramaika will be considered.

Tsaratanana region (including the important pegmatites of Berere and Betanimena). Berere (Fig. 1, loc. 13) is situated some 250 km north of Tananarive. The pegmatites in this general area are emplaced in migmatites and pyroxene-amphibolites and according to Giraud (1957a) three main types can be distinguished: muscovite pegmatites, biotite-muscovite or two-mica pegmatites; and biotite pegmatites. The muscovite pegmatites show a regular zoning and are the principal carriers of beryl and columbite; the uraniferous niobotantalites, xenotime, monazite, malacon and bismutite are of secondary importance. In the two-mica pegmatites the internal structure is well developed and is characterised by a distinct graphic zone. These pegmatites carry smaller amounts of beryl and columbite than the muscovite pegmatites, and the columbite is commonly replaced by euxenite and ampangabeite; allanite, xenotime, monazite and zircon are also present. In the biotite pegmatites, beryl is seldom present and columbite is less prominent, but fergusonite or samarskite are locally important.

Andriamena–Ankazobe region. Andriamena (Fig. 1, loc. 13) is situated some 170 km north of Tananarive and the pegmatite region stretches south to Befanamo and south-west towards Ankazobe. The individual deposits are generally smaller in size than in the Tsaratanana region and are emplaced in migmatitic gneisses or pyroxene-amphibolites. Giraud (1957b) records biotite pegmatites containing uraniferous niobotantalites and monazite, biotite and muscovite pegmatites carrying some beryl and niobotantalites, and muscovite pegmatites with larger amounts of beryl and columbite.

A well-known deposit from this area is the Befanamo pegmatite, the type locality of Madagascar thortveitite or befanamite. The yellow to greenish-grey thortveitite is closely associated with smoky quartz, encrusting green gem beryl or is intergrown with monazite. Thortveitite is a rare mineral and since its discovery in the eluvium of this pegmatite in 1918, only 40 kg have been recovered.

Sahatany region. Hundreds of mineralised pegmatites are scattered along the Sahatany valley (Fig. 1, loc. 14) north-west and west of the imposing quartzitic Ibity mountain, some 30 km south of the town of Antsirabe in central Madagascar. The pegmatites have been known for a long time as important deposits of excellent gemstones such as coloured beryl and tourmaline, kunzite, blue topaz, and many coloured quartz varieties. The rare minerals rhodizite and pollucite have also been obtained from this region. The pegmatites are generally small and emplaced in mica schist, quartzite and limestone. They are usually composed of quartz, biotite, microcline, muscovite and often black tourmaline in addition to well-developed sodium and lithium phases. Most of them are typical lithium pegmatites.

Ampandramaika–Malakialina region. The central part of this vast pegmatite area (Fig. 1, loc. 15) is situated some 275 km south-west of Tananarive. The many important pegmatites of this region were originally exploited for industrial mica, gem beryl and tourmaline but, during the 1950s, the main beryl and columbite mining of Madagascar appears to have been concentrated here. The mineralised pegmatites are of varied composition and often exhibit well-developed internal structures. Among the more important deposits only the Marivolanitra, Ampandramaika and Malakialina pegmatites will be considered in this account. Of these the Marivolanitra pegmatite is one of the most important in Madagascar; it lies some 10 km north of the town of Ampandramaika in a terrain of mica schist. The main beryl mineralisation is confined to the albitic zone and is closely associated with plumose muscovite and quartz. The intermediate zone contains large phosphate nodules principally of triplite with associated complex manganese–iron phosphates such as graftonite, barbosalite, bermanite, lipscombite, mitridatite and tavorite.

The Ampandramaika pegmatite is a large body in the form of a steeply inclined sheet with well-developed zoning emplaced in mica schist (Guigues, in Besairie, 1965).

The Malakialina pegmatites are the source rocks for a large amount of beryl mainly concentrated at the quartz core margin, and giant beryl crystals up to 270 tons in weight have been recovered.

Mineralogy and geochemistry

One of the most characteristic features is the widespread occurrence of the so-called uraniferous niobotantalites such as betafite, ampangabeite, euxenite, samarskite, fergusonite; these give a specific geochemical pattern to the Madagascar pegmatite province. In addition, the rare mineral bastnäsite is common in some pegmatites and hydrothermal deposits; uranothorianite is prominent in the basic pegmatites in the south. In common with the geochemical trend of the Mozambique Belt, the pegmatites are rich in rare earths. In addition, scandium, yttrium, titanium, thorium, niobium and uranium may be considered as type elements of this pegmatite province.

Lithium mineralisation is widespread in Madagascar, but much less prominent, on the whole, than in typical pegmatitic lithium provinces. Although numerous smaller deposits (for instance, the Sahatany Valley) carry lithium, quantitatively the amount of such lithium minerals as lepidolite, spodumene and amblygonite is negligible.

The distribution of the rarer *rubidium* and *caesium* follows that of potassium and consequently the bulk of these elements is found in the microcline-perthites and micas, especially the lithium-bearing ones. The coloured alkali beryls carry lithium and caesium. In addition to pollucite, the unique caesium- and rubidium-bearing mineral rhodizite appears to be of common occurrence in some pegmatites where it is typically associated with rubellite, albite, microcline and quartz.

Beryllium is a common element in the principal pegmatite fields of Madagascar and beryl, bertrandite, chrysoberyl, rhodizite and hambergite have all been recorded. Beryl is, however, the main beryllium mineral mined for industrial purposes and as a gemstone. The excellent coloured beryls, especially the aquamarines, the rose morganites and the yellow beryls, are well known. Occasionally unusual polychrome crystals of gem quality are found. In the Sahatany Valley pegmatites beryllium is associated with boron in the rare minerals rhodizite and hambergite.

Boron mineralisation is represented by the common black tourmaline found in most pegmatites and by the coloured and polychrome varieties mainly associated with the sodium-lithium pegmatites, and the characteristic brown tourmaline which is often observed in the basic pegmatites of southern Madagascar. The coloured tourmalines are frequently of excellent gem quality.

The *rare earths* are concentrated in various uraniferous niobotantalites and also in monazite, xenotime, bastnäsite and allanite.

The element *scandium* is of special geochemical interest and an extensive study undertaken by B.R.G.M. showed considerable enrichment of this element in a variety of minerals from several pegmatite areas (Phan *et al.*, 1967). Thus garnet, beryl, tourmaline, muscovite and biotite contain up to some tenths of a per cent scandium, a significant concentration compared with the average content of 5 ppm in the earth's lithosphere. Apart from thortveitite, the most interesting scandium-bearing minerals are the uraniferous niobotantalites, columbites and a specific scandian ixiolite previously known from Mozambique (Von Knorring, 1965a) with a scandium content of some 4% (Table 2, analysis 29).

Tin is a very rare element in the pegmatites of Madagascar, and Behier (1960) records only one pegmatite with cassiterite. As regards alluvial cassiterite, only traces have been observed from various parts of the island. Some tin is, however, present in the columbite-tantalites and especially in the scandian ixiolites.

Titanium in pegmatites is concentrated in rutile or occasionally in sphene and more commonly in various uraniferous niobotantalites, ilmenite and niobian and tantalian ilmenites.

Zircon, and the variety cyrtolite, are common in pegmatites. The zircons associated with the thortveitite-bearing pegmatites at Befanamo have a high hafnium/zirconium ratio. Hafnian zircons, such as those found in Alto Ligonha area and in South-West Africa, have not been recorded.

Thorium mineralisation is common in southern Madagascar, where many extensive pyroxenite pegmatites are mined for thorianite.

Niobium appears to be by far the dominant element in the niobium-tantalum minerals. The most widespread minerals in this group are perhaps the uraniferous niobotantalites, betafite, euxenite, fergusonite, samarskite, ampangabeite, the columbite-tantalites and the niobian and tantalian rutiles. Rather common minerals like manganotantalite and microlite, type minerals of lithium pegmatites, are extremely rare and tapiolite has not been recorded. In the specific group of scandian ixiolites that I have investigated, both niobium- and tantalum-dominant members have been observed. Compared with the Alto Ligonha region, the Madagascar pegmatite province shows a distinct paucity of the element tantalum.

Summary and conclusions

When considering the geochemical characteristics of the three major pegmatite belts of equatorial and southern Africa—the Kibaran Belt affected by the 1100 ± 200 m.y. orogeny and the Damaran and Mozambique Belts of 550 ± 100 m.y. orogenesis—there are some similarities and differences amongst the pegmatites; yet a certain geochemical pattern emerges, irrespective of the age differences of the orogenic belts, reflecting perhaps the fundamental geochemical variations in the earth's crust. In the Kibaran and Damaran zones, the most outstanding feature is the ubiquitous presence of tin throughout the entire length of some 2000 miles, comparable only to the tin belt of south-east Asia (Burma–Siam–Malaya–Indonesia) where tin is associated with granites of Mesozoic age. The heavy elements niobium and tantalum and, to a smaller extent, the rare earths zirconium and hafnium, are constant associates of tin; and tungsten is another element occasionally associated with tin-bearing pegmatites in the Kibaran–Damaran Belts.

In the pegmatites of the Mozambique Belt, on the other hand, tin is virtually absent but the rare earths titanium, thorium and uranium are characteristic and more representative. In particular, a distinct enrichment of scandium has been observed in the pegmatites of Mozambique and Madagascar. The light elements lithium and beryllium are variable in the Kibaran Belt and are predominant in the pegmatites of South-West Africa, Mozambique and Madagascar. Apart from tin, the general geochemical pattern of these three southern regions is very similar. Pegmatites in the northern part of the Mozambique Belt are particularly low in the type elements lithium and beryllium, but their content of rare-earths conforms with the rest of the belt.

From these geochemical considerations of the various pegmatite regions two main pegmatite belts can be deduced: a western or the 'tin belt' following the Kibaran-Damaran trend; and an eastern or the 'rare-earths belt' associated principally with the Mozambique Belt.

REFERENCES

ALTMANN, J. 1961. L'occurrence de brazilianite, augelite, frondelite et lithiophilite dans la pegmatite de Buranga. *Schweiz. miner. petrog. Mitt.*, **41**, 407.

BAKER, B. H. 1963. Geology of the Baragoi area. *Rep. geol. Surv. Kenya*, **53**.

BARNES, J. W., and MEAL, P. F. 1959. Mineralised pegmatites in Uganda. *Bull. geol. Surv. Uganda*, **4**.

BEHIER, J. 1957. Les minéraux phosphates de la pegmatite de Marivolanitra, Madagascar. *C.C.T.A. east centr. and southern reg. Comm. Geol.*, Tananarive, **1**, 237.

—— 1960. Contribution a la minéralogie de Madagascar. *Annls. géol. Madagascar*, **29**.

BELLASIS, J. W. M., and VAN DER HEYDE, C. 1963. Operations at Kamativi Mines Limited, Southern Rhodesia. In *Pegmatites in southern Rhodesia*, p. 47. Inst. Mining Metall., Salisbury, 47.

BESAIRIE, H. 1965. Géologie économique de la Sous-Prefecture D'Ambatofinandrahana. *Doc. Serv. géol. Malgache*, **170**.

BETTENCOURT DIAS, M. 1961. Os pegmatitos do Alto Ligonha. *Bol. Est. Not. Trab., Serv. Geol. Min. Mozambique*, **27**, 17.

BURGER, A. J., VON KNORRING, O., and CLIFFORD, T. N. 1965. Mineralogical and radio-metric studies of monazite and sphene occurrences in the Namib desert, South-West Africa. *Miner. Mag.*, **35**, 519.

CAHEN, L., and SNELLING, N. J. 1966. *The geochronology of equatorial Africa*, Amsterdam.

CLIFFORD, T. N. 1966. Tectono-metallogenic units and metallogenic provinces of Africa. *Earth Planetary Sci. Let.*, **1**, 421.

COMBE, A. D., and GROVES, A. W. 1932. The geology of south-west Ankole. *Mem. geol. Surv. Uganda*, **2**.

DANA, E. S. 1892. *Descriptive mineralogy*, 6th ed., New York.

DE KUN, N. 1965. *The mineral resources of Africa*, Amsterdam.

GALLAGHER, M. J., and GERARDS, J. F. 1963. Berlinite from Rwanda. *Miner. Mag.*, **33**, 613.

—— and HAWKES, J. R. 1966. Beryllium minerals from Rhodesia and Uganda. *Bull. geol. Surv. G.B.*, **25**, 59.

GEVERS, T. W. 1936. Phases of mineralisation in Namaqualand pegmatites. *Trans. geol. Soc. S. Afr.* **39**, 331.

GIRAUD, P. 1957a. Le champ pegmatitique de Berere à Madagascar. *C.C.T.A. east centr. and southern reg. Comm. Geol.*, Tananarive, **1**, 125.

—— 1957b. Les principaux champs pegmatitiques de Madagascar. *C.C.T.A. east centr. and southern reg. Comm. Geol.*, Tananarive, **1**, 139.

HUTCHINSON, R. W., and CLAUS, R. J. 1956. Pegmatite deposits, Alto Ligonha, Portu-guese East Africa. *Econ. Geol.*, **51**, 757.

KENNEDY, W. Q. 1965. Introduction to the ninth annual report on scientific results, session 1963–64. *9th Ann. Rep. Res. Inst. African Geol., Univ. Leeds*, 5.

LINDBERG, M. L., and PECORA, W. T. 1955. Tavorite and barbosalite, two new phosphate minerals from Minas Gerais, Brazil. *Am. Miner.*, **40**, 952.

McKIE, D. 1957. A synopsis of mineral parageneses in the complex pegmatites of Tanganyika. *C.C.T.A. east centr. and southern reg. Comm. Geol.*, Tananarive, **1**, 159.

NICKEL, E. H., KARPOFF, B. S., MAXWELL, J. A., and ROWLAND, J. F. 1960. Holmquistite from Barraute, Quebec. *Can. Miner.*, **6**, 504.

PHAN, K. D., FOISSY, B., KERJEAN, M., MOATTI, J., and SCHILTZ, J. C. 1967. Le scandium dans les minéraux et les roches encaissantes des certaines pegmatites malgaches. *Bull. Bur. Rech. géol. min.*, **3**, 77.

QUADRADO, R., and LIMA DE FARIA, J. 1966. High hafnium zircon from Namacotche Alto Ligonha, Mozambique. *Garcia de Orta (Lisbon)*, **14**, 311.

REEDMAN, A. J. 1967. Geological environment and genesis of the tungsten deposits of Kigezi district, south-western Uganda. Unpublished Ph.D. thesis, Univ. Leeds.

SAFIANNIKOFF, A., and VAN WAMBEKE, L. 1961. Sur un terme plombifère du groupe pyrochlore-microlite. *Bull. Soc. fr. Minér. Crist.*, **84**, 382.

SAHAMA, TH. G., VON KNORRING, O., and LEHTINEN, M. 1968. Cookeite from the Muiane pegmatite, Zambezia, Mozambique. *Lithos*, **1**, 12.

SAMPSON, D. N. 1962. The mica pegmatites of the Uluguru mountains. *Bull. geol. Surv. Tanganyika*, **35**.

THOREAU, J., and BASTIEN, G. 1954. Les phosphates de pegmatites du Ruanda occidental. *Bull. Acad. roy. Belg.*, **25**, 1595.

VARLAMOFF, N. 1957. Consideration sur la zonéographie et le zonage interne des pegmatites Africaines. *C.C.T.A. east centr. and southern reg. Comm. Geol.*, Tananarive, **1**, 185.

VON KNORRING, O. 1965a. Mineralogical studies. Niobium–tantalum minerals. *9th Ann. Rep. Res. Inst. African Geol., Univ. Leeds*, 42.

—— 1965b. Notes on some pegmatite minerals from Rwanda. *Bull. Serv. geol. Rwanda*, **2**, 11.

—— and CLIFFORD, T. N. 1960. On a skarn monazite occurrence from the Namib desert near Usakos, South-West Africa. *Miner. Mag.*, **32**, 650.

—— and HORNUNG, G. 1961. Hafnian zircons. *Nature, Lond.*, **190**, 1098.

—— and HORNUNG, G. 1965. Pegmatite investigations in east Africa. *9th Ann. Rep. Res. Inst. African Geol., Univ. Leeds*, 44.

—— and MROSE, M. E. 1966. Bertossaite a new mineral from Rwanda. *Can. Miner.*, **8**, 668.

—— SAHAMA, TH. G., and SAARI, E. 1964. A note on euclase from Muiane mine, Alto Ligonha, Mozambique. *C.R. Soc. géol. Finlande*, **36**, 143.

—— SAHAMA, TH. G., and SAARI, E. 1966. A note on the properties of manganotantalite. *C.R. Soc. géol. Finlande*, **38**, 47.

—— SAHAMA, TH. G., and LEHTINEN, M. 1969a. Ferroan wodginite from Ankole, south-west Uganda. *Bull. geol. Soc. Finland*, **41**, 65.

—— VORMA, A., and NIXON, P. H. 1969b. Rankamaite, a new tantalum mineral from Kivu, Congo. *Bull. geol. Soc. Finland*, **41**, 47.

—— SAHAMA, TH. G., and LEHTINEN, M. 1969c. Scandian ixiolite from Mozambique and Madagascar. *Bull. geol. Soc. Finland*, **41**, 75.

VORMA, A., SAHAMA, TH. G., and HAAPALA, I. 1965. Alkali position in the beryl structure. *C.R. Soc. géol. Finlande*, **37**, 119.

R. BLACK and M. GIROD

9   Late Palaeozoic to Recent
igneous activity in West Africa
and its relationship to
basement structure

ABSTRACT. *Three major phases of Phanerozoic igneous activity in West Africa are briefly described: (i) post-Visean–pre-Jurassic dolerites of Mauritania, Mali, Guinea and Ivory Coast; (ii) late Palaeozoic– Tertiary alkaline Younger Granite ring-complexes of Niger, Nigeria and Cameroons; and (iii) Tertiary–Recent alkali basalts of Senegal, Hoggar, Aïr, Nigeria and Cameroons. The nature and distribution of these eruptive rocks are shown to be closely related to the structure of the basement. Whereas the first phase, a typical tholeiitic association, is restricted to the West African Craton, the latter two phases displaying alkaline affinities only occur in the relatively unstable belt of Katangan–Damaran orogenesis (500-700 m.y.). In contrast to the tholeiites which were injected during the gentle down-warping of the Taoudenni basin, the alkaline rocks tend to occupy zones of pronounced epeirogenic uplift, a pre-existing system of transcurrent faults having facilitated the rise of magma. It is suggested that the genesis of the tholeiites and the Younger Granite ring-complexes may be related to events in the mantle leading to the break-up of Gondwanaland. In the case of the Tertiary to Recent vulcanism, magma is thought to have been generated locally by pressure release in arched zones.*

In this chapter some of the most important features of West African late Palaeozoic to Recent igneous activity are reviewed in the light of modern work on the structure of the basement complex.

Formation of the West African stable platform

It is now generally accepted that the African continent, bordered by the Alpine Atlas chain to the north, the Hercynian Mauritanides to the north-west, and the Cape Coast folding at its southern tip, consisted of three ancient cratons in late Precambrian–early Palaeozoic times, namely the West African, Congo and Kalahari Cratons, bounded by zones affected by thermo-tectonic events during Katangan-Damaran (Pan-African) cycle 500-700 m.y. ago (Kennedy, 1964; Rocci, 1965; Cahen and Snelling, 1966; see Clifford, this volume, p. 14).

The *West African Craton*, whose centre is gently downwarped to form the Taoudenni basin (Fig. 1), has been unaffected by orogenesis for over 1600 m.y. Its limits are clearly defined by geochronological data, and are often marked by faults, and by thrusting of more recent orogenic belts towards the craton (see Figs. 1 and 3). Recently, it has been recognised that the lowest platform deposits on the West African Craton are Precambrian in age, whereas those

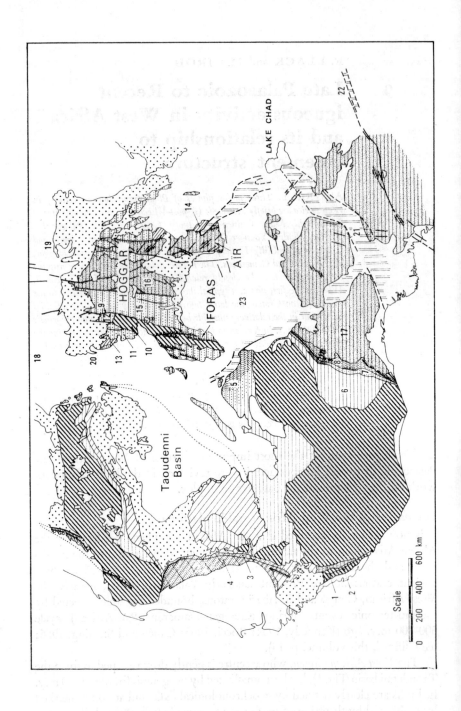

Scale

0   200   400   600 km

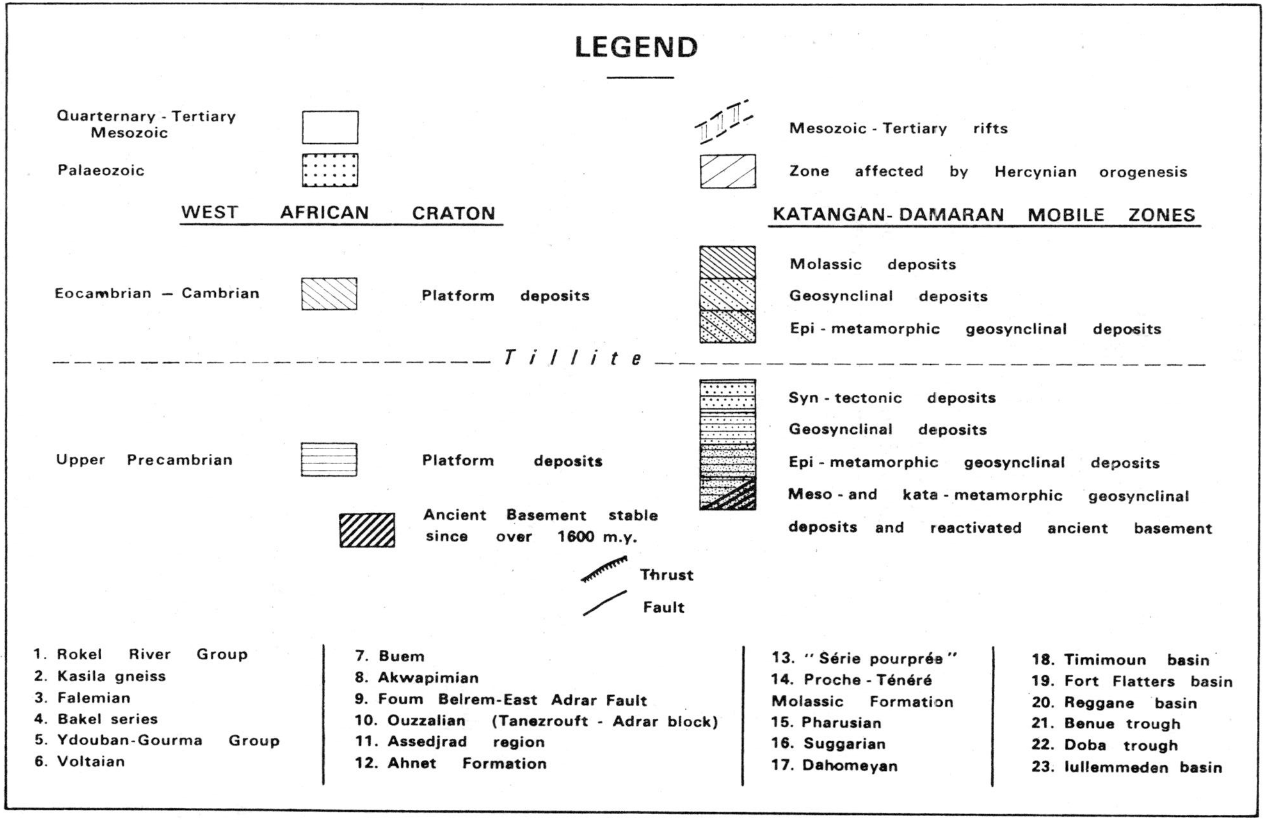

LEGEND

Quarternary - Tertiary
Mesozoic

Palaeozoic

WEST AFRICAN CRATON

Eocambrian — Cambrian   Platform deposits

----------- T i l l i t e -----------

Upper Precambrian   Platform deposits

Ancient Basement stable since over 1600 m.y.

Mesozoic - Tertiary rifts

Zone affected by Hercynian orogenesis

KATANGAN-DAMARAN MOBILE ZONES

Molassic deposits

Geosynclinal deposits

Epi - metamorphic geosynclinal deposits

Syn - tectonic deposits

Geosynclinal deposits

Epi - metamorphic geosynclinal deposits

Meso - and kata - metamorphic geosynclinal deposits and reactivated ancient basement

Thrust

Fault

1. Rokel River Group
2. Kasila gneiss
3. Falemian
4. Bakel series
5. Ydouban-Gourma Group
6. Voltaian
7. Buem
8. Akwapimian
9. Foum Belrem-East Adrar Fault
10. Ouzzalian (Tanezrouft - Adrar block)
11. Assedjrad region
12. Ahnet Formation
13. "Série pourprée"
14. Proche - Ténéré Molassic Formation
15. Pharusian
16. Suggarian
17. Dahomeyan
18. Timimoun basin
19. Fort Flatters basin
20. Reggane basin
21. Benue trough
22. Doba trough
23. Iullemmeden basin

FIG. 1. Structural map of West Africa.

overlying the more recent orogenic belt in the Hoggar to the east (Fig. 1) are of uppermost Cambrian or Lower Ordovician age (Legrand, 1964; Beuf *et al.*, 1968a). This recognition raised the possibility that the horizontal deposits lying on the craton may be correlative with the folded and metamorphosed formations in peripheral geosynclinal belts (Black, 1966, 1967).

In the Taoudenni basin, a tillite separates the underlying Precambrian sediments from formations showing lithological affinities with Silurian and later sediments, but which have either been attributed to the Cambrian–Ordovician (Zimmermann, 1960) or to the Eocambrian–Lower Cambrian. The pre-tillite sequence displays internal unconformities and a pattern of thickness variations which differ from those of the post-tillite formations. Moreover, the former pass laterally into geosynclinal sediments flanking both western and eastern margins of the craton (Fig. 1). At the western margin, for example, the Rokel River Group of Sierra Leone (Fig. 1, no. 1) was deformed and locally metamorphosed in pre-Silurian times during the Katangan–Damaran Orogeny (Allen, 1968), which also rejuvenated, probably by upthrusting, the coastal band of Kasila gneiss (Fig. 1, no. 2). Further north, in Senegal and Mauritania, where subsidence continued into the Palaeozoic, the folded Falemian sequence and its metamorphic equivalent the Bakel Series (Fig. 1, nos 3 and 4) have been correlated with the *post-tillite* deposits of the Taoudenni basin. Whereas radiometric data in Senegal suggest the metamorphism to be essentially Caledonian (Bassot *et al.*, 1963), the presence of folded and thrust Devonian further north in Mauritania has led Sougy (1962) to consider the orogeny as Hercynian. Recent structural studies have shown the superposition of several phases of folding, the last Hercynian phase petering out southwards into Senegal (Bassot, 1966).

Along the eastern margin of the West African Craton, between Timbuctu and Gao, the Ydouban-Gourma Group (Fig. 1, no. 5) represents the filling of a geosyncline, folded, partly metamorphosed, and intruded by granite at the close of the Precambrian. Reichelt (1966) has shown these deposits to be the lateral equivalent of the pre-tillite platform deposits of the Taoudenni basin. Further south in Ghana, Togo and Dahomey, the Voltaian-Buem-Akwapimian sequence (Fig. 1, nos. 6-8) may also represent the same passage from platform deposits in the west to geosynclinal deposits in the east; the intensity of folding and of metamorphism increases eastwards towards the most mobile part of the belt (Grant, 1967). The highly metamorphosed formations, formerly assigned to the Dahomeyan (Fig. 1, no. 17), which extend over the greater part of Dahomey, Nigeria, and the northern Cameroons, are intruded by abundant syn- and post-orogenic granites of Katangan-Damaran age (Bonhomme, 1962; Jacobson *et al.*, 1964; Hurley *et al.*, 1966; Lasserre, 1964). These formations are now thought to include both geosynclinal sediments of Upper Precambrian age, and remobilised ancient basement, the latter often displaying metamorphism in the granulite facies (Black, 1966). The general structural picture in Ghana, Togo and Dahomey shows westward-directed overfolding and thrusting of the younger Katangan-Damaran Orogenic Belt towards the craton.

In the Hoggar–Iforas–Aïr regions (Fig. 1), the basement is divided into distinct tectonic and stratigraphical units by major north-south faults. The Foum Belrem–East Adrar Fault (Fig. 1, no. 9), which can be traced for over 900 km (Karpoff, 1965), separates two contrasting structural domains:

To the west, the Tanezrouft–Adrar Belt of Ouzzalian rocks (Lelubre, 1967) (Fig. 1, no. 10) consists of granulite facies gneisses and granites, which have yielded Rb-Sr whole rock isochron ages around 2850 m.y., and have been unaffected by thermo-tectonic rejuvenation since 1700 m.y. ago (Eberhardt *et al.*, 1963; Ferrara and Gravelle, 1966). Structurally this belt formed a rigid block, separated from the West African Craton to the southwest by a geosyncline filled by the Ydouban-Gourma Group. To the north, the Ouzzalian rocks plunge beneath the Ahnet Formation (Fig. 1, no. 12) comprising a considerable thickness of quartzites (3500 m) overlain by stromatolitic limestones and capped by calc-alkaline volcanics (Arene, 1968). Lastly, all these formations are unconformably overlain by a platform deposit ('Série pourprée' or Nigritian) composed of sandstones, greywackes, conglomerates, jaspers and shales (Reboul *et al.*, 1962) (Fig. 1, no. 13); the presence of a tillite (Fabre *et al.*, 1962) and of intercalated ignimbrites, related to volcanic rocks dated at 550 ± 30 m.y. by Rb-Sr whole rock studies, indicates an Eocambrian or Cambrian age for this series and suggests a correlation with the post-tillite formations of the Taoudenni basin (Caby, 1967a and b).

To the east of the Foum Belrem–East Adrar Fault (Fig. 1, no. 9), the Katangan-Damaran orogenic zone is composed of alternating tectonic blocks of Suggarian and Pharusian rocks. The Suggarian, a highly metamorphosed complex comparable to the Dahomeyan, probably comprises both Upper Precambrian sediments and zones of reactivated ancient basement. The Pharusian, generally metamorphosed in the greenschist facies, is composed of polygenetic conglomerates, locally attaining a thickness of several thousand metres (Lelubre, 1952), rhyolites and andesites with intercalated quartzites and limestones overlain by flysch; recently, an unconformity has been discovered within the sequence (Gravelle, 1965; Bertrand *et al.*, 1966). The lithology of the Pharusian indicates an unstable zone of deposition in proximity to a rising Suggarian mountain chain that was undergoing rapid erosion. As both the Suggarian and Pharusian have been invaded by abundant syn- and post-orogenic granites, whose radiometric ages range between 500 and 650 m.y., the two units have been tentatively interpreted as representing closely succeeding orogenic cycles within the same metamorphic belt (Black, 1966; Eberhardt *et al.*, 1963; Lay and Ledent, 1963; Boissonnas *et al.*, 1964; Lay *et al.*, 1965; Picciotto *et al.*, 1965; Boissonnas *et al.*, in preparation). Indeed the destruction of the chain had actually already started with the deposition of the Pharusian and was completed with the deposition of the Proche-Ténéré tillite-bearing molassic formation (Fig. 1, no. 14) and the 'Séries Intermédiaires' of central Hoggar. As the last Pharusian orogenic granites have been dated at 560 m.y.(*ibid.*) there is an interval of about 50 m.y. to allow for the peneplanation of the Hoggar before the deposition overlying (Upper Cambrian–Ordovician) cover.

To conclude, West Africa attained its identity as a stable platform at a much younger date than was previously thought; it is of Lower Palaeozoic age.

Instability of the Katangan–Damaran Orogenic Belt during the Phanerozoic

Since Lower Palaeozoic times the zones affected by orogenesis on the western and eastern sides of the West African Craton have continued to display a degree of instability (Kennedy, 1965). In contrast to the quiescent conditions that prevailed on the craton with the broad shallow down-warping of the Taoudenni basin, Hercynian orogenesis to the west was followed by the formation of coastal Mesozoic and Tertiary basins whilst, to the east, the Katangan–Damaran Orogenic Belt was, throughout the Phanerozoic, the site of important vertical movements largely determined by a prominent pre-existing fault pattern.

Focussing our attention on the Aïr (Fig. 1), Fig. 2 shows that the region was affected by two types of major faulting which reflect the Katangan–Damaran Orogeny (Black *et al.*, 1967). In late Precambrian times the chain was uplifted by movements of more or less rigid blocks displaced along conjugate inwardly-dipping thrust planes inclined towards the axial zone of the Aïr (Tafadek and Aouzegueur Thrusts). Rupture along these planes can be explained by an initial stress distribution with the direction of maximum compressional stress east-west, thus coinciding with the direction of the lateral compression which caused folding in depth, and a minimum vertical stress due to the counteraction of lithostatic load by isostasy. Once isostatic equilibrium was reached, following erosion of the chain and peneplanation and deposition of the Proche-Ténéré molassic formation, the stress field was modified: the direction of minimum stress became horizontal and perpendicular to the east-west stresses which continued to play a predominant role. The new regime induced north-west-trending sinistral wrench faults accompanied by a few complementary east-north-east and north-east-trending dextral faults. It seems therefore that continuity prevailed in the direction of orogenic forces which produced folding in depth, uplift by conjugate thrusting, and finally sinistral rotation of the massif by transcurrent faulting.

This pattern of pre-Ordovician faulting, although partly masked by later superposed faulting, extends over most of the Katangan–Damaran Orogenic Belt to the east of the West African Craton (Fig. 1). The major north-south faults that partition the Hoggar are thought to have been pre-existing high-angle thrusts that were active during the rise of fault-bounded blocks of Suggarian (Fig. 1, no 16) and the deposition of the Pharusian (Fig. 1, no. 15). Transcurrent faulting similar to that of the Aïr is extensively

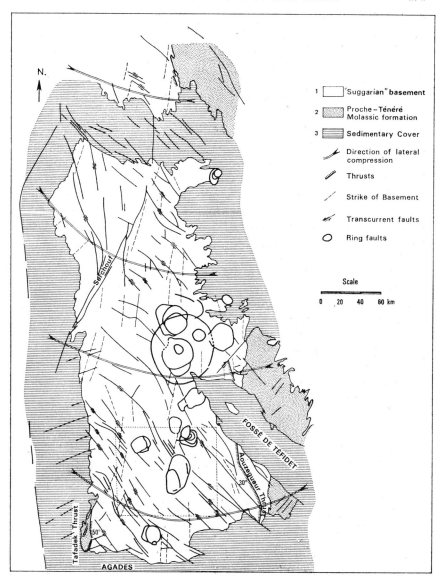

FIG. 2. Tectonic map of the Aïr. Rectangular area in the south is the area of Fig. 7.

developed in central and eastern Hoggar. In Dahomey, Togo and Ghana to the south, thrusts and reverse faults mark the western limit of the Katangan–Damaran Orogenic Belt. The major faults recorded in Nigeria and the Cameroons are north-easterly and east-north-easterly-trending dextral wrench faults that are complementary to the north-westerly sinistral faults that traverse the Aïr.

Many of these fractures have suffered renewed movements at intervals throughout the Phanerozoic and have influenced the localisation of zones of subsidence and intermittent uplift. The Hoggar–Iforas–Aïr massif is surrounded by Palaeozoic basins, some of which are over 5000 m deep; for example, Reggane (Fig. 1, no. 20), Timimoun (Fig. 1, no. 18) and Fort Flatters (Fig. 1, no. 19). The centres of maximum subsidence avoid the prolongation of the stable Tanezrouft–Adrar cratonic zone (Fig. 1, no. 10), and lie on either side of the major north-south faults that cut the Hoggar basement. The first stages in the development of individual basins can be traced back to the Ordovician with activity along the Foum Belrem–East Adrar Fault (Beuf et al., 1968b), and have been accentuated by Caledonian and important Hercynian epeirogenic movements, accompanied by renewed vertical displacements along north-south faults (Heybroek, 1963; Reboul et al., 1962). Caledonian uplift of the Aïr block resulted in the removal of the Ordovician cover, so that the Lower Devonian lies directly on the basement. During the Hercynian movements much of the Hoggar and the eastern portion of the Aïr acted as uplifted blocks. Additional evidence for this uplift is indicated by seismic results which show that the Conrad discontinuity, at − 19 km to the north of the Hoggar, rises to − 11 km in the centre of the crystalline massif (Merlet, 1962). The southern provenance of Lower Palaeozoic sediments, the thinning of formations to the west of the Aïr as they are traced southwards, and the absence of Palaeozoic formations in the south suggest that most of Nigeria and the Cameroons formed an emerged land mass in Palaeozoic and early Mesozoic times.

Mesozoic deformation and faulting have been important over the entire zone affected by the Katangan-Damaran Orogeny. Besides reactivation of north-south faults both to the north (Heybroek, 1963) and south (Joulia, 1957) of the Hoggar–Iforas–Aïr Massif, and the deepening of the northern Sahara Mesozoic basins, the pre-Ordovician wrench faults have played a role in localising new zones of subsidence. In Nigeria geophysical studies suggest that the Benue trough (Fig. 1, no. 21), filled by over 5000 m of Cretaceous sediments, was initially an Albian rift valley bound by east-north-easterly faults, a direction which coincides with that of pre-existing transcurrent faulting (Cratchley and Jones, 1965). The pre-Maestrichian trough to the south of the Iforas (Radier, 1953) may also be a rift structure localised on ancient north-westerly wrench faults and may link up with the lower Niger Cretaceous basin (King, 1950). In southern Chad the Doba trough (Fig. 1, no. 22), masked by Tertiary cover (Louis, 1962), also lies on the prolongation of a zone of ancient east-north-easterly wrench faulting (the Ngaoundéré Fault). The Cenomanian-Turonian marine transgression, which led to the coalescence of the Benue Sea and the Mediterranean, submerged parts of the Hoggar and the Aïr, between land masses formed by Taoudenni–Iforas to the west and Djado–Chad to the east (Faure, 1966). Following an important phase of Senonian folding and associated north-north-easterly faulting in Nigeria, the sea retreated from the Benue and eastern Niger to form a basin centred on the Iforas, where marine conditions continued into the Tertiary (Greigert, 1966). The Tefidet–Termit graben to the east of the Aïr, localised on ancient north-westerly-trending wrench faults, was also probably formed at the close of the Cretaceous or in the early Tertiary (Faure, 1966).

The Tertiary and Quaternary history was marked by pronounced uplift of the central Saharan crystalline massifs (Kilian, 1928; Lelubre, 1952), accompanied by renewed movements along ancient faults (Birot and Dresch, 1955; Conrad, 1968). The rise of the Aïr–Jos Plateau axis (see Fig. 3), which coincides with the distribution of the Younger Granites, separated the Iullemmeden basin (Fig. 1, no. 23) to the west from the Chad basin to the

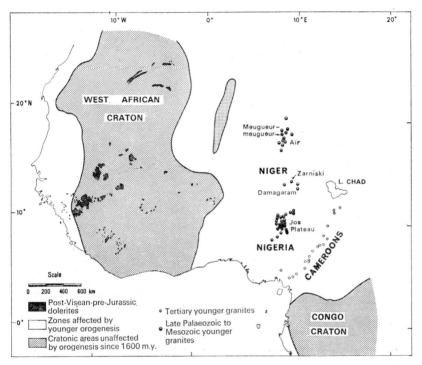

FIG. 3. Map of West Africa showing the distribution of post-Visean–pre-Jurassic dolerites and of the late Palaeozoic–Tertiary Younger Granites.

east (Furon, 1965). Uplift of a horst in the southern Cameroons, and of the Cameroon Highlands, also occurred as a counterpart to the deepening of the adjacent Niger delta and Chad basin.

Late Palaeozoic to Recent igneous activity

The difference in behaviour between the stable craton and the adjacent younger metamorphic belts is reflected in the nature and distribution of late Palaeozoic to Recent eruptive rocks. Since its consolidation as a platform, West Africa has been the site of three major phases of igneous activity:

1. Post-Visean–pre-Jurassic dolerites of Mauritania, Mali, Guinea and Ivory Coast (Fig. 3).

2. Late Palaeozoic–Tertiary ring-complexes of Niger, Nigeria and Cameroons (Fig. 3).

3. Tertiary–Recent vulcanism of Senegal, Hoggar, Aïr, Nigeria and Cameroons (Fig. 4).

The first group is a typical tholeiitic dolerite-granophyre association and is restricted to the West African Craton, whereas the other two groups

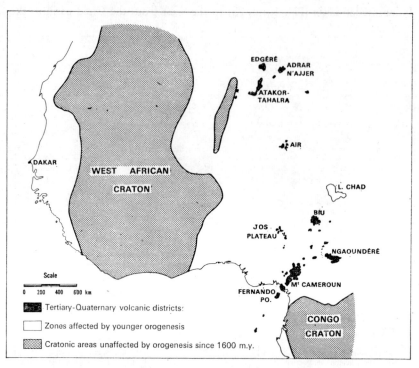

FIG. 4. Map of West Africa showing the distribution of Tertiary–Recent vulcanism.

of pronounced alkaline affinities occur only in zones affected by younger orogenesis. The Cretaceous intrusions and vulcanism associated with the development of the Benue trough will not be discussed.

POST-VISEAN–PRE-JURASSIC DOLERITES

Dolerites are extensively developed on the West African Craton (Fig. 3). Although several periods of injection are known to have occurred in the Precambrian (Machens, 1966; Knopf, in press), the major phase of emplacement is younger. The dolerite forms sheets, sills and dyke-swarms that crop out around the margin of the Taoudenni basin and are overlain by formations ascribed to the 'Continental Intercalaire'. Individual sills locally attain a thickness of several hundred metres and can be traced laterally for over 50 km. The intrusions are found in rocks of various ages including Lower

Palaeozoic, Devonian and Visean horizons (Dars, 1961). The post-Visean–pre-Jurassic age limits have been fixed by stratigraphical relationships.

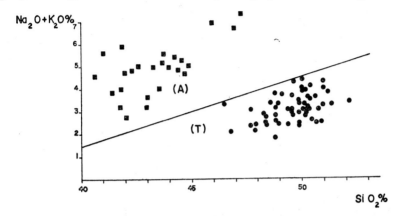

Fig. 5. Variation diagram showing ratios of total alkalies to silica (weight %) for the post-Visean–pre-Jurassic dolerite–granophyre association (T) and for the Tertiary–Recent alkali olivine basalt–trachyte association (A).

The dolerites form a very homogeneous group and display very similar mineralogical, textural and chemical features over distances exceeding 1500km. They are associated with granophyres, which also form dykes and pipes

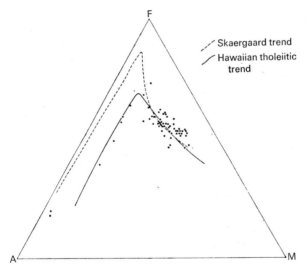

Fig. 6. AFM variation diagram for the post-Visean–pre-Jurassic dolerite–granophyre association.

(*ibid.*). Amongst the thicker sills, examples of gravity differentiation have been recorded (Barrère and Slansky, 1965).

All these rocks belong to the tholeiitic series. The dolerites frequently contain low-calcium clinopyroxenes (pigeonite) and sometimes orthopyroxene and quartz. Published chemical analyses (Dars, 1961) show low total alkalies to silica ratios, the values lying within the limits determined for the Hawaiian tholeiites (Fig. 5). Furthermore, the dolerite-granophyre trend shows the characteristic iron enrichment of the tholeiite association (Fig. 6).

Villemur (1967) has pointed out that the Taoudenni basin (Fig. 1), as it is known at present, started its development in the Devonian, and gentle down-warping continued during and after the Carboniferous. There may be therefore a direct relationship between the formation of the basin and the emplacement of the dolerites. In this connection, it is interesting to note the rarity of dolerites in eastern Ivory Coast and Upper Volta, especially along the upwarp separating the Taoudenni and Voltaian basins.

Close structural and petrological analogies can be drawn between the dolerite-granophyre association of West Africa and that of the Karroo Period of southern Africa (see Cox, this volume, p. 211). It is significant that in both cases the tholeiites are chiefly situated on the old cratons (West African, Congo, Kalahari Cratons) and particularly in down-warped zones (Taoudenni, Congo and Kalahari basins).

LATE PALAEOZOIC–TERTIARY YOUNGER GRANITE RING–COMPLEXES

The Younger Granites are high-level non-orogenic intrusions (Fig. 3). In Nigeria and Niger, they form a group of over 60 massifs lying within a narrow north-south strip, 1500×200 km, defined by longitudes 8° and 10° E. and latitudes 8° and 21° N. Radiometric age determinations on rocks from Nigeria have given a wide range of results, but the most reliable of these indicate a Jurassic age of about 160 m.y. (Darnley et al., 1962; Jacobson et al., 1964). There are, however, reasons for believing that the more northerly complexes of the Aïr (see Fig. 3) are rather older, and recent unpublished K-Ar determinations have yielded ages around 295 m.y. (Schetcheglov, 1965). It seems therefore that along the ninth meridian, magmatism occurred over a long time interval and may in part be contemporaneous with the tholeiitic activity on the craton. A few isolated small ring-complexes, probably of Cretaceous age, have been described in the Hoggar (Rémy, 1960).

In the Cameroons and Chad (Fig. 3), another group of over twenty small plutons, some of which have yielded Tertiary isotopic ages (Lasserre, 1966), have a north-easterly alignment coincident with that of the Fernando Po and São Thomé volcanic islands. It is probable that the calderas with associated ignimbrites of Tibesti which lie on the same alignment, represent the volcanic super-structures of this group of plutons (Vincent, 1963 and this volume, p. 301).

Detailed field studies of these complexes have demonstrated a succession of magmatic activity from vulcanism to plutonism and have revealed a range of structural and tectonic processes associated with the high-level emplacement of granite by cauldron subsidence (Jacobson et al., 1958; Rocci, 1960; Turner, 1963; Black, 1963; Black et al., 1967). The initial stage in the development of the complexes was marked by the extrusion of vast quantities of acid lavas and welded tuffs, now only partially preserved as a result of subsidence along ring-faults. The rhyolites generally fall into two distinct groups which represent different modes of eruption. The earlier rhyolites are typical of sporadic activity from simple central-vent volcanoes or from groups of elongated vents along ring-fractures. The later, por- phyritic rhyolites are believed to have been extruded as a result of surface cauldron subsidence. Intrusive late rhyolites also occur as dykes, sheets and shallow sub-surface cauldrons within the earlier volcanic accumulations. With one exception (Bilète), where a thin undated conglomerate and water- laid tuffs underlie the volcanics (Black *et al.*, 1967), extrusive rhyolites directly overlie the basement, thereby providing direct evidence for the emplacement of Younger Granites in zones of uplift which had already been stripped of their sedimentary cover in late Palaeozoic and Jurassic times.

Ring-dykes are the most striking features of the province. The majority range from 8 to 20 km in diameter but there are many exceptions, notably the perfectly circular Meugueur-meugueur gabbro ring-dyke (Figs. 2 and 3), with a radius of 33 km (Black *et al.*, *ibid.*), and the small complete ring of peralkaline granite within the Zarniski Complex whose span does not exceed 2 km (Black, 1963). In outline they are either circular or polygonal, the latter form predominating among the narrow dykes cutting the basement and the former in broad ring-dykes within the massifs. For most of the poly- gonal ring-dykes there is clear evidence that the form has been controlled by pre-existing wrench faults and jointing in the basement (Fig. 7). The width is also variable and whilst in most cases intrusion is believed to have been per- missive, locally block stoping in a wide fracture zone has taken place. Elliptical and crescent-shaped intrusions are common and some resulted by subsidence of blocks bounded by two or more ring-fractures of different radii. Granites and syenites also appear as simple stocks and bosses but doubt- less form part of ring-dykes at greater depth. Cone-sheets have been re- cognised but are not common. They are intrusive into the rhyolites but generally predate the intrusion of the granites by which they are frequently truncated.

Ring-fracturing, which accompanied the emplacement of ring-dykes, has exerted a major control on the structure and form of individual com- plexes. Thus the initial ring-dyke frequently defines a centre of repeated igneous activity and encloses a number of granitic intrusions emplaced concentrically. Intrusions frequently partly overlap one another to form chain-like alignments, igneous activity having ended in each centre before the initiation of a new centre. This fact, together with the rarity of intrusions in the basement separating individual massifs, and the similar sequences in successive centres, suggest that each centre marked the formation of a new and separate magma chamber (Turner, 1963). Although a general tendency towards a southern shift of centres is revealed in Nigeria, there are many exceptions and it is impossible to relate linear groupings of complexes to regional tectonic features.

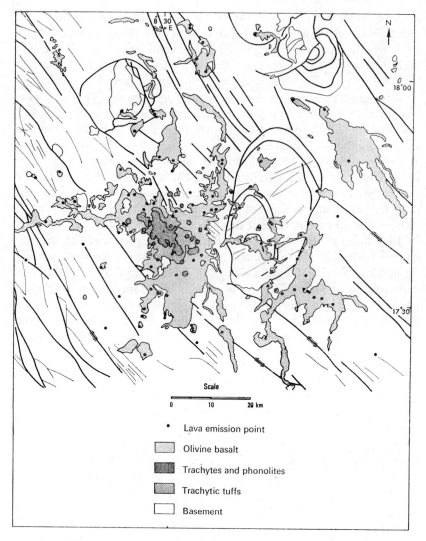

FIG. 7. Map of southern Aïr showing the relationship between the location of lava emission points and faulting (pre-Ordovician wrench faults, late Palaeozoic–Mesozoic ring-faults). See inset area of Fig. 2.

It has been pointed out that, in Nigeria, the Younger Granites display a unified joint pattern; the major joints trending in a north-westerly direction (Turner, in press). The same observation has been made in the Aïr, but there joints have an east-north-easterly orientation which coincides with the planes of weakness that are complementary to the north-westerly wrench faults and which were under tension during the emplacement of the Younger Granites. Microgabbro, rhyolite and peralkaline microsyenite dykes cutting the basement follow the same east-north-easterly trend.

Turner (1963) has distinguished two types of ring-dyke which are believed to have resulted from distinct sets of processes. The initial peripheral ring-dykes are generally vertical and were emplaced in a ring-fracture that probably reached the surface and which had already taken part in the formation of a caldera during the volcanic cycle. The later internal ring-dykes, on the other hand, often display outward-dipping contacts and, at a higher level, they feed a sub-horizontal sheet extending across the complex. These are plutonic structures and fit admirably the model suggested by Anderson (1936). Almost all the rock types of the suite appear as ring-dykes, but the peripheral variety which often initiated an evolving petrographic series within a centre, is favoured by nordmarkites (quartz syenites, microsyenites) in the north, and by the group of hastingsite granite-porphyries and granites in the southern part of the province.

The most striking petrographical feature of the province is the overwhelmingly acid character of the suite and the uniformity of facies found in all areas. Over 95% of the rocks can be classified as rhyolites, quartz-syenites, peralkaline granites, hastingsite granites and biotite granites, with anorthosite and olivine gabbro forming the remaining 5% (Table 1). The order of intrusion of plutonic phases varies from one complex to another but as a general rule quartz-syenite preceded peralkaline granite and (fayalite-hedenbergite-) hastingsite granite was followed by biotite granite. In several of the Aïr massifs, the plutonic sequence leading to peralkaline granite starts with the high-level emplacement of anorthosite.

The overall unity of the acid and intermediate rocks is emphasised by the importance of perthite. With the exception of the albite-riebeckite granites, which have notably high soda content, soda and potash values are not particularly high for alkaline rocks and in terms of molecular proportions there is generally only a slight excess of soda over potash. Lime is consistently low and is reflected in the nature of the plagioclase which rarely contains more than 15% of the anorthite molecule. The peralkaline character of many of the rocks is essentially due to a deficiency of alumina relative to alkalies. The coloured minerals are very varied and provide the chief distinctions between the main rock types. Their nature (Borley, 1963; Fabries and Rocci, 1965) is to a large extent determined by the total alkalies to alumina ratio which gives rise to two trends: (i) a *miaskitic series* characterised by iron-calcium and iron minerals (hedenbergite, hastingsite, biotite); and (ii) an *agpaiitic series* containing iron-soda minerals (aegirine, riebeckitic arfvedsonite, aenigmatite, astrophyllite and rarely narsarsukite). The very low magnesia/iron ratio is shown in the nature of the femic minerals: fayalite, hedenbergite, soda-iron amphiboles and iron-rich micas. The suite is characterised by a high fluorine content, fluorite, topaz, cryolite and thomsenolite appearing as accessory minerals, and an abundance of rare-earth and radio-active elements entering into pyrochlore, fergusonite, monazite, xenotime, allanite and euxenite. The Younger Granite province is one of the richest tin and columbite producing regions of the world. Cassiterite is generally found in greisens associated with the biotite granites but is also present as a fine-grained accessory mineral in some of the biotite granites. Niobium is particularly

TABLE 1

Estimate of areal distribution (in km²) of principal rock types in the Younger Granite province of Niger, Nigeria and Cameroons (see Fig. 3)

	Anorthosite	Gabbro	Quartz–syenite	Hastingsite granite	Biotite granite	Peralkaline granite	Rhyolite	Totals
AÏR	379	91	653	467	171	1964	392	4117
Areal %	9	2	16	11	4	48	10	
DAMAGARAM	—	—	152	2	3	896	978	2031
Areal %	—	—	8	—	—	44	48	
NIGERIA	—	35	320	520	3716	839	1245	6675
Areal %	—	1	4	8	56	12	19	
CAMEROONS	—	27	184	98	88	225	66	688
Areal %	—	4	27	14	13	33	9	
GRAND TOTALS	379	153	1309	1087	3978	3924	2681	13511
Areal %	3	1	10	8	29	29	20	

abundant in soda-rich granites and crystallises in the form of columbite in the biotite granites and pyrochlore in the peralkaline granites.

The gabbros and anorthosites have similar mineralogy and are composed of plagioclase mainly in the andesine-labradorite range, olivine with a composition around Fa_{35}, titaniferous augite, amphibole and biotite. The anorthosites which are generally extremely coarse-grained are very similar to those described from Gardar, Greenland (Upton, 1964; Bridgwater and Harry, 1968).

The variation diagrams shown in Figs. 8 and 9 illustrate the essential chemical relationships between the main rock types of the province and

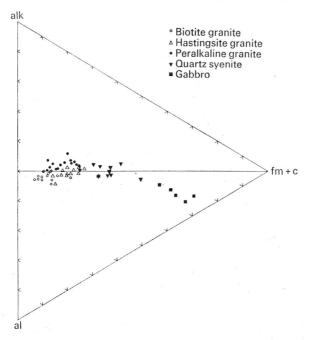

FIG. 8. Niggli variation diagram (alk, al, (fm+c)) of Younger Granites.

show the significant diminution of lime, iron and magnesia from syenite to granite. Fig. 8 shows the close balance between alkalies and alumina which prevails in the acid and intermediate rocks and suggests that differentiation has tended towards an alumina-deficient fraction represented by the peralkaline granites and an alumina-rich fraction represented by the biotite granites. In the AFM diagram (Fig. 9), it is interesting to note separate fields for syenites and peralkaline granites on the one hand, and for syenites, hastingsite granites and biotite granites on the other; the peralkaline granites and the hastingsite granites fall in the same field.

The Younger Granites belong to the gabbroic-alkaline association (Sheinmann et al., 1961). The syenites and peralkaline granites display a trend similar to that observed in more evolved sequences in oceanic islands and two main types of genetic hypotheses have been advanced. Upton (1960), discussing the Gardar province, maintains that they have developed by fractionation of an alkali olivine basalt magma in a narrow deep crustal and sub-crustal chamber of great vertical extent. Fractionation gave rise to a lower horizon of olivine gabbro with concentrations of anorthosite, and an upper syenite magma, which by further fractionation gave peralkaline granites. However, Bailey and Schairer (1966) have suggested that syenite liquids represent an optimum fractional melting composition of olivine basalt and that peralkaline granites are also primary magmas produced by partial melting of an oversaturated syenite; in their opinion the associated

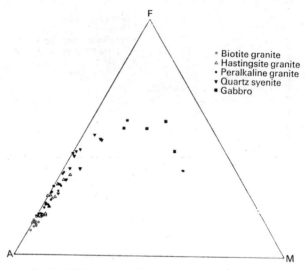

FIG. 9. AFM variation diagram of Younger Granites.

sub-aluminous granites (hastingsite granites and biotite granites) have resulted from sialic contamination.

In the Younger Granite province, the overwhelming predominance of acid rocks lends support to the Bailey and Schairer hypothesis involving the formation of primary intermediate and acid magmas by partial fusion of crustal and sub-crustal rocks, a process favoured by cauldron subsidence. The existence at depth of immense basaltic magma reservoirs seems unlikely in view of the rarity of associated basaltic lavas in a terrain sliced by pre-Ordovician wrench faults which provided readily-opened conduits for the rise of basalts, as shown by the Tertiary–Recent vulcanism (Fig. 7). The partial fusion of sialic material in the case of the sub-aluminous granites would also serve to explain the presence of cassiterite both in the Younger Granites and in the Nigerian basement pegmatites, and its absence in similar igneous provinces (New Hampshire, Oslo, Gardar). This view is confirmed by the discovery of isotopic lead ratios yielding Katangan-Damaran ages in feldspar from Younger Granites (Tugarinov, 1968).

TERTIARY—RECENT VULCANISM

Tertiary to Recent alkali vulcanism in West Africa is limited to zones affected by younger orogenesis on either side of the West African Craton (Fig. 4). To the east, the main districts occur in the Hoggar (Atakor–Tahalra, Adrar n'Ajjer, Edjéré), southern Aïr, Jos Plateau, Biu, southern Cameroons, Ngaoundéré and Tibesti. To the west, lavas of the same age and composition have been described from the Dakar region (Debant, 1961) and in the Cape Verde Islands (Berthois, 1950).

In the Hoggar the extrusion of lavas in Tertiary–Quaternary times has been accompanied by important epeirogenic movements. Contour maps of the region show the altitude of the Pharusian and Suggarian basement to be at its highest beneath the volcanic districts (Fig. 10). In Atakor, the basement has been raised by over 1000 m, presumably since early Pliocene (Girod, 1968). This observation also applies to southern Aïr (500 m of uplift), the Jos Plateau (1000 m of uplift), Biu (Carter et al., 1961), the southern Cameroon horst (Diebold, 1960), Ngaoundéré and Tibesti (see Fig. 4).

The rise of magma has been facilitated by the presence of pre-existing deep-seated wrench faults (Fig. 7). In the Hoggar (Girod, 1968), Aïr (Black et al., 1967) and Cameroons (Sarcia and Sarcia, 1952), the emission points of lavas are generally aligned along north-westerly and east-north-easterly faults. The volcanoes on the Jos Plateau occur along the more recent north-north-easterly faults. Both in Niger and Nigeria volcanic craters are also located on Younger Granite ring-faults and on their intersection with pre-Ordovician transcurrent faults in the basement (Fig. 7).

Epeirogenic uplift has been locally accompanied by the formation of grabens and it is significant that in southern Aïr and Ngaoundéré the volcanic districts are situated in arched zones at the head of the Tefidet–Termit and Doba Rifts (see Fig. 1).

The structural similarity of these regions is accompanied by a magmatic similarity which is particularly evident for the lavas extruded on either side of the West African Craton. They all belong to the alkali olivine basalt-trachyte association, and are characterised by the predominance of basic lavas (basanites, basanitoids, ankaratrites), by the scarcity of intermediate types (hawaiite, mugearites) and by the peralkaline nature of some of the most differentiated members (peralkaline phonolites, comendites, pantellerites). The agpaiitic tendency is expressed by the presence of minerals such as aegirine, arfvedsonite, kataphorite, aenigmatite, eudialyte and lavenite.

The vulcanism of Mt. Cameroon (Gèze, 1943) is rather more diversified and includes lavas which are particularly deficient in silica and sometimes potassic. Jérémine (1943) has described nephelinites, leucitites and hauynophyres whose closest equivalents are to be found in East Africa (Denaeyer and Chellinck, 1965). The presence in this region of large central-vent volcanoes with the development of calderas possibly marks a transition between the Younger Granite ring-complexes and the simple olivine basalt-trachyte series (Dumort, in press).

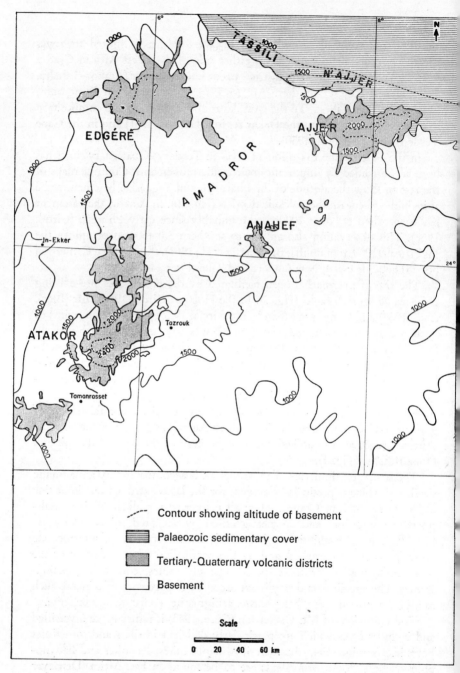

Fig. 10. Map of Tertiary–Recent volcanic districts of central Hoggar showing altitudes of the basement.

Basalts almost invariably are related to strato-volcanoes of strombolian type. Individual cones attain a height of over 200 m and have generally emitted a single flow whose thickness rarely exceeds 20 m. They contain numerous ultrabasic nodules as well as large euhedral intratelluric crystals of kaersutite, ferrisalite, zircon and ferripleonaste (Girod, 1968; Wright, 1968). The trachytes and phonolites were generally highly viscous and formed extrusions, plugs and cumulo-domes, all striking features in the landscape.

The contrast between the alkaline character of all these lavas occurring mainly in zones of uplift and the older tholeiites present in the down-warped Taoudenni basin is clearly shown in Fig. 5. The 140 analyses of West African

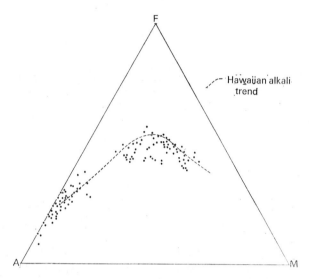

Fig. 11. AFM variation diagram of Tertiary–Recent lavas of West Africa.

Tertiary–Quaternary lavas plotted on an AFM diagram (Fig. 11) fall closely on the Hawaiian alkali basalt trend.

It appears that the Tertiary–Quaternary alkali vulcanism, in the relatively unstable Katangan-Damaran Belt of orogenesis to the east of the craton, has developed only in zones of pronounced epeirogenic uplift where deep-seated pre-existing faults provided readily opened conduits for the rise of sub-crustal magma.

Conclusions

In West Africa, there is a close relationship between the nature and distribution of late Palaeozoic to Recent igneous activity and the structure of the basement. Whereas the tholeiites are restricted to the West African Craton and have been irrupted during the down-warping of the Taoudenni basin, the alkali Younger Granites and alkali vulcanism occurring in the relatively unstable belt of Katangan-Damaran orogenesis, tend to occupy zones of pronounced epeirogenic uplift, a pre-existing system of transcurrent faults having facilitated the rise of magma.

Modern experimental and theoretical work suggest that whilst the genesis of tholeiites and alkali basalts may be explained by partial fusion of the upper mantle in different pressure-temperature environments (Yoder and Tilley, 1962; Kushiro and Kuno, 1963), the genesis of the gabbroic-alkaline series, which includes sub-aluminous granites, probably involves partial fusion of crustal as well as sub-crustal material. Although it is impossible at present to derive a theoretical model for the location and depth of melting, it may be significant that cratonic areas generally display a low geothermal gradient when compared to more recent orogenic belts (Makarenko et al., 1968).

The tholeiites and the alkali Younger Granites were probably partly contemporaneous, and it is tempting to relate their genesis to events in the mantle leading to the break-up of Gondwanaland (see Cox, this volume, p. 228). The comparison of structural maps of Brazil and West Africa, using the computed fit of Bullard et al. (1965), shows a unified system of east-north-easterly dextral and north-westerly sinistral wrench faults, and the presence of a Mesozoic peralkaline granite at Cabo (Brazil) which lies on the prolongation of the Aïr–Nigeria alignment (Almeida and Black, in press). Our current prejudice is to relate the zone of high heat flow along the ninth meridian and the formation of the basin on the craton to a convection system in the mantle. The Tertiary Younger Granites in the Cameroons are aligned on a more recent zone of high heat flow indicated by the Fernando Po and São Thomé 'oceanic' islands ('Cameroon line').

For the Tertiary to Recent olivine basalt-trachyte association, a genesis of magma by pressure release in arched zones as advocated by Bailey (1964) appears to fit well with field observations.

Acknowledgements

We wish to express our gratitude to Prof. M. Roques for having critically read the manuscript and for his helpful suggestions. We thank Madame D. Hilt of the Centre de Recherches sur les Zones Arides (C.N.R.S., Paris) for drawing the figures.

REFERENCES

ALLEN, P. M. 1968. The stratigraphy of a geosynclinal succession in western Sierra Leone, West Africa. Geol. Mag., 105, 62.

ALMEIDA, F. F. M. DE, and BLACK, R. (in press). Comparaison structurale entre le N.E. du Brésil et l'Ouest-Africain. Symp. Continental Drift, Montevideo, 1967.

ANDERSON, E. M. 1936. The dynamics of the formation of cone-sheets, ring-dykes and caldron subsidence. Proc. r. Soc. Edinb., 56, 128.

ARENE, J. 1968. Stratigraphie et évolution structurale du Précambrien dans la région de l'Adrar Ahnet (Sahara central). C.R. Acad. Sci. Paris, 266, 868.

BAILEY, D. K. 1964. Crustal warping—a possible tectonic control of alkaline magmatism. *J. geophys. Res.*, **69**, 1103.

—— and SCHAIRER, J. F. 1966. The system $Na_2O-Al_2O_3-Fe_2O_3-SiO_2$ at 1 atmosphere, and the petrogenesis of alkaline rocks. *J. Petrology*, **7**, 114.

BARRERE, J., and SLANSKY, M. 1965. Notice explicative de la carte géologique 2,000,000[e] de l'Afrique occidentale. *Mém. Bur. Rech. géol. min.*, **29**.

BASSOT, J. P. 1966. Etude géologique du Sénégal oriental et de ses confins guinéo-maliens. *Mém. Bur. Rech. géol. min.*, **40**.

—— BONHOMME, M., ROQUES, M., and VACHETTE, M. 1963. Mesures d'âges absolus sur les séries précambriennes et paléozoïques du Sénégal oriental. *Bull. Soc. géol. Fr.*, **5**, 401.

BERTHOIS, L. 1950. Contribution à la connaissance de l'archipel du Cap Vert. *Publ. Junta Invest. Colon. (Lisbon)*, **7**.

BERTRAND, J. M. L., BOISSONNAS, J., CABY, R., GRAVELLE, M., and LELUBRE, M. 1966. Existence d'une discordance dans l'Antécambrien du 'fossé pharusien' de l'Ahaggar occidental (Sahara central). *C.R. Acad. Sci. Paris*, **262**, 2197.

BEUF, S., BIJU-DUVAL, B., MAUVIER, A., and LEGRAND, PH. (1968a). Nouvelles observations sur le 'Cambro-Ordovicien' du Bled el Mass (Sahara central). *Bull. Serv. Carte géol. Algérie*, **38**, 39.

—— BIJU-DUVAL, B., CHARPAL, O. DE, GARIEL, O., BENACEF, A., BLACK, R., ARENE, J., BOISSONNAS, J., CACHAU, F., GUERANGE, B., and GRAVELLE, M. (1968b). Une conséquence directe de la structure du bouclier africain: l'ébauche des bassins de l'Ahnet et du Mouydir, au Paléozoïque inférieur. *Bull. Serv. Carte géol. Algérie*, **38**, 105.

BIROT, P., and DRESCH, J. 1955. Une faille du Quaternaire récent dans la plaine d'Amguid. *C.R. Soc. géol. Fr.*, **11-12**, 209.

BLACK, R. 1963. Note sur les complexes annulaires de Tchouni-Zarniski et de Gouré. *Bull. Bur. Rech. géol. min.*, **1**, 31.

—— 1966. Sur l'existence d'une orogénie riphéenne en Afrique occidentale. *C.R. Acad. Sci. Paris*, **262**, 1046.

—— 1967. Sur l'ordonnance des chaînes métamorphiques en Afrique occidentale. *Chron. Mines Rech. min.*, **364**, 225.

—— JAUJOU, M., and PELLATON, C. 1967. Notice explicative sur la carte géologique de l'Aïr, à l'échelle du 1 : 500,000. *Dir. Mines Géol.*, *Niger*.

BOISSONNAS, J., DUPLAN, L., MAISONNEUVE, J., VACHETTE, M., and VIALETTE, Y. 1964. Etude géochronologique et géochimique des roches du compartiment suggarien du Hoggar central (Algérie). *Annls. Fac. Sci. Univ. Clermont*, **25** (8), 73.

—— BORSI, S., FABRE, J., FABRIES, J., FERRARA, G., and GRAVELLE, M. (in preparation). On the Lower Cambrian age of two late granites of west central Ahaggar (Algerian Sahara).

BONHOMME, M. 1962. Contribution à l'étude géochronologique de la plateforme de l'Ouest-Africain. *Annls Fac. Sci. Univ. Clermont*, **5** (5).

BORLEY, G. D. 1963. Amphiboles from the younger granites of Nigeria. Part I, Chemical classification. *Miner. Mag.*, **33**, 358.

BRIDGWATER, D., and HARRY, W. T. 1968. Anorthosite xenoliths and plagioclase megacrysts in Precambrian intrusions of South Greenland. *Medd. Grønland*, **185** (2).

BULLARD, E. C., EVERETT, J. E., and SMITH, A. G. 1965. The fit of continents around the Atlantic. *Phil. Trans. r. Soc. Lond.*, **258**, Ser. A, 41.

CABY, R. 1967a. Existence de Cambrien à faciès continental ('Série pourprée', Nigritien) et importance du volcanisme et du magmatisme de cet âge au Sahara central (Algérie). *C.R. Acad. Sci. Paris*, **264**, 1386.

—— 1967b. Un nouveau fragment du craton ouest africain, dans le nord-ouest de l'Ahaggar (Sahara algérien); ses relations avec la série à stromatolites; sa place dans l'orogénie du Précambrien supérieur. *C.R. Acad. Sci. Paris*, **265**, 1452.

CAHEN, L., and SNELLING, N. J. 1966. *The geochronology of equatorial Africa*, Amsterdam.

CARTER, J. D., BARBER, W., TAIT, E. A., and JONES, G. P. 1963. The geology of parts of Adamawa, Bauchi and Bornu Provinces in north-eastern Nigeria. *Bull. geol. Surv. Nigeria*, **30**.

CONRAD, G .1968. L'évolution continentale post-hercynienne du Sahara algérien. Unpublished thesis, Univ. Paris.

CRATCHLEY, C. R., and JONES, G. P. 1965. An interpretation of geology and gravity anomalies of the Benue Valley, Nigeria. *Geophys. Pap. Overseas geol. Survs.*, **1**.

DARNLEY, A. G., SMITH, G. H., and CHANDLER, T. R. D. 1962. The age of fergusonite from the Jos area, northern Nigeria. *Miner. Mag.*, **33**, 48.

DARS, R. 1961. Les formations sédimentaires et les dolérites du Soudan occidental (Afrique de l'Ouest). *Mém. Bur. Rech. géol. min.*, **12**.

DEBANT, P. 1961. Les roches volcaniques récentes de la feuille Ouakam au 1 : 20,000. *Dipl. Etudes sup., Univ. Dakar*.

DENAEYER, M. E., and CHELLINCK, F. 1965. Recueil d'analyses des laves du fossé tectonique de l'Afrique centrale. *Annls. Mus. roy. Afr. centr.*, **49**.

DIEBOLD, P. 1960. Notes on the geology of southern Cameroons. *C.C.T.A. west-centr. reg. Comm. Geol.*, Kaduna, 16.

DUMORT, J. C. (in press). Carte géologique de reconnaissance du Cameroun au 1/500,000e. Feuille Douala-Ouest, avec notice explicative. *Publ. Dir. Mines Géol. Cameroun*.

EBERHARDT, F., FERRARA, G., GLANGEAUD, L., GRAVELLE, M., and TORGIORGI, R. 1963. Sur l'âge absolu des séries métamorphiques de l'Ahaggar occidental dans la région de Silet-Tilehaouine (Sahara central). *C.R. Acad. Sci. Paris*, **256**, 1126.

FABRE, J., FREULON, J. M., and MOUSSU, H. 1962. Présence d'une tillite dans la partie inférieure de la 'Série pourprée' de l'Ahnet (Nord-Ouest du Hoggar, Sahara central). *C.R. Acad. Sci. Paris*, **255**, 1965.

FABRIES, J., and ROCCI, G. 1965. Le massif granitique de Tarraouadji (République du. Niger). Etude et signification pétrogénétique des principaux minéraux. *Bull. Soc. fr. Minér. Crist.*, **88**, 319.

FAURE, H. 1966. Reconnaissance géologique des formations sédimentaires post-Paléozoïques du Niger oriental. *Mém. Bur. Rech. géol. min.*, **47**.

FERRARA, G., and GRAVELLE, M. 1966. Radiometric ages from western Ahaggar (Sahara) suggesting an eastern limit for the West African Craton. *Earth Planetary Sci. Let.*, **1**, 319.

FURON, R. 1965. Matériaux pour l'étude de la 'houle crustale' et de la méga tectonique du socle africain. *Rev. Géogr. phys. Géol. dynam.*, **7**(1), 21.

GEZE, B. 1943. Géographie physique et géologie du Cameroun occidental. *Mém. Mus. Hist. nat.*, **17**.

GIROD, M. 1968. Le massif volcanique de l'Atakor (Hoggar, Sahara algérien). Etude pétrographique, structurale et volcanologique. Unpublished thesis, Univ. Paris.

GRANT, N. K. 1967. Complete late Pre-cambrian to early Palaeozoic orogenic cycle in Ghana, Togo and Dahomey. *Nature, Lond.*, **215**, 609.

GRAVELLE, M. 1965. Problèmes de la géologie de l'Antécambrien dans la région de Silet (Ahaggar occidental, Sahara occidental, Sahara central). *C.R. Soc. géol. Fr.*, **7**, 233.

GREIGERT, J. 1966. Description des formations cretacées et tertiaires du Bassin des Iullemmeden (Afrique occidentale). *Dir. Mines Géol. Niger*, **2**.

HEYBROEK, P. 1963. Note on the structural development of the El Biod high and Amguid spur. *Rev. Inst. fr. Pétrole*, **10**, 1363.

HURLEY, P. M., RAND, J. R., FAIRBAIRN, H. W., PINSON, W. H. Jr., POSADAS, V. C., and REID, J. B. 1966. Continental drift investigations. *14th Ann. Rep. Dept. Geol. Geophys., M.I.T.* (for 1966), 3.

JACOBSON, R. R. E., MACLEOD, W. N., and BLACK, R. 1958. Ring-complexes in the Younger Granite Province of northern Nigeria. *Mem. geol. Soc. Lond.*, **1**.

—— SNELLING, N. J., and TRUSWELL, J. F. 1964. Age determinations on the geology of Nigeria with special reference to the older and younger granites. *Overseas Geol. Mineral. Resources*, **9**, 168.

JEREMINE, E. 1943. Contribution à l'étude pétrographique du Cameroun occidental. *Mém. Mus. Hist. nat.*, **17**.

JOULIA, F. 1957. Sur l'existence d'un important système de fractures intéressant le Continental Intercalaire à l'W. de l'Aïr (Niger). *Bull. Soc. géol. Fr.*, **6**, 137.

KARPOFF, R. 1965. Les grandes époques de fracture et de bombement au Sahara central. *Bull. Soc. géol. Fr.*, **7**, 469.

KENNEDY, W. Q. 1964. The structural differentiation of Africa in the Pan-African (±500 m.y.) tectonic episode. *8th Ann. Rep. Res. Inst. African Geol., Univ. Leeds*, 48.

—— 1965. The influence of basement structure on the evolution of the coastal (Mesozoic and Tertiary) basins of Africa. In *Salt basins around Africa*, p. 7. Inst. Petrol., London.

KILIAN, C. 1928. Sur la structure du Sahara Sud-Constantinois et Central. *C.R. Soc. géol. Fr.*, 71.

KING, L. C. 1950. Speculations upon the outline and the mode of disruption of Gondwanaland. *Geol. Mag.*, **89**, 353.

KNOPF, D. (in press). Les dolérites de Côte d'Ivoire. *West Afr. Assoc.*, Abidjan.

KUSHIRO, I., and KUNO, H. 1963. Origin of primary basalt magmas and classifications of basaltic rocks. *J. Petrology*, **4**, 75.

LASSERRE, M. 1964. Etude géochronologique par le méthode strontium–rubidium de quelques échantillons en provenance du Cameroun. *Annls. Fac. Sci. Univ. Clermont*, **25** (8), 53.

—— 1966. Confirmation de l'existence d'une serie de granites Tertiaires au Cameroun. *Bull. Bur. Rech. géol. min.*, **3**, 141.

LAY, C., and LEDENT, D. 1963. Mesures d'âge absolu de minéraux et de roches du Hoggar (Sahara central). *C.R. Acad. Sci. Paris*. **265**, 3113.

—— LEDENT, D., and GRÖGLER, N. 1965. Mesures d'âge absolu de zircons du Hoggar (Sahara central). *C.R. Acad. Sci. Paris*, **265**, 3113.

LEGRAND, PH. 1964. Découverte de nouveaux gisements fossilifères dans les grès inférieurs du Tassili N'Ajjer. *C.R. Soc. géol. Fr.*, **1**, 14.

LELUBRE, M. 1952. Recherches sur la Géologie de l'Ahaggar central et occidental (Sahara central). *Bull. Serv. Carte géol. Algérie*, **22**.

—— 1967. Chronologie du Précambrien au Sahara central. Abstract. *Int. Meeting geol. Ass. Can.*, Kingston, 49.

LOUIS, P. 1962. Interprétation géologique d'une partie de la carte gravimétrique du Bassin de Logone (République du Tchad). *C.R. Acad. Sci. Paris*, **264**, 3732.

MACHENS, E. 1966. Sur l'âge des dolérites 'récentes' du Liptako (République du Niger). *Bull. Bur. Rech. géol. min.* **1**, 113.

MAKARENKO, F. A., POLAK, B. G., and SMIRNOV, J. B. 1968. Geothermal field on the U.S.S.R. Territory. *Int. geol. Congr.*, **23**(5), 67.

MERLET, J. 1962. Note relative aux phases sismiques observées entre 100 et 200 km dans le massif du Hoggar. *C.R. Acad. Sci. Paris*, **252**, 3441.

PICCIOTTO, E., LEDENT, D., and LAY, C. 1965. Etude géochronologique de quelques roches du socle cristallophyllien du Hoggar (Sahara central). *Sci. Terre*, **10**(3-4), 481.

RADIER, H. 1953. Contribution à l'étude stratigraphique et structurale du détroit soudanais. *Bull. Soc. géol. Fr.*, **6**, 677.

REBOUL, C., MOUSSU, H., and LESSARD, L. 1962. Notice explicative de la carte géologique au 1 : 500,000 du Hoggar (Sahara central). *Bur. Rech. géol. min.*

REICHELT, R. 1966. Métamorphisme et plissement dans le Gourma et leurs âges (République du Niger). *C.R. Acad. Sci. Paris*, **263**, 589.

REMY, J. M. 1960. Les manifestations éruptives du Sud-Est de l'Amadror en Ahaggar. *Rev. Géogr. phys. Géol. dynam.*, **3**(2), 95.

ROCCI, G. 1960. Le massif de Tarraouadji, étude géologique et pétrographique. *Notes Bur. Rech. géol. min.*, Dakar, **6**.

—— 1965. Essai d'interprétation de mesures géochronologiques. La structure de l'Ouest Africain. *Sci. Terre*, **10** (3-4), 461.

H

SARCIA, J., and SARCIA, J. A. 1952. Volcanisme et tectonique dans le nord-est Adamaoua (Cameroun français). *Bull. volc.*, **2**, 129.

SCHETCHEGLOV, A. D. 1965. Conclusions des experts géologues soviétiques concernant les problèmes du développement des travaux géologiques en République du Niger. *Arch. Dir. Mines Géol. Niger.*

SHEINMANN, Y. M., APEL'TSIN, F. R., and NECHAEVA, E. A. 1961. Shchelochnye intruzii, ikh razmeshchenie i svyazannaya s nimi mineralizatsiya. *Geologiya Mestorozh. redk Elem.*, **12** (Gosgeoltekhizdat).

SOUGY, J. 1962. West African fold belt. *Bull. geol. Soc. Am.*, **73**, 871.

TUGARINOV, A. 1968. Age absulu et particularités génétiques des granites du Nigéria et du Cameroun septentrional. *C.R. UNESCO Coll. Granites Ouest-Africain (Paris)*, 119.

TURNER, D. C. 1963. Ring-structures in the Sier-Fier Younger Granite Complex, northern Nigeria. *Q. J. geol. Soc. Lond.*, **119**, 345.

—— (in press). Structure of the Jos Plateau, northern Nigeria. *Symp. Continental Drift, Montevideo* 1967.

UPTON, B. G. J. 1960. The alkaline igneous complex of Kûngnât Fjeld, South Greenland. *Medd. Grønland*, **123**(4).

—— 1964. The geology of Tugtutôq and neighbouring islands, south Greenland. *Medd. Grønland*, **169**(3).

VILLEMUR, J. R. 1967. Reconnaissance géologique et structurale du nord du Bassin de Taoudenni. *Mém. Bur. Rech. géol. min.*, **51**.

VINCENT, P. M. 1963. Les volcans tertiaires et quaternaires du Tibesti occidental et central (Sahara du Tchad). *Mém. Bur. Rech. géol. min.*, **23**.

WRIGHT, J. B. 1968. Oligoclase-andesine phenocrysts and related inclusions in basalts from part of a Nigerian Cenozoic Province. *Miner. Mag.*, **283**, 1024.

YODER, H. S., and TILLEY, C. E. 1962. Origin of basalt magmas: an experimental study of natural and synthetic rock systems. *J. Petrology*, **3**, 342.

ZIMMERMANN, M. 1960. Nouvelle subdivision des séries antégothlandiennes de l'Afrique occidentale (Mauritanie, Soudan, Sénégal). *Int. geol. Congr.*, **21** (8), 26.

K.G. COX

10   Tectonics and vulcanism
       of the Karroo Period
       and their bearing on the
       postulated fragmentation
       of Gondwanaland

ABSTRACT. *The sedimentary, tectonic, and volcanic evolution of southern Africa during the Karroo Period (Upper Carboniferous–Lower Jurassic) is summarised, and it is suggested that the pre-existing pattern of basement cratons and orogenic belts exerted a profound influence on the development of the Karroo rocks. With few exceptions the tectonic phenomena can be ascribed to a tensional origin, and are correlated with a postulated convective uprise of mantle materials responsible for the fragmentation of the super-continent of Gondwanaland. Gross inhomogeneity of the crustal rocks apparently allowed strain to be concentrated along zones of basement weakness, and it is suggested that surface tensional features such as zones of igneous activity, rift valleys and mid-ocean rises do not mark the actual sites of upwelling mantle currents, except by coincidence.*

Introduction

The close of the Karroo Period in southern Africa was marked by what must have been one of the most spectacular volcanic episodes the earth has ever seen. Although a reconstruction of the original extent of the lavas is difficult, it is likely that something like 2,000,000 km² of the continent was either covered by them or affected by the host of dykes and sills which accompanied their eruption. Even today after a long period of erosion and post-Karroo sedimentation, there still remains about 140,000 km² of lava outcrop assignable to the Karroo period. The lavas are not only extensive but they are locally extremely thick, particularly in the easternmost outcrops (see Fig. 1) where the volcanic rocks pass beneath Cretaceous and Tertiary sediments along a monoclinal flexure extending for 1200 km from Lupata to Natal. In the southern part of this zone, surface mapping has suggested that the volcanic succession is at least 9 km thick (Du Toit, 1929; Cox et al., 1965) and even greater thicknesses have been postulated from geophysical evidence (Hales and Gough, 1962).

Not only was the vulcanism on a massive scale but the rocks produced were of considerable variety and interest. Associated with this period are intrusive complexes which include gabbros, granites, quartz-syenites, nepheline-syenites, ijolites, and carbonatites. The most important extrusive rocks are basalts and rhyolites but there are substantial amounts of glassy,

211

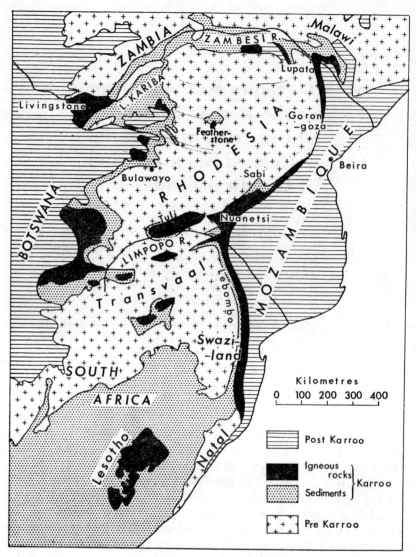

FIG. 1. Geological map of south-eastern Africa showing the distribution of rocks of the Karroo System and principal localities mentioned in text.

olivine-rich, tholeiites, locally termed limburgites, as well as minor amounts of alkalic types such as nephelinite and shoshonite.

The Karroo province provides a wealth of interesting problems, both of a petrological and of a more generally geological nature. It is the latter type of problem that I wish to discuss here, particularly the relation of the Karroo rocks to the structure of the basement. This question has been discussed previously (Cox et al., 1965, pp. 116–130) but new evidence has since come to light so that a fuller interpretation is worthwhile. It is also pertinent to discuss the postulated disruption of Gondwanaland during Mesozoic times,

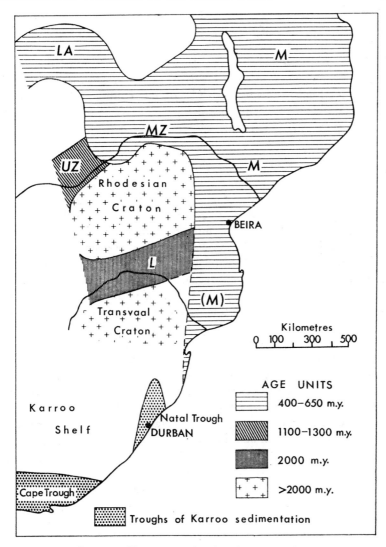

FIG. 2. Outline map of basement rocks of south-eastern Africa (largely after Vail, 1965), and sedimentary troughs of South Africa (after Woodward, 1966). *M*, Mozambique Belt; *(M)*, postulated southerly extension of Mozambique Belt beneath Karroo and younger cover; *MZ*, mid-Zambesi Belt; *LA*, Lufilian Arc; *UZ*, upper Zambesi Belt; *L*, Limpopo Belt.

and the possible relationship of vulcanism to that event. The main part of this chapter will therefore be devoted to a consideration of geological evidence which is relevant to these questions.

Structure of the basement

Vail (1965) has summarised the present state of knowledge regarding orogenic belts in the basement rocks of southern Africa (see Fig. 2). I should like to draw particular attention to the Mozambique and Limpopo Belts, and also to the gneiss belts of the various parts of the Zambesi Valley. Even a casual comparison of Figs. 1 and 2 will suggest that in some areas, particularly for example, around the edges of the Rhodesian Craton, the present distribution of the Karroo rocks is related to basement structure. The relationship is a fundamental one, for it is not only the present distribution of the rocks but also the original distribution of zones of sedimentation and eruptive activity that shows evidence of basement control.

Karroo sedimentation

In the interpretation of the igneous rocks, any tectonic information which can be gleaned for the period preceding the eruptive activity is of great value. Many of the tectonic features associated with the vulcanism were present at an earlier stage and had important effects on the development of the sedimentary rocks of the Karroo Period. Taking the correlations of Bond (1952) and the time scale of Harland *et al.* (1964) the period over which the tectonism accompanying the igneous activity was foreshadowed stretches from the Permian about 280 m.y. ago to the end of the Trias about 200 m.y. ago. The igneous activity may thus be thought of as the culmination of a long train of events.

In this context the sedimentation of the Limpopo and Zambesi valleys is of particular interest since, as a result of strong faulting and warping, the details of basement geology are well seen, and the effects of basement control on sedimentation are particularly clear. The wide outcrop of Karroo sediments in South Africa is less informative, however, because the basement is concealed over a large area (Fig. 1). A generalised stratigraphic sequence of Karroo sediments is given in Table 1.

TABLE 1

Generalised Karroo stratigraphy

Karroo Basin (*South Africa*)	Northern Area (*Transvaal, Rhodesia, Zambia and Malawi*)		
Basalts	Basalts	}	Upper Karroo
Stormberg Series	Stormberg Series		
Beaufort Series			
Ecca Series	Ecca Series ~~~*	}	Lower Karroo
Dwyka Series			
Pre-Karroo	Pre-Karroo ~~~*		

* The Beaufort and Dwyka Series are either absent or much reduced throughout the northern area.

THE ZAMBESI VALLEY

The part of the Zambesi Valley which trends north-east in the vicinity of Lake Kariba forms a well-documented example of a trough of Karroo sedimentation, although in this particular area there was apparently little or no subsequent vulcanism. The thickness of the Karroo sediments here, and in other parts of the Zambesi Valley (see Table 2), contrasts with the thin sequences preserved on the Rhodesian Craton to the south. A well-developed sedimentary trough was present in the Kariba area during Lower Karroo times and it was not until Upper Karroo times that the sediments transgressed onto the Rhodesian Craton (Bond, 1952); Gair (1956) has suggested that the trough developed as the sediments accumulated.

THE LIMPOPO VALLEY

Sedimentation in this region and its structural continuation in the Sabi area of south-east Rhodesia (Fig. 3) has been discussed by Visser (1961), Swift et al. (1953) and Cox et al. (1965). Visser concluded that the area marked a trough of Karroo sedimentation and that the sediments did not extend over the Zoutpansberg to the south.

North of the Limpopo a relatively thick sequence of sediments is preserved in the Bubye Coalfield but thins markedly to the north as it passes onto the Messina Block (Table 2 and Fig. 3). The latter is a fault block coinciding almost exactly with the central zone of the Limpopo Orogenic Belt

TABLE 2

Variations in thickness of Karroo sediments

Locality and Reference	Lower Karroo	Upper Karroo	Total	
S. Malawi (Habgood, 1963)	800 m	3000 m	3800 m	Zambesi
Gwembe (NE) (Gair, 1956)	3000 m	900 m	3900 m	Valley
Gwembe (SW) (Taverner-Smith, 1962)	2500 m	1000 m	3500 m	and environs
Magnet Mine (Worst, 1962)	—	300 m	300 m	Rhodesian
Robb's Drift (Worst, 1962)	—	250 m	250 m	Craton
Featherstone (Worst, 1962)	—	50 m	50 m	
Mazunga (Worst, 1962)	?	?	300 m	
Massabi (borehole T10); (Worst, 1962)	?	?	300 m	Messina Block
Nuanetsi shelf area (Cox et al., 1965)	?	?	0-50 m	
Bubye Coalfield (Cox et al., 1965)	400 m	300 m	700 m	Limpopo Trough
N. Transvaal (Van Eeden et al., 1955)	400 m	300 m	700 m	

(Cox et al., 1965, p. 81) and over much of its upper surface the sediments are from 0 to 30 m thick, though they reach about 300 m at Massabi and Mazunga in the western part of the block. Relatively thick sediments are also preserved in the Sabi area of Rhodesia (Swift et al., 1953), but in general from the Limpopo–Sabi region north-westwards towards the centre of the Rhodesian

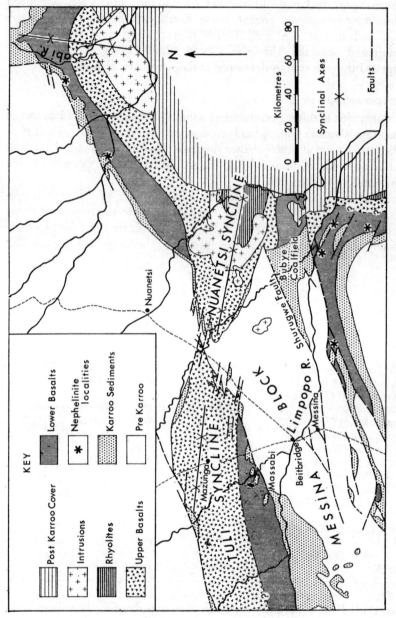

Post Karroo Cover

Intrusions

Rhyolites

Upper Basalts

Lower Basalts

Nephelinite
localities

Karroo Sediments

Pre Karroo

Sabi R.

Nuanetsi

NUANETSI SYNCLINE

Bubye
Coalfield

Shurugwe Fault

Limpopo R.

Messina

Beitbridge

Massabi

Mazungo

TULI SYNCLINE

BUBYE BLOCK

MESSINA

N

Kilometres

0   20   40   60   80

Synclinal Axes

Faults

FIG. 3. Geological map of the Limpopo–Nuanetsi–Sabi area.

Craton the sediments become very thin and only the upper beds are present. Because of erosion, however, original thicknesses can only be determined where the overlying basalts are still preserved, such as in the area north of Bulawayo (for example, Magnet Mine and Robb's Drift of Table 2), and at Featherstone some 300 km to the north-east of Bulawayo (Fig. 1).

RELATION TO BASEMENT STRUCTURE

From the evidence given above it can be inferred that the Rhodesian Craton acted as a positive area during the period of Karroo sedimentation, that is during much of the Permian and Trias. Sediments only accumulated in any appreciable thickness in the areas floored by the surrounding younger orogenic belts. In the large area of Karroo rocks further south in South Africa, however, the basement is concealed, though the stratigraphic evidence given by Woodward (1966) suggests that a similar control may have been exerted. Woodward has distinguished a Karroo 'shelf' area, which may be a southerly continuation of the Transvaal Craton (Fig. 2), and two sedimentary troughs. The general pattern is similar to that seen further north, and the parallelism of the axis of the Natal trough with the postulated southern extension of the Mozambique Belt is probably not fortuitous.

Karroo volcanics

MODE OF ERUPTION AND LOCATION OF ERUPTIVE ZONES

There is a general concensus of opinion that the great majority of the volcanics, which overlie the sediments (Table 1), were erupted relatively quietly from fissures. The direct evidence of this is, as usual in these cases, rather sparse although a few examples of flows being fed by dykes are known. Volcanic necks plugged by agglomerate and basalt are common in the general vicinity of south-west Lesotho (Gevers, 1929; Stockley, 1947) and there is some interbedding of tuffs within the lower parts of the basalt sequence. Pyroclastics have also been reported from the Sabi area of Rhodesia (Swift et al., 1953). In general, however, explosive vulcanism appears to have been only a minor and local early stage of the igneous activity. It is possible that the intrusive complexes of South-West Africa (Korn and Martin, 1954) and south-east Rhodesia (Cox et al., 1965) may have given rise to surface vulcanism of central type but the present erosion level is too deep for traces of this to be preserved.

The distribution of dolerite dykes from which the basalts were probably erupted is by no means uniform and several prominent swarms exist (see Vail, this volume, p. 348). Worst (1962) has suggested that all the basalts in Rhodesia, including those lying on the craton north of Bulawayo and at Featherstone (Fig. 1), were erupted from fissures lying in the Nuanetsi–Sabi area to the south-east and the Livingstone area of the Zambesi Valley to the north-west. This is based on an apparent absence of dykes within the craton, and reservations have been expressed about this hypothesis (Cox et al., 1967). Flow distances of 250 km are involved and yet the Featherstone basalts retain the quenched textures typical of the lower basalts of the Nuanetsi–Sabi area. Nevertheless a growing body of evidence supports the idea that plateau basalts are capable of flowing over such great distances as, for example, in the Deccan Traps of India (West, 1959) and the Columbia River basalts of America (Macdonald, 1967; Swanson, 1967). Detailed mapping of individual flows in the area between Bulawayo and Lake Kariba would be helpful in the present case, though the provenance of the Featherstone flows is likely to remain doubtful.

In contrast to the well-developed dyke swarms of the Nuanetsi–Sabi, Gorongoza, Lower Zambesi and Shire Highlands (southern Malawi) areas in the north (Fig. 1), the basalts of the inland parts of southern Africa were probably erupted from a diffuse zone of dykes with trends in several directions (Vail, this volume, p. 348). This applies particularly to the Lesotho region and the surrounding parts of the Karroo basin, where, in addition, dolerite sills (the Karroo dolerites) were emplaced in vast numbers. To the north-east of that area however a very strong meridional dyke swarm is encountered from which the majority of the eruptive rocks of the Lebombo Range probably issued.

RELATION TO BASEMENT STRUCTURE

In the northern part of south-eastern Africa the dykes seem to be largely confined to, and show a marked parallelism with, the regional orogenic zones. The Gorongoza dykes, for example, are parallel to the trend of the Mozambique Belt, and many of the dykes in the Limpopo–Nuanetsi–Sabi zone are parallel to the Limpopo Orogenic Belt. It should be emphasised that in these zones it is apparently the broad regional trend of the orogenic belts which has the controlling influence, rather than local basement trends to which the dykes are sometimes strongly discordant. As a consequence, where the basement structure is complicated and discontinuously exposed it becomes difficult to determine the existence of any control over dyke orientations. The Shire Highlands (Cholo) dyke swarm of southern Malawi (Evans, 1965), for example, is about 150 km wide and can be traced for about 120 km along the strike of the dykes in a north-easterly direction until it fades out in the Zomba area (Woolley and Garson, this volume, p. 241). Throughout much of its extent the swarm lies almost at right angles to the local basement trends, yet one may at least suspect that basement control has been operative because much of the basement beyond Zomba has a strong trend in the same direction as the dyke swarm.

The situation is much clearer with regard to the dykes located at the north-eastern corner of the Rhodesian Craton; there, about 100 km south-west of the river, the dyke swarm, and the basement trends, run south-eastwards parallel to the edge of the craton and to the course of the lower Zambesi.

In the southern part of Africa the most important eruptive zone lies along the Lebombo Monocline and although there is no direct evidence of basement control the parallelism of the monocline with the general course of the

Mozambique Belt down the eastern side of the continent is marked, and one would assume that the belt continues southwards, east of the Lebombo, beneath a cover of Cretaceous and later sediments (compare Figs. 1 and 2). The possibility that the Natal Trough of Karroo sedimentation is related to that belt has been mentioned previously.

The cratonic area to the west of the Lebombo, which is the domain of the Karroo 'shelf' sediments, the Karroo dolerites, and the Lesotho basalts with the attendant multi-directional dykes, is evidently quite different from the Rhodesian Craton in a tectonic sense. It did not have such a strongly positive character during the period of sedimentation, and it was extensively permeated by basic magma. Woodward (1966) has pointed out that the southern limit of the dykes and sills coincides closely with the hinge separating the Cape Trough from the Karroo Shelf (Fig. 2), and there is some evidence that a similar situation exists at the edge of the Natal Trough. He attributes the absence of basic intrusives in the Cape Trough to the continuation of the compressive stress that had earlier been responsible for the formation of the Cape Fold Belt.

RELATION OF ERUPTIVE ZONES TO SEDIMENTARY TROUGHS

In the northern part of the Karroo province there was, as we have seen, a strong degree of basement control on the location of both sedimentary troughs in the earlier part of the Karroo period and the later eruptive zones. However, the relationship between these latter features is itself often obscure. The Limpopo zone offers one unequivocal example of a coincidence between the two, and the mid-Zambesi Valley is a clear example of a sedimentary trough which did not become an eruptive zone. Elsewhere, however, if the eruptive rocks are well developed, for example in the Lebombo Monocline or the Livingstone area of the Zambesi, the evidence relating to the sedimentary history is largely hidden.

STRATIGRAPHY AND TIME RELATIONS OF THE KARROO LAVAS

The studies of Karroo sedimentation discussed above are essentially concerned with the deformation of the sub-Karroo surface during the period of sedimentation. A consideration of the stratigraphy of the lavas can extend this study into younger geological periods. Stratigraphic successions of selected areas are given in Table 3.

A definition of what a 'Karroo' lava is would be helpful at this stage, and it will become clear that only an arbitrary one is possible. Igneous activity was in evidence in one part or another of southern Africa throughout almost the entire time-range from 200 to 100 m.y., that is from the late Trias to well into the Cretaceous (see Woolley and Garson, this volume, p. 242). Relatively few radiometric age determinations have been made on the extrusive rocks but Jurassic ages are given by rhyolites and basalts from the Lebombo and Nuanetsi (Hales, 1960; Manton, 1968; Snelling, 1967) and from the basalts of the lower Zambesi at Lupata (Flores, 1964). On the other hand the alkaline lavas overlying the Jurassic volcanics at Lupata give Cretaceous ages (*ibid.*) and the basalts of the Kaokoveld in South-West Africa are also Cretaceous (Siedner and Miller, 1968).

In this context, therefore, it would be convenient to restrict the term 'Karroo' to those lavas of southern Africa which are of Jurassic age. It is by no means certain whether this definition could be made less arbitrary by relating it to some regional tectonic event, producing a widespread unconformity between Karroo and later volcanics. An unconformity does separate these two groups at Lupata but it does not necessarily represent a logical top to the Karroo succession. For example, Monkman (in Cox *et al.*, 1965) mapped a number of quite striking unconformities within the rhyolite succession at Nuanetsi and these almost certainly are of only local significance.

The Karroo lavas are predominantly basaltic except for the succession along the Lebombo Monocline where the upper part of the sequence contains great thicknesses of rhyolite. Rhyolites are also seen overlying the basalts at

TABLE 3

Generalised sequences of Mesozoic lavas (thicknesses given in metres where known)

(*a*) *Central part of continent*

NORTH Victoria Falls (Zambesi Valley)	Tuli Syncline (Limpopo Valley)	Transvaal (Springbok Flats)	SOUTH Lesotho
Olivine-poor basalts (1000 m)	Olivine-poor basalts Olivine-rich basalts (total 400 m)	Olivine-poor basalts (400 m)	Olivine-poor basalts (1500 m)

(*b*) *Eastern part of continent*

NORTH Lupata (Zambesi Valley)		Nuanetsi, Sabi and Northern Lebombo	SOUTH Southern Lebombo (Swaziland and Mozambique)
Phonolites, rhyolites, trachytes	} Lupata Series	Rhyolite Group Olivine-poor basalts Olivine-rich basalts	Upper Basalts and alkalic lavas Rhyolite Group
Rhyolite Basalt	} Karroo	Nephelinites (total 6700 m)	Olivine-poor basalts (estimated *c.* 9000 m)

Lupata (Dixey and Campbell Smith, 1929). Small amounts of nephelinite are found at the base of the succession in the Sabi area of Rhodesia (Swift *et al.*, 1953) and immediately south of the Limpopo at the northern end of the Lebombo (Rogers, 1926; Lombaard, 1952). An occurrence of shoshonite has been reported by Vail *et al.* (1969), and there are rare occurrences of trachyte (Stockley, 1947; Assunçao *et al.*, 1962).

The most abundant basalt type is a tholeiite either devoid of olivine or containing it in very small amounts. Mineralogically the tholeiitic nature is shown by the frequent occurrence of groundmass pigeonite (Lombaard,

1952; Cox and Hornung, 1966; Cox et al., 1967) and chemically it is confirmed by the importance of normative hypersthene, often accompanied by normative quartz. If we exclude the Limpopo–Nuanetsi–Sabi region, discussed below in more detail, this description probably applies to the great majority of the Karroo basalts, though locally, for example near the base of the Lesotho succession, a few more olivine-rich flows are known (Stockley, 1947; Cox and Hornung, 1966). The basalts are usually mildly feldspar-phyric, though often the phenocrysts are small and not very abundant.

The rocks of the Limpopo–Nuanetsi–Sabi zone contrast strongly with the monotonous successions elsewhere. Here, at the intersection of the Lebombo and Limpopo structural lines, there is a thick development of glassy olivine-rich basalts usually termed *limburgites*. The term has been in use for these rocks for many years (see Mennell, 1913) but is something of a misnomer since the type limburgites from the Kaiserstuhl are highly undersaturated rocks whereas the Karroo ones are conspicuously hypersthene normative.

The olivine-rich basalts are abundant within the lower part of the sequence in the Limpopo area and a fairly sharp dividing line can be drawn between them and the olivine-poor and olivine-free basalts which overlie them. These two groups are designated the Lower Basalts and the Upper Basalts in Fig. 3 and it is possible to draw an approximate boundary between them over most of the region. In addition to the occurrences of the olivine-rich basalts within that area, it should be noted that they extend southwards down the Lebombo for some 120 km (Lombaard, 1952), though their stratigraphic position within the lava sequence is not known. The small outlier at Featherstone in the middle of the Rhodesian Craton also includes olivine-rich rocks. The nephelinites, found very locally at the base of the lava succession, are indicated by asterisks in Fig. 3.

The stratigraphic development of the lavas of the Limpopo–Nuanetsi–Sabi zone is somewhat complex and a brief structural digression will therefore be made as a preliminary part of the discussion.

The important structural elements in this area include the *Nuanetsi Syncline* which plunges to the east and contains an estimated 7-8000 m of volcanic rocks at its eastern end where it disappears beneath the conglomerates and sandstones of the Malvernia Beds (Cox, 1963). The northern limb of the syncline swings into a monoclinal flexure which runs towards the Sabi River to the north-east. The monocline is modified here by the north-south *Sabi Syncline* but continues beyond this, out of the area covered by Fig. 3.

The *Tuli Syncline* is a fold of much smaller amplitude than was formerly thought by Cox et al. (1965) and it is now clear from the outcrop pattern established by Vail et al. (1969), and from the borehole near Mazunga reported by Worst (1962), that the rocks over much of its area are nearly horizontal. According to Worst (*ibid.*) the total thickness of Karroo rocks near Mazunga is little more than 400 m. Thus despite some fairly large displacements of the outcrop pattern along the south side of the syncline it is probable that none of the faults shown in this area has a substantial throw.

Hence the structural distinction made previously between the Tuli Syncline and the Messina Block (Cox *et al.*, 1965) is probably not an important one. As is indicated in Fig. 4(c) the Messina Block has an upper surface which slopes gently to the north and the Tuli Syncline consists largely of an area of almost flat-lying Karroo rocks. The axis of the syncline, running somewhat to the north of Mazunga, appears to be a minor prolongation of the axis of the Nuanetsi Syncline, superimposed on the northern extension of the Messina Block (Fig. 3).

South of the Messina Block lies the *Northern Transvaal Fault Zone* where moderate thicknesses of northward-dipping Karroo rocks are preserved (Fig. 3). South of the Bubye Coalfield this faulted monoclinal structure swings through a right angle and joins the northern end of the Lebombo Monocline.

Points of interest in this general area include the great thicknesses of volcanic rocks involved, and the very rapid lateral changes in the succession. At the east end of the Nuanetsi Syncline, for example, the Lower Basalts are estimated to be 2150 m thick (Cox *et al.*, *ibid.*, p. 85) but they are overstepped westwards and cut out in the area north of the Bubye Coalfield. They reappear along the south side of the Tuli Syncline but must be less than 80 m thick since this is the total thickness of basalt encountered in a bore-hole near Mazunga (Worst, 1962). We have no detailed information for the basalt stratigraphy south of the Messina Block in the Northern Transvaal Fault Zone, but the descriptions of Rogers (1926), Van Eeden *et al.* (1955) and Wilke (1965) suggest that most of the outcrops should be assigned to the Lower (olivine-rich) Basalts. The eastern part of the fault zone is one of fairly high dips and in one locality Van Eeden *et al.* (1955) estimated that about 1400 m of basalts are present. Thus there is every indication that the Messina Block maintained its positive character during the eruption of the Lower Basalts which are thin or absent upon it and much thicker to the south and to the east.

Variations in thickness in the Upper Basalts can rarely be established because of truncation by the erosion surface. Nevertheless, where the overlying rhyolites are present in the Nuanetsi Syncline, it is possible to demonstrate that the sequence thins westwards, and at the edge of the Messina Block (at the north-eastern termination of the Shurugwe Fault) has only a fraction of the thickness seen 30 km further east. It can be inferred that the Upper Basalts are generally transgressive over the more restricted Lower Basalts.

A second point of interest concerns the limited occurrences of nephelinites, which are found at the base of the sequence. Presumably these rocks are overstepped by the Lower Basalts in much the same way that the latter are overstepped by the Upper Basalts. This tantalising glimpse of nephelinites suggests that much of petrological interest may be hidden beneath the basaltic succession further to the east.

Faulting

Faults which displace Karroo rocks are very widespread in southern Africa and have been reviewed by Vail (1967) who interprets the fault-pattern as a modified southward extension of the East African Rift System. Much of the faulting follows the zones of Karroo sedimentation and eruptive activity and, as a result, the present outcrop of Karroo rocks round the margins of the Rhodesian Craton is often fault-bounded.

There is a wide range in the age of the faults and many developed over long periods. A number of major faults are later than the extrusion of the basalts, including that marking the southern boundary of the Bubye Coalfield (Fig. 3) and the Deka Fault (Lightfoot, 1913) along the south-east margin of the basalts in the Livingstone area of the Zambesi Valley (Fig. 1). The northern margin of this outcrop is also strongly faulted near the west end of Lake Kariba (Taverner-Smith, 1960) and many other examples of post-basalt faulting could be quoted.

Slightly older faults are represented by those that cut the Karroo sediments but which fail to displace the basalts or displace only the lower flows. Minor examples of this type are reported from the Zambesi Valley (*ibid.*) and from Lesotho (Stockley, 1947) while the Shurugwe Fault, shown marking the northern boundary of the Bubye Coalfield in Fig. 3, is a major structure of this age.

In categorising the ages of faults in this fashion we are to a large extent only taking note of the latest movements on the fault plane. There is some evidence, for example, that pre-Karroo faults have been reactivated during the Karroo Period. Both Hitchon (1958) and Taverner-Smith (1960) regard this as a likelihood in the Zambesi Valley and an excellent example is afforded by the history of the Tshipise Fault in the Northern Transvaal Fault Zone (Van Eeden *et al.*, 1955; Van Zyl, in Söhnge, 1945). In some instances there is also evidence that movements took place, possibly semi-continuously, during the period of Karroo sedimentation. This has been suggested for the Zambesi (Gair, 1959) and Luano Valleys (Gair, 1960), the latter an area of Karroo rocks slightly to the north of the area shown in Fig. 1. The Shurugwe Fault (Fig. 3) also provides excellent evidence of activity during this period, though whether it existed as a fault scarp or as a monoclinal flexure is debatable.

The great majority of Karroo faulting is probably of normal type. Admittedly faults are usually regarded as normal until proved otherwise, but the attitudes of quite a number of fault planes have been reported and these are all normal. No case of reverse faulting has been noted as far as I am aware. The Dowe–Tokwe Fault, shown on Fig. 3 as passing in an east-west direction about 5 km south of Messina, is a dextral wrench fault with a horizontal displacement of approximately 1 km. Its age, however, is unknown since it cuts only basement rocks. It may be continuous with the Shurugwe Fault referred to previously, but this is not in itself evidence of Karroo age since as noted above there is evidence of rejuvenation of pre-Karroo faults in this area.

Warping

The Karroo rocks rest nearly horizontally over much of the interior of the African continent, but they plunge steeply in the east beneath a cover of

Cretaceous and Tertiary sediments, the fringing deposits of the Mozambique channel. In this zone is found the Lebombo Monocline (Fig. 1), the most impressive of all the Karroo structures. It runs almost straight in a general north-south direction for 700 km and the difference of structural level from one side to the other is extremely large. Precisely how large is a matter of some debate since there is a lack of detailed mapping and the possible influence of strike faulting is unknown. Du Toit (1929) estimated that the sub-Karroo surface was warped down to the east by at least 9 km, while the geophysical studies of Hales and Gough (1962) suggested a figure of about 13 km. Whatever the exact amount, it is clear that the structure is one of great magnitude and that a substantial proportion of the crust to the east of the monocline is made up of Karroo and the overlying Cretaceous lavas.

Du Toit (1929) argued that the monocline was formed during the eruption of the lavas, and a similar interpretation has been placed on the time-relations between deformation and eruption in the Nuanetsi Syncline (Cox et al., 1965). The latter has been interpreted as a volcano-tectonic structure which owes its origin to the eruption of enormous amounts of lava from a dyke swarm parallel to the synclinal axis. The implied transfer of materials from below to the surface seems to be a sufficient reason for the formation of an intense downwarp in the sub-Karroo surface. Thus this fold has an entirely tensional origin.

The location of the Lebombo Monocline is probably controlled by basement structure, in this case the postulated southerly extension of the Mozambique Belt (Fig. 2). The Nuanetsi Syncline (Fig. 3), on the other hand, appears to represent a completely different type of basement control, possibly unique amongst the Karroo structures. It has been suggested (ibid.) that the shape of the combined Transvaal and Rhodesian Cratons was such that, if they were stressed as a whole, a strain concentration resulting in profound fracturing and eruptive activities would be produced in the re-entrant angle marked by the Nuanetsi area. A rather close analogy is found in the coastal structure of east Greenland described by Wager (1947).

Elsewhere in the Karroo outcrop deformational structures comparable in magnitude with the Lebombo Monocline and Nuanetsi Syncline are not found, though future investigations may reveal that the broad syncline in the Karroo and Cretaceous volcanic rocks astride the Zambesi south-east of Lupata has much in common with them. The broad and gentle synclinal structure of the Kariba section of the mid-Zambesi will be discussed in association with its faulting in the next section.

The relationship between faulting and warping

The Northern Transvaal Fault Zone and the Kariba section of the mid-Zambesi show a relationship between faulting and warping which is one of the most crucial aspects of Karroo tectonics. The structure of the northern Transvaal is well known as a result of the work of Söhnge (1945), Söhnge et al. (1948), Van Zyl (1950), Van Eeden et al. (1955) and Visser (1961). Part

PLATE 1. (*above*) The Drakensberg at Mont–aux–Sources on the Natal–Lesotho border. A 1300 m succession of horizontal Karroo basalt flows overlies sediments exposed in foreground. (*below*) The Maluti Mountains, Lesotho. Horizontal Karroo basalt flows form a dissected plateau 3000 m above sea level.

of the fault zone is shown in the southern part of Fig. 3 and it appears in diagrammatic form in Fig. 4(c). The Karroo rocks dip consistently to the north at an angle of about 12° and are repeatedly displaced along normal faults hading to the south. It has been calculated that the cumulative throw of these faults is at least 13 km (Van Zyl, in Söhnge, 1945), and Visser (in Van Eeden et al., 1955) suggests that faulting and tilting of the beds were simultaneous. The almost exact compensation of the dip by the throw of the faults makes it clear that the whole structure is concerned with the horizontal extension of the crust, which has been effected by the rotation of fault-bounded blocks. However clear this aspect of the structure may be, the reason for the constant hade direction of the faults requires explanation, for under ideal conditions of horizontal tensional stress conjugate fault sets should be formed. This point will be examined in more detail after a consideration of other fault zones.

The Karroo rocks of the Kariba section of the mid-Zambesi have been studied by Gair (1956, 1959) and Taverner-Smith (1960, 1962). The rocks are disposed in a broad faulted syncline with a steeper north-western limb and a very gentle south-eastern limb. The axis of the syncline lies slightly south-east of the Zambesi and is roughly parallel to it (Fig. 1); a diagrammatic cross-section is given in Fig. 4(d).

The faulting of the mid-Zambesi Syncline is more variable than that of the northern Transvaal but a very well-developed series of rotated fault blocks is present on the north-western limb. This structure seems to be almost exactly analogous to the Northern Transvaal Fault Zone, though the compensation of the dip by the throw of the faults is not quite so perfect. On the south-eastern limb of the fold there is some evidence of a similar type of faulting, with the faults hading to the south-east, but the structure is complicated by basement horsts.

Fault zones showing a common direction of hade and the rotation of fault-bounded blocks have been described from Madagascar by Cliquet (1958) who used the term 'factory roof' to describe this type of structure; fault zones of this general type are referred to as 'half grabens' by Badgley (1965). Coastal Natal (Fig. 1) at the south end of the Lebombo Monocline shows a somewhat similar structure described by Beater and Maud (1960), Maud (1961) and Hardie (1962); here the eastward dip of the rocks is offset by faults throwing down inland. Du Toit (1929) has described the same phenomenon on a minor scale from the Lebombo, but to what extent the strong Natal structures continue northwards within the Lebombo volcanic succession must await future investigation. The outcrop pattern of the rhyolite-basalt contact in Swaziland (Urie and Hunter, 1963) is at least suggestive of the presence of major strike faults throwing down to the west, that is, offsetting the strong eastward dip of the lavas.

It seems therefore that faulted monoclinal structures of the type discussed above are of great importance in the understanding of the tectonics of continental margins, particularly those which are suspected of having been disrupted by continental drift. For example, a structure of this type is spectacularly developed along the northern margin of the Gulf of Aden and has been described in outline by Wissmann *et al.* (1942).

A possible origin of faulted monoclines

It has been argued above that fault zones of the type under discussion are associated with crustal elongation effected by rotation of fault blocks. A more detailed hypothesis is required to account for the uniformity of hade direction, and the association with monoclinal flexuring.

On a relatively small scale, fault patterns showing uniformity of hade are developed antithetically to major faults dipping in the opposite direction. A discussion of this type of faulting, defined by Cloos (1928), is given by De Sitter (1956), and is interpreted as being mainly due to the reduction in the angle of dip of the master fault with depth. Taverner-Smith (1960) has interpreted the faulting of the north-western limb of the mid-Zambesi Syncline in this way, though admitting that the master fault is not always present or may be undetected where it runs through basement rocks. The scale of the phenomenon is not however favourable to this hypothesis. The sections given in Figs. 4(c) and 4(d) are diagrammatic, but they do show the hades of the faults with some degree of reality. For the observed surface faults, which form zones of constant hade direction some 50 km wide, to intersect a curved master fault plane hading in the opposite direction the sections would have to be extended to depths of similar magnitude, that is well into the mantle. This factor, coupled with the lack of evidence that master faults actually exist in most of the zones, suggests that the interpretation should proceed without them.

If we interpret the faulting as tensional in origin, and associated with crustal extension, then it is also logical to regard the Karroo sedimentary troughs of the Zambesi and Limpopo Valleys as tensional features. Faulting, downwarping and sedimentation seem in a general way to be contemporaneous; and downwarping and faulting continued during the volcanic period. Thus there is no room for compressive phases of tectonism in the history of the northern part of the Karroo area.

Crustal extension carries with it an implication of crustal thinning, a concept given considerable attention by Du Toit (1937, p. 325). In the Limpopo area, a trough of sedimentation, roughly coinciding with the Northern Transvaal Fault Zone, developed in a tensional environment (the Shurugwe Fault being active during that period) and subsequently passed into a phase of crustal extension by fracturing. The sedimentary trough may then be regarded as a manifestation of earlier crustal extension and thinning accomplished by plastic flow. Such troughs may be thought of as 'necks' developing on a crustal scale, analogous to the necks formed in experimentally stretched metals.

This hypothesis enables an explanation to be found for the constant hade direction of the fault zones. Fig. 4(a) shows a hypothetical section of crust which has developed a neck as a result of plastic flow. The broken lines indicate the directions of relative sub-horizontal tension (P_{min}) and maximum sub-vertical stress (P_{max}). Heavy lines with arrows indicate directions of potential shear planes. On the flanks of the neck it will be seen that the conjugate shear directions are no longer symmetrical relative to the horizontal and

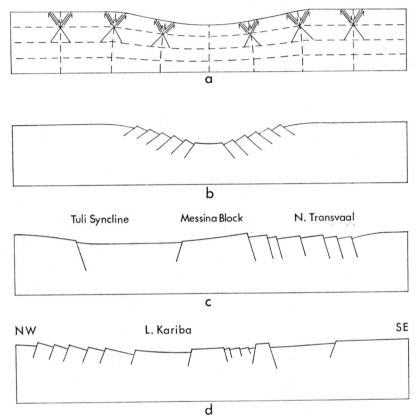

FIG. 4. (a) Diagrammatic section through crust undergoing 'necking': broken sub-horizontal lines show direction of P_{min} (tension), broken sub-vertical lines show direction of P_{max}; potential shear directions marked by arrows. (b) As above showing the development of faulting on shear planes of lowest dip. (c) Diagrammatic section across the Limpopo zone showing faulting and warping of pre-Karroo surface; section is 175 km long. (d) Diagrammatic section across the Kariba section of the mid-Zambesi: the north-western part is after Gair (1959); the south-eastern part is based on the 1:1,000,000 Geological Map of Southern Rhodesia (1961); scale as in Fig. (c) above.

vertical, and this is perhaps a sufficient reason for one set of shear planes to be developed rather than the other. For a given amount of movement on fault planes the development of the faults which dip at a relatively low angle will result in a larger amount of crustal extension. Fig. 4(b) indicates faulting of this type in association with a crustal neck, and it seems a reasonable inference that structures of the type illustrated in Figs. 4(c) and 4(d) might develop in this way when the crust undergoing necking is inhomogeneous.

The pattern of crustal extension

In Fig. 5 the distribution of zones of crustal extension in south-eastern Africa is indicated. The zones marked *A*, *E* and *F* represent the mid-Zambesi, Limpopo and Lebombo structures respectively, while *B*, *D* and *C* represent the lower Zambesi, Gorongoza and Shire Highlands (Cholo) dyke swarms. From the foregoing evidence, these can be taken as essentially contemporaneous zones of extension marked either by crustal necking or by dyke intrusion or by both. The most striking feature of the pattern is its polygonal nature,

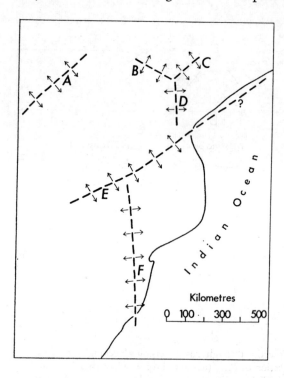

Fig. 5. Zones of crustal extension in south-eastern Africa; area covered is the same as in Fig. 1: *A*, mid-Zambesi; *B*, dykes parallel to lower Zambesi; *C*, Shire Highlands (Cholo) dykes; *D*, Gorongoza dykes; *E*, Limpopo zone; and *F*, Lebombo Monocline.

which is consistent with a general expansion of this part of the African continent during Karroo times. A detailed analysis would include several more important lineaments, but this would not detract from the polygonal character of the pattern. It should also be clear that the pattern owes its geometry to the structure of the basement rocks.

If we accept these ideas and examine them in the light of the continental drift hypothesis we must bear in mind that we are considering a relatively rigid crust which undergoes necking along lines of weakness dictated by basement structure. The necking must take place over long periods of time and must proceed approximately simultaneously in different areas. A few necks will become deformed to the point of complete rupture which will be accompanied by a massive transgressive vulcanism of the type seen in the Lebombo and its environs. Other necks will not develop beyond the point of initial trough formation (e.g. the mid-Zambesi) or of relatively mild vulcanism (e.g. northern Transvaal).

A partial reconstruction of Gondwanaland is given in Fig. 6. The position chosen for Madagascar is somewhat arbitrary since it could equally well be placed nearer to Africa, to the south-west of the position shown. The zones marked *A*, *B* and *C* represent postulated lines of complete disruption (Fig. 6). *A* can be dated as an early Jurassic break corresponding with the period of maximum deformation of the Lebombo Monocline. Volcanic activity however, probably continued in the Mozambique channel (the strait separating Madagascar from Africa) into the Cretaceous and presumably gave rise to widespread vulcanism now concealed beneath the Cretaceous and Tertiary sediments of Mozambique, except at Lupata. Basalts are interbedded with

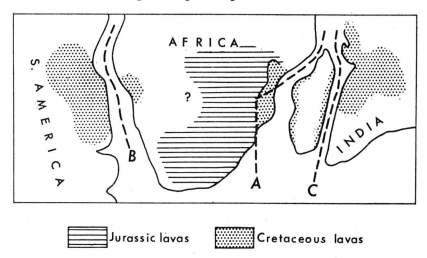

Jurassic lavas   Cretaceous lavas

FIG. 6. Zones of complete crustal disruption in part of Gondwanaland. Areas of Jurassic and Cretaceous lavas are indicated.

marine Cretaceous rocks in western Madagascar (Haughton, 1963) though no Jurassic (Karroo) lavas are developed.

The course of the line of postulated Jurassic disruption can be seen in more detail in Fig. 5 where it is represented by the Lebombo Monocline (*F*) and the postulated north-eastern projection of the Limpopo Zone (*E*). The relationship of the line to basement structures can be seen from inspection of Figs. 1 and 2. It seems possible that, for a long distance to the north-east of the Sabi area in Rhodesia, the Mozambique Belt structures are overprinted on earlier Limpopo structures, and it is the latter which have had a controlling influence on Mesozoic tectonic events. This is suggested by the way in which the Karroo structures north-east of the Sabi continue the line of the Limpopo Belt while cutting diagonally across the exposed Mozambique Belt. The trend of the coastline north-east of Beira may also be dictated by the orientation of Limpopo zone structures within the Mozambique Belt.

In accordance with the ideas of Siedner and Miller (1968), zone *B* (Fig. 6) is a rather younger zone of disruption contemporaneous with the outpourings of the Cretaceous basalts of the Kaoko area of South-West Africa and the Serra Geral Formation of Brazil. Siedner and Miller (*ibid.*) have obtained ages of about 125 m.y. for the Kaoko lavas which correspond remarkably closely with the dates obtained from the Serra Geral basalts (Creer *et al.*, 1965; Amaral *et al.*, 1966; McDougall and Rüegg, 1966; Melfi, 1967).

C may be a younger break still if it is contemporaneous with the Deccan Trap lavas of India, which are usually taken to be of very late Cretaceous age (Krishnan, 1949). Future radiometric dating will show to what extent the Deccan rocks match the Cretaceous basalts, preserved locally on the east coast of Madagascar, and it is hoped that further detailed work on volcanic activity will aid in the understanding of the earlier history of the fragmentation of Gondwanaland. For the moment it is worth noting that the evidence discussed strongly supports the hypothesis (Du Toit, 1937; King, 1962) that the main fragmentation of the southern continents took place in the Cretaceous, an idea recently endorsed by Hallam (1967) on faunal evidence. The Mozambique channel disruption may be of early Jurassic age but in terms of continental displacement it did not develop into an important zone.

The mechanism of continental drift

It is often assumed in studies of continental drift that surface features such as rift valleys and mid-oceanic rises mark the sites of up-welling convection currents in the mantle. Wilson (1967) and Hurley (1968) provide two examples of recent reviews accepting this view, which originates in the work of Dietz (1961) and Hess (1962). On the other hand, many workers in Africa have been impressed by the dependence of young tectonic features on the pre-existing structure of the basement, and I have discussed examples of this sort of control in some detail in the previous pages. It is not easy, however, to adhere to both of these ideas for they seem to contain elements that are mutually contradictory.

For example, are we to postulate that each zone of extension shown in Fig. 5 is generated by its own convective up-welling? It is surely more reasonable to presume that a general sub-crustal flow has acted upon an inhomogeneous crustal plate which has responded by necking along lines of weakness. All that is required is a general stretching mechanism and a crust of sufficient strength to transmit stresses laterally over modest distances so that strains can become concentrated along certain zones; for example, the Limpopo zone. We may decide, because of the generally meridional trend of the major zones of necking (Fig. 6), that the general stress had an important east-west component, but if we accept the premise that strains are concentrated and resolved along existing lines of weakness, we cannot proceed very far in reconstructing the general stress pattern. We are very much in the position of contemplating a tortoiseshell broken into its component polygonal plates by some unknown force. Reconstructing the shell is one thing; deciding exactly how it came to be broken is quite another.

Considering a feature such as the Mid-Atlantic Ridge, many geologists would accept that it originated as a fracture across a combined South American/African continent. Moreover the pattern of the ridge today is still a more or less faithful copy of the shape of that original fracture. The evidence that the fracture was controlled by basement structures, in much the same way as the coast of south-eastern Africa, is however persuasive (Kennedy, 1965). If we follow the logic of the previous arguments we must conclude that the original zone of fracture was unlikely to have marked the position of a convective up-welling in any exact way, and that like other zones of necking it was a localised response to a more generalised stress pattern. If the original fracture did not mark the position of a convective uprise it follows that the Mid-Atlantic Ridge today probably does not either. Moreover, the features observed along mid-ocean rises do not necessarily demand the existence of convective up-wellings for their formation. McKenzie (1967),

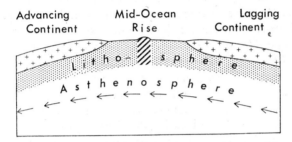

FIG. 7. Relation of drifting continents and mid–ocean rise to convective flow of uniform direction. The zone of concentrated strain is shown by heavy diagonal ornament. Arrows show direction of sub-crustal flow.

for example, has argued that neither the anomalously high heat flow nor the gravity anomalies need reflect special conditions in the upper mantle. The alternative is to regard the role of the mid-ocean rises as an essentially passive one concerned with the filling of the rift left between two crustal blocks which are undergoing relative separation. Morgan (1968) has given an account of how such a process might operate.

The structural evidence considered above can be reconciled with the convection hypothesis if it is postulated that convective flow passes beneath mid-ocean rises with a uniform travel direction as illustrated in Fig. 7, rather than rising and spreading beneath the rise. The requirement for the initiation of necking and its subsequent propagation as a mid-oceanic rise is a general tension, which unidirectional sub-crustal flow will produce providing that the whole continent is not transported at the same velocity as the flow. It is possible that the continental mass of Gondwanaland was itself responsible for the creation of thermal instability in the underlying mantle, and that this was eventually expressed as a broad zone of convective uprise having a strong radial element in the pattern of flow. This pattern would prevent the bodily translation of the whole mass and could induce necking in a great variety of directions consistent with the generally expansional nature of the tectonic features illustrated in Fig. 5.

Fig. 7 shows two continental fragments designated as 'lagging' and 'advancing' respectively, the latter being the fragment furthest from the centre of the flow pattern. In Gondwanaland it is possible that the centre of the uprising convection coincided roughly with southern Africa and Antarctica for it is in these areas that the Jurassic magmatism is prevalent (for Antarctic age determinations, see McDougall, 1963), while Cretaceous magmatism is characteristic of India, Madagascar, and Brazil. The latter units may have had the role of advancing continents relative to the lagging continent of Africa.

The feasibility of this hypothesis depends almost entirely on the mechanical properties of the continental and new oceanic crust and their ability to transmit stress over considerable distances. I have already briefly mentioned the factors which appear to have controlled the location of the Nuanetsi Syncline (p. 224) and these seem to imply rigidity of the continental crust over distances of perhaps some hundreds of kilometres. In a more general way the concentration of Mesozoic tectonic events in Africa along the orogenic belts of a variety of ages surrounding cratonic blocks implies that the latter were substantially rigid. These arguments, based on geological observations, refer to the continental crust, but supporting evidence suggesting that newly-created oceanic crust can also act in a rigid way has recently been provided by the studies of Oliver and Isacks (1967), Morgan (1968) and Le Pichon (1968). These studies collectively represent a revival of the concept of the rigid lithosphere overlying a mobile asthenosphere (Daly, 1940) and are supported by evidence from crustal geometry and seismology.

REFERENCES

AMARAL, G., CORDANI, U. G., KAWASHITA, K., and REYNOLDS, J. H. 1966. Potassium-argon dates of basaltic rocks from southern Brazil. *Geochim. et cosmochim. Acta*, **30**, 159.

ASSUNÇAO, A. F. T. DE, COELHO, A. V. P., and ROCHA, A. T. 1962. Petrologia das lavas dos Libombos. *Mem. Junta Invest. Ultramar (Lisbon)*, **99**.

BADGLEY, P. C. 1965. *Structural and tectonic principles*, New York.

BEATER, B. E., and MAUD, R. R. 1960. The occurrence of an extensive fault system in S.E. Zululand and its possible relationship to the evolution of a part of the coastline of southern Africa. *Trans. geol. Soc. S. Afr.*, **63**, 51.

BOND, G. 1952. The Karroo System of Southern Rhodesia. *Int. geol. Congr.*, **19** (*Symp. Gondwana*), 209.

CLIQUET, P. L. 1958. La tectonique profonde du sud du Bassin de Morondava. *C.C.T.A. east centr. and southern reg. Comm. Geol.*, Tananarive, **1**, 199.

CLOOS, H. 1928. Uber antithetische Bewegungen. *Geol. Rdsch.*, **19**, 246.

COX, K. G. 1963. Malvernia Beds. *Trans. geol. Soc. S. Afr.*, **66**, 341.

—— and HORNUNG, G. 1966. The petrology of the Karroo basalts of Basutoland. *Am. Miner.*, **51**, 1414.

Cox, K. G., Johnson, R. L., Monkman, L. J., Stillman, C. J., Vail, J. R., and Wood, D. N. 1965. The geology of the Nuanetsi Igneous Province. *Phil. Trans. r. Soc. Lond.*, **257**, Ser. A, 71.

—— Macdonald, R., and Hornung, G. 1967. Geochemical and petrographic provinces in the Karroo basalts of southern Africa. *Am. Miner.*, **52**, 1451.

Creer, K. M., Miller, J. A., and Smith, A. Gilbert. 1965. Radiometric age of the Serra Geral Formation. *Nature, Lond.*, **207**, 282.

Daly, R. A. 1940. *Strength and structure of the earth,* New York.

De Sitter, L. U. 1956. *Structural geology,* New York.

Dietz, R. S. 1961. Continent and ocean basin evolution by spreading of the sea floor. *Nature, Lond.*, **190**, 854.

Dixey, F. and Campbell Smith, W. 1929. The rocks of the Lupata gorge and the north side of the lower Zambezi. *Geol. Mag.*, **66**, 241.

Du Toit, A. L. 1929. The volcanic belt of the Lebombo—a region of tension. *Trans. r. Soc. S. Afr.*, **18**, 189.

—— 1937. *Our wandering continents.* Edinburgh.

Evans, R. K. 1965. The geology of the Shire Highlands. *Bull. geol. Surv. Malawi,* **18**.

Flores, G. 1964. On the age of the Lupata rocks, lower Zambesi River, Mozambique. *Trans. geol. Soc. S. Afr.*, **67**, 111.

Gair, H. S. 1956. A summary of the structure and tectonic history of the mid-Zambesi Valley. *C.C.T.A. east centr. reg. Comm. Geol.*, Dar-es-Salaam, 123.

—— 1959. The Karroo System and coal resources of the Gwembe district, north-eastern section. *Bull. geol. Surv. N. Rhodesia,* **1**.

—— 1960. The Karroo System of the western end of the Luano valley. *Rep. geol. Surv. N. Rhodesia,* **6**.

Gevers, T. W. 1929. The volcanic vents of the western Stormberg. *Trans. geol. Soc. S. Afr.*, **31**, 43.

Habgood, F. 1963. The geology of the country west of the Shire River between Chikwawa and Chiromo. *Bull. geol. Surv. Nyasaland,* **14**.

Hales, A. L. 1960. Research at the Bernard Price Institute of Geophysical Research, University of the Witwatersrand, Johannesburg. *Proc. r. Soc. Lond.*, **258**, Ser. A, 1.

—— and Gough, D. I. 1962. The gravity survey of the Republic of South Africa. II. Isostatic anomalies and crustal structure. *Handb. geol. Surv. S. Afr.*, **3**, 355.

Hallam, A. 1967. The bearing of certain palaeozoogeographic data on continental drift. *Palaeogeog. Palaeoclim. Palaeoecol.*, **3**, 201.

Hardie, L. A. 1962. The fault pattern of coastal Natal: an experimental reproduction. *Trans. geol. Soc. S. Afr.*, **65**, 203.

Harland, W. B., Smith, A. Gilbert and Wilcock, B. (Eds.). 1964. *The Phanerozoic time-scale,* Geol. Soc. Lond.

Haughton, S. H. 1963. *Stratigraphic history of Africa south of the Sahara,* Edinburgh.

Hess, H. H. 1962. History of the Ocean Basins. In *Petrologic studies: a volume in honor of A. F. Buddington,* p. 599. (Eds. A. E. J. Engel, H. L. James, and B. F. Leonard), Geol. Soc. Am.

Hitchon, B. 1958. The geology of the Kariba area. *Rep. geol. Surv. N. Rhodesia,* **3**.

Hurley, P. M. 1968. The confirmation of continental drift. *Scient. Am.*, **218**, 53.

Kennedy, W. Q. 1965. The influence of basement structure on the evolution of the coastal (Mesozoic and Tertiary) basins of Africa. In *Salt basins around Africa,* p. 7. Inst. Petrol., London.

King, L. C. 1962. *The morphology of the earth,* Edinburgh.

Korn, H., and Martin, H. 1954. The Messum Igneous Complex in South-West Africa. *Trans. geol. Soc. S. Afr.*, **57**, 83.

Krishnan, M. S. 1949. *Geology of India and Burma,* Madras.

Le Pichon, X. 1968. Sea-floor spreading and continental drift. *J. geophys. Res.*, **73**, 3661.

Lightfoot, B. 1913. The geology of the north-western part of the Wankie coalfield. *Bull. geol. Surv. S. Rhodesia,* **4**.

LOMBAARD, B. V. 1952. Karroo dolerites and lavas. *Trans. geol. Soc. S. Afr.*, **55**, 175.

MACDONALD, G. A. 1967. Forms and structures of extrusive basaltic rocks. In *Basalts*, p. 1. (Eds. H. H. Hess and A. Poldervaart), Vol. 1, New York.

MANTON, W. I. 1968. The origin of associated basic and acid rocks in the Lebombo-Nuanetsi Igneous Province, southern Africa, as implied by strontium isotopes. *J. Petrology*, **9**, 23.

MAUD, R. R. 1961. A preliminary review of the structure of coastal Natal. *Trans. geol. Soc. S. Afr.*, **64**, 247.

McDOUGALL, I. 1963. Potassium-argon age measurements on dolerites from Antarctica and South Africa. *J. geophys. Res.*, **68**, 1535.

—— and RÜEGG, N. R. 1966. Potassium-argon dates on the Serra Geral Formation of South America. *Geochim. et cosmochim. Acta*, **30**, 191.

McKENZIE, D. P. 1967. Some remarks on heat flow and gravity anomalies. *J. geophys. Res.*, **72**, 6261.

MELFI, A. J. 1967. Potassium-argon ages for core samples of basaltic rocks from southern Brazil. *Geochim. et cosmochim. Acta*, **31**, 1079.

MENNELL, F. P. 1913. *A manual of petrology*, London.

MORGAN, W. J. 1968. Rises, trenches, great faults, and crustal blocks. *J. geophys. Res.*, **73**, 1959.

OLIVER, J., and ISACKS, B. 1967. Deep earthquake zones, anomalous structures in the upper mantle, and the lithosphere. *J. geophys. Res.*, **72**, 4259.

ROGERS, A. W. 1926. Notes on the north-eastern part of the Zoutpansberg district. *Trans. geol. Soc. S. Afr.*, **28**, 33.

SIEDNER, G., and MILLER, J. A. 1968. K-Ar age determinations on basaltic rocks from South-West Africa and their bearing on continental drift. *Earth Planetary Sci. Let.*, **4**, 451.

SNELLING, N. J. 1967. Age determination unit. In *Ann. Rep. Inst. geol. Sci.* (for 1966), p. 142.

SÖHNGE, P. G. 1945. The geology of the Messina copper mines and surrounding country. *Mem. geol. Surv. S. Afr.*, **40**.

—— Le ROEX, H. D., and NEL, H. J. 1948. The geology of the country around Messina. Explanation of sheet 46. *Geol. Surv. S. Afr.*

STOCKLEY, G. M. 1947. *Report on the geology of Basutoland*, Maseru.

SWANSON, D. C. 1967. Yakima basalt of the Tieton River area, south-central Washington. *Bull. geol. Soc. Am.*, **78**, 1077.

SWIFT, W. H., WHITE, W. C., WILES, J. W., and WORST, B. G. 1953. The geology of the Lower Sabi coalfield. *Bull. geol. Surv. S. Rhodesia*, **40**.

TAVERNER-SMITH, R. 1960. The Karroo System and coal resources of the Gwembe district, south-western section. *Bull. geol. Surv. N. Rhodesia*, **4**.

—— 1962. Karroo sedimentation in a part of the mid-Zambezi Valley. *Trans. geol. Soc. S. Afr.*, **65**, 43.

URIE, J. G., and HUNTER, D. R. 1963. The geology of the Stormberg volcanics. *Bull. geol. Surv. Swaziland*, **3**.

VAIL, J. R. 1965. An outline of the geochronology of the late-Precambrian formations of eastern central Africa. *Proc. r. Soc. Lond.*, **284**, Ser. A, 354.

—— 1967. The southern extension of the East African Rift System and related igneous activity. *Geol. Rdsch.*, **57**, 601.

—— HORNUNG, G., and COX, K. G. 1969. Karroo basalts of the Tuli syncline, Rhodesia, *Bull. volc.*, **33**, 398.

VAN EEDEN, O. R., VISSER, H. N., VAN ZYL, J. S., COERTZE, F. J., and WESSELS, J. T. 1955. The geology of the eastern Zoutpansberg and the Lowveld to the north. Explanation of sheet 42. *Geol. Surv. S. Afr.*

VAN ZYL, J. S. 1950. Aspects of the geology of the northern Zoutpansberg area. *Annls. Univ. Stellenbosch.*, **26**, Ser. A.

VISSER, H. N. 1961. The Karroo System in northern Transvaal. *C.C.T.A. southern reg. Comm. Geol.*, Pretoria, 115.

WAGER, L. R. 1947. Geological investigations in East Greenland. Part IV. The stratigraphy and tectonics of Knud Rasmussens Land and the Kangerdlugssuaq region. *Medd. Grønland*, **134** (5).

WEST, W. D. 1959. The source of the Deccan trap flows. *J. geol. Soc. India*, **1**, 44.

WILKE, D. P. 1965. Magnesite deposits north of the Zoutpansberg, Transvaal. *Bull. geol. Surv. S. Afr.*, **44**.

WILSON, J. T. 1967. Rift valleys and continental drift. *Trans. Leicester lit. phil. Soc.*, **61**, 22.

WISSMAN, H. V., RATHJENS, C., and KOSSMAT, F. 1942. Beiträge zur Tektonik Arabiens. *Geol. Rdsch.*, **33**, 221.

WOODWARD, J. E. 1966. Stratigraphy and oil occurrences in the Karroo basin, South Africa. *Proc. 8th Commonw. Mining metall. Congr.*, **5** (Petrol.), 45.

WORST, B. G. 1962. The geology of the Mwanesi range and the adjoining country. *Bull. geol. Surv. S. Rhodesia*, **54**.

A. R. WOOLLEY and M. S. GARSON

II Petrochemical and tectonic relationship of the Malawi carbonatite-alkaline province and the Lupata-Lebombo volcanics

ABSTRACT. *The Lebombo Monocline can be traced continuously northwards into the African Rift System (see King, this volume, p. 263). The igneous rocks along this structure at Lebombo, Nuanetsi, Sabi, Lupata and southern Malawi (Chilwa Alkaline Province), which are closely related temporally and spatially to the monoclinal and rift structures, are compared petrochemically and are shown to become distinctly more alkaline northwards; in that direction the number of carbonatite and nephelinite centres increases, and an alkaline and normal rock series coexist. It is suggested that faulting to the north tapped successively deeper levels in the mantle, and that the partial melting of the mantle at varying depths was responsible for the multiple rock series. However, some degree of crustal melting, in some cases of nephelinised or fenitised rocks, is envisaged to explain the range and relative abundance of the rocks of the Chilwa Alkaline Province.*

Mesozoic igneous activity

INTRODUCTION

The carbonatite centres in southern Malawi and the adjacent parts of Mozambique belong to the Chilwa Alkaline Province, a dominantly alkaline group of igneous intrusions of Upper Jurassic to Cretaceous age, which post-date the Stormberg volcanic episode at the close of Karroo sedimentation (Dixey *et al.*, 1937). A correlation has been suggested between these rocks and the alkaline lavas of the Lupata Series of the lower Shire–Zambesi area (*ibid.*, p. 55; Garson, 1962, p. 28).

The rhyolitic and basaltic lavas and dolerites of Stormberg age in the lower Shire area and Zambesi Valley were regarded by Dixey (1929) as the equivalents of the volcanic rocks in the Lebombo area of Mozambique (Portugese East Africa) described by Du Toit in the same year. Recently, Vail (in press) has drawn attention to the petrographic similarities between the Chilwa Alkaline Province, the Lupata volcanics and the Nuanetsi Igneous Province of Rhodesia which forms a northward continuation of the Lebombo Monocline; he believes that all these rocks belong to one broad magmatic episode which occurred mainly in the Stormberg period but extended into the Cretaceous.

CHILWA ALKALINE PROVINCE

Seventeen centres of carbonatite intrusion have been mapped in Malawi; seven are in the form of volcanic necks or ring-structures, and the remainder occur as dyke-like bodies infilling fissures (Garson, 1965). Together they form an irregular eastern chain of igneous bodies near the eastern border of southern Malawi, and a western belt within the Shire Rift Valley (Fig. 2).

In the eastern belt, the main carbonatite centres at Chilwa Island, Tundulu, Nkalonje and Songwe comprise early ring-structures of sövite and associated feldspathic breccias within aureoles of fenitised gneiss. Later plugs and ring-dykes of nepheline syenite, and ankeritic and sideritic carbonatite are associated with plugs and cone-sheets of trachyte, phonolite, ijolite, lamprophyre and nephelinite. At Namangali Hill the main vent rocks are leucotrachytes and phonolites followed by sövite.

In the western belt the carbonatite and agglomerate vent at Kangankunde differs from the other main centres in its virtual lack of associated silicate intrusives apart from a few small dykes and plugs of alnöite and carbonatised nephelinite. The carbonatites here include early sideritic types and later ankeritic and strontianite-rich carbonatites. Nepheline syenite and phonolite are associated with some of the dominantly feldspathic vents in the Lake Malombe area (Fig. 2).

The large ring-structure at Muambe in the lower Zambesi area comprises feldspathic agglomerate and ring-dykes of sideritic and manganiferous carbonatite intruded into updomed and fenitised Karroo sediments and Stormberg lavas (Dias, 1961).

The alkali granite and syenite intrusions of the Chilwa Alkaline Province form prominent steep-sided mountains at Zomba, Mlanje, Michese and Machemba. The Zomba Massif is formed of a ring-structure of separate intrusions of quartz syenite, perthosite, alkali granite and microgranite in that order of age (Bloomfield, 1965, p. 28). To the north of Zomba the Chikala–Chaone–Mongolowe–Chinduzi line of connected ring-structures is orientated east-west and includes early volcanics caught up in ring-dykes of perthosite, and later pulaskite and nepheline syenite. These rocks are cut by late vents of feldspathic agglomerate. The ring-structures at Junguni and Mauze in Malawi (Fig. 2), and at Chiperone in Mozambique consist of varieties of foyaite and microfoyaite, while at Milange and Morrumbala Mountain there are intrusions of microgranite, alkali granite, syenite and nepheline syenite (Coelho, 1959a, p. 15). The ring-structures at Salambidwe and Gorongosa, believed by Coelho (*ibid.*, p. 100) to belong to the Chilwa Series, are characterised by early gabbroic rocks followed by ring-intrusions of granitic rocks, and at Salambidwe also by ring-dykes of pulaskitic syenite and nepheline syenite (Garson, 1965, p. 97).

The Stormberg doleritic dyke-swarm was closely paralleled by the later alkali dyke-swarms of Chilwa Alkaline Province age. The youngest of these comprises a swarm of sölvsbergite dykes radiating from and connecting the Salambidwe and Zomba ring-structures (Fig. 2). A similar swarm of sölvsbergite and riebeckite microgranites is associated with the alkali granite intrusions at Michese and Mlanje.

LUPATA VOLCANICS

The succession of volcanic rocks and sediments in the Lupata Gorge of the Zambesi River (Figs. 1 and 2) was first established by Dixey (Dixey and Campbell Smith, 1929, p. 241), and the stratigraphy was described in more detail by Flores (1964). The following succession along the gorge is based on Dixey and Flores:

Conglomeratic sandstones (Sena Sandstones)		
Alkaline lavas (Lupata)	300 m	Cretaceous
Conglomeratic sandstones and tuffs	150 m	
UNCONFORMITY		
Columnar rhyolite	80 m	
Arkosic sandstones	80–100 m	Stormberg
UNCONFORMITY		(Jurassic)
Basalts	600 m	
Arkosic sandstones		

Intruding these alkaline lavas are dykes of microfoyaite and a mass of nepheline syenite about a mile long (Dixey and Campbell Smith, 1929, p. 259) which closely resembles petrographically the foyaites at Songwe (Dixey *et al.*, 1937, p. 55).

Rhyolites similar to the flow at Lupata Gorge occur near Sena and at the southern rim of the basalt flows extending into Malawi. An age determination (Table 1) shows that the Lupata rhyolite is of Upper Stormberg age (Flores, 1964, p. 115) so that it is evident, as originally thought by Dixey, that the main rhyolites are a late phase of the Stormberg activity equivalent to the rhyolitic flows of the Lebombo Monocline far to the south.

In the Sena area the Sena Sandstones are intruded by pipes of olivine nephelinite. A similar group of nephelinite and phonolite plugs cuts Stormberg basalts just south of the Pompue River and east of Gorongosa (Fig. 2). The age of these pipes is uncertain and may be as late as Tertiary (Teale and Campbell Smith, 1923, p. 234). However they have strong petrographic similarities with small, late, pipe-like intrusions at the Tundulu, Chilwa Island and Muambe carbonatite centres (Fig. 2), and they may therefore all belong to the Chilwa Alkaline Province.

SABI–NUANETSI–LEBOMBO VOLCANICS

The volcanics along the Lebombo Monocline comprise essentially lower and upper groups of basaltic flows separated by a thick series of rhyolites, all dipping eastwards beneath Cretaceous and Tertiary sediments of the Mozambique geosyncline (Du Toit, 1929; Cox, this volume, p. 220). These Stormberg volcanics can be traced from Natal, through Swaziland, the Transvaal, Mozambique and the Nuanetsi area of Rhodesia. North-eastwards

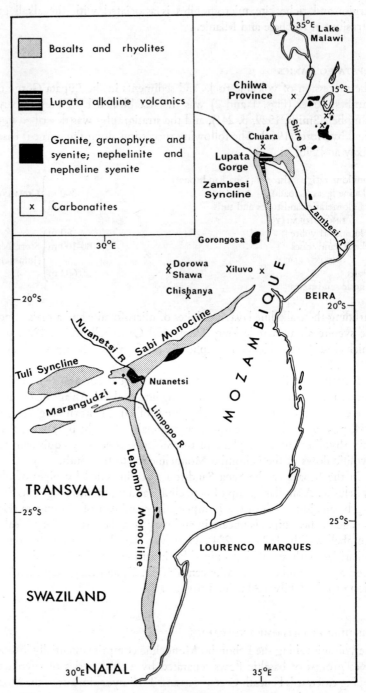

FIG. 1. Locality map of the Chilwa Alkaline Province and the volcanic suites of Lupata, Sabi, Nuanetsi and Lebombo.

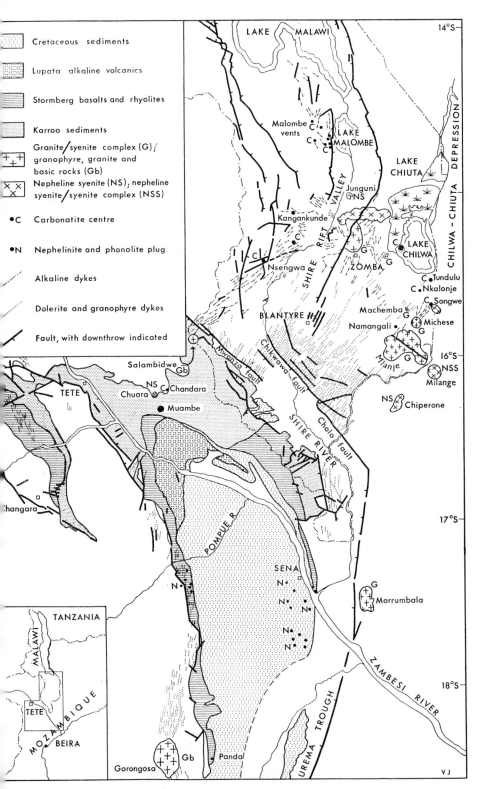

FIG. 2. Structural geology map of southern Malawi and adjacent parts of Mozambique. Based on geological maps of Malawi (1966), Mozambique (Dias, 1956) and in Vail (1965b).

I

the volcanic belt extends through the Sabi area of Rhodesia thence across Mozambique to the Lupata Gorge (Fig. 1).

At the base of the succession, limburgites and nephelinites occur in the Limpopo River area, while in the Nuanetsi and Sabi areas small amounts of nephelinite and allied rocks were erupted at an early stage, succeeded by limburgitic and picritic lavas and then by normal tholeiitic lavas (Cox et al., 1965, p. 84). At the eastern edge of the Nuanetsi Syncline the total thickness exceeds 6700 m, comprising: rhyolitic extrusives, 1650 m; upper basalts, 3000 m; olivine-rich group, 2000 m.

In the southern part of the Lebombo Monocline the upper basaltic group includes intercalated hyperalkaline and calc-alkaline rhyolitic flows (Assunçao et al., 1962, p. 64). In the same area there are minor intrusions of porphyritic nepheline syenite (Du Toit, 1929, p. 208) which may correspond to the alkaline rocks cutting the Lupata alkaline lavas.

The first phase of plutonic activity in the Nuanetsi area was the emplacement of olivine and quartz gabbro sheets in the Northern Ring, and the Masukwe and Dembe Complexes, followed by the intrusion of a thick, slightly discordant sheet of granophyre (Cox et al., 1965). This granophyre is similar chemically to the rhyolitic lavas and is probably contemporaneous with them (ibid., p. 104). Igneous activity continued with the intrusion into the main granophyre and older rocks of a series of acid igneous rocks in the form of ring-dykes, cone-sheets and plugs within nine ring-complexes. In some of the complexes nordmarkite preceded granitic rocks. Nepheline syenite forms the core of the Marangudzi Complex while early carbonatite and ijolite occur at Chishanya, Dorowa and Shawa (Fig. 1). The age of the carbonatite centre at Xiluvo south of Gorongosa is uncertain.

Dyke swarms of basic and acid rocks are associated with the ring-complexes and Stormberg lavas.

Age determinations

A considerable number of age determinations (Table 1) have recently been made on igneous rocks in Central and South Africa, so that it is now possible to establish the pattern of the main phases of the Mesozoic igneous activity discussed in this paper.

Within the period of Stormberg igneous activity the basic rocks of the Nuanetsi ring-structure were evidently intruded during part of the phase of outpouring of basaltic lavas in the Lebombo Monocline. The ijolite at Shawa, at present the oldest dated intrusion of the area, and the malignitic and pulaskitic rocks at Marangudzi, probably correspond in age with the early nepheline basalts of the Limpopo and Sabi area at the base of the Stormberg volcanics. In this connection the correlation of the nepheline syenites in the Nuanetsi area with the Lupata alkaline volcanics shown by Cox et al. in their tabulation of the Karroo volcanic cycle in the Lupata–Nuanetsi–Lebombo zone is clearly wrong (ibid., p. 209). The equivalence in age of the Lupata Gorge and Lebombo rhyolites is confirmed; the nearest

Age determinations

Locality	Rock type	Mineral dated	Method	Reference	Age (m.y.)
		CHILWA ALKALINE PROVINCE			
Zomba Ring-Structure	Quartz syenite	Hornblende and pyroxene	K-Ar	Geol. Map Malawi (1966)	105±12
Zomba Ring-Structure	Microphyric syenite	Hornblende	K-Ar	Geol. Map Malawi (1966)	115±12
Zomba Ring-Structure	Alkali granite	Hornblende	K-Ar	Geol. Map Malawi (1966)	128±13
Chambe Ring-Structure, Mlanje Massif	Perthosite	Biotite	K-Ar	Cahen and Snelling (1966)	116±6
Nanchidwa, Mlanje Massif	Quartz syenite	Biotite	K-Ar	Cahen and Snelling (1966)	128±6
Chinduzi Ring-Structure	Pulaskite	Biotite	K-Ar	Cahen and Snelling (1966)	116±6
Chaone Ring-Structure	Nepheline syenite or perthosite	Zircon	Pb-α	Bloomfield (1961)	138±14
Kangankunde Vent	Carbonatite	Phlogopite	K-Ar	Cahen and Snelling (1966)	123±6
Tundulu Ring-Structure	Sövite	Biotite	K-Ar	Snelling (1965)	133±7
Chilwa Island	Sövite	Biotite	K-Ar	Snelling (1965)	136±7
		LUPATA ALKALINE VOLCANICS			
Lupata Gorge	Kenyte (?)	Anorthoclase	K-Ar	Gough et al. (1964)	106±7
Lupata Gorge	Kenyte (?)	Anorthoclase	K-Ar	Gough et al. (1964)	110±5
Lupata Gorge	Kenyte (?)	Whole rock	K-Ar	Flores (1964)	115±10
Lupata Gorge	'Basalt' (phonolite?)	Feldspar	K-Ar	In Vail (1965a)	130
		STORMBERG IGNEOUS ROCKS			
South Africa and Mozambique	Karroo dolerites	—	K-Ar	McDougall (1963)	154—190
Lebombos Mountain	Rhyolite	—	Rb-Sr	Hales (1960)	160
Lupata Gorge	Rhyolite	Whole rock	K-Ar	Flores (1964)	166±10
Grootvlei, Transvaal	Dolerite dyke	Plagioclase	K-Ar	McDougall (1963)	161
Sasolberg, Transvaal	Dolerite	Whole rock	K-Ar	McDougall (1963)	173
Masukwe and Dembe-Divula Ring Complexes	Granite	—	Rb-Sr	Brock (1968)	177±4 (168)*
Marangudzi Ring-Complex	Foyaite	Biotite	K-Ar	Gough et al. (1964)	182±10
Marangudzi Ring-Complex	Granite	Whole rock	K-Ar	Gough et al. (1964)	186±10
Marangudzi Ring-Complex	Gabbro	Biotite	K-Ar	Gough et al. (1964)	187±10
Marangudzi Ring-Complex	Gabbro	Biotite	K-Ar	Gough et al. (1964)	192±10
Marangudzi Ring-Complex	Pulaskite	Hornblende	K-Ar	Gough et al. (1964)	195±10
Marangudzi Ring-Complex	Malignite	Biotite	K-Ar	Gough et al. (1964)	196±10
Shawa Carbonatite	Ijolite	Biotite	Rb-Sr	Nicolaysen et al. (1962)	208±16

* Age according to the decay constant $\lambda = 1.475 \times 10^{-11} \text{yr}^{-1}$.

igneous counterparts are the main granophyre sill and the granites at the
Masukwe and Dembe–Divula Ring-Complexes.

The age-determinations in Table 1 show that the major unconformity in
the Lupata Gorge above the Stormberg rhyolites represents a maximum
pause in igneous activity of about 30 m.y. (see however p. 246). This is confirmed also by the age determinations of the igneous equivalents forming the
Chilwa Alkaline Province.

The oldest dated rocks of the Chilwa Alkaline Province are the sövites of
the first ring-structure at Tundulu and Chilwa Island, and the perthosites of
the Chikala–Chaone ring-structures north of Zomba (Garson, 1962, p. 27).
The later great acid and alkaline intrusions were coeval with the alkaline
volcanics in the Lupata Gorge. The equivalence of the ages of the Mlanje
Massif and the Lupata volcanics has been substantiated by Briden (1967,
p. 375) who showed that they have identical palaeomagnetic poles of Lower
Cretaceous age. It would be useful now to have age determinations of the
Muambe carbonatite centre and the adjacent intrusion of nepheline syenite at
Chuara to determine if, as seems likely, these were feeders for the volcanics.
Age determinations on the plugs of nephelinite and phonolite in Mozambique
and the similar late plugs at the Malawi carbonatite centres would serve to
define the upper limits of the alkaline activity.

Relationship between Mesozoic igneous activity and rift tectonics

It was first recognised by Dixey (1937, 1956) that the southern end of the
great African Rift System was largely developed in Jurassic times along lines
of crustal weakness developed in the Cambrian or Precambrian. Dixey (*ibid.*)
also described the thousand mile overlap of Jurassic and Quaternary structures in the Rukwa–Nyasa–Shire–Lebombo Rift and the contemporaneous
development of the Lebombo Monocline with the earlier structures.

The present study of Mesozoic igneous activity in Central and southern
Africa demonstrates its close relationship with the early rifting at the close of
Karroo sedimentation and the change in composition of the eruptive rocks
during later stages of the rifting.

KARROO SEDIMENTARY BASINS

According to Dixey the rift disturbances in Central Africa were heralded by
crustal movements which produced ridges and zones of subsidence or
troughs (1930, p. 57). In the lower Shire and Limpopo areas, Karroo sedimentation occurred under conditions of steady downwarp concomitant with
the elevation of the surrounding areas of Precambrian rocks (Habgood,
1963, p. 42; Cox *et al.*, 1965, p. 124). The occurrence of similar rocks along
the edge of the Lebombo and Sabi Monoclines indicates lines of downwarp
during the same period (*ibid.*, p. 124). In general these sedimentary areas are
closely associated with the older orogenic belts encircling the Rhodesian
Craton (Vail, in press).

STORMBERG IGNEOUS ACTIVITY

Following the deposition of the Karroo sediments, downwarping culminated in a stage of intense fracturing which led to the injection of doleritic dyke-swarms and the outpouring of thick flows of basalt.

Lebombo Monocline

The dip of the volcanics is eastward at between 5° and 20°, reaching 30° to 40° and occasionally as much as 60° in the axis of the monocline. The volcanics flatten out to the east where they pass unconformably beneath Cretaceous beds (Du Toit, 1929). The maximum crustal bending probably occurred during the main period of rhyolitic vulcanicity. The volcanics are cut by a dyke-swarm which strikes parallel to the monocline, and in which the majority of the dykes dip at high angles to the west suggesting that they are axial planar to the monoclinal structure. This pattern is similar to that of the dyke-swarm which cuts the Skaergaard Intrusion (Wager and Deer, 1939, p. 19; Cox, this volume, p. 224). Another significant feature is that on the western edge of the monocline the narrow slices between basic dykes downthrow to the west.

Du Toit (1929) believed that the sequence of events started with pressure from the south-south-west which resulted in a period of east-west tension in the Lebombo area. This produced a line of weakness running north-south associated with a great magmatic wedge of basic material and viscous upper acid differentiates partly derived by assimilation. West of the wedge, basalts and limburgites were erupted from numerous parallel fractures while the acid wedge immediately to the east supported the overlying crust. Basaltic injection and strike-faulting continued on the west side until the acid reservoir was finally tapped and rhyolites were discharged from fissures concomitant with crustal warping. A narrow zone of granophyre was produced during renewed tensional conditions until the acid magma reservoir was virtually exhausted and basalts, perhaps of a more alkaline type, were emitted together with a few sporadic eruptions of hyperalkaline rhyolite. Finally alkaline rocks were intruded in the Little Lebombo area (for an alternative hypothesis see Cox, this volume, p. 228).

This period of closely associated vulcanism and earth movements is taken to be the first phase in the creation of the African Rift System.

Sabi Monocline and Lake Chilwa—Chiuta Depression

Between the Limpopo and Nuanetsi Rivers the Lebombo Monocline is deflected into the Nuanetsi Syncline (Fig. 1) which was also developed during the Stormberg volcanic period (Cox et al., 1965, p. 125). The north-eastern extension termed the Sabi Monocline is similarly diverted into the Zambesi Syncline. The monocline as a whole is parallel to the Lake Chilwa–Chiuta Depression (see Figs. 1 and 2) which passes, to the north-north-east, into a major fault zone in Mozambique within which there are narrow troughs of Karroo sediments. In this depression Stormberg lavas are largely absent, possibly owing to erosion, although there is a spectacular swarm of dolerite dykes orientated parallel to the length of the depression. These dykes are mainly vertical or have high westward dips, a similar attitude to that of the Lebombo swarm suggesting that the same stress field operated in both cases. Possible remnants of flows may occur on the Mlanje Massif where acidic volcanics are invaded by syenite (Garson and Walshaw, in press), and north of Zomba where basic volcanics are down-faulted and nephelinised in the Chikala ring-structure (Stillman and Cox, 1961).

A section across the Lake Chilwa–Chiuta depression eastwards from the edge of the Shire Rift reveals a striking resemblance to the monoclinal structure to the south. The eastern edge of the rift is marked by a series of faults downthrowing to the west. The acidic rocks of Zomba Mountain occur in the highest area of the gneissic platform which is tilted to the east. The platform is downwarped in the area of Lake Chilwa and levels out eastwards below colluvial and alluvial deposits.

Shire Rift and Lower Zambesi River

Although the Shire Rift is probably influenced to some extent by ancient lines of weakness, the faults bounding the rift were initiated in Stormberg times. In the southern part of the rift system the Mwanza, Chikwawa and Cholo Faults are important features parallel to the Zambesi River (Fig. 2). In places the faults are infilled by dolerite.

The basaltic flows within the Zambesi Valley were erupted dominantly from fissures, volcanic centres being comparatively rare and small. In Malawi, the flows attained a thickness of 1200 m prior to being intruded by numerous sills and dykes of dolerite. The unconformity between the rhyolitic flow at the Lupata Gorge and the underlying basalt may indicate earth movements of some magnitude. Likewise the deep erosion following the eruption of this rhyolite has left a mere 80 m thickness compared with over 1750 m in the Nuanetsi area. Because of this, it cannot now be established if the upper group of alkaline basalts and hyperalkaline rhyolites found in the Lebombo Monocline was erupted in Stormberg times in the Zambesi area. The break between the Lupata rhyolite and the overlying alkaline lavas (Table 1) may not therefore represent as long a pause in igneous activity as 30 m.y.

Nuanetsi Syncline and Zambesi Syncline

The Nuanetsi Syncline has been compared by Cox et al. (1965, p. 128) to the Kangerdlugssuaq area in Greenland (Wager, 1947), and the Zambesi Syncline apparently has a similar origin. In all three areas a monocline paralleled by dyke-swarms changes direction significantly (Fig. 1) and at the point of the re-entrant angle a subsidiary, apparently tensional, structure is found (Cox, this volume, p. 224). The formation of the Nuanetsi Syncline was attributed to the extrusion of enormous quantities of lava which led to subsidence of the original land-surface; the fold was therefore referred to as a volcano-tectonic feature. A similar explanation is probably valid for the Zambesi Syncline (Fig. 1).

The disposition of the Karroo ring-complexes in the Nuanetsi area, and of the Cretaceous carbonatitic and alkaline intrusions and flows in the Zambesi Syncline, is probably significant and tectonically related to the intersection of the monoclinal structure by major directions of tension. At Nuanetsi the ring-complexes are mainly situated on a straight line parallel to the east-north-east Limpopo trend of faulting. The age relations suggest that the complexes were the source of some of the thick basaltic flows, and that some of the rhyolites are the effusive equivalents of the acid intrusives. In the Lupata area the intersecting line is the north-west-trending Zambesi Rift Valley bounded by the major Mwanza, Chikwawa and Cholo Faults and a series of faults in the vicinity of Tete (Fig. 2). These faults bisect the angle defined by the Chilwa–Chiuta Depression, parallel to the Malawi dolerite dyke-swarm, and the basalt and rhyolite sequence to the south of Lupata which follows the northward continuation of the Sabi Monocline (see Fig. 1). A comparable area of similar age is the Rufunsa area of Zambia where Bailey (1961, p. 280) has shown that the Upper Jurassic–Cretaceous carbonatites are sited at the intersection of two major fault trends.

CHILWA ALKALINE PROVINCE

The close association between the igneous activity of the Chilwa Alkaline Province and renewed rifting is shown by both the eastern and western belts of igneous centres in Malawi, and also by the already quoted intersection of the Zambesi Syncline and the rift fault-zone.

In the eastern belt, the alkaline dykes that are aligned parallel to the earlier Stormberg dolerite dyke-swarm indicate renewed tensional conditions (see Fig. 2). In general these dykes link the carbonatitic centres which are themselves roughly aligned parallel to the Lake Chilwa–Chiuta Depression. At some of these centres there is evidence of strong updoming prior to the intrusion of carbonatite, and the Mlanje Massif is similarly emplaced in updomed Basement Complex gneisses. It appears that the granitic and syenitic plutons extending from Morrumbala to the Songwe area (Fig. 2) were intruded in high positions along a ridge on the eastern edge of the Lake Chilwa–Chiuta Depression and Urema Trough. The Zomba Massif and the ring-structures to the north were emplaced under similar conditions. The sequence of intrusions indicates that the igneous centres north of Zomba were emplaced progressively to the west, as faulting became more pronounced towards the Shire Rift Valley, concomitant with further down-warp in the Lake Chilwa area. This is compatible with Dixey's idea of a 'rise to the rift' during arching (Dixey, 1956, p. 6).

At the Tundulu centre, wrench-faulting along north-south to north-north-east to south-south-west directions occurred before the intrusion of beforsite, alnöite and melanephelinite of the third ring-structure. Sölvsbergite dykes were intruded into wrench-faults at a later stage. Similar sölvsbergite and riebeckite microgranite dykes are radially disposed towards the ring-structures at Michese and the Mlanje Massif.

In the western belt (Fig. 2) the Lake Malombe vents are situated adjacent to the fault zone delineating the western edge of the main rift valley. Further south the Kangankunde and Kapiri vents occur on lines of parallel faulting associated with the rifting while the small Chaumbwi vent infills a small fissure parallel to the main rift valley fault in the area. The Nsengwa Complex occurs at the intersection of a main rift valley fault and a later zone of transcurrent faulting infilled by a swarm of sölvsbergite dykes 'connecting' the Zomba and Salambidwe ring-structures. As in the eastern belt of igneous centres, carbonatitic intrusion has occurred during and after periods of rift faulting.

UREMA SEDIMENTARY TROUGH

The zone of parallel faults defining the eastern edge of the Urema Trough (Fig. 2) is the southern extension of the main eastern rift fault of Malawi (Mouta, 1957, p. 29). It is evident therefore that the southern end of the African Rift was initiated in Stormberg times as a monocline with some tensional faulting, and later developed into a trough-like feature bounded to the west by the monocline and to the east by a zone of Cretaceous to Tertiary faults. The intrusions related to this feature comprise the nephelinite plugs near Sena, and the phonolite and nephelinite plugs south of the Pompue River and east of Gorongosa. The Urema Trough is parallel to the post-Karroo Mozambique geosyncline and is believed by Dixey to be related to its development (Dixey, 1956, p. 6).

Petrochemistry

THE CHILWA ALKALINE PROVINCE

Over fifty chemical analyses of igneous silicate rocks of the Chilwa Alkaline Province are now available and these are plotted in the quartz-nepheline-kalsilite system (Fig. 3), together with five new analyses of lavas and a nepheline syenite from the Lupata Gorge (Table 2). The plotted points spread along the low temperature trough in a similar distribution to that found by Bowen (1937) for a suite of East African rhyolites, trachytes and phonolites, suggesting that crystal-liquid equilibrium was the dominant influence on their evolution. However, in the undersaturated part of the system, Bowen's analyses clustered close to the potash side of the trough, and a few were outside the trough; this is an effect that Bowen identified (*ibid.*, p. 13) and which may be due to the presence of anorthite. The tendency towards enrichment in potash is considerably enhanced in the Lupata and Chilwa rocks, and is probably caused, in part at least, because these are plutonic and porphyritic rocks in which some degree of crystal separation has occurred. There appears to be, in terms of Fig. 3, a distinct separation of the syenitic and granite rocks.

In Fig. 4 the available analytical data have been plotted on a Harker diagram. The wide range of silica values of these rocks lends itself to the Harker plot, while the low magnesia values exclude the meaningful application of one of the differentiation indices as abscissa. It is immediately apparent from Fig. 4 that the suite subdivides into three distinct groups: granites, syenites and trachytes; nepheline syenites and phonolites; and nephelinites. The nephelinites are somewhat isolated from the nepheline syenites in terms of silica, but extrapolation of the trends defined by the nepheline syenites passes close to the nepheline points. A comparison with data from similar

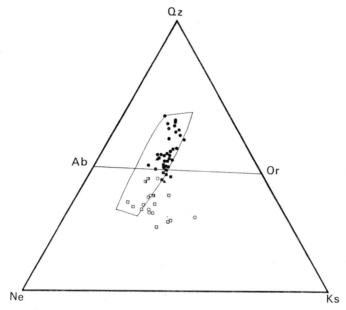

Fig. 3. Plot of Chilwa Alkaline Province and Lupata alkaline rocks in the quartz-kalsilite-nepheline system. ○ Lupata alkaline volcanics.

Chilwa Alkaline Province: ☐ foyaite and phonolite; ◧ pulaskite; ▲ perthosite; ■ syenite and trachyte (Salambidwe); ● granite and syenite (Zomba and Mlanje).

provinces (p. 257), suggests that there is usually a continuous range of compositions from nephelinite through to nepheline syenites. The distinctive nature of the granite-syenite suite is particularly apparent with regard to MgO, CaO and $(FeO + Fe_2O_3)$. For these oxides there is a sudden increase concomitant with a slight increase in silica. This increase is quite contrary to all the usual differentiation trends and suggests strongly that there is not a continuous differentiation sequence from nepheline syenite to syenite and granite. The Al_2O_3 trend is also broken, as is that for Na_2O, but there is no apparent displacement of the K_2O values. It would appear therefore, that we are dealing with two distinct magma series which differentiated independently.

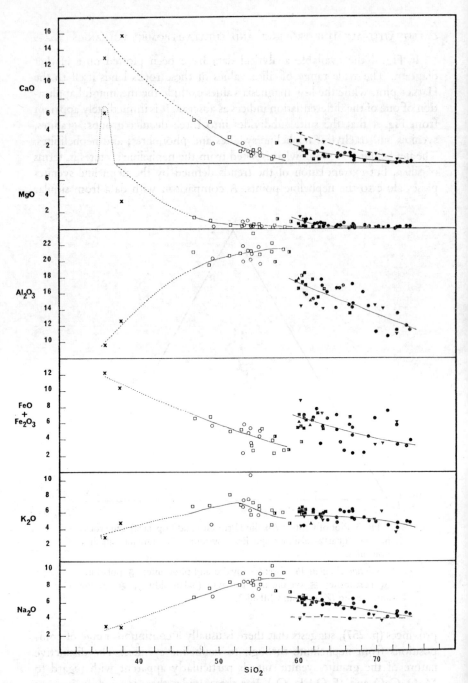

FIG. 4. Harker diagram of rocks of the Chilwa Alkaline Province and the alkaline volcanics of the Lupata Gorge. ○ blairmorite, kenyte, phonolite and nepheline syenite from the Lupata Gorge.

Chilwa Alkaline Province: × melanephelinite; □ foyaite, microfoyaite and phonolite; ▣ pulaskite (Chikala); ◼ syenite, pulaskitic syenite, porphyritic trachyte, sölvsbergite (Salambidwe); ● granite, quartz syenite, and syenite (Mlanje); ▼ granite, microgranite, syenite, and perthosite (Zomba); ▲ perthosite and microsyenite.

The dashed trend lines are for Zomba Mountain—where these trends are distinctive.

Within the syenite-granite sequence the Salambidwe, Mlanje and Zomba Mountain plots show some difference in their trends. This is particularly apparent for the Na_2O and CaO values of Zomba, which are distinguished by the dashed lines on Fig. 4; Zomba is distinctly poorer in soda but richer in lime. Salambidwe is rather lower in lime and magnesia than the other two. Although the analytical data on these complexes are not abundant enough to be statistically incontravertible it appears (Fig. 4) that either the parent magmas were slightly different in composition, or that there was independent, and distinctive, differentiation within each complex from a common parental magma.

A number of Chilwa complexes, in particular Chikala, Chaone and Mongolowe, are built of perthosite, pulaskite and foyaite. These rocks should extend across the boundary between the two magma series on Fig. 4. Unfortunately there are no analyses available of foyaite or perthosite from these three complexes, although the three analyses of pulaskite from Chikala (Bloomfield, 1965, p. 119) do extend across the syenite-nepheline syenite boundary in terms of Qz-Ne-Ks (Fig. 3). The situation is complicated by the fact that the perthosites are earlier than the nepheline syenites and have, to some extent, been nephelinised by them. From the intimate association of perthosite, foyaite and pulaskite in these complexes one would expect them to define a single differentiation series. In terms of MgO and CaO against SiO_2 the pulaskites appear to form part of the syenite-granite, rather than the nepheline syenite series, though in terms of the other oxides they occupy an intermediate position.

THE LUPATA VOLCANICS

The alkaline lavas that comprise part of the Lupata Series are phonolites, kenytes, analcite kenytes, and blairmorites. Although described in considerable petrographic detail by Dixey and Campbell Smith (1929, p. 252), no chemical data were given in that account. However, Coelho (1959b, p. 37) gives seven analyses of phonolites, trachytes, and a rhyolite from Lupata, but none of these rocks are described as containing analcite. The blairmorites, which contain analcite phenocryts up to 1 cm in size, are extremely rare rocks. The suite of Lupata rocks collected by Dixey and described by him and Campbell Smith is now housed in the British Museum (Natural History), and the analyses presented with this account were made on that material (Table 2). The details of the analysed rocks are not duplicated here; full references for each rock are given at the foot of Table 2.

A modal analysis made on the blairmorite hand-specimen gave: analcite phenocrysts, 25·4%; nepheline phenocrysts, 14·4%; vesicles, 1·7%; matrix, 58·5%. The analysis of blairmorite (Table 2, analysis 4(d)) is difficult to compare with that of the type rock (MacKenzie, 1914) because the Canadian rock has 71% modal analcite which tends to considerably inflate the Na_2O. The phonolites, kenyte, and blairmorite given in Table 2 have notably higher K_2O values than the average phonolite of Nockolds (1954, p. 1024)

TABLE 2
Analyses of Lupata Gorge alkaline volcanics

	(a)	(b)	(c)	(d)	(e)
SiO_2	55·2	55·1	54·0	49·3	54·1
Al_2O_3	19·7	21·3	19·9	20·0	20·9
Fe_2O_3	2·7	3·3	2·9	5·5	1·9
FeO	2·5	1·2	2·6	0·2	1·7
TiO_2	0·40	0·25	0·65	1·7	0·80
CaO	1·4	1·7	1·5	2·5	1·3
MgO	0·35	0·40	0·50	0·3	0·40
P_2O_5	0·05	0·05	0·10	0·30	0·10
MnO	0·35	0·25	0·35	0·25	0·15
Na_2O	9·7	7·7	8·6	6·9	6·9
K_2O	6·5	5·9	6·4	4·7	10·8
CO_2	0·10	0·20	0·05	1·2	0·10
$H_2O +$	1·35	2·50	2·90	5·5	1·00
$H_2O -$	0·25	0·35	0·30	1·3	0·15
F	0·10	0·05	0·05	0·1	0·10
Total	100·65	100·25	100·8	99·75	100·4

Norms

	(a)	(b)	(c)	(d)	(e)
c	0·0	0·0	0·0	2·5	0·0
or	38·4	34·9	37·8	27·8	45·2
ab	15·5	31·4	18·3	35·9	0·0
an	0·0	7·4	0·0	10·4	0·0
lc	0·0	0·0	0·0	0·0	14·6
np	26·9	17·0	26·2	4·4	25·7
nc	0·2	0·5	0·1	2·9	0·2
ac	7·8	0·0	4·8	0·0	5·5
ns	1·6	0·0	0·0	0·0	0·8
di	5·8	0·6	5·8	0·0	4·9
ol	1·9	0·5	1·2	0·5	0·4
mt	0·0	4·0	1·8	0·0	0·0
hm	0·0	0·6	0·0	5·5	0·0
il	0·8	0·5	1·2	0·9	1·5
ru	0·0	0·0	0·0	1·2	0·0
ap	0·1	0·1	0·2	0·7	0·2
$H_2O +$	1·3	2·5	2·9	5·5	1·0
$H_2O -$	0·2	0·3	0·3	1·3	0·1
others	0·1	0·0	0·0	0·1	0·1
Norm total	100·6	100·3	100·6	99·6	100·2

(a) Phonolite. Domue Ridge about 6 miles from Chiganga. B.M.1929,173,28. (Dixey and Campbell Smith, 1929, p. 252 [N447]).

(b) Phonolite. Between Nkomadzi and Gorufa. B.M.1929,173,69. (*ibid.*, p. 253 [N721]).

(c) Kenyte. Between Penze and Chirisa. B.M.1929,173,47. (*ibid.*, p. 254 [N690]).

(d) Blairmorite with fresh analcite phenocrysts. Between Gorufa and Sungo. B.M.1929,173,72. (*ibid.*, p. 256 [N724]).

(e) Biotite aegirine-augite foyaite. Along track between Penze and Chirisa. B.M. 1929,173,50. (*ibid.*, p. 259; Dixey *et al.*, 1937, p. 55 [N696]).

Analyst, A. A. Moss.

For accurate localisation of rocks see geological map in Dixey and Campbell Smith (1929).

and this tendency is even more marked in the foyaite which is more enriched in potash than the average nepheline syenite of Juvet type (*ibid.*).

The tendency already pointed out for Chilwa and Lupata under-saturated rocks to lie on the potash side of the low-temperature trough in the Qz-Ne-Ks triangle can be explained by the crystallisation and separation, perhaps by flotation, of primary analcite crystals, thus enriching the liquid in potassium. An analysis of the matrix of the blairmorite (analcite and nepheline phenocrysts removed by hand picking) gave values of $K_2O = 7\cdot0\%$ and $Na_2O = 3\cdot4\%$ which contrasts with whole rock values of $4\cdot7\%$ and $6\cdot9\%$ respectively. This indicates potassium enrichment and sodium impoverishment in the liquid phase. The widespread availability of the analcite crystallising magma is supported by the presence of pseudoanalcite-bearing dykes at Tundulu. The general similarity of the Lupata volcanics to the Chilwa foyaites is apparent on Fig. 3 and is brought out on the Harker diagram (Fig. 4), on which the only notable differences are the high K_2O of the Lupata foyaite and the low K_2O of the blairmorite. It seems possible that the foyaite results from the removal of analcite, and the blairmorite represents an analcite concentrate.

A COMPARISON OF THE CHILWA, LUPATA AND LEBOMBO–NUANETSI–SABI SUITES

The close petrochemical similarity between the phonolite-kenyte-blairmorite lavas of the Lupata Gorge and the foyaite suite of the Chilwa Alkaline Province has already been pointed out. Significantly analcite is an important constituent of many of the phonolite dykes of the Tundulu Complex (Garson, 1962, p. 135). The pseudomorphs in these dykes, originally described as pseudoleucite phonolites, are now regarded as analcite not leucite. That analcite can crystallise as a primary mineral is amply proved by the blairmorites of the Lupata Gorge, and the presence of such rare rock types in the Tundulu Complex and amongst the Lupata lavas is strong evidence that they are comagmatic. Another suite of dykes found both among the Lupata lavas and at Tundula are the spherulitic tinguaites described in detail by Dixey *et al.* (1937, p. 25). These tinguaites are characterised by radiating, spherical groups of natrolite needles, which weather to circular depressions commonly 3-4 cm across; similar rocks occur west of Lake Malombe (*ibid.*, p. 25).

The strongly alkaline rocks of the Lebombo–Nuanetsi Province have not been studied in detail, and reference to these rocks are widely scattered. However, Cox *et al.* (1965, p. 195), on the basis of the few available analyses, suggested that there are upper and lower alkaline groups, and Assunçao *et al.* (1962, p. 68) point out that the few analyses of alkaline rocks from the Lebombo Monocline are from the eastern side of the structure; so these presumably are part of the upper alkaline group. The alkaline rocks described from the Lebombo–Nuanetsi–Sabi fall into three groups which comprise: (*i*) lavas described as tephrite, phonolite, nepheline basalt and nephelinite; (*ii*) plugs and dykes of nepheline dolerite, phonolite, olivine nephelinite and ijolite; and (*iii*) stocks of nepheline syenite. Descriptions of these rocks and a

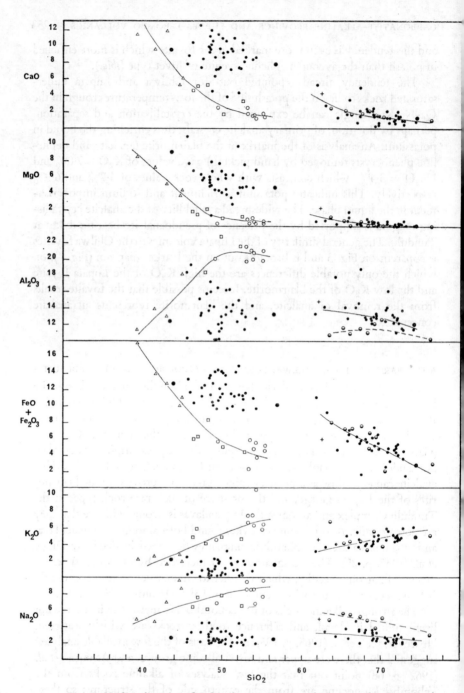

FIG. 5. Harker diagram of Lebombo, Nuanetsi, and Lupata volcanic rocks.
○ lupata alkaline volcanics; □ nepheline syenite (Marangudzi); tephrite,
nepheline syenite and phonolite (Lebombo); △ nephelinite, ijolite, and
nepheline basalt (Sabi Coalfield); nepheline basalt (Kruger National Park);
• basalt and dolerite; alkaline and calcalkaline rhyolite (Lebombo and
Nuanetsi); ◕ Malawi dolerite and Lupata Gorge rhyolite; ◑ hyperalkaline
rhyolite (Lebombo); + quartz trachyte (Lebombo);

The dashed trend lines are for the hyperalkaline rhyolites—where they are
distinctive.

few analyses are given by Young (1920, pp. 106 and 108), Teale and Camp-bell Smith (1923), Lombaard (1952, p. 190), Swift et al. (1953, pp. 40 and 44), Assunçao et al. (1962) and Cox et al. (1965).

The analyses of the Stormberg alkalic rocks together with the new analy-ses and Coelho's data (1959b) of the slightly younger Lupata rocks are plotted as hollow symbols on Fig. 5. Although there is some scattering of the points, the MgO plots in particular show a relatively smooth curve suggesting strongly that these rocks can be considered as a single differentiation series. The plots for Na_2O, $Fe_2O_3 + FeO$ and CaO are also relatively compact, and the Lupata analyses, shown as circles, fit in well with the general trend. Although somewhat similar curves would result for provinces of similar rock types it has been found that there are usually distinct differences from province to province. Collected analyses from the Rungwe Province of southern Tanzania (Harkin, 1960), and from eastern Uganda, for instance, define rather different trends. The trends of CaO, MgO and $Fe_2O_3 + FeO$ agree well with those of the foyaites of the Chilwa Alkaline Province, but the Chilwa rocks are distinctly richer in K_2O. The whole suite of Chilwa, Lupata and Lebombo–Nuanetsi alkaline rocks have been combined in the silica–total alkalies diagram of Fig. 6b (hollow symbols), where the chemical difference is apparent between the alkaline rocks and the basalts with which they are associated.

The rhyolites, which lie above the Stormberg basalts in the Lupata Gorge, can be traced southwards almost continuously into South Africa. This great development of Karroo rhyolite is restricted to the line of the Lebombo 'hinge', for in southern Africa the Stormberg volcanics are mainly tholeiitic basalts. A large number of chemical analyses of these rocks have been made, and are listed by Cox et al. (1965, p. 198), and plotted on Figs. 5 and 6b. Assunçao et al. (1962, p. 61) were able to distinguish a hyperalkaline group of rhyolites, which differ from the more normal alkaline and calc-alkaline groups in having higher Na_2O and lower Al_2O_3. The hyperalkaline group (ibid.) containing normative acmite are differentiated on Figs. 5 and 6b; from these figures the distinctive nature of this group with regard to total alkalies, soda and alumina is apparent. Cox et al. (1965) found that the alka-line-calc-alkaline group of rhyolites, referred to by them as the rhyodacite-rhyolite series (ibid. p. 201), are chemically indistinguishable from the ana-lysed granites and granophyres of the Nuanetsi area; this similarity is brought out on the silica-alkalies diagram (Fig. 6b). However, it is also apparent from Fig. 6b that the hyperalkaline rhyolites, for which there are no plutonic equivalents at Nuanetsi, are closely comparable to the Chilwa Alkaline Province syenite-alkali granite series, represented particularly by rocks from Zomba and Mlanje Mountains, and are probably, therefore, extrusive equivalents of this series.

The published analyses of basalts from the Nuanetsi-Lebombo region are tabulated by Cox et al. (ibid., p. 188) and these have also been plotted on Figs. 5 and 6b, as have the only two dolerites from Malawi that have so far been analysed. The dolerites lie almost exactly along the trends, often rather

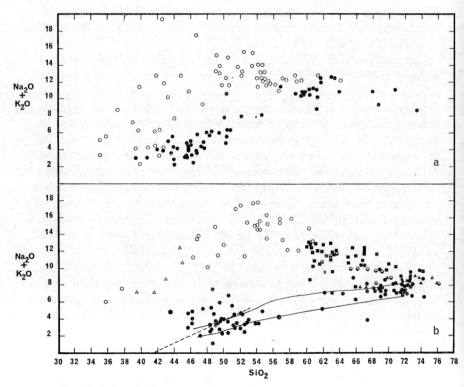

FIG. 6. Silica-alkalies diagram for (a) the Neogene Province of northern
Tanzania and southern Kenya, and (b) the Chilwa—Lupata–Lebombo–Nuanetsi
Province.

(a) ○ feldspathoid-bearing lavas; ● other lavas. (b) ○ foyaite, phonolite,
and tephrite of Chilwa, Lupata, and Lebombo Provinces; □ nepheline
syenite (Marangudzi); △ nephelinite, ijolite, and nepheline basalt (Sabi
Coalfield, and Kruger National Park); • basalt, dolerite, and calcalkaline
rhyolite (Lebombo and Nuanetsi); ● Malawi dolerites; ◐ hyperalkaline
rhyolite (Lebombo); ■ granite, syenite, trachyte and perthosite (Chilwa
Province); ▲ granophyre and granite (Nuanetsi and Lebombo); ◰ nordmarkite (Nuanetsi); × quartz trachyte (Lebombo).

Solid boundary lines separate the fields for tholeiite, high-alumina basalt, and
alkali olivine basalt of central and south-western Japan (Kuno, 1968); dashed line
separates fields of tholeiitic and alkalic lavas of Hawaii (Macdonald and Katsura,
1964).

poorly defined, of the basalts although one is more basic and the other more
acid than the basalts. It is probable that the Malawi dolerite dyke-swarm
includes a complete spectrum of rocks between the two specimens analysed.
In the Tundulu area alone it is possible to distinguish, on petrographic
criteria, five types of dolerite dyke (Garson, 1962, p. 41), so that there is an
urgent need for a geochemical study of this swarm. The basalts described by
Cox et al. (1965, p. 186) from Nuanetsi are tholeiites, but a number of the
rocks from the Lebombo Monocline area, analysed by Assunçao et al. (1962),
are alkaline basalts as indicated by a lack of normative hypersthene; the

majority of these rocks contain normative nepheline. The big spread in the values of the total alkalies of the basalts of the province as a whole is apparent on Fig. 6b. On this figure are shown the lines separating tholeiites, high-alumina basalts, and alkaline olivine basalts for central and south-western Japan (Kuno, 1968) and the dashed line separates the fields of tholeiites and alkaline olivine basalts of Hawaii (Macdonald and Katsura, 1964). On Fig. 6b many of the points which plot above the lines delimiting Hawaiian tholeiites and Japanese high-alumina basalts are undoubted tholeiites as they are hypersthene normative, though these rocks are often abnormal in being enriched in potassium (Cox *et al.*, 1965, p. 191; Cox *et al.*, 1967, p. 1472). However, the chemical data of Assunçao *et al.* (1962) indicate the presence of true alkaline basalts in the Lebombo sequence. Of the two Malawi dolerites which have been analysed one is a tholeiite and the other is an alkaline dolerite.

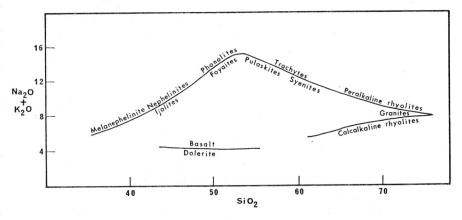

FIG. 7. Generalised silica-alkalies diagram for the Chilwa–Lupata–Sabi–Nuanetsi–Lebombo Province based on Fig. 6b.

It is noteworthy that there are few intermediate types between the strongly alkaline suite of nephelinites and phonolites and the more alkaline basalts. Similarly the basalts are completely isolated from the rhyolitic rocks, there being no analyses with SiO_2 values between 57% and 62%, except in the alkaline series. From the abundance of basalt and rhyolite analyses which are now available it is very unlikely indeed that this gap is due to inadequate sampling, so it reflects a real dearth of rocks of intermediate composition. Although it seems very reasonable to consider the whole of the Lebombo–Nuanetsi, Lupata and Chilwa rocks as part of a single, comagmatic province it is apparent from the variation diagrams that there are a number of distinct rock series within the province; and this is emphasised in Fig. 7.

In several African provinces a basaltic and a nephelinitic series of rocks coexist, together with carbonatites, for instance in the Neogene Province of northern Tanzania and Kenya (Saggerson and Williams, 1964) (Fig. 6a) and in eastern Uganda (King and Sutherland, 1966; Varne, 1968). In contrast, the Lebombo–Nuanetsi–Lupata–Chilwa Province also includes a great abundance of rhyolitic and granitic rocks and a suite of syenites (Fig. 6b), which present further genetic problems. The double trend for the Neogene Province corresponds to the alkali olivine basalt–trachybasalt–alkali trachyte and ankaratrite–melanephelinite–nephelinite–phonolite series determined by Saggerson and Williams (1964). The Neogene Province also includes carbonatites as evidenced by the carbonatite lavas which have been extruded from Oldoinyo Lengai (Dawson, 1962).

Discussion

The Lebombo–Nuanetsi–Sabi–Lupata–Chilwa area, when contrasted with the type Karroo volcanic province of South Africa, provides a superb example of the effect of tectonic environment upon the chemistry of an igneous province. If it is assumed that the volcanic suite which extends along the Lebombo monoclinal structure into southern Malawi is a manifestation of Karroo vulcanism and if, furthermore, the Chilwa Province is considered to be a late pulse of this Karroo activity then the differences in the more northerly rocks are found to correlate closely with the different tectonic regime.

The Karroo vulcanism of South Africa is essentially tholeiitic (Walker and Poldervaart, 1949). As the Karroo volcanics are traced into the Lebombo structure large volumes of rhyolite appear in the succession, and this development of rhyolites is entirely confined to the Lebombo Monocline (Cox et al., 1965, p. 205); a similarly restricted outcrop applies to the alkaline rocks, including alkali basalts. The alkaline rocks become more important northwards reaching their greatest development in the Lupata area and southern Malawi. This change in the character of the igneous rocks is matched by a change in the tectonics from a region of little or no deformation (South Africa), through the Lebombo region of warping and some faulting (see Cox, this volume, p. 225) to an area of major dislocation associated with the rift valleys. The fact that rocks of tholeiitic type only are developed where crustal deformation is slight while the alkaline character of the rocks, and the presence of carbonatites, increases in importance as faulting becomes more profound, can be fitted quite closely to petrogenetic schemes suggested by Kuno (1959, 1968), Yoder and Tilley (1962), and Kushiro and Kuno (1963). Kuno (1959, p. 60) suggested that alkaline olivine basalts were derived from greater depths than tholeiites and he correlated the depth of origin of the magmas with the depth to deep-focus earthquake epicentres beneath Japan. Yoder and Tilley (1962, p. 509) also considered that alkali basalt magmas are derived from greater depths than tholeiite magmas, while Kushiro and Kuno (1963) classify the basaltic rocks into three broad groups (A, B and C) based on their 'mantle norms'. Their group A includes tholeiites and high-alumina basalts, and some alkali basalts; group B includes some alkali basalts, basanite and leucite basalt; while group C comprises nephelinite, nepheline basalt, and melilite basalt. They suggest that group A magmas originate at shallow depths, group *B* at intermediate depths, and group *C* at the greatest depths within the mantle (*ibid.*, p. 84).

If this general hypothesis of the correlation of depth of magma generation with magma series is applied to the area under discussion it suggests that in South Africa magma generation was probably restricted to high mantle levels, so producing a single rock series of tholeiitic affinities. Where, however, deep faulting prevailed, lower as well as higher levels in the mantle were tapped and two or even three primary magmas and their differentiation products could reach the surface. A petrogenetic scheme for the Lebombo–Nuanetsi–Sabi–Lupata–Chilwa Province based on this hypothesis and including the rock suites defined on Fig. 7 is shown in Fig. 8. This system differs from the hypothesis of Cox *et al.* (1965, p. 212) which invokes 'cyclical

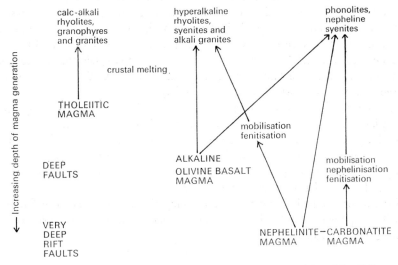

Fig. 8. Diagram summarising suggested origins and relationships of the Chilwa
—Lupata–Sabi–Nuanetsi—Lebombo igneous suite.

rise, culmination, and fall of the geoisotherms through a layered earth structure' to explain changes in rock types with time; instead it is suggested here that the mantle was tapped to varying depths by faulting, thus closely correlating rock suites with tectonics.

A number of difficulties are apparent with regard to some of the relationships among the rock suites depicted on the generalised silica-alkalies diagram (Fig. 7). The great bulk of the rhyolites is more than could be expected from differentiation of a tholeiitic basalt. It seems very probable that some degree of crustal melting must be envisaged to account for the calc-alkali rhyolites, a conclusion reached by Cox *et al.* (1965, p. 206). However, Manton (1968, p. 35), as a result of strontium isotope work on acid and basic rocks of the Nuanetsi and Lebombo area, was not able to suggest a simple model of magma genesis but did conclude that the acid rocks were derived from the mantle. The hyperalkaline lavas are rarer and may be accounted for by differentiation of the more alkaline basalts, although the plutonic equivalents of these rhyolites, in Mlanje and Zomba Mountains, are large complexes and would require the differentiation of vast volumes of basalt.

On the silica–alkalies diagram it appears that there is a continuous series from the foyaites through pulaskitic rocks to the syenites. However, it is apparent on the Harker diagram of Malawi rocks (Fig. 4) that there is a break between the foyaites and the syenites, with the pulaskites occupying a rather variable position in between. The syenites and foyaites do not appear, therefore, to define a single differentiation series. Although the evidence from Malawi is incomplete, it appears from the Lebombo–Nuanetsi–Sabi–Lupata data (Fig. 5) that there is a complete series from the nephelinites through to the phonolites. It seems probable that either the phonolites are products of differentiation of a nephelinite magma, or they represent mobilised crustal rocks nephelinised by a nephelinitic, or possibly a carbonatitic, magma. The syenites may also be the products of mobilisation of fenites, a mechanism which has been suggested previously to explain the syenites and nepheline syenites of East Africa (King and Sutherland, 1960, p. 719; King, 1965, p. 97) and of Malawi (Woolley, 1969).

The detailed petrogenesis of carbonatites is still obscure, but the close spatial relationship between carbonatites and the undersaturated alkaline rocks, particularly the ijolites, is amply documented. We consider that the carbonatites are associated with the differentiation of a melanephelinite magma produced by partial melting deep in the mantle, as already suggested. The close association of carbonatites with rift valleys would appear, therefore, to be determined by the rifts marking the lines of very deep faulting which tap the deepest levels of the mantle where nephelinitic and carbonatitic magmas are produced.

REFERENCES

Assuncao, A. F. T. de, Coelho, A. V. P., and Rocha, A. T. 1962. Petrologia das lavas dos libombos, Moçambique. *Mem. Junta Invest. Ultramar (Lisbon)*, **99**.

Bailey, D. K. 1961. The Mid-Zambezi-Luangwa Rift and related carbonatite activity. *Geol. Mag.*, **98**, 277.

Bloomfield, K. 1961. The age of the Chilwa Alkaline Province. *Rec. geol. Surv. Malawi*, **1**, 95.

—— 1965. The geology of the Zomba area. *Bull. geol. Surv. Malawi*, **16**.

Bowen, N. L. 1937. Recent high-temperature research on silicates and its significance in igneous geology. *Am. J. Sci.*, **33**, 1.

Briden, J. C. 1967. A new palaeomagnetic result from the Lower Cretaceous of East-Central Africa. *Geophys. J. astron. Soc.*, **12**, 375.

Brock, A. 1968. Palaeomagnetism of the Nuanetsi Igneous Province and its bearing upon the sequence of Karroo igneous activity in southern Africa. *J. geophys. Res.*, **73**, 1389.

Cahen, L., and Snelling, N. J. 1966. *The geochronology of equatorial Africa*, Amsterdam.

Coelho, A. V. P. 1959a. Primeiro reconhecimento petrografico da Serra da Gorongosa (Moçambique). *Bol. Serv. Ind. Ser. Geol. Mozambique*, **25**.

COELHO, A. V. P. 1959b. Reconhecimentos petrograficos sumarios dos maciços da Lupata, Morrumbala, Chiperône-derre e Milange. *Bol. Serv. Ind. Ser. Geol. Mozambique*, **26**,

COX, K. G., JOHNSON, R. L., MONKMAN, L. J., STILLMAN, C. J., VAIL, J. R., and WOOD, D. N. 1965. The geology of the Nuanetsi Igneous Province. *Phil. Trans. r. Soc. Lond.*, **257**, Ser. A, 71.

—— MACDONALD, R., and HORNUNG, G. 1967. Geochemical and petrographic provinces in the Karroo basalts of southern Africa. *Am. Miner.*, **52**, 1451.

DAWSON, J. B. 1962. Sodium carbonate lavas from Oldoinyo Lengai, Tanganyika. *Nature, Lond.*, **195**, 1075.

DIAS, M. B. 1956. Explanation of the structural map of Mozambique. *Bol. Serv. Ind. Ser. Geol. Mozambique*, **18**.

—— 1961. Geologia do Monte Muambe. *Bol. Serv. Ind. Ser. Geol. Mozambique*, **27**, 43.

DIXEY, F. 1929. The Karroo of the lower Shire-Zambezi area. *Int. geol. Congr.*, **15**(2), 120.

—— 1930. The geology of the ower Shire-Zambezi area. *Geol. Mag.*, **67**, 49.

—— 1937. The pre-Karroo landscape of the Lake Nyasa region, and a comparison of the Karroo structural directions with those of the rift valley. *Q. J. geol. Soc. Lond.*, **93**, 77.

—— 1956. The East African Rift System. *Bull. colon. Geol. Mineral Resources*, Suppl. **1**.

—— and CAMPBELL SMITH, W. 1929. The rocks of the Lupata Gorge and the north side of the lower Zambezi. *Geol. Mag.*, **66**, 241.

—— CAMPBELL SMITH, W., and BISSET, C. B. 1937. The Chilwa Series of southern Nyasaland. *Bull. geol. Surv. Nyasaland*, **5** (revised 1955).

DU TOIT, A. L. 1929. The volcanic belt of the Lebombo—a region of tension. *Trans. r. Soc. S. Afr.*, **18**, 189.

FLORES, G. 1964. On the age of the Lupata rocks, lower Zambezi River, Mozambique. *Trans. geol. Soc. S. Afr.*, **67**, 111.

GARSON, M. S. 1962. The Tundulu carbonatite ring-complex in southern Nyasaland. *Mem. geol. Surv. Nyasaland*, **2**.

—— 1965. Carbonatites in southern Malawi. *Bull. geol. Surv. Malawi*, **15**.

—— and WALSHAW, R. D. (in press). The geology of the Mlanje area. *Bull. geol. Surv. Malawi*.

GOUGH, D. I., BROCK, A., JONES, D. L., and OPDYKE, N. D. 1964. The palaeomagnetism of the ring complexes at Marangudzi and the Mateke Hills. *J. geophys. Res.*, **69**, 2499.

HABGOOD, F. 1963. The geology of the country west of the Shire River between Chikwawa and Chiromo. *Bull. geol. Surv. Nyasaland*, **14**.

HALES, A. L. 1960. Research at the Bernard Price Institute of Geophysical Research, University of Witwatersrand, Johannesburg. *Proc. r. Soc. Lond.*, **258**, Ser. A, 1.

HARKIN, D. A. 1960. The Rungwe volcanics at the northern end of Lake Nyasa. *Mem. geol. Surv. Tanganyika*, **2**.

KING, B. C. 1965. Petrogenesis of the alkaline igneous rock suites of the volcanic and intrusive centres of eastern Uganda. *J. Petrology*, **6**, 67.

—— and SUTHERLAND, D. S. 1960. Alkaline rocks of eastern and southern Africa. *Sci. Prog.*, **48**, 298, 504, 709.

—— and SUTHERLAND, D. S. 1966. The carbonatite complexes of eastern Uganda. In *Carbonatites*, p. 73. (Eds. O. F. Tuttle and J. Gittins), New York.

KUNO, H. 1959. Origin of Cenozoic petrographic provinces of Japan and surrounding areas. *Bull. volc.*, **20**, 37.

—— 1968. Differentiation of basalt magmas. In *Basalts*, p. 623. (Eds. H. H. Hess and A. Poldervaart), New York.

KUSHIRO, I., and KUNO, H. 1963. Origin of primary basalt magmas and classification of basaltic rocks. *J. Petrology*, **4**, 75.

LOMBAARD, B. V. 1952. Karroo dolerites and lavas. *Trans. geol. Soc. S. Afr.*, **55**, 175.

MACDONALD, G. A., and KATSURA, T. 1964. Chemical composition of Hawaiian lavas. *J. Petrology*, **5**, 82.

MACKENZIE, J. D. 1914. The Crowsnest volcanics. *Bull. geol. Surv. Mus. Can.*, **4**, 19.

McDOUGALL, I. 1963. Potassium-argon age measurements on dolerites from Antarctica and South Africa. *J. geophys. Res.*, **68**, 1535.

MANTON, W. I. 1968. The origin of associated basic and acid rocks in the Lebombo-Nuanetsi Igneous Province, southern Africa, as implied by strontium isotopes. *J. Petrology*, **9**, 23.

MOUTA, F. 1957. L'éffrondrement de l'Uréma, extrême sud des 'Rift Valleys' au Moçambique (Afrique Orientale Portugaise). *Bol. Serv. Ind. Ser. Geol. Mozambique*, **24**, 29.

NICOLAYSEN, L. O., BURGER, A. J., and JOHNSON, R. L. 1962. The age of the Shawa carbonatite complex. *Trans. geol. Soc. S. Afr.*, **65**, 293.

NOCKOLDS, S. R. 1954. Average chemical compositions of some igneous rocks. *Bull. geol. Soc. Am.*, **65**, 1007.

SAGGERSON, E. P., and WILLIAMS, L. A. J. 1964. Ngurumanite from southern Kenya and its bearing on the origin of rocks in the northern Tanganyika alkaline district. *J. Petrology*, **5**, 40.

SNELLING, N. J. 1965. Age determinations on three African carbonatites. *Nature, Lond.*, **205**, 491.

STILLMAN, C. J., and COX, K. G. 1961. The Chikala Hill syenite-complex of southern Nyasaland. *Trans. geol. Soc. S. Afr.*, **63**, 99.

SWIFT, W. H., WHITE, W. C., WILES, J. W., and WORST, B. G. 1953. The geology of the Lower Sabi Coalfield. *Bull. geol. Surv. S. Rhodesia*, **40**.

TEALE, E. O., and CAMPBELL SMITH, W. 1923. Nepheline-bearing lavas and intrusive rocks from south of the Zambezi River, with a note on an outcrop of Karroo lavas in the Buzi Valley, Portuguese East Africa. *Geol. Mag.*, **60**, 226.

VAIL, J. R., 1965a. Estrutura e geocronologia da parte oriental da Africa Central com referencias a Moçambique. *Bol. Serv. Geol. Min. Moçambique*, **33**, 13.

—— 1965b. Aspects of the stratigraphy and the structure of the Umkondo System in the Manica belt of Southern Rhodesia and Mozambique, and an outline of the regional geology. *Trans. geol. Soc. S. Afr.*, **68**, 13.

—— (in press). Mesozoic igneous activity in Central Africa. *Int. geol. Congr.*, **22**.

VARNE, R. 1968. The petrology of Moroto Mountain, eastern Uganda, and the origin of nephelinites. *J. Petrology*, **9**, 169.

WAGER, L. R. 1947. Geological investigations in East Greenland. Part IV. The stratigraphy and tectonics of Knud Rasmussens Land and the Kangerdlugssuak region. *Medd. Grønland*, **134**(5).

—— and DEER, W. A. 1939. Geological investigations in East Greenland. Part III. The petrology of the Skaergaard Intrusion, Kangerdlugssuak. *Medd. Grønland*, **105**.

WALKER, F., and POLDERVAART, A. 1949. Karroo dolerites of the Union of South Africa. *Bull. geol. Soc. Am.*, **60**, 591.

WOOLLEY, A. R. 1969. Some aspects of fenitization with particular reference to Chilwa Island and Kangankunde, Malawi. *Bull. Brit. Mus. (Miner.)*, **2**(4).

YODER, H. S., and TILLEY, C. E. 1962. Origin of basalt magmas: an experimental study of natural and synthetic rock systems. *J. Petrology*, **3**, 342.

YOUNG, R. B. 1920. The rocks of a portion of Portuguese East Africa. *Trans. geol. Soc. S. Afr.*, **23**, 98.

B. C. KING

12   Vulcanicity and rift tectonics in East Africa

ABSTRACT. *The East African Rift System and its associated vulcanism are among the most remarkable geological phenomena in the world. While it appears inescapable that they are related to a pattern of earth behaviour of a major order, the details of rift faulting are closely determined by older, mostly Precambrian, structures.*

In southern Africa 'rifts' in part controlled, but more largely followed the deposition of the Karroo, and while in some cases they were rejuvenated in later times, the evidence in East Africa points clearly to mid-Tertiary and later dates for the rift faulting there. Perhaps the most notable feature of the rifts is the extraordinary intensity of faulting (ranging from Tertiary to Recent) within their limits and the almost complete absence of faults outside.

There seems to be a correlation between vulcanism and topographic culminations in the 'rift system' of which the Kenya 'dome' is the most obvious example. A pattern has often been sought in the vulcanism of the rifts, both in its relation to tectonics and in terms of composition. Neither is clear, although alkaline, sometimes strongly alkaline, compositions are characteristic. Notable in Tanzania and Kenya are the early but sporadic manifestations of highly undersaturated basic volcanics, but even more are the later (mid-Tertiary) vast 'plateau' outpourings of phonolites and the great development of trachytes, and lavas and pyroclastics, in the form of huge broad volcanoes of late Tertiary, or early Pleistocene age. Recent vulcanism is marked by numerous central volcanoes along the median zone of the rift, mostly trachytic, but with basic associates, and often characterised by spectacular calderas.

Introduction

Within eastern Africa the rift system extends some 2300 miles, from the neighbourhood of the Limpopo River in the south to the Red Sea in the north. Here, despite considerable variation in pattern, the typical Rifts are graben about 30-50 miles in width. Northwards the Red Sea Rift extends the system a further 1400 miles but, like the branch rift into the Gulf of Aden, it is a much wider structure (up to 300 miles in width). The Levantine rift, which continues from the Gulf of Akaba northwards some 700 miles to the Taurus Mountains, again consists of narrower troughs resembling those of eastern Africa.

Although the most characteristic rifting is of Tertiary and later age, a similar pattern of faulting is recognisable in several sectors dating back to late Karroo or early Jurassic times. The fact that there are no older crustal structures or lineaments in this great belt of such magnitude and persistence suggests that the system has been determined by a major deep-seated crustal

However

or sub-crustal mechanism. The occurrence of distinctive volcanic associations along many parts of the rift system is also in accord with this conclusion. Nevertheless, when the detailed pattern of dislocations is studied there is a notable correlation between fault directions and older Precambrian structural trends, a correspondence that was first clearly appreciated by Dixey (1956). The coincidence of rift faulting with Precambrian trends and the extended history of rift movements suggested to Dixey and also to McConnell (1951) that the fundamental cause was a stress system that had operated from Precambrian times in eastern Africa. This, however, is a questionable inference, for the older trends that are followed by the rift faults belong to Precambrian orogenic belts of widely differing ages and to structures of differing kinds, such as fold trends, foliations, faults or mylonite zones. It is rather that the structures of the old orogenic belts constituted a crustal 'grain' of greatly varying regularity and trend, by which the course of later faults was largely directed.

The profound dislocations that affected the Karroo in southern Africa are well documented, the vertical displacements being particularly obvious from the varying altitudes of the Stormberg lavas (Cox, this volume, p. 223). Throws of 10,000 ft or more are demonstrable on some faults, but there is also evidence that effective bevelling by erosion occurred simultaneously (Dixey, 1956, p. 9). The general uplift within the south-eastern and eastern margins of the continent was intermittently active from Karroo times on-wards and the complementary coastal and offshore downwarp determined the pattern of Cretaceous and Tertiary marine sedimentation. The uplift is represented by the highlands and plateaus of eastern Africa and with it the ancestral watershed between the Indian Ocean and Atlantic drainages; Dixey has appropriately referred to this major feature as the 'rising rim of Africa'.

Northwards from the Limpopo the Karroo is largely confined to a number of structural troughs (Cox, this volume, p. 214) which have much in common with the later rifts; the present morphological resemblance is, however, largely due to erosion of the less resistant infill of Karroo sediments; the troughs have alignments which were evidently largely determined by Precambrian trends (see De Swardt, 1965a). The Zambesi contrived to main-tain its course to the Indian Ocean against the rising rim of Africa which lifted the Batoka basalts from 3500 ft above sea-level at the Victoria Falls to 6500 ft above sea-level on the Rhodesian plateau.

Profound excavation and planation within the Karroo troughs and across their margins was achieved by the beginning of the Cretaceous and produced what Dixey (1939) terms 'valley-floor peneplanes'; extensive deposits of early Cretaceous 'dinosaur beds' on these surfaces point to their former much more widespread occurrence in the troughs. Early Cretaceous sediments, generally resting on eroded Karroo, occur in the Rukwa and other troughs of southern Tanzania and in the Nyasa-Shire trough, thus testifying not only to the age of the earlier movements in these troughs, but also to a period of stability that intervened before the Tertiary movements.

The recognition of the nature and importance of the earlier troughs was one of Dixey's major contributions to an understanding of the evolution of the East African Rift System. He showed that the southern rift valleys were late- or post-Karroo troughs rejuvenated by Quaternary faulting; indeed, in the case of the Malawi Rift the greater part of the eastern escarpment was determined by the Mesozoic displacements.

Alkaline igneous centres occur in association with the Mesozoic rifts and troughs. The most notable of these are the Chilwa Series of Malawi, comprising a group of relatively large alkali granite-syenite ring-complexes, together with numerous much smaller carbonatite centres (Garson, 1965; Woolley and Garson, this volume, p. 238). It is probable that volcanic activity was associated with the Chilwa centres, but it may be doubted if they were surmounted by major volcanoes. Similar carbonatite complexes occur around the junction of the rifts in northern Malawi, north-eastern Zambia and southern Tanzania.

The main rift system of East Africa

Throughout its length the rift system is characterised by Tertiary to Recent tectonic and, in many sectors, associated volcanic activity. On a continental scale it is essentially a north-south feature, yet its structural elements only exceptionally show this trend. The coincidence of detailed structures with older Precambrian trends is, however, very striking. Equally, it seems likely that the 'rejuvenation' of earlier troughs by later rifting merely reflects the fact that both sets of dislocations were determined by the ancient Precambrian grain.

Major faults showing strong parallelism and regularity are generally found where the Precambrian grain is also regular and persistent and trends approximately north-south. This is true of large parts of the Eastern Rift in Kenya. Similar persistence and continuity is also shown where a pronounced Precambrian grain is markedly oblique to the north-south trend, as in the case of the Rukwa and other troughs of south Tanzania and of the Lake Albert Rift.

Where the Precambrian structures depart considerably from north-south the rift valley may be defined by a pattern of *en echelon* faults, following the older grain, which relay the effective displacement along a more nearly north-south course. Examples are shown by the Albert Nile towards the Sudan border and to the south of Ruwenzori. Again where the Precambrian grain is complex and variable in direction the rift fault pattern is correspondingly irregular, as in eastern Tanzania.

It seems clear that the bifurcation of the system into an Eastern and a Western Rift is the result of deflection of the fracture systems around the Tanganyika 'Shield' or Craton (Fig. 1); this consists of a gneissic complex and meta-sedimentary or meta-volcanic systems (Dodoman, Nyanzian, Kavirondian) with prevalently east-west trends. The south-west of the craton is sharply limited by the Ubendian and associated formations, with their parallel and strongly marked folds, mylonite belts and lines of intrusions;

these have determined the series of north-west to south-east faults bounding troughs and horsts that constitute the rift structures of northern Lakes Nyasa, Rukwa and southern Lake Tanganyika (Fig. 1). Towards Lakes Kivu and Edward the rift continues to follow the Ubendian Belt but northwards it swings to the north-east following the grain of the Toro System and the Uganda 'basement'. Conforming also to this grain is the well-defined Lake Albert Rift, which is sited *en echelon* to the Lake Edward Trough, separated from it by the horst of Ruwenzori. Northwards again the relaying series of *en echelon* faults of the Albert Nile continue to follow the 'basement' grain. The Western Rift terminates abruptly against the strongly marked north-west to south-east trends of the Madi Series of the Sudan border and the parallel structures of the Aswa Shear Belt of northern Uganda.

The Eastern or Gregory Rift Valley in Kenya approximates in many places to a classic graben (Fig. 1); its north-south trend and general regularity are evidently related to the equally regular structures of the Mozambique Orogenic Belt. Northwards, towards the Ethiopian border, however, it splays out, again apparently largely controlled by older structures, to form a broad zone up to about 200 miles in width (Fig. 1), within which only a narrow and relatively insignificant trough passes from the southern end of Lake Rudolph and by way of Lake Stephanie to link with the great rift valley of Ethiopia (see Gass, this volume, p. 286).

Southwards in Tanzania the Eastern Rift rapidly 'degenerates' into a broad zone of faults of varying persistence and trend which define a series of tilted blocks (James, 1956; Dundas, 1965; Pallister, 1965). Control by older structures is, however, still evident. The zone of faulting, over 200 miles in northern Tanzania, is commensurate with the width of the Mozambique Belt and the fault trends, though widely varying, conform nevertheless with the complexities of structural trends in the Precambrian. More persistent and strongly marked older structures at the margin of the Mozambique Belt determine, for example, the Usungu Trough in the south, while the horst blocks of the Pare and Usambara Mountains and associated troughs reflect more regular 'Mozambiquian' trends in north-eastern Tanzania (Fig. 1). The subsidiary rifts or troughs that branch west-wards from the main rift, namely the troughs of Kavirondo, Musoma, Speke Gulf and Lake Eyasi, cut across the margin of the Mozambique Belt and appear to be controlled by the transverse trends within the Tanganyika 'Shield'.

Two aspects of the morphology of the rifts that evidently have genetic significance are the rise from the surrounding land surface to the rift shoulders and the variations in altitude of floor and margins. Gregory (1921), in his classic description of the Kenya Rift Valley, recognised that it consists essentially of a downfaulted belt or graben along the crest or axis of an elongated arch; he thereby developed the notion that the rift was the consequence of the tensional stresses set up by the process of arching and that the down-faulted strip is thus analogous mechanically to the keystone of an architectural arch. Wayland (1930), however, concluded that the arch was more satisfactorily interpreted as a compressional structure, the central graben representing in effect an inverted keystone held down by the same compressive forces, explaining thereby, among other things, the negative gravity anomalies normally characteristic of the rift floors. This hypothesis is now largely discredited in view of the almost ubiquitous evidence for tensional, normal faulting associated with the rift valleys.

Traced along their length the rift valleys show great variations in altitude, but the fact that elevations and depressions of the floor are matched by corresponding elevations and depressions of the shoulders shows that the overall displacements are everywhere of the same general order. The tendency for the rifts to assume similar proportions, in terms both of width and height, is presumably related to such common factors as crustal thickness and strength as well as to the general magnitude of the primary stresses.

In many places the present altitudes of rift floors and shoulders do not indicate the actual displacements, since the floors locally are infilled by hundreds or thousands of feet or sediments and/or volcanics, and the rift shoulders similarly have great thicknesses of volcanics surmounting the 'basement'. Nevertheless, in general terms, the floor of the Eastern Rift ranges from below sea-level in the Red Sea to 7000 ft in central Ethiopia, below 1000 ft to the south of Lake Rudolf, 7000 ft around Nakuru in central Kenya and again to below 2000 ft at Lake Magadi near the Kenya–Tanzania border (Figs. 1 and 2). Elevations of the shoulders and flanks of the rift, corresponding to those of the floor, constitute the Ethiopian and Kenya 'domes'. Significantly these are both regions of volcanic activity in the Eastern Rift. Elsewhere similar culminations in the rift system occur around Lakes Edward and Kivu in the Western Rift and in southern Tanzania and northern Malawi (Fig. 1). Again these are centres of volcanic activity. The association of vulcanism with regions of uplift is not confined to the rift zone, but also characterises, for example, the volcanic areas of Hoggar and Tibesti in the Sahara (this volume: Black and Girod, p. 185; Vincent, p. 303).

Erosion surfaces

Erosion surfaces assume considerable significance in establishing the nature and sequence of movements within the greater part of the region of the rift valley system.

The pattern of the river systems in Uganda and north-western Tanzania shows clearly that the original drainage was towards the Atlantic, across the site of the Western Rift (Wayland, 1930). The watershed was approximately along the line of the Eastern Rift, although, owing to disturbance by tectonic and volcanic activity, the earlier drainage pattern here has been largely obscured. The difference of situation of the two rift valleys in relation to the drainage history accounts for the much greater sedimentation in the Western Rift and its more extensive occupation by lakes. The sediments of the Western Rift accumulated almost continuously from its inception during the Miocene to the reversal of drainage in the Pleistocene. By contrast, the Eastern Rift is

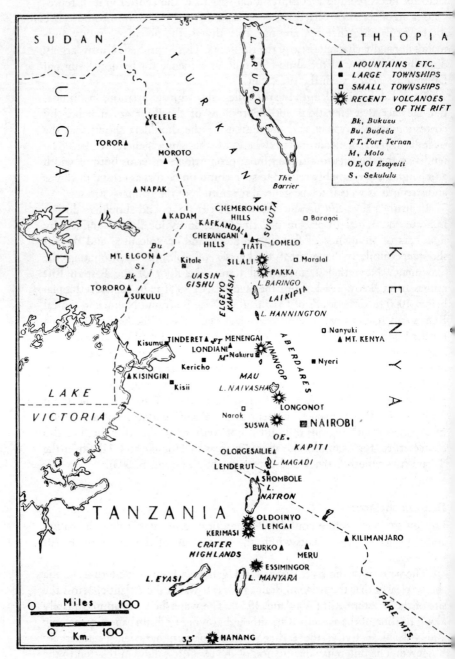

FIG. 2. Locality map of the volcanic area of the Eastern Rift in Kenya and Tanzania.

an internal drainage system with limited inflow and a number of shallow saline lakes.

The confusion in the recognition and correlation of erosion surfaces has largely arisen from the assumption that they were everywhere almost flat, with little or no relief, and that correlation is possible by projections based on concordance of summit levels. In the tectonically active regions, the dislocation of the surfaces by warping and faulting has been underestimated. De Swardt (1965b) affirms that there are in Uganda only two surfaces of major significance, marked by deep weathering and lateritisation. The *upper surface* (Buganda surface; P II of Wayland; or mid-Tertiary of Dixey) forms closely concordant flat-topped relics on the schists and gneisses of central Uganda, where it approximates to a peneplane, but shows broad elevations over more resistant formations with a relief up to 1000 ft or more in south-western and western Uganda. The *lower surface* (P III of Wayland; end-Tertiary of Dixey) is widely expressed over northern and eastern Uganda. In central Uganda it is 500-1000 ft below the upper surface and extends southwards as broad valley shoulders set below the Buganda surface; it ascends with the valleys themselves until, in watershed regions between the river systems, it almost converges with the upper surface, as for example around the north-western shores of Lake Victoria. Both upper and lower surfaces rise towards the Western Rift; the rise is, however, much greater towards Lake Edward, where the upper surface reaches more than 8000 ft, than towards Lake Albert, while it is still less towards the Albert Nile (see Bishop and Trendall, 1967). The rise to the rift is more marked for the upper surface, so that the interval between the two becomes progressively greater towards the west; tilting was therefore initiated after the formation of the earlier upper surface and continued after the development of the later lower surface. Nevertheless, the drainage maintained its westward course and reversal did not occur until valleys had been cut below the lower surface. Lake Victoria, which was a consequence of the reversal, has submerged not only part of the lower surface but also the later valley incisions.

In western Kenya also only two lateritised surfaces are represented; but, since uplift was towards the Eastern Rift rejuvenating the upper courses of the original drainage and so accelerating erosion, the older surface is preserved only as remnants standing now at widely divergent levels, notably the Cherangani Hills (up to 11,000 ft), the Chemerongi Hills and the Kisii Hills (Fig. 2). The last named show flat-topped summits, morphologically similar to the outliers of the Buganda surface, which slope westwards from over 8000 ft to 5000 ft within 30 miles or so. In addition, numerous hills which rise above volcanics of the Uasin Gishu Plateau (Fig. 2) and above the general level of the basement plateau to the north and north-east of Kitale, are eroded residuals of the older surface. In eastern Uganda, erosion consequent upon upwarping towards the Eastern Rift has removed the older surface beyond regions underlain by the more resistant formations of the Buganda Series.

The lower lateritised surface can be traced continuously from eastern Uganda into western Kenya; not only does it truncate the lowest volcanics and penetrate as broad valleys into the volcanics of Mounts Elgon and Kisingiri, but it extends above the Nandi Scarp to form a lateritised planation, truncating alike the volcanics of the Uasin Gishu Plateau to the edge of the Elgeyo Escarpment and the basement of the Kitale area. On the Kericho Plateau it is similarly a modification of the surface of the lavas and rises into the Kisii Hills as shoulders to the valleys.

Sediments or tuffs, containing Lower Miocene faunas, are found at many localities below or within the lowest volcanics in eastern Uganda and western Kenya, while the earliest sediments of the Western Rift are of similar age. An extensive sub-Miocene erosion surface has been extrapolated by contouring from these various occurrences (see Pulfrey, 1960). In fact, however, the visible surface is a landscape of greatly diverse and often high relief that reflected the effects of current tectonic and paravolcanic activity resulting in rapid erosion of monoclinal upwarps and fault lines to form scarps, sections of which are preserved under Elgon and Kadam (Fig. 2), and complementary sedimentary accumulation in troughs, valleys and downwarps. Relatively planar elements in the landscape approximate to dislocated sectors of the older surface which have suffered little modification, as under the western flanks of Elgon and Kadam or around Napak. Even the Kitale plain, regarded as part of the sub-Miocene bevel (*ibid.*, p. 7), shows considerable irregularity beneath the lavas (see Sanders, 1964). Only in eastern Kenya is it possible to recognise a sub-Miocene surface of planar character (Saggerson and Baker, 1965).

These features suggest that the upper surface of Uganda (the Buganda surface) must, like the surfaces over the Kisii and Cherangani Hills, be older than mid-Tertiary (or Miocene), the age that has commonly been ascribed to it; indeed, it may well date back to the Cretaceous (see Bishop and Trendall, 1967, p. 407).

The lower surface cannot as yet be dated with any precision, although it is broadly 'late Tertiary', but it is likely to prove an important 'horizon' in a regional demarcation of tectonic and volcanic events in and around the Eastern Rift.

The Western Rift

The Western Rift System has been reviewed recently by Macdonald (1965). Hopwood and Lepersonne (1953) showed that in the Albert–Lower Semliki part of the rift the oldest sediments are Lower Miocene. The initial downwarp of the rift was thus at least as early, depressing the upper (Buganda) erosion surface and, presumably, severing the direct connection of the Uganda drainage with the Congo. The floor of the rift sank continuously, for the sediments are all of shallow water type. In Lake Albert drilling and geophysical work have established a maximum thickness of some 8000 ft of sediments. Older faulting can be recognised, as near the Murchison Falls, where faults are overlapped by sediments, but the major, visible faulting is Pleistocene, and later than the formation of the main erosion surfaces. The total amount of displacement increases from hundreds of feet near the Victoria Nile to over 10,000 ft in the Semliki area, the increase being marked by a rise of the shoulders from 2800 ft to 5500 ft, while the floor declines to 6000 ft below sea-level. On the Congo side of the rift the shoulders rise to around 10,000 ft indicating an even greater displacement of the north-western boundary fault. No more than a 100 ft or so of later Pleistocene sediments occur, reflecting the effect of reversal of the drainage. The fault pattern in the Albert Rift is comparatively simple, but numerous small faults in the sediments have been recorded (see Bishop, 1965) and continuance of movement to the present day is shown by the small scarp produced by the 1967 earthquake.

In the Albert Nile sector to the north the sediments are thinner, but 1300 ft have been proved by drilling and up to 3000 ft inferred geophysically. The faults here are also mostly north-east to south-west, but have an *en echelon* pattern which relays the movements northwards. Pleistocene faulting is evident where sediments are thrown against basement.

The continuation of the Lake Albert graben south-westwards along the Semliki River is confirmed by the form of the belt of high negative anomalies. To the south-east of the Semliki, Ruwenzori forms a parallel horst block rising to 17,000 ft. The upper surface is arched from north-east to south-west and tilted to the south-east, and despite its altitude it may well represent the uplifted Buganda surface.

Further south is the offset Lake Edward Trough, which is shallow at its north-eastern end, but deepens towards the Rwanda border; from Lake George it is defined by a series of faults which form an *en echelon* pattern.

In the Birunga-Bufumbira region eight major volcanoes have been built up in the rift along a trend which is almost at right angles to that of the trough itself (Holmes and Harwood, 1937). They are steep-sided cones, rising to a maximum of well over 14,500 ft and are composed of lavas ranging from leucitites or nephelinites to trachytes, but are characterised by a high potash/soda ratio. The two westernmost volcanoes are still active. Nyamlagira (10,010 ft) erupted in 1938–42 and its double caldera, $1\frac{1}{2}$ miles wide and 1000 ft deep, is unique among African volcanoes in containing a lava lake; Niragongo (11,385 ft) erupted in 1948. The volcanoes may have commenced activity in Pliocene times, but the bulk of the lavas are ascribed to the later part of the Pleistocene. There are also several hundred smaller cones, mostly formed of scoria. The culmination of the floor of the Western Rift is Lake Kivu (4790 ft) which, as a result of a volcanic accident, drains into Lake Tanganyika and thence to the Congo. South of Lake Kivu (Fig. 1) there are lava fields largely composed of alkali (olivine) basalts and basanites, but trachytes and trachyandesites are also represented.

In western and south-western Toro, Uganda, within and marginal to the rift zone, is a unique region of numerous explosion craters, often so closely spaced that they are contiguous or even overlap (Combe, 1930). They are mostly surrounded by insignificant rings of ejectamenta, but issuing from the Katunga crater there is a flow composed of a highly undersaturated basic lava with kalsilite. This mineral is also recorded in the lavas of the main Birunga field (Fig. 1) and, indeed, has proved to be rather widespread in the pyroclastics of Toro and Ankole (Holmes, 1942). Formations originally recorded by Combe as carbonated pyroclastics have more recently been interpreted as carbonatite lavas (Von Knorring and Du Bois, 1961).

Southwards from Lake Kivu the Western Rift swings into the 400 mile north-north-west to south-south-east belt occupied by Lake Tanganyika (Fig. 1). The structures here have not as yet been closely studied, but appear to be relatively simple. The lake is not comparable to Lake Albert, for it has a negligible sedimentary infill and its floor is well below sea-level. The lake itself occupies sub-parallel graben defined by various pairs of faults and, towards its southern end, the rift structures extend over a width of about 100 miles, involving parallel graben and horsts, into the rejuvenated post-Karroo Rukwa Trough (Fig. 1).

The Livingstone Fault bounding the eastern side of the Nyasa Trough continues the line of the fault to the east of the Rukwa Trough and at Mbeya Mountain and the Livingstone Mountains the shoulders rise to over 8000 ft, some 2000 ft higher than the Mbozi block on the opposite side of the rift. Between Mbeya Mountain and the Livingstone Mountains the transverse Usangu Trough joins the Rukwa–Nyasa Trough and it is here that the Rungwe volcanics form extensive highlands. This complex volcanic region shows the association of moderately alkaline basalts, phonolites and trachytes with more strongly alkaline nephelinites; lavas are accompanied by pyroclastics and although activity was at its maximum during the Pleistocene, it may have begun during the Pliocene, and there is evidence of tuff eruptions during the historic period (Harkin, 1959).

The Eastern Rift

In the northern part of eastern Africa lies the great Abyssinian Plateau, an enormous mass of volcanics rising to over 13,000 ft and extending eastwards into Somalia (Gass, this volume, p. 286). It is traversed by the rift graben widening to the north-east as the Afar depression, the bounding faults of which diverge to the Red Sea and Gulf of Aden. The first movements in the early Tertiary involved uplifts of the region by a maximum of 8000 ft accompanied by faulting. The volcanics mostly overlie various Mesozoic sediments, which rest, in turn, on the Precambrian. The older basalts were succeeded by over 3000 ft of later basalts before the Middle Tertiary. Major faulting and warping form the main graben and the Afar depression, and were contemporaneous with general uplift of the plateau. This was followed in the Quaternary by further vulcanism which continued with the development of basaltic volcanoes (Mohr and Rogers, 1965), including the numerous craters of the Red Sea. Although this is by far the greatest volcanic region of eastern Africa, it has been very inadequately investigated.

By contrast the Eastern Rift of Kenya and northern Tanzania has been examined in comparative detail, largely by the systematic work of the Geological Surveys in those territories (Figs. 2 and 3).

The older alkaline complexes of south-eastern Uganda may be considered as the earliest manifestations of igneous activity associated with the Eastern Rift although they lie well to the west of its main course. Extending over about 40 miles in a north-north-east to south-south-west line they comprise the centres of Sukulu (about 22 miles diameter), Tororo and Bukusu, formed of carbonatites with associated alkali syenites (including ijolites) and fenites,

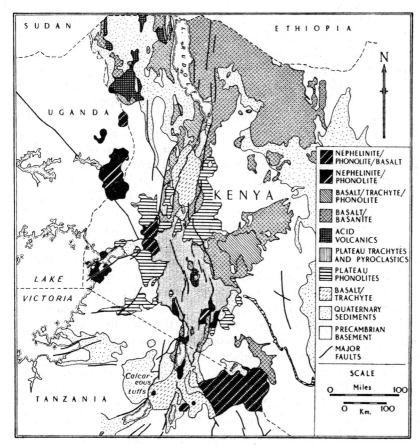

FIG. 3. The distribution of the principal types and associations of volcanics in the Kenya–Tanzania sector of the Eastern Rift. (Map compiled by L. A. J. Williams, whose permission to use it is gratefully acknowledged.)

and the centres of Sekululu and Budeda, which have been revealed by partial removal of the Elgon volcanics. It is evident from the configuration of the surface under the Elgon volcanics (Fig. 1) that these complexes produced relatively abrupt domes, originally 1000 ft or more high. Although the radiometric ages obtained so far show a wide scatter they do suggest that the complexes may be no older than early Tertiary, rather than Mesozoic as formerly assumed.

In general terms the Eastern Rift is a north-south trough which roughly

K

bisects the oval dome of the Kenya Highlands (Fig. 2). Corresponding to the greatest height of the rift floor, over 6000 ft between Nakuru and Naivasha at the centre of the dome, the lava plateaus of the rift shoulders rise to 8000 or 9000 ft in the west, culminating at over 10,000 ft in the Mau area, and from 5000 to 6000 ft in the east, but stepping up successively to around 8500 ft on the Kinangop Plateau and 13,000 ft at the crest of the Aberdare Range (Fig. 2).

The rift appears to have been initiated in the early Miocene as a downwarp with only minor faults; thereafter faulting, vulcanism and episodes of fluviatile or lacustrine sedimentation have continued to the present day. The most characteristic feature of the rift here is that it is a sharply-defined zone of more or less intense faulting, for the adjacent regions are largely or totally devoid of such dislocations. In general, faulting has developed inwards from the rift margins, so that in places there is a later, narrow central graben. The major faults have displacements of several thousands of feet and often provide evidence of repeated or long-continued movement. Uplift of the shoulders has occurred, as well as subsidence of the floor. The latest episodes of faulting have consisted of close-set, sub-parallel 'grid' faults, which generally displace the surface by no more than hundreds and often only tens of feet, producing a succession of narrower or wider horsts and troughs (Baker, 1965a and b).

There are great structural variations from one section of the rift to another. In southern Kenya and northern Tanzania the western margin is marked by the Nguruman Fault (Fig. 1) with a throw of at least 5000 ft in basement rocks. Later Tertiary basalts, banked against the fault scarp, were faulted to form a lower scarp-bound terrace above the floor of the rift. Northwards, the margin of the rift is lower and is essentially an eroded downwarp in Pleistocene trachytic tuffs and lavas. East of the Mau, the scarp, although again of greater height, is largely modified by a mantling of late tuffs, and descends to a subdued feature where the Kavirondo Trough joins the main rift. Thence to the north the western margin is downwarped, step-faulted and strongly dissected, but, with the decline in the floor to less than 4000 ft towards Lake Baringo (Fig. 2), it becomes an increasingly prominent feature and passes into the Kamasia Range rising to 8000-9000 ft. Here, however, the side of the rift is relayed farther to the west by the Elgeyo Fault with a throw of upwards of 5000 ft; this great displacement develops within a few miles from a series of minor step-faults. The Kamasia Range is essentially a block tilted to the west to the Kerio Valley at the foot of the Elgeyo escarpment (Fig. 2), and with a step-faulted eastern slope in which rather steep eastward dips are largely offset by repeated faults which northwards, however, become dominated by a single major fault, the Kamasia Fault. Thence, the Kamasia fault passes into several faults of smaller throw, while the Elgeyo Escarpment continues as an impressive feature surmounted by the Cherangani Hills which rise to 11,000 ft.

Farther to the north the rift becomes less well defined and from a width of 30–40 miles, widens out to almost 200 miles. The western margin is relayed by a splay fault, branching at a large angle from the Elgeyo Fault and carrying most of the displacement in passing the Chemerongi Hills; a similar splay structure following the Turkana Escarpment (Fig. 1), and marking the Kenya-Uganda border, is probably an eroded downwarp rather than a fault. Of these structures the Turkana downwarp was evidently formed at an early stage, for it was eroded and locally buried by the Miocene lavas from Moroto Mountain and Yelele in eastern Uganda, whereas the main movement of the Elgeyo Fault was in late Tertiary times.

The structures of the eastern margin of the rift are generally different from those of the western, so that the rift itself is commonly asymmetric both in structure and morphology. From northern Tanzania almost to the latitude of Nairobi (Fig. 2), the eastern side of the rift is a step-faulted downwarp, marked by fault strips which are tilted away from the floor. West of Nairobi and for some distance northwards the eastern margin is largely defined by a single fault scarp and the rift as a whole approximates to a simple graben. Northwards the faults which define the Kinangop Plateau die out and the main wall of the rift is relayed to the east by the fault marking the western side of the Aberdares which continues into the faults of the Laikipia Escarpment (Fig. 2).

The Laikipia sector is marked by erosion scarps etched along or across a pattern of north-south faults which are relayed successively eastwards in a northerly direction; the fault-bound strips in general step down into the rift but are variously tilted towards or away from it. Farther to the north, from Maralal to the Ethiopian border, the eastern margin is a broad downwarp, with comparatively minor faulting, towards the central trough.

The Kavirondo Trough (Shackleton, 1951) appears to have originated as a transverse downwarp or lag, faulted along part of its northern margin, probably in association with early arching along the site of the main rift, and contemporaneous with the formation of the west-facing downwarp or lag under Elgon and Kadam and the east-facing downwarp of the Turkana Escarpment to the north (Fig. 2). These movements dislocated the older (? late Mesozoic) lateritised erosion surface which subsquent erosion modified to form the sub-volcanic (sub-Miocene) surface.

Vulcanism in eastern Uganda and the Kavirondo Trough

Among the earliest volcanic activity of the region was the formation of the large, low-angle volcanoes of eastern Uganda and the Kavirondo Trough. At many localities fossiliferous sediments or tuffs, below or within the basal volcanics, indicate an early Miocene age; radiometric ages (around 19 m.y.) are in agreement. The sediments accumulated in basins, troughs and valleys, including the marginal depressions to basement domes, such as those associated with the early stages of the Napak and Kisingiri Volcanoes, or the para-volcanic depressions underlying the northern flank of Elgon (Fig. 2).

The volcanoes are typically composed of nephelinites and melanepheli-nites, varieties with melilite being common on Elgon, Kadam, Napak, Kisingiri and Tinderet (Fig. 2), while basalts and basanites predominate on

Moroto and also occur on Yelele (Nixon and Clark, 1967); basanites and phonolites form the later eruptions of Tinderet. Phonolites, trachyandesites and trachytes are represented only to a minor extent on the volcanoes of eastern Uganda and not at all at Kisingiri. Whereas Elgon, Kadam and Napak are built up predominantly of coarse pyroclastics, which also contribute in the Kisingiri and Tinderet Volcanoes, Moroto and Yelele are almost entirely formed of lavas.

All the volcanoes are considerably dissected, erosion of Napak revealing a central plug of carbonatites, surrounded by ijolites and fenitised basement. Kisingiri (Fig. 2) is similarly dissected but shows a more complex history, for early ijolites and carbonatites, with associated fenites, occupy large parts of the central dome area of the volcano; they were exposed by erosion before the extrusion of the volcanics. The central vent cuts eccentrically a circular mass of uncompahgrite and ijolite and is occupied by agglomerates, breccias and tuffs together with later carbonatites. Toror is now only represented by intrusive formations: a central carbonatite and agglomerate, with feldspar rock and related potash trachytes, and dykes and plugs of phonolite (Fig. 2).

The Tinderet Volcano in its early stages formed a pyroclastic, nephelinitic cone, against which the plateau phonolites of Kericho and Uasin Gishu (Fig. 2) were banked and with which they were partly intercalated. An Upper Miocene fauna in associated tuffs at Fort Ternan agrees with radiometric ages around 14 m.y. for the phonolites. Activity continued at Tinderet with nephelinite agglomerates, basanites, etc. to build a broad low-angle cone. The adjacent volcano of Londiani, lying near the junction of the Kavirondo Trough with the main rift, commenced activity after the plateau phonolites with associated nephelinites, phonolitic nephelinites and trachyte tuffs. It was largely built up alongside Tinderet and continued after partial erosion of the older cone. Both Londiani and Tinderet are largely mantled by deep red soils and laterites of the 'lower' or late Tertiary surface which extend across the plateau phonolites and on to basement beyond. Pleistocene eruptions of tuffs and basalts from minor centres, either overlie the laterite or rest in erosion hollows.

East of Kisingiri are the carbonatite complexes of the Ruri Hills and Homa Mountain (King et al., 1966); these post-date nearby ijolite-carbonatite complexes and consist of numerous small carbonatite cone-sheets or ring-dykes, with associated breccias and agglomerates of fenitised basement rocks. They are sited on abrupt domes of the basement 1500-2000 ft high and no more than a few miles across; although they were very near-surface phenomena, the associated extrusive activity appears to have been on a minor scale. Radiometric ages suggest that these centres were late Tertiary, post-dating the Kisingiri volcanics, and that some activity extended into the Pleistocene.

The Kavirondo Trough was emphasised by later faulting which displaced the plateau phonolites near Kisumu and the older part of the Tinderet Volcano, while a great part of the Kisingiri Volcano has been dropped down between parallel east-north-east to west-south-west faults which attain maximum throws of 1000-2000 ft adjacent to the volcano itself.

Vulcanism in the Eastern Rift

Radiometric ages so far available confirm faunal evidence that the earliest volcanics of the rift are not older than Middle or even Upper Miocene. Sequences vary greatly in different parts of the rift and the following, therefore, represents a very generalised succession (see Baker, 1965b):

QUATERNARY
{
Trachytes, phonolites, rhyolites
Basalts and basanites
}
{
mainly central volcanoes
and minor volcanic cones
}

PLIOCENE-MIOCENE
{
Trachytes and trachytic pyroclastics: including large trachytic volcanoes
Later Tertiary basalts
Trachytes, phonolites, nephelinites: including older central volcanoes
Main plateau phonolites
Early basalts, basanites, nephelinites: some central volcanoes
}

Although most volcanic events were of local significance, some were represented widely; the distribution of the main volcanic associations is shown in Fig. 3 (see also Williams, 1965). The main plateau phonolites were evidently erupted as vast floods, presumably from numerous fissures, over the greater part of the Kenya dome; radiometric ages range from 11 to 14 m.y. The early members of the succession very largely filled in the incipient rift, so that the phonolites spread far over the shoulders. After major phases of rift-faulting, the later volcanics were mostly confined within the rift itself. Trachytic pyroclastics and lavas were also very widely distributed during later Pliocene and Pleistocene times.

Basalts, often with basanites and nephelinites, are among the earliest members of volcanic sequences, but also appear at intervals throughout the succession. They commonly build thick and extensive flood lava piles and are often associated as early and late phases in large trachytic volcanoes, but rarely form any but minor volcanic cones. There was a notable tendency for vulcanism, like tectonic activity, to affect progressively narrower zones of the rift system, and the latest phases include a series of large volcanoes, and many smaller cones, along the central belt or graben of the rift.

All the volcanics are alkaline and soda-rich, but two genetic series have been distinguished: one strongly alkaline and nepheline-bearing (ankaratrite-melanephelinite-nephelinite-phonolite); the other mildly alkaline and without modal nepheline (alkali basalt-trachybasalt-alkali trachyte-soda rhyolite) (Saggerson and Williams, 1964; King and Sutherland, 1960). Chemical data are at present limited, but in conjunction with petrographic studies they suggest that there may be a number of genetic trends lying between a strongly alkaline series and one that ranges from basalt ('transitional' or even tholeiitic) to trachyandesite and alkali rhyolite (Fig. 4). It also appears likely that within the series there are hiatuses corresponding to compositions that are sparingly represented. The phonolites and trachytes were extruded in enormous volumes, equal to those of the basaltic lavas.

In northern Tanzania the Tertiary extrusives are chiefly basalts, together variously with basanites, tephrites, nephelinites and phonolites, which extend along and across the general region of the rift. They are seen, for example, in the Crater Highlands and as horst blocks within the main rift (Figs. 2 and 3), which is here only clearly delimited on its western side by a major fault of up to 2000 ft throw. The older, 3000 ft fault which defines the northern side of the Eyasi Trough (Fig. 1) is partly buried by volcanics. The numerous large central volcanoes are divided into older and younger groups according to their relations with the main faulting; this trends around north-south in the southern extension of the Kenya Rift, but swings eastwards into

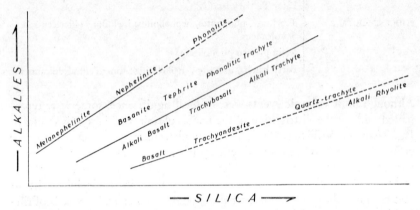

FIG. 4. Relationships among the volcanic rocks of the Eastern Rift expressed diagrammatically in terms of the proportions of alkalies and silica.

a pattern of graben, horsts and tilted blocks trending north-north-west to south-south-east and thence becomes nearly east-west. The older volcanoes include those of the Crater Highlands, such as Lemagarut, Oldeani (10,460 ft) and the giant caldera of Ngorongoro, in which trachytes rather than basaltic lavas predominate and pyroclastics are widespread. In the east, the older volcanoes are formed chiefly of basalts and trachybasalts.

The younger volcanoes, of Pleistocene to Recent age, are mostly aligned along the main rift; Mosonik and Kerimasi bury its western wall, while Oldoinyo Lengai (9443 ft) and Hanang lie close to it; Essimingor and Burko are farther to the east (James, 1959). They are notable for their strongly alkaline and undersaturated character; nephelinites, melanephelinites and nephelinitic phonolites are typical, and pyroclastics rather than lavas build steep-sided cones. Blocks of ijolite, jacupirangite, biotite-pyroxenite, fenite and carbonatite are found in the agglomerates and tuffs, while Mosonok and Kerimasi have extrusive carbonatites and Oldoinyo Lengai (Dawson, 1962) has erupted carbonatite ash and natro-carbonatite flows in recent years (Fig. 2).

Over the border into Kenya the earlier central volcanoes are continued within the rift by the eroded cones of Shombole (3200 ft of nephelinite lavas and tuffs), Lenderut (1000 ft of trachyandesite, trachyte, basalt and basanite) and Olorgesaille (3000 ft of trachytes, basalts with pyroclastics surmounted by nephelinites). The Olorgesaillie volcanics are stepped up by a series of Pleistocene faults on to the eastern flanks of the rift, where, to the south of Nairobi, they partly overlie and are partly equivalent to a succession of tuffs, trachytes, phonolites and basalts which succeed the Kapiti phonolite; the latter rests on basement and is of Upper Miocene age (13·4 m.y.) (Williams, 1967). Burying the step-faults are the trachytes, tephrites and phonolites of the later Ol Esayeiti centre.

North of Nairobi, volcanics up to about 5000 ft thick continue along the eastern flanks of the rift; the Kapiti phonolites are succeeded by basalts, tuffs and more basalts which are sharply uptilted to the fault scarp of the Aberdare Range at 13,000 ft (Fig. 2). On the western flanks of the rift nephelinites are succeeded by basalts along the Nguruman Escarpment, while northwards in the Narok and Mau areas (Fig. 2) the plateau phonolites are overlain southwards by basalts, but more widely by sheets of welded tuff. Within the rift itself the older volcanics are succeeded by great developments of Pleistocene trachytic pyroclastics and lavas, with intercalated lacustrine sediments. On the east they also form the intermediate Kinangop Plateau below the Aberdares; on the western side they are downwarped into the rift from the vicinity of Narok. Comparable trachytes (2000 ft) occupy the rift floor and flank the earlier volcanoes in the Magadi–Natron area. Over much of the central and southern parts of the rift, Pleistocene to Recent sediments and tuffs obscure the earlier volcanics, but a great number of volcanoes and minor cones, ranging in age to Recent, protrude through or surmount the sediments. These include the caldera volcanoes of Suswa, Longonot and Menengai, all predominantly trachytic, aligned along the centre of the rift. Numerous domes and flows of rhyolite and comendite occur in the Naivasha area.

The volcanics of the eastern flanks of the rift are mantled by laterites and thick lateritic soils, which cover the Tertiary basalts up to the crest of the Aberdare Range where they are faulted down and pass under the Pleistocene tuffs of the Kinangop Plateau. Eastwards the laterites are overlain by the volcanics of Mount Kenya (Fig. 2). On the western side of the rift the laterites which formed alike on the Kericho plateau phonolites and adjacent basement are similarly covered by Pleistocene tuffs.

North of the Aberdares the Laikipia sector is dominated by plateau phonolites (11-12 m.y.) which are underlain and overlain by basalts; these are fault-stepped westwards into the rift and overlain by later 'grid' faulted (Hannington) phonolites. Traced northwards the older basalts and phonolites wedge out and are overlapped by upwards of 3000 ft of late Tertiary basalts and over 1000 ft of trachytes with associated basalts. These are downwarped and stepped-down by numerous minor faults into the rift, the tilting being increasingly marked in the older formations. The trachytes form a strongly dissected terrain overlooking the central Suguta trough.

On the western side of the rift, north of Molo, the Uasin Gishu Plateau phonolites are surmounted by the upper Tinderet volcanics and are gradually downfaulted into the trough where they are lapped against and overlain by tuffs and the series of younger (Hannington) phonolites. Northwards the Elgeyo escarpment (8500 ft) shows basement overlain by about 1500 ft of plateau phonolites, locally interleaved with sediments of possible early Miocene age. In the Kamasia range (Fig. 2) over 3000 ft of sediments, tuffs, basalts, basanites and phonolites, accumulated in the early rift trough and are interposed between basement and the plateau phonolites, the latter here being surmounted by trachytes. This entire sequence is exposed in the Kamasia fault scarp. On the eastern flanks of the Kamasia Range (8500 ft) numerous faults step down the phonolites and trachytes, bringing in successively younger formations of basalts and phonolites towards the central trough around Lake Baringo (3500 ft) (Fig. 2). In total, therefore, some 10,000 ft of volcanics and sediments may be inferred within this part of the rift, compared with 1500 ft on the shoulders. As well as the basal sediments, fossiliferous sediments and tuffs of Pliocene, Lower Pleistocene (Martyn, 1967) and Upper Pleistocene age are recorded among the lavas; radiometric ages of the latter confirm the range and continuity of the succession from Miocene to Recent.

Northwards again the downfaulted Miocene phonolites and basalts are mantled by a number of large, low-angle strongly dissected trachytic volcanoes of Pliocene age. These volcanoes form the eastern part of the Tiati Range and the regions to the north and north-east. Along the central zone of the rift between Lake Baringo and Lake Rudolf there is a series of large, predominantly trachytic late Pleistocene to Recent volcanoes—Karossi, Pakka, Silali, and Emoruongongolak—which are characterised by large craters or calderas (Fig. 2). All show steam jets or other signs of recent or waning activity. Immediately south of Lake Rudolf the complex volcanic area of the Barrier (Fig. 2) includes Teleki's volcano, which was active in 1894. Minor cones and extensive floods of Recent basalts occur in the central graben.

In northern Kenya to the west and east of Lake Rudolf the volcanic region widens with the rift system, which is no longer so clearly defined by major fault scarps. Tilting of the faulted segments often reveals basement, and the volcanics are in many places a comparatively thin veneer. Basalts are extensive, especially to the south and east of Lake Rudolf, but to the west and north-west a great diversity of basalts, basanites, nephelinites, phonolites, 'andesites' and rhyolites are represented. These are mostly Tertiary and overlie Miocene sediments.

The greatest of the African volcanoes, Mounts Kenya (17,058 ft) and Kilimanjaro (19,340 ft), lie to the east of the main Rift Valley (Fig. 2). The activity of Mount Kenya dates back to the latest Pliocene, but continued into the Pleistocene. Basalts and phonolites, together with pyroclastics of similar compositions, constitute most of its bulk, but late basalt flows and ashes were erupted from parasitic vents. Kilimanjaro, some 50 miles across, has been built by eruption from three major centres; it is presumed to be largely

Pleistocene and still shows fumarolic activity. The lavas include basalts nephelinites, basanites and trachybasalts, but highly alkaline trachytes and phonolites are particularly abundant among the later flows.

Conclusions

The East African Rift System is a structure of a major order in the earth's crust. Its history extends from at least early Miocene to the present day, while in the south it has partly rejuvenated similar rifts of early Jurassic age. The trends of rift faults are closely dependent on the older Precambrian 'grain'. The faulting is mostly normal and tensional; only to this extent are the rifts dilational phenomena. They are therefore unlike the so-called 'rifts' of the ocean floors.

Arching over the sites of the rifts is characteristic, and intermittent uplift of the rift shoulders has occurred concomitant with subsidence of rift floors. Typical rifts show rather regular widths between 30 and 50 miles and vertical displacements of 5000-10,000 ft, which are independent of their absolute altitude and probably reflect relations between thickness and strength of the crust and imposed stresses. Activity, both tectonic and volcanic, has tended to migrate towards the central graben during rift history and the latest movements commonly took place along numerous close-set faults of small displacement. These features suggest a sequence of secondary adjustments to strains induced by primary dislocations.

Vulcanism is not a direct consequence of rifting, for extensive sectors of the rift are devoid of volcanic activity and volcanic centres rarely, if ever, show a close association with faults. Nevertheless, the coincidence of vulcanism in East Africa with parts of the rift system shows that there is a relation, ship of some kind. Indeed, volcanic regions are associated with culminations or broad topographic domes in the rift system.

The volcanic rocks range from mildly to strongly alkaline and include highly undersaturated types. There is no evidence of simple sequences or cycles, but in the Eastern Rift the first appearance of particular compositions on a major scale may be of genetic significance. The earliest (Miocene) volcanics are highly undersaturated nephelinites and/or alkali basalts. Widespread phonolites followed in later Miocene. Trachytes were abundantly represented in the Pliocene, whilst alkali rhyolites are Pleistocene to Recent in age and of limited distribution. Alkali basalts are, indeed, of frequent occurrence throughout the volcanic sequences, while central volcanoes ranging in age from Miocene to Recent, and correspondingly more or less dissected, testify to the association between nephelinite lavas and intrusive ijolite-carbonatite.

Highly characteristic of the volcanic province is the great development of phonolite and trachyte, comparable in extent with basalts; the proportions are strikingly different from those of similar rock types in the volcanics of the ocean basins. The total volume of volcanic rocks associated with the Eastern Rift is probably at least 150,000 cubic miles (600,000 km^3).

Acknowledgements

Adequate recognition of source material for a largely appreciative and synthetic account is not possible, but reference is conveniently made to the very comprehensive bibliographies that were prepared for the UMC/UNESCO Symposium held in Nairobi in 1965. I am also much indebted to the work, as yet largely unpublished, of members of my research teams in East Africa and I thank B. Collins in particular for the preparation of the maps.

REFERENCES

BAKER, B. H. 1965a. The Rift System in Kenya. *Rep. UMC/UNESCO Seminar on the E. African Rift System (Nairobi)*, 82.
—— 1965b. An outline of the geology of the Kenya Rift Valley. *Rep. UMC/UNESCO Seminar on the E. African Rift System (Nairobi)*, 1.
BISHOP, W. W. 1965. Quaternary geology and geomorphology of the Albertine Rift Valley, Uganda. *Spec. Pap. geol. Soc. Am.*, **84**, 291.
—— and TRENDALL, A. F. 1967. Erosion-surfaces, tectonics and volcanic activity in Uganda. *Q. J. geol. Soc. Lond.*, **122**, 385.
COMBE, A. D. 1930. Volcanic areas of Bunyaruguru and Fort Portal. *Ann. Rep. geol. Surv. Uganda (for 1929)*, 16.
DAWSON, J. B. 1962. Sodium carbonate lavas from Oldoinyo Lengai, Tanganyika. *Nature, Lond.*, **195**, 1075.
DE SWARDT, A. M. J. 1965a. Rift faulting in Zambia. *Rep. UMC/UNESCO Seminar on the E. African Rift System (Nairobi)*, 105.
—— 1965b. Lateritisation and landscape development in parts of equatorial Africa. *Rep. UMC/UNESCO Seminar on the E. African Rift System (Nairobi)*, 134.
DIXEY, F. 1939. The early Cretaceous valley-floor peneplain of the Lake Nyasa region and its relation to Tertiary rift structures. *Q. J. geol. Soc. Lond.*, **95**, 75.
—— 1956. The East African Rift System. *Bull. colon. Geol. Mineral Resources*, Suppl. **1**.
DUNDAS, D. L. 1965. Review of rift faulting in Tanzania. *Rep. UMC/UNESCO Seminar on the E. African Rift System (Nairobi)*, 95.
GARSON, M. S. 1965. Summary of present knowledge of the rift system in Malawi. *Rep. UMC/UNESCO Seminar on the E. African Rift System (Nairobi)*, 94.
GREGORY, J. W. 1921. *The rift valleys and geology of East Africa*. London.
HARKIN, D. A. 1959. The Rungwe volcanics. *Mem. geol. Surv. Tanganyika*, **2**.
HOLMES, A. 1942. A suite of volcanic rocks from south-west Uganda containing kalsilite. *Miner. Mag.*, **26**, 197.
—— and HARWOOD, H. F. 1937. The petrology of the volcanic area of Bufumbira. *Mem. geol. Surv. Uganda*, **2**, Pt. 2.
HOPWOOD, A. T., and LEPERSONNE, J. 1953. Présence de formations d'âge miocène inférieur dans le fossé tectonique du Lac Albert et de la Basse Semliki (Congo belge). *Annls. Soc. géol. Belge*, **77**, 83.
JAMES, T. C. 1956. The nature of rift faulting in Tanzania. *C.C.T.A. east centr. reg. Comm. Geol.*, Dar-es-Salaam, 81.
—— 1959. Carbonatites and rift valleys in E. Africa. Abstract. *Int. geol. Congr.* **20**, *As. Serv. geol. Afr.*, 325.
KING, B. C., SUTHERLAND, D. S., COLLINS, B., DIXON, J. A., LE BAS, M. J., CLARKE, M. C. G., FLEGG, A., and FINDLAY, A. L. 1966. *In* Volcanism in eastern Africa. *Proc. geol. Soc. Lond.*, **1629**, 16.
—— and SUTHERLAND, D. S. 1960. Alkaline rocks of eastern and southern Africa. *Sci. Prog.*, **48**, 298, 504, 709.

MACDONALD, R. 1965. The status of rift valley studies in Uganda. *Rep. UMC/UNESCO Seminar on the E. African Rift System (Nairobi)* 52.

MARTYN, J. E. 1967. Pleistocene deposits and new fossil localities in Kenya. *Nature, Lond.* **215**, 476.

McCONNELL, R. B. 1951. Rift and shield structures in E. Africa. *Int. geol. Congr.*, **18(14)**, 199.

MOHR, P. A., and ROGERS, A. S. 1965. Status of geological and geophysical studies and resumé of the geology of Ethiopia. *Rep. UMC/UNESCO Seminar on the E. African Rift System (Nairobi)*, 47.

NIXON, P. H., and CLARK, L. 1967. The alkaline centre of Yelele and its bearing on the petrogenesis of other eastern Uganda volcanoes. *Geol. Mag.*, **104**, 455.

PALLISTER, J. W. 1965. The rift system in Tanzania. *Rep. UMC/UNESCO Seminar on the E. African Rift System (Nairobi)*, 86.

PULFREY, W. 1960. Shape of the sub-Miocene erosion bevel in Kenya. *Bull. geol. Surv. Kenya*, **3**.

SAGGERSON, E. P., and BAKER, B. H. 1965. Post-Jurassic erosion-surfaces in eastern Kenya and their deformation in relation to rift structure. *Q. J. geol. Soc. Lond.*, **121**, 51.

—— and WILLIAMS, L. A. J. 1964. Ngurumanite from southern Kenya and its bearing on the origin of rocks in the northern Tanganyika alkaline district. *J. Petrology*, **5**, 40.

SANDERS, L. D. 1964. Geology of the Eldoret area. *Rep. geol. Surv. Kenya*, **64**.

SHACKLETON, R. M. 1951. A contribution to the geology of the Kavirondo Rift Valley. *Q. J. geol. Soc. Lond.*, **106**, 345.

VON KNORRING, O., and DU BOIS, C. G. B., 1961. Carbonatite lava from Fort Portal area in western Uganda. *Nature, Lond.*, **192**, 1064.

WAYLAND, E. J. 1930. Rift valleys and Lake Victoria. *C. R. Int. geol. Congr.*, **15(6)**, 323.

WILLIAMS, L. A. J. 1965. Petrology of volcanic rocks associated with the Rift System in Kenya. *Rep. UMC/UNESCO Seminar on the E. African Rift System (Nairobi)*, 33.

—— 1967. In *Nairobi: City and Region*, p. 1. (Ed. W. T. W. Morgan).

I. G. GASS

13   Tectonic and magmatic evolution of the Afro-Arabian dome

ABSTRACT. *In Lower Tertiary times, the junction area of the Red Sea, Gulf of Aden and Ethiopian Rifts became the site of intense magmatic and tectonic activity that has continued to the present day. The variety of magmatic products and the style of deformation have altered radically during this period: it appears that alkali basalts were erupted during an early phase when vertical uplift was dominant; zones where the sialic crust was attenuated became the sites of peralkali vulcanism; whereas the 'oceanic' crust in the Red Sea and Gulf of Aden, formed during the lateral separation of sialic blocks in late Tertiary times, is of tholeiitic character. It is suggested that both the tectonic history and the volcanic evolution are the result of an isolated lithothermal event in the upper mantle; the theoretical thermo-dynamic evolution of such a phenomenon is compared with the actual tectono-magmatic development of the area.*

Introduction

The region under examination is easily recognised, for it is the area, covered almost entirely by Tertiary and Recent volcanic rocks, that forms the elevated land masses of Ethiopia, Somalia and the south-west corner of Arabia and trisected by the great rift systems of the Red Sea, the Gulf of Aden and Ethiopia (Fig. 1). In Lower Tertiary times, after a long period of tectonic stability and magmatic quiescence extending from late Precambrian times, this region became the site of intense magmatic and tectonic activity that has continued, virtually without interruption, to the present day. However, throughout the last 70 m.y., the type of magmatic activity and the style of deformation have changed radically. It is the purpose of this article to review both the sequence of magmatic events and the tectonic history to see if these phenomena are related in time and space, and if so, to establish the nature of the relationship.

The dominant magmatic activity of the area has been the extrusion of basic lavas, and although silicic eruptives are abundant, they can usually be recognised as differentiation products of a basaltic parent. As well as the alkaline and tholeiitic associations, originally recognised by Kennedy (1933), a third, intermediate between true tholeiites and true alkali basalts, which produced peralkali differentiates on fractionation (Coombs, 1963), is also abundant. The abundance of peralkali silicic rocks within the region has long been recognised (Prior, 1903) and has led to its acceptance as one of the largest, if not the largest, peralkali provinces in the world (Shackleton, 1954).

Tectonic deformation commenced in the early Tertiary and until the end

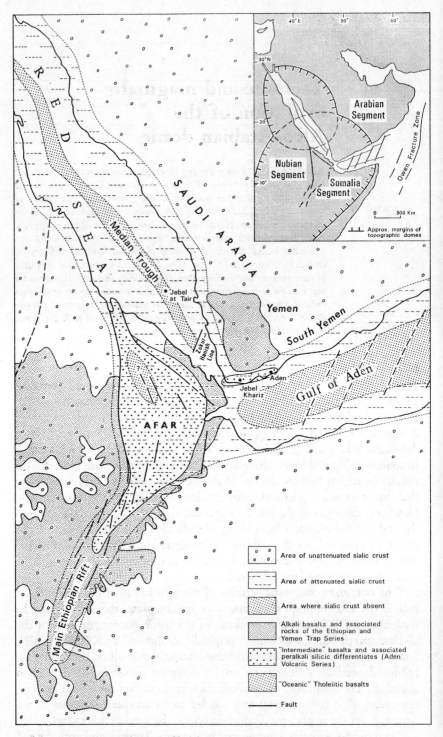

FIG. 1. Distribution of basalt types and major structural elements.

of the Eocene period the region was the site of extensive vertical uplift that has produced the major topographic feature of the Afro-Arabian dome (Dainelli, 1943; Beydoun, 1964; Mohr, 1963). Attenuation of the sialic crust across the dome, in response to the vertical uplift, caused it to be trisected into three distinct crustal segments by zones of weakness (Cloos, 1939): (1) the Arabian segment; (2) Egypt–Sudan–Ethiopia (Nubian) segment; and (3) the Horn of Africa–Somaliland (Somalia) segment (see Fig. 1, inset). Each of the zones of weakness, the 'proto' Red Sea, the Gulf of Aden and the Ethiopian Rift, was bounded by crustal flexures associated with tensional faulting and became the site of further tectonic activity when, in the Miocene, the style of deformation changed to one of predominantly lateral displacement.

Since that time some 50 km of new oceanic crust have been formed in the southern Red Sea (Girdler, 1958; Drake and Girdler, 1964) whereas the width of oceanic crust in the Gulf of Aden formed during this period is as much as 200 km (Laughton, 1966); this means that the Arabian peninsula must have moved horizontally away from the Nubian and Somalia segments. Although no obvious oceanic crust is present in the Ethiopian Rift (see p. 293), there is evidence (Gibson, in press; Gass and Gibson, 1969) that the Somalia segment is moving north-eastwards with respect to the Nubian segment which, so far as can be determined, has remained relatively stationary with respect to Eurasia since the Middle Miocene.

On a regional scale it appears that the variety of magmatic products can be correlated with the style of deformation: first, preceding and concurrent with vertical uplift alkali basalts were erupted; second, the zones of crustal attenuation are the site of volcanic activity that is characterised by the abundance of peralkali silicic differentiates; and third, areas of new ocean crust in the Red Sea, the Gulf of Aden, and possibly in the northern Afar depression, are of low potassium tholeiitic basalts. There are exceptions to this regional correlation, in particular alkali basalts of very recent age are found in the attenuated zones and on the flanks of the dome. The spatial extent of these provinces is shown in Fig. 1 and the pertinent data are given in Table 1.

The pre-rift alkali basalt association

The term 'Trap Series' has been applied and is here used, for the dominantly basaltic, Lower Tertiary volcanics which covered some 750,000 km^2 in Ethiopia and some 30,000 km^2 in south-west Arabia, and which in the extensive and uniform individual flows are similar to flood basalts elsewhere. The data presented hereunder are culled mainly from the works of Mohr (1962, 1963) who in turn has drawn extensively on the publications of earlier workers in Ethiopia, notably Blanford (1869), Comucci (1928, 1932, 1933a, and b, 1948, 1950), Dainelli (1943), Duparc (1930) and Hieke-Merlin (1950, 1953). Information on the Yemen is sparse (Roman, 1926; Lamare, 1930; Lipparini, 1954; Shukri and Basta, 1954; Gass and Mallick, 1968) and mainly descriptive.

TABLE 1.

TIME SCALE	M.Y	REGIONAL TECTONICS	REGIONAL MAGMATISM	CRUSTAL SEGMENTS			RIFT ZONES		
				NUBIAN	SOMALIA	ARABIAN	GULF OF ADEN	RED SEA	ETHIOPIAN RIFT
RECENT	0.5	HORIZONTAL NORTHWARD MOVEMENT OF CRUSTAL SEGMENTS AT VARIOUS RATES	"OCEANIC" THOLEIITIC VULCANISM	continuing uplift greatest at margins	continuing uplift greatest at margins	basaltic vulcanism in Saudi Arabia; continuing uplift greatest at margins	basaltic vulcanism in south Arabia; widening of "oceanic" zone concomitant with tholeiitic vulcanism	widening of median trough concomitant with injection & extrusion of tholeiitic basalt	tholeiitic vulcanism in N. Afar, incipient crustal separation
PLEISTOCENE	1.5		"PERALKALI" VULCANISM						peralkali vulcanism within rift zone
PLIOCENE	7.0			alkali shield vulcanism	alkali shield vulcanism		peralkali vulcanism on S. Arabia coast		rift widening by downwarp associated with normal faulting
MIOCENE	25	VERTICAL UPLIFT; FORMATION OF AFRO-ARABIAN DOME & PROTO RIFT ZONES	ALKALINE TRAP SERIES VULCANISM	marginal alkali vulcanism in Sudan		block faulting near margins	separation of sialic crust	separation of sialic crust	
OLIGOCENE	37			trap vulcanism in Ethiopia & Sudan; younging southwards	trap vulcanism in Somalia	trap vulcanism in S.W. Arabia	attenuation of sialic crust by downwarp & normal faulting	attenuation of sialic crust by downwarp & normal faulting	
EOCENE	54	NORTHWARD MOVEMENT OF UNIFIED AFRO-ARABIAN PLATE							
PALAEOCENE	65				folding of Mesozoic sediments	folding of Mesozoic sediments			
UPPER CRETACEOUS	100								
PRE-UPPER CRETACEOUS		EPEIROGENIC MOVEMENTS							
PRECAMBRIAN	600								

Time correlation between magmatic and structural events.

Before the volcanic activity, the area was a peneplaned surface cut into Mesozoic sediments (Mohr, 1963). It was on to this regionally flat surface that the Trap Series, amounting to some 3500 m in Ethiopia and 2000 m in the Yemen, was extruded. The series is thickest in Ethiopia but thins southwards into Kenya, northwards and westwards into the Sudan and rapidly eastwards into Somalia. In the Yemen the maximum thickness is in the southwest part of that country and the series 'thins' northwards and eastwards into Saudi Arabia. There is little doubt that areas now occupied by the Red Sea, the Gulf of Aden and the Ethiopian Rift were covered by rocks of this series, and indeed the Ethiopian Rift is, in places, floored by basalts of the Trap Series. Although extensively fractured by subsequent faulting, particularly in the south Yemen (Gass and Mallick, 1968), and downwarped along the margins of the rift zones, the original regional disposition of the Trap Series was approximately horizontal.

In Ethiopia, Mohr (1963) substantiated the subdivision made by earlier workers and divided the Trap Series into two units: a lower unit ranging in thickness from 200 to 1200 m, characterised by a few thick, extensive, and petrographically uniform flows of aphyric alkali basalt erupted from fissures; and an upper unit up to 2600 m thick, of more widely varying, generally porphyritic types that were erupted from central vent volcanoes which, although they contain abundant silicic and ultramafic varieties, are of alkaline affinity throughout. The subdivision is petrological and not structural and although on a regional scale the two units vary in age, some temporal overlap is indicated. Lavas of the lower unit are thought to have issued from fissures and, although none can now be recognised, it is suggested by Mohr (ibid.) that these lineaments were roughly north-south; this meridianal alignment is also apparent among the central volcanoes of the younger group. According to Mohr (ibid.) the period of Trap vulcanism ranged from the Oligocene to the lowest Miocene; the uplift that formed the Afro-Arabian dome did not continue after Upper Eocene times, and the central vent vulcanism did not become generally active until the period of uplift was over. This is disputed by other workers (Gibson, Azzaroli; personal communications) but a Lower Tertiary age for the volcanic activity is generally accepted.

Roman (1926) describes the Trap Series of the Yemen, considered by Lipparini (1954) as being of Eocene age, as an indisputably alkaline sequence with sodium greatly in excess of potassium and in which porphyritic trachyandesites and trachydolerites are the dominant rock types. Those who have studied the Yemen Trap Series (Roman, 1926; Lamare, 1930; Shukri and Basta, 1954; El-Hinnawi, 1964) have all emphasised the pronounced similarity of the series to the coeval volcanics of Ethiopia. It is however evident from the descriptions of the Yemen rocks that they are most closely comparable to the highly porphyritic upper part of the Ethiopian sequence. Whether the underlying aphyric unit, present in Ethiopia, is represented in the Yemen is not known, but it is at least suggestive that the lowest volcanics of the Yemen Trap Series are more alkaline and contain fewer intermediate members (Roman, 1926). There is a clear correlation, in time and composition, between the Yemen and the Ethiopian Trap Series so that the chemistry of the sequence can be considered as a whole.

Data on the chemistry of the Trap Series comes mainly from Ethiopia; analyses presented by Dainelli (1943), Comucci (1943) and Hieke–Merlin (1953) are given and discussed by Mohr (1963). Only isolated analyses of specimens from the Yemen are available (Roman, 1926; Shukri and Basta, 1954; and El-Hinnawi, 1963). It is however clear that in both areas the series is alkaline on the normative classification of Yoder and Tilley (1962) provided that the effect of excessive oxidation is taken into account. Mohr (1963) noted that the average for Ethiopian basalts is closely comparable to Nockold's (1954) average for world alkali basalts, and emphasised that the series has such typical alkali association traits as low TiO_2, high H_2O and, particularly, high Na/Si and ferric/ferrous ratios. In Ethiopia, the aphyric alkali basalts which form the lower part of the Trap Series are low in MgO in keeping with the absence of modal olivine (see Table 2). The upper part of the sequence, on average richer in MgO, is more typically alkaline. However, as a wider variety of generally porphyritic rock types is present, it is clear that these are the products of more extensive, high-level fractionation probably in a magma chamber under a central vent volcano. The isolated analyses from the Yemen Trap Series often lack accompanying field and petrographic data and are, in many cases, of an obviously less reliable nature. However, the better analyses are averaged in Table 2 and show the alkali nature of the series and further emphasise the similarity of the Yemen Traps to those in Ethiopia.

Igneous activity in the zones of crustal attenuation

In places, a period of magmatic quiescence followed the eruption of the Trap Series in both Ethiopia and in Arabia (Dainelli, 1943; Gass and Mallick, 1968), although vertical uplift, and the crustal attenuation that formed the 'proto' Red Sea, the Gulf of Aden and the Ethiopian Rift, seems to have continued without pause. When volcanic activity recommenced it was of two types which were broadly contemporaneous but differed markedly in the composition of their products and their eruptive mechanisms from each other, and also from the preceding Trap Series.

In the median zones of the Red Sea and the Gulf of Aden, separation of the sialic crust took place, and new ocean floor of tholeiitic basalt was formed. Elsewhere, in the zones of attenuation where the sialic crust was thin but still present, the vulcanism was of the intermediate type that leads on fractionation to peralkali differentiates. In volume this latter association, often termed the Aden Volcanic Series, constitutes about 1/500 of the Trap Series and must be considered as essentially a transitional stage between the voluminous Trap volcanics and the undoubtedly abundant tholeiitic basalts of the central Gulf of Aden and the Red Sea. It is this intermediate association that is now discussed.

Although seismic evidence (Knott et al., 1967; Gouin, in Mohr, 1963; Laughton, 1967) indicates that the sialic crust is attenuated along the margins of the Red Sea and the Gulf of Aden and also within the Ethiopian Rift, only two zones of peralkali vulcanism have been convincingly identified and studied in any detail. These are the line of Miocene-Pliocene central vent volcanoes along South Arabian coast between the entrance of the Red Sea and Aden (Gass et al., 1965; Gass and Mallick 1968; Cox et al., 1970); and the Pliocene to Recent volcanoes within the Ethiopian Rift (Mohr, 1963; Gibson, in press). It is significant that in both of these regions the volcanic centres lie along lines of structural significance; the Arabian centres are on an east-west line parallel to the northern margin of the Gulf of Aden whereas those in Ethiopia are within the seismically active Wonji Fault Belt or along the northern margin of the Somaliland horst (Gouin, in Mohr, 1963) (see Fig. 1).

It should be recalled however that in both Africa and Arabia the silicic volcanics occurring towards the top of the Trap Series also have marked peralkaline affinities and although they are in many ways petrochemically similar to the Aden Volcanic Series, they are separate from it in both time and space; the reason for the community of character is discussed later.

Dickinson et al. (1969) have shown that specimens from the South Arabian centres of the Aden Volcanic Series, including the most silicic varieties, have Sr^{87}/Sr^{86} ratios of 0·704–0·707, which strongly suggests that they are of mantle origin.

Theoretically, it is reasonable to assume that the original melt produced by partial fusion of the upper mantle was basaltic and, in turn, that the least evolved surface rocks would be aphyric basalts. Average compositions of aphyric basalts from the Aden Volcanic Series in Ethiopia and from Jebel Khariz in the south coast of Arabia are presented in the Table 2. Once the effect of excessive oxidation is removed the aphyric basalt average, considered to represent the most primitive magmatic liquid, is mildly hypersthene normative in Ethiopia and mildly nepheline normative in Arabia (Table 2, analyses 4a and 5a). Mohr (1963), comparing the basalts of the Aden Volcanic Series with those of the underlying Trap Series, notes that the former is enriched in Fe_2O_3, Mn and P and lower in Ti, K and H values. In Arabia, K is the only major element that shows any consistent variation between the two series and is invariably lower in the Aden Volcanic Series. It seems probable that the diminution of potassium accounts for the change from the dominantly alkaline Trap Series to the mildly alkaline Aden Volcanic Series.

However, silicic rocks form a large proportion of the products of the Aden Volcanic Series in Arabia, although types ranging from olivine basalt through intermediate varieties to peralkali rhyolite are present; in the Ethiopian Rift, silicic rocks are abundant and basalts occur, but rocks of intermediate composition are rare or absent. In the case of the Arabian centres, for which 60 analyses and detailed field relationships are available, the abundance imbalance has been ascribed to the fact that the basaltic primary magma has been extensively fractionated before eruption and that central vent volcanoes are particularly inefficient at erupting a representative selection of the available magmatic liquids (Gass and Mallick, 1968; Cox et al., 1970).

The tholeiitic association of the Red Sea and Gulf of Aden

Geophysical studies indicate that sialic crust is absent under most of the Gulf of Aden and under the southern part of the Red Sea median trough (Laughton, 1966, 1967; Drake and Girdler, 1964). At present the most acceptable explanation is that concomitant with the Upper Tertiary separation of Arabia from Africa, new oceanic crust was generated in the axial zone of both rifts. This process is considered to have operated since the early Miocene in the Gulf of Aden (Laughton, 1967), but only during the last 5 m.y. in the Red Sea (Vine, 1966). To explain the creation of new oceanic crust and the presence of magnetic strip anomalies in these areas and elsewhere it has been suggested (Hess, 1962; Vine and Matthews, 1963) that basaltic material was injected along the axial zone of mid-ocean rises, probably as dyke-like feeders leading to submarine lava flows. Logically, with time, the older intrusions will move aside to make way for subsequent injections in the axial zone; the once adjacent continental masses would thus be moved further apart as more ocean floor was created.

Laughton (1966, 1967) shows that the main feature of the Gulf of Aden is a central east-west zone of rough topography which is associated with an earthquake epicentre belt, linear magnetic anomalies and high heat flow. This zone is considered to be equivalent to the crest regions of mid-ocean ridges. The east-west alignment is however displaced by a series of north-easterly structures that Laughton (*ibid.*) interprets as transform faults along which movement of separate parts of the axial zone is accommodated.

Magnetic, seismic, gravity and heat flow data indicate that the southern part of the 50 km wide Red Sea median trough is floored by oceanic crust. The number of magnetic strip anomalies suggest that the initial separation of the sialic masses of Africa and Arabia took place in the Red Sea some 5 m.y. ago (Vine, 1966). Although magnetic anomalies, and thus ocean crust, are not present in the northern part of the Red Sea, the trough continues northwards as a bathymetric feature and is probably linked with the Jordan Rift (Freund, 1965). To the south the oceanic trough loses its identity in the shallow water surrounding the volcanic islands of the Zukur and Hanish groups (Fig. 1) which are thought to lie on a transform fault or faults which, it should be noted, are parallel to similar structures in the Gulf of Aden (Gass and Gibson, 1969).

There is increasing evidence (for example, Engel *et al.*, 1965; Cann and Vine, 1966) that the volcanic rocks of the deep ocean floors, the abyssal basalts, intruded and extruded by the mechanism just described, have characteristic major oxide and trace element compositions. On the classification of Yoder and Tilley (1962) they are dominantly over-saturated tholeiites, carrying abundant normative hypersthene and usually normative quartz. In the major oxides, abyssal basalts are high in SiO_2 and CaO, low in TiO_2 and particularly low in K_2O; in trace elements, they are impoverished in the large ion elements such as rubidium, thorium, uranium, barium (Gast, 1968).

Dredge hauls from the Gulf of Aden have brought up a variety of mafic and ultramafic rocks in various stages of alteration, but also unaltered aphyric basalts. Eight of these basalts have been analysed by Cann, and the average (Table 2, analysis 7) is typical for abyssal basalts elsewhere (Cann, personal communication); petrochemical data thus substantiate the oceanic nature of the crust within the Gulf of Aden.

In the Red Sea, specimens of volcanic glass recovered from depressions in the median trough (Chase, 1969) have major oxide compositions close to those of abyssal basalts (Table 2, analyses 8 and 9). These specimens are considerably altered and might not be acceptable alone as evidence of the oceanic character of the Red Sea median trough; however the volcanic island of Jebel at Tair, which lies within the median trough, is formed of distinctly tholeiitic lavas (Gass et al., in preparation) (Table 2, analysis 10) and, although these are considerably more evolved than the abyssal basalts, they are nevertheless of 'oceanic' character and are comparable in their major oxide chemistry to oceanic island tholeiites. Confirmatory evidence has recently been provided by Schilling (in press), who notes that the chondritic relative rare-earth element abundance pattern of the dredge samples from the median trough is strictly analogous to that of abyssal basalts elsewhere, whilst the pattern for Jebel at Tair basalts is closely comparable to those of mid-ocean islands.

There is therefore little doubt that the floor of most of the Gulf of Aden and the southern part of the Red Sea median trough is formed of abyssal basalts belonging to a tholeiitic association that is significantly different in its chemistry from the alkali and mildly alkali suites of the Trap Series and the Aden Volcanic Series.

Recent volcanoes in the central part of the northern Afar depression (see Fig. 1) have been investigated and described by Tazieff et al. (in press). Chemically, the products of these volcanoes are tholeiitic and most closely comparable with the 'oceanic' basalts of the Red Sea and Gulf of Aden (Table 2, analysis 6). As the area is one of active east-west dilation, witnessed by the abundance of north-south tensional faults and basic dykes, it is possibly of embryonic 'oceanic' character. Although the lavas are low in potassium, this element is more abundant than in the ocean floor tholeiites but this is not entirely surprising as the magma has had to pass through thousands of metres of potassic evaporites (Holwerda and Hutchinson, 1968).

Tectono-magmatic relationships

The coincidence in time and space of the extensive Trap vulcanism with the vertical uplift which formed the Afro-Arabian dome can hardly be fortuitous. That doming, rifting and magmatism are the expressions of the same major process has been previously proposed by Shackleton (1954) and Bailey (1964, p. 1105). Coeval regional doming and extensive vulcanism also occurred in East Africa, Tibesti and the Hoggar (this volume: King, p. 263; Vincent, p. 301; Black and Girod, p. 185). The generation of basaltic magmas, and regional uplift, are thought to have originated by processes within the upper mantle, and the spatial coincidence of both the doming and the vulcanism suggest that the primary mechanism could well be a localised thermal disturbance. In this regard, Elder (1966), using the term penetrative convection, has, on experimental evidence, postulated that there are isolated lithothermal systems involving both heat and mass transfer. These systems are discrete portions of the mantle, hotter than their surroundings, that can exist either as isolated, roughly equidimensional 'blobs' or as rising thermal plumes; in their space-form they have been compared to salt diapirs (Elder, *ibid.*; Harris, this volume, p. 422).

The theoretical thermodynamic evolution of a rising lithothermal system is now compared with the actual tectono-magmatic evolution of the Afro-Arabian dome (Fig. 2). Thermal activity of this type would cause partial melting in the upper mantle and thereby make basaltic magma available, and would increase the thermal gradient thus necessitating the downward movement of the main phase boundaries in the mantle. The lowering of the phase boundaries would result in an increase in volume as the low temperature, high pressure minerals inverted to their less dense, high temperature equivalents. The increase in volume would be most easily relieved by vertical uplift, in this case producing the Afro-Arabian dome, with the consequent attenuation of the sialic crust and the formation of the 'proto' rift zones of the Red Sea, Gulf of Aden and Ethiopia.

With the rising of the lithothermal system through the upper mantle, partial melting would occur in response to the elevation of the thermal gradient. Both Hess (1960, p. 81) and McBirney (1963, p. 6351) have suggested that as more heat is required to extend the zone of melting vertically than to raise the temperature of the rock still below the melting range, a body of magma will extend laterally at a far greater rate than it will grow vertically. The first body of magma generated is likely therefore to have a roughly space-tabular form (see Fig. 2a). As the overlying crust and mantle were updomed, fractures would occur in this brittle carapace and give ready access to ascending magma. The first melts would be produced in a zone where the melting point curve for peridotite or garnet peridotite intersected the thermal gradient. In the case of the Afro-Arabian dome this seems to have been at depths of about 60 km for the products of this first phase are alkaline (Kushiro, 1965, 1968; Kushiro and Kuno, 1963). The tendency for the upper part of the Trap Series to become less alkaline, noted by Mohr (1963) and implicit in the presence of abundant peralkali silicic differentiates (p. 291), could be accounted for either by the upward movement of the primary magma-genetic zone or, more probably, by a pause in the upward movement of the magma and fractionation in a lower pressure regime on the way to the surface (O'Hara, 1965). The latter mechanism seems more likely as McBirney (1967) has shown that once formed, a zone of partial fusion would tend to migrate downwards rather than upwards.

As isostatic equilibrium appears to be maintained throughout doming (Bullard, 1936), the volume changes in the mantle must be compensated by uplift; linear zones of crustal attenuation form to relieve the tensional stresses. It follows that these zones would be areas of mass deficiency and lines of structural weakness, factors facilitating intense magmatic injection. Logically, the attenuated zones would be the sites of most intense injection and vulcanism and this is, in fact, the case at the present time in the Ethiopian

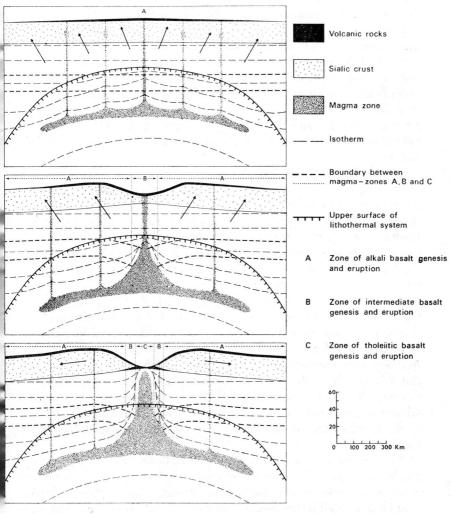

Volcanic rocks

Sialic crust

Magma zone

Isotherm

Boundary between magma-zones A, B and C

Upper surface of lithothermal system

A Zone of alkali basalt genesis and eruption

B Zone of intermediate basalt genesis and eruption

C Zone of tholeiitic basalt genesis and eruption

FIG. 2. Schematic diagram illustrating the postulated distribution of magma-genetic zones and related eruptive rocks during the structural evolution of the area.

Rift and Afar depression, where active fault belts are also the sites of the most active vulcanism (Mohr, 1963; Tazieff *et al.*, in press).

Injection of magma along the axial zones of weakness must distort the isothermal surfaces from their original, near-horizontal attitude, to give a region of high thermal gradient adjacent to and above the injection zone; the situation envisaged is similar to that demonstrated by the electric analogue experiments of McBirney (1963, p. 6328). With the increased thermal gradient and the lowering of pressure in the injection zone, partial fusion under the areas of crustal attenuation would extend higher into the mantle than elsewhere. Experimental evidence (Kushiro, 1965) suggests that magma generated in such a low pressure regime would tend to be more silicic than that created at greater depth. It is postulated that the basaltic magma generated at this stage in the tectono-magmatic cycle would result in a rock, intermediate between alkali and tholeiitic varieties, that leads on fractionation to peralkali silicic differentiates (Coombs, 1963). This stage is shown in Fig. 2b, and is envisaged for the Recent volcanics that lie along the active Wonji Fault Belt of the Ethiopian Rift. Nevertheless intermediate basaltic magmas of this type can be produced after crustal separation has occurred (for example, zone B in Fig. 2c) and this is thought to apply to the Miocene-Pliocene volcanoes that lie along the south coast of Arabia, between Aden and the Red Sea.

Repeated injection along the zones of crustal attenuation would result in the moving apart of the bordering sialic blocks and eventually the host rock for the injections would be the previously emplaced basaltic dykes. By this stage the sialic crystalline basement would be entirely separated by new basaltic crust. The effect of repeated injection would steepen, still further, the thermal gradient over the original attenuated belts so that the zone of partial melting might well extend to within 10 km of the surface (Fig. 2c); it is in this low pressure environment that tholeiitic basalts appear to be generated (Kushiro, 1965; Oxburgh and Turcotte, 1968; Gast, 1968; McBirney and Gass, 1967). This stage commenced in the Miocene and has continued ever since so far as the Gulf of Aden is concerned. In the Red Sea, sialic separation took place about 5 m.y. ago (Vine, 1966) whereas in the northern part of the Afar depression the strongly tholeiitic character of the central basalts suggests that this stage has only just been reached (Tazieff et al., in press). Once this stage has been reached (Fig. 2c), the process continues as long as a primary magma source is available because the high thermal gradient caused by the injections themselves ensures the production of magma high in the mantle.

The question now arises whether the sialic blocks, fractured by the updoming and separated by basaltic injection, will continue to separate in response to the emplacement of basalt alone or whether some other mechanism must be invoked. It is therefore pertinent to note that from Cretaceous times onwards, Africa and Arabia have been moving north-eastwards, possibly in response to the generation of new oceanic crust along the ridges of the south Atlantic and Indian Oceans (McElhinny et al., 1968). It is within this major 'conveyor belt' system that the Afro-Arabian lithothermal event occurred, and its effect has been to split the earlier single continental mass of Afro-Arabia into segments and, by injection of basalts along the fracture lines, to generate ocean crust in the Gulf of Aden and the Red Sea, thus accentuating the north-easterly movement of the Arabian peninsula.

Throughout the foregoing discussion the upper surface of the lithothermal system has been treated as a uniform sphere. This is unlikely for there is distinct geomorphological evidence that a secondary topographic dome exists athwart the Red Sea with its major elevation at about 22° N. (Fig. 1, inset). This, together with crustal separation only in the central part of the Red Sea median trough suggests a subsidiary thermal 'plume'. Furthermore, since coeval vulcanism and rifting in East Africa are associated with major updoming, it is tentatively suggested that the African Rift System may be the combined result of a number of lithothermal systems, each with its own regional updoming and resultant axial fracturing (Bailey, 1964) which coalesces or overlaps at the dome margins. The difference between the Afro–Arabian dome and others in Africa is that the former was succeeded by the formation of basaltic oceanic crust in the floor of the ever-widening rifts. This process in turn, could be related to the position of the dome near the margin of the Afro-Arabian continent, so that the lines of weakness were accentuated by ocean-floor spreading already operating from the Carlsberg Ridge in the north-west Indian Ocean (Le Pichon and Hertzler, 1968). The Carlsberg Ridge mechanism combined with the spreading effect of the developing lithothermal system could well explain why the Gulf of Aden is widest to the east. Speculating on the future tectonic development of the area, it is probable that as the separation of Arabia from Africa is virtually complete, oceanic crust will continue to be formed in the Gulf of Aden and the Red Sea and that the Afar depression and the westward continuation of the Gulf of Tadjura will become the site of 'oceanic' tholeiitic vulcanism.

Acknowledgements

In this essay the information used is, perforce, very largely that of others; I have attempted to make this evident by quoting, as fully as possible, the sources of the data. In many cases recent works of synthesis written in English have been consulted rather than the original publications, most commonly presented in Italian. I would therefore stress the invaluable contribution made by Italian geologists whose researches, particularly in Ethiopia and Somalia, are outstanding. I gratefully acknowledge the generosity of Dr J. R. Cann in making unpublished analytical data on the Gulf of Aden basalts available, and that of Professor H. Tazieff and his co-workers, Professor J. G. Schilling, Dr P. A. Mohr and Dr R. L. Chase for data in pre-print form. My colleagues, Drs Dorothy H. Rayner and Tom N. Clifford, are thanked for their constructive criticism of the text, and Mr R. C. Boud for cartographic assistance. Finally, the theme of this essay, an attempt at a regional correlation of basalt types with major structural elements, is 'Kennedian' in its approach and reflects, albeit imperfectly, W. Q. Kennedy's influence as a teacher, friend and colleague.

REFERENCES

BAILEY, D. K. 1964. Crustal warping—a possible tectonic control of alkaline magmatism. *J. geophys. Res.*, **69**, 1103.

BEYDOUN, Z. R. 1964. The stratigraphy and structure of the eastern Aden Protectorate. *Overseas Geol. Mineral Resources*, Suppl. **5**.

BLANFORD, W. T. 1869. On the geology of a portion of Abysinnia. *Proc. geol. Soc. Lond.*, **25**, 401.

BULLARD, E. C. 1936. Gravity measurements in East Africa. *Phil. Trans. r. Soc. Lond.*, **235**, Ser. A, 445.

CANN, J. R., and VINE, F. J. 1966. An area on the crest of the Carlsberg ridge: petrology and magnetic survey. *Phil. Trans. r. Soc. Lond.*, **259**, Ser. A, 198.

CHASE, R. L. 1969. Basalt from the axial trough of the Red Sea. In *Hot brines and recent heavy metal deposits in the Red Sea*. (Eds. E. T. Degens and D. A. Ross), New York.

CLOOS, H. 1939. 'Hebung-Spaltung-Vulcanismus. *Geol. Rdsch.*, **30**, 405.

COMUCCI, P. 1928. Contributo allo studio delle rocce effusive della Dancalia. *Mem. Soc. Toscana. Sci. Nat.*, **39**, 93.

—— 1932. Alcune rocce dei dintorni di Addis Ababa (Abissinia). *Proc. Verb. Soc. Toscana Sci. Nat.*, **41**, 100.

—— 1933a. Note petrografiche sulle rocce raccolte dalla spedizione Cerulli nell' Ethiopia Occidentale. In *Ethiopia Occidentale*, p. 227. (Ed. E. Cerulli), Vol. 2 Rome.

—— 1933b. Rocce dello Iemen raccolte dalla missione de S.E. Gasparini. *Periodico Miner.*, **4**, 89.

—— 1948. *Le rocce della regione di Jubdo (Africa Orientale)*, Acad. naz. Lincei, Roma.

—— 1950. *Le vulcaniti del Lago Tana (Africa orientale)*, Acad. Naz. Lincei, Roma.

COOMBS, D. S. 1963. Trends and affinities of basaltic magmas and pyroxenes as illustrated on the diopside-olivine-silica diagram. *Min. Soc. Am. Spec. Pap.*, **1**, 227.

COX, K. G., GASS, I. G., and MALLICK, D. I. J. 1970. The structural evolution and volcanic history of the Aden and Little Aden volcanoes. *Q.J. geol. Soc. Lond.*, **124**, 283.

DAINELLI, G. 1943. *Geologia dell'Africa Orientale*, Rome.

DICKINSON, D. R., DODSON, M. H., GASS, I. G., and REX, D. C. 1969. Correlation of initial [87]Sr/[86]Sr with Rb/Sr in some late Tertiary volcanic rocks of South Arabia. *Earth Planetary Sci. Let.*, **6**, 84.

DRAKE, C. L., and GIRDLER, R. W. 1964. A geophysical study of the Red Sea. *Geophys. J.*, **8**, 473.

DUPARC, L. 1930. Sur les basaltes et les roches basaltiques du plateau Abyssin. *Schweiz. miner. petrog. Mitt.*, **10**, 1.

ELDER, J. W. 1966. Penetrative convection: its role in volcanism. *Bull. volc.*, **29**, 327.

EL-HINNAWI, E. E. 1964. Petrochemical characters of African volcanic rocks. Part 1. Ethiopia and Red Sea Region (including Yaman and Aden). *Neues. Jb. Miner.*, **3**, 65.

ENGEL, A. E. J., ENGEL, C. G., and HAVENS, R. G. 1965. Chemical characteristics of oceanic basalts and the upper mantle. *Bull. geol. Soc. Am.*, **76**, 719.

FREUND, R. 1965. A model of the structural development of Israel and adjacent areas since Upper Cretaceous times. *Geol. Mag.*, **102**, 189.

GASS, I. G., and GIBSON, I. L. 1969. The structural evolution of the rift zones in the Middle East. *Nature, Lond.*, **221**, 926.

—— and MALLICK, D. I. J. 1968. Jebel Khariz: an Upper Miocene strato-volcano of comenditic affinity on the South Arabian coast. *Bull. volc.*, **32**, 33.

—— MALLICK, D. I. J., and COX, K. G. 1965. The Royal Society volcanological expedition to the South Arabian Federation and the Red Sea. *Nature, Lond.*, **205**, 952.

—— MALLICK, D. I. J., and COX, K. G. (in preparation). The volcanic islands of the Red Sea.

GAST, P. W. 1968. Trace element fractionation and the origin of tholeiitic and alkaline magma types. *Geochim. et cosmochim. Acta*, **32**, 1057.

GIBSON, I. L. 1967. Preliminary account of the volcanic geology of Fant-ale, Shoa. *Bull. geophys. Obs. Addis Ababa*, **10**, 59.

—— (in press). The structure and volcanic geology of an axial portion of the main Ethiopian Rift. *Tectonophys.*

GIRDLER, R. W. 1958. The relationship of the Red Sea to the East African Rift System. *Q. J. geol. Soc. Lond.*, **114**, 79.

HESS, H. H. 1960. Stillwater Igneous Complex, Montana. *Mem. geol. Soc. Am.*, **80**.

—— 1962. History of ocean basins. In *Petrologic studies: a volume in honor of A. F. Buddington*, p. 599. (Eds. A. E. J. Engel, H. L. James, and B. F. Leonard), Geol. Soc. Am.

HIEKE-MERLIN, O. 1950. I basalti dell'Africa Orientale. *Inst. Geol. Miner. Univ. Padova*, **17**, 1.

—— 1953. Le vulcaniti acide dell'Africa Orientale. *Soc. miner. Ital.*, **9**, 135.

HOLWERDA, J. G., and HUTCHINSON, R. W. 1968. Potash-bearing evaporites in the Danakil area, Ethiopia. *Econ. Geol.*, **63**, 124.

KENNEDY, W. Q. 1933. Trends of differentiation in basaltic magmas. *Am. J. Sci.*, **25**, 239.

KNOTT, S. T., BUNCE, E. T., and CHASE, R. L. 1967. Red Sea seismic reflection studies. *Pap. geol. Surv. Can.*, **66-14**, 33.

KUSHIRO, I. 1965. The liquidus relations in the systems forsterite-$CaAl_2SiO_6$-silica and forsterite-nepheline-silica at high pressures. *Yb. Carneg. Instn.*, **64**, 103.

—— 1968. Compositions of magmas formed by partial zone melting of the earth's upper mantle. *J. geophys. Res.*, **73**, 619.

—— and KUNO, H. 1963. Origin of primary basalt magmas and classification of basaltic rocks. *J. Petrology*, **4**, 75.

LAMARE, P. 1930. Les manifestations volcaniques post-Crétacées, de la Mer Rouge et des pays limitrophes. In: Etudes géologiques—Ethiopie, Somalie et Arabie méridionale. *Mem. Soc. géol. Fr.* 6(N.S.), 21.

LAUGHTON, A. S. 1966. The Gulf of Aden. *Phil. Trans. r. Soc. Lond.*, **259**, Ser. A, 150.

—— 1967. The Gulf of Aden, in relation to the Red Sea and the Afar depression of Ethiopia. *Pap. geol. Surv. Can.*, **66-14**, 78.

LE PICHON, X., and HERTZLER, J. R. 1968. Magnetic anomalies in the Indian Ocean and sea-floor spreading. *J. geophys. Res.*, **73**, 2101.

LIPPARINI, T. 1954. Contributi alla conoscenza geologica del Yemen. *Bull. Serv. Geol. Ital.*, **76**, 93.

McBIRNEY, A. R. 1963. Conductivity variations and terrestrial heat-flow distribution. *J. geophys. Res.*, **68**, 6323.

—— 1967. Genetic relations of volcanic rocks of the pacific ocean. *Geol. Rdsch.*, **57**, 21.

—— and GASS, I. G. 1967. Relations of oceanic volcanic rocks to mid-ocean rises and heat flow. *Earth Planetary Sci. Let.*, **2**, 265.

McELHINNY, M. W., BRIDEN, J. C., JONES, D. L., and BROCK, A. 1968. Geological and geophysical implications of paleomagnetic results from Africa. *Rev. Geophys.*, **6**, 201.

MOHR, P. A. 1962. *The geology of Ethiopia*, Addis Ababa.

—— 1963. The Ethiopian Cainozoic lavas—a preliminary study of some trends: spatial, temporal, and chemical. *Bull. geophys. Obs. Addis Ababa*, **3**, 103.

NOCKOLDS, S. R. 1954. Average chemical compositions of some igneous rocks. *Bull. geol. Soc. Am.*, **65**, 1007.

O'HARA, M. J. 1965. Primary magmas and the origin of basalts. *Scott. J. Geol.*, **1**, 19.

OXBURGH, E. R., and TURCOTTE, D. L. 1968. Mid-ocean ridges and geotherm distribution during mantle convection. *J. geophys. Res.*, **73**, 2643.

PRIOR, G. T. 1903. Contributions to the petrology of British East Africa, comparison of volcanics from the Great Rift Valley with rocks from Pantellaria, the Canary Islands, Ascension, St Helena, Aden and Abyssinia. *Miner. Mag.*, **8**, 228.

ROMAN, D. 1926. Studii Petrografice in Yemen (Reguinea-Hodeida-Saana). *Annls. geol. Inst. Rumania*, **11**, 207.

SCHILLING, J. G. (in press). Origin of the Red Sea floor: rare earth evidence. *Bull. volc.*

SHACKLETON, R. M. 1954. The tectonic significance of alkaline igneous activity. In *The tectonic control of igneous activity*, 1st Inter. Univ. geol. Congr., Univ. Leeds, p. 21.

SHUKRI, N. M., and BASTA, E. Z. 1954. Petrography of the alkaline volcanic rocks of Yaman. Egyptian University scientific expedition of S.W. Arabia. *Bull. Inst. Egypt*, **36**, 130.

TAZIEFF, H., MARINELLI, G., BARBERI, F., and VARET, J. (in press). Géologie de l'Afar Septentrional. *Bull. volc.*

VINE, F. J. 1966. Spreading of the ocean floor: new evidence. *Science*, **154**, 1405.

—— and MATTHEWS, D. H. 1963. Magnetic anomalies over ocean ridges. *Nature Lond.*, **199**, 947.

YODER, H. S., and TILLEY, C. E. 1962. Origin of basalt magmas: an experimental study of natural and synthetic systems. *J. Petrology*, **3**, 342.

PLATE 1 and OVERLAY. The mountainous Tibesti Massif from Egheï (Lybia) to the Chad plains. *Abbreviations*: A, Abéki; EK, Emi Koussi; T, Toôn; Ti, Tieroko; TN, Trou Natron; To, Tousside; V, Voôn: and Yé, Yéeza.

Facing page 301

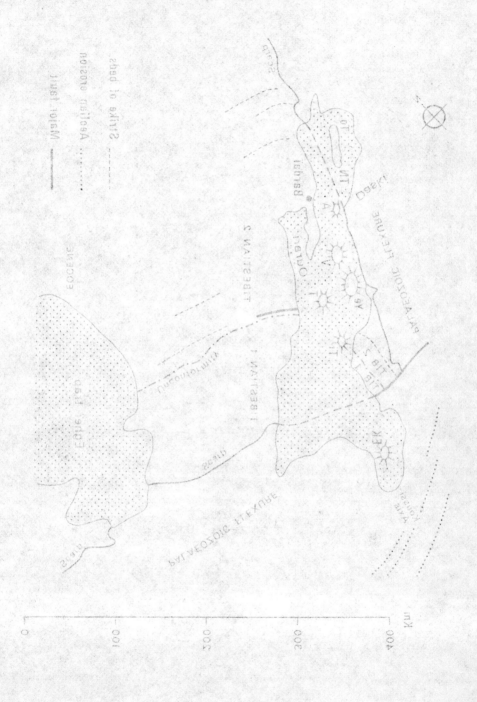

Strike of beds ⟶
Aeolian erosion ⋯⋯⋯
Major fault ⟶

EOCENE

Elgie Tap

TIBESTIAN 2

Bardaï

Ouari

Dagsï

Scarp

TIBESTIAN 1

Unconformity

Scarp

PALAEOZOIC RELATIVE

PALAEOZOIC RELATIVE

Scarp

EK

Koussi
Axis

km

0
100
200
300
400

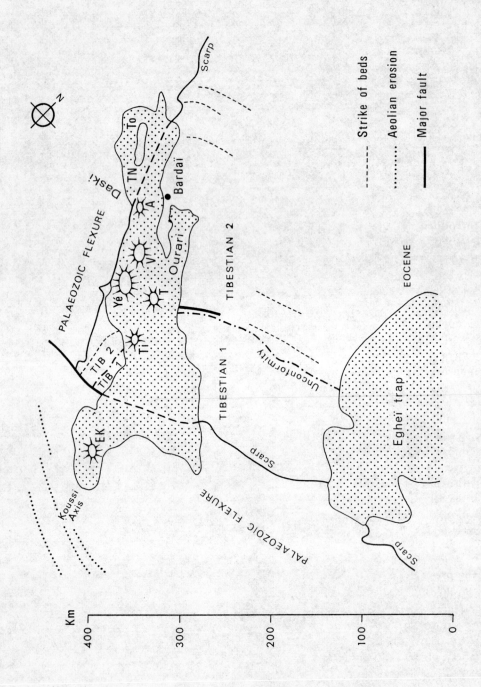

Km

400

300

200

100

0

N

PALAEOZOIC FLEXURE

Daskï

TN

To

A

V

Yé

T

Ti

TIB 2

TIB 1

EK

Koussi
Axis

Ourari

Bardaï

Scarp

TIBESTIAN 2

TIBESTIAN 1

Unconformity

PALAEOZOIC FLEXURE

Scarp

Eghei trap

EOCENE

Scarp

Scarp

– – – – Strike of beds

· · · · · · Aeolian erosion

——— Major fault

P. M. VINCENT

14   The evolution of the Tibesti
Volcanic Province, eastern
Sahara[1]

ABSTRACT. *Vulcanism in the Tibesti Province was characterised by the eruption of vast quantities of basalt and rhyolitic ignimbrite. Overlying plateau basalts are shield volcanoes of basalt and ignimbrite, both having their own calderas. Eruptive processes, the relationship between vulcanism and tectonics, and the development of acid magmas by anatexis, are discussed.*

Introduction

The Tibesti Massif is a mountainous, triangular area of approximately 100,000 km², which lies in the extreme north of the Chad Republic, half-way between Lake Chad and the Mediterranean (see Plate 1). Although the entire massif stands in marked topographic contrast to the surrounding low-lying country, it is in a smaller triangular area, composed of volcanic rocks and occupying an area of about 30,000 km², that the highest peaks (e.g. Emi Koussi, 3415 m) occur. All the high summits are volcanic, but the accumulation of volcanic products is not totally responsible for the topographic eminence of the area as the surface on which the volcanoes lie rises in places to 2000 m.

An account of work undertaken up to 1960, together with a comprehensive bibliography, has been given by Vincent (1963) in a memoir primarily concerned with work carried out since 1954 in west and central Tibesti. A summary of work undertaken in 1957 on the vulcanism of the whole massif is given by Gèze *et al.* (1959); and a geological map on the scale 1 : 1,000,000, with explanatory text, and covering an area of 500,000 km² of the Chad part of the Sahara was published by Wacrenier *et al.* in 1958. This paper reiterates some of the earlier conclusions (Vincent, 1963), but incorporates the results of three subsequent investigations which, until now, have remained unpublished.

The pre-volcanic formations

The Precambrian rocks of Tibesti consist of two tectonically and petrographically distinct groups separated by a major unconformity; these groups are termed the Lower and Upper Tibestian (Fig. 1). The Lower Tibestian occurs only in the north-east and south-east of Tibesti and includes a great variety of mesozonal metamorphic rocks, trending between 25° N and

[1] The editors are very grateful to Dr R. Black and Mrs E. Nutt for translating the original French text of this paper into English.

60° N, which are often migmatised and injected by syn-tectonic granodiorites (Wacrenier, 1956). In contrast, the Upper Tibestian occupies a larger
area and consists of a succession of phyllites and arkosic quartzites resting on a
thick succession of conglomerates in which examples of all the rocks in the
Lower Tibestian occur. This conglomerate lies unconformably on the Lower
Tibestian and the whole Upper Tibestian sequence is weakly metamorphosed.
In western Tibesti the strike of these strata is very regular and close to north-
north-east. After the folding of this sequence, numerous small, discordant,
granitic bodies were intruded to the north of the present volcanic region;
these masses were the precursors of the great Araye Batholith which extends
northwards into Libya.

It would appear that the Tibesti area has acted as a stable block since
Lower Tibestian times for it controlled the sedimentation during the Upper
Tibestian and the subsequent fold patterns.

Palaeozoic formations rest unconformably on a perfectly peneplaned and
weathered Precambrian surface. The stratigraphic succession established by
Bizard et al. (1955) and Bonnet et al. (1955) is applicable to Tibesti. The
predominance of sandstone facies in this sequence reflects the proximity of a
North African shoreline, and work carried out by petroleum geologists, on
the basins around the massif, has confirmed that marine transgression came
from the north or north-west and that the detritus was derived from the
south. Had there been no subsequent tectonic activity, the situation would
have been one of a stable basin gently inclined towards the north, mantled
successively by Palaeozoic and Mesozoic formations.

Only Cambro-Ordovician deposits are present on the periphery of the
Tibesti Massif; they are several hundred metres thick and consist mainly of
fine-grained to conglomeratic, white or fawn-coloured sandstones, often
cross-bedded and with rare intercalations of siltstones and siliceous shales.
A traverse from the plains to the east and west, towards the massif, shows a
rapid change from a very shallow to a steeper dip as the massif is approached;
this flexure is rectilinear (see the overlay of Plate 1).

The Tibesti plateau has the shape of an elongated 'U' open to the north-
east (Fig. 1). Three sides of the 'U' are formed by the escarpments of Cambro-
Ordovician sandstones the inclination of which is particularly spectacular on
the south-west and south-east flanks where the Precambrian of the plateau
reaches its highest elevations. The Palaeozoic is absent in the inner part of
the 'U' and patches of Nubian sandstone, of probable Lower Cretaceous age,
lie directly on the Precambrian. The first upwarp or swell which resulted in
the erosion of thick Palaeozoic formations is therefore thought to be of pre-
Cretaceous age. To the south-east, outside the area of Fig. 1, localised, intense
earth movements of Hercynian age occurred, for Nubian sandstone of a
probable Triassic age overlies folded Upper Devonian and Carboniferous
strata.

Indications of the Libyan Eocene transgression are visible north and
north-west of Aozou (Fig. 1). The deposits are only a few tens of metres thick
and contain fossiliferous limestones, marls and some gypsiferous horizons. It is

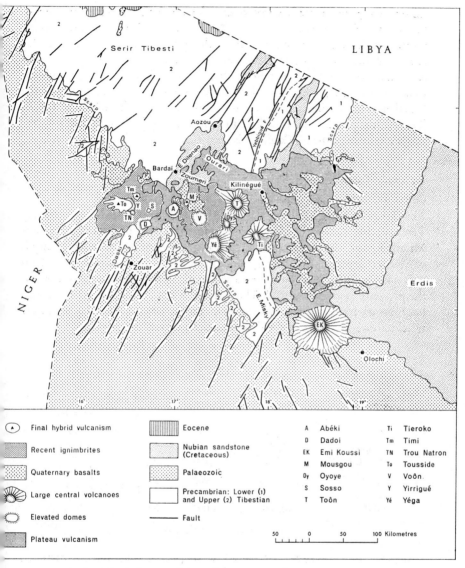

FIG. 1. A geological map of the Tibesti region; after Gèze *et al.* (1959) and Vincent (unpublished work).

believed that this transgression extended southwards as far as the Palaeozoic scarps and that subsequent regression was connected with a slight tilting of the block towards the north-north-east, which has raised these deposits to a height of 500 m. The earliest basalts of Egheï (Plate 1) overlie the Middle Eocene (Lelubre, 1946) and are identical with the earliest basalts of Tibesti.

Summarising the structural history of the Tibesti region before the Tertiary vulcanism it is clear that the basement below the volcanic field is part of a vast Precambrian block, roughly rectangular in shape, which underwent positive movements over a long period. In Palaeozoic times, uplift reached its maximum after the Carboniferous Period, resulting in the erosion of the sedimentary cover. After the Nubian sandstones were deposited, but before the Lutetian, the massif was cut into elongated slices by north-east- and north-north-east-trending faults which follow the Precambrian trends. Subsequent uplift took place along the Emi Koussi axis and, after the Lutetian transgression, the block was tilted towards the north-north-east (Fig. 1). It is pertinent to note here that the centre of the vulcanism is at the steepest corner of the block and it is clear that the two facts are interrelated.

Volcanic formations

INTRODUCTION

The Tibesti Massif has been the site of intense vulcanism since the early Tertiary; the earliest basalts of Egheï, to the north of Tibesti, post-date Middle Eocene marine sediments, and a major part of the volcanic activity is of Quaternary age. The petrographic groups which constitute the Tibesti volcanics are listed in Table 1, and a schematic stratigraphic cross-section for the province is given in Fig. 2. The surface area of outcrop given in Table 1 has been accurately calculated from maps on the scale of 1 : 200,000, but the usual uncertainty arises in the determinations of average thickness.

TABLE 1

Volcanic series*	Surface area in km^2	Volume in km^3	Volume %	Average value for SiO$_2$%
SH	380	38	1·2	58·6
SN3–SN4	1,400	25	0·8	46·5
SC (I, II, III)	14,730	1,150	37	72
SN2	4,500	900	29 {25 / 4}	{50·2 / 58·4}
SN1	10,000	1,000	32	46·2

Total 3,100 approx.

* SN, Black Series (basalts, andesites); SC, Pale Series (ignimbrites, rhyolites); and SH, Hybrid Series (trachyandesites predominant). See Fig. 2 for stratigraphic sequence of rock-types.

In the older part of the volcanic sequence (Fig. 2), rocks of intermediate composition are absent; rhyolites (Formation SC I–III) and basalt-andesites (Formations SN1 and SN2) forming 37% and 61% respectively of the total volume of the rocks erupted (Fig. 3a). Moreover the histogram for the basalt-andesite formations shows two peaks corresponding to two distinct associations and it must be stressed that, had the curve been drawn for the whole of Tibesti and for its Libyan continuation of the Egheï (Plate 1), the SN1 peak would have been even more dominant, for the SN2 association is only known in the region of Emi Koussi (Fig. 1). A histogram of all recent volcanic products (SN3, SN4 and SH) including all Quaternary activity but excluding the ignimbrites of Tousside, shows the presence of basic rocks with low silica content and also of trachyandesites (Fig. 3b), but these rock types represent only 2% of the whole.

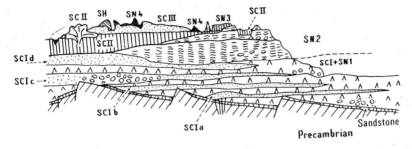

FIG. 2. A schematic section of the Tibesti volcanic formations. SN, Black Series (basalts and andesites): SN1, plateau vulcanism; SN2, Hawaiian shield vulcanism; SN3 and SN4, Recent vulcanism. SC, Pale Series (ignimbrites and rhyolites): SCI, II and III, lower, middle and upper series respectively. SH, Final Hybrid Series (trachyandesites dominant).

THE PLATEAU VULCANISM: THE COMPOSITE TRAPS

Under this heading have been grouped the 'première série noire' (SN1) or the First Black Series, its differentiation products and the early ignimbrites (SCI) (see Table 1 and Fig. 2). The First Black Series was first studied in Ourari (Fig. 1) where it is essentially basaltic, whereas the early ignimbrites

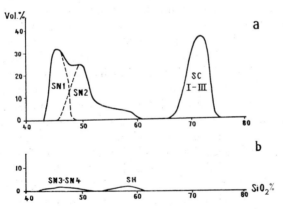

FIG. 3. Frequency curves for: (a) vulcanism up to, and including, SCIII; and (b) more recent vulcanism (see Fig. 2). The ordinate shows the relative volumes.

of SCI were examined in the south where these rocks predominate. Recent mapping has shown the succession is much more complex to the east where three ignimbrite sheets are seen intercalated in the basaltic succession.

The First Black Series is composed essentially of flood basalts with olivine and titaniferous augite in varying abundance. Chemically these rocks contain

L

normative nepheline but this mineral has not been observed as a modal constituent. A small differentiated alkaline suite of little volumetric importance plays an important morphological role forming spectacular plugs of alkaline trachyte, trachy-phonolite and phonolite; these have been called vulcanism of 'Atakor type'. The whole series belongs to the alkaline olivine basalt association of Kennedy (1938). Using the Jung and Brousse diagrams (1958) the series can be classified as a weakly sodic association (Vincent, 1963). The rhyolites are similar to those of other ignimbrite sheets of Tibesti in that they are alkaline or peralkaline, with sodic metasilicates and characterised by the abundance of sanidine phenocrysts; silica is present mostly in the form of tridymite.

THE LARGE CENTRAL VOLCANOES

This general group includes: (a) shield volcanoes of Hawaiian type and the acid vulcanism of the calderas; (b) large elevated domes; and (c) some less-voluminous rhyolitic volcanoes which are generally in the form of 'cumulo-domes'. These centres are all older than the Voon ignimbrite (SCIII) (see Figs. 1 and 2).

One of the characteristic features of the central part of the Tibesti Massif is the presence of *Hawaiian type shield volcanoes*. Four of these occur together and are still easily recognisable despite erosion; they are Tieroko, Toôn, Oyoye and Yéga (Fig. 1). The last three form a marked alignment of 100 km along the prolongation of the north-north-east-trending Yebbigué Fault (Fig. 1). All these volcanoes are morphologically very similar, with diameters of between 40 and 60 km, slopes of low angle between 10° and 15° and slightly convex profiles. The lavas are olivine-poor basalts, pyroxene andesites and some rare dacites without notable quartz. The dominant rock type is a labradorite-rich porphyritic basalt, in which the labradorite phenocrysts attain lengths of 20 mm, the rock also contains glass, olivine and Ca-poor augite. The ore minerals magnetite and ilmenite are abundant. These lavas represent the Second Black Series (SN2) and belong entirely to the tholeiitic volcanic association, being characterised by the presence of normative hypersthene, usually with diopside and quartz. One of the fundamental characteristics of this association is its potassic tendency for the average K_2O is 1·59%, and this corresponds to 10% normative orthoclase. The norm for an average labradorite-rich basalt calculated from 14 analyses shows 5% quartz. This high value is due in part to a high Fe_2O_3/FeO ratio but it must be noted that this oxidation is probably primary for magnetite is very abundant in well-developed crystals; this is a constant feature of the porphyritic varieties and must be of genetic significance. Associated acid vulcanism has produced alkaline rhyolites poor in ferromagnesian minerals.

Two *large elevated domes* are known in western Tibesti, namely Abéki and Dadoi (Fig. 1); Bounai in central Tibesti is another but is, as yet, largely unstudied. The formation of Abéki started before the emission of the Second Black Series (SN2) in the area and continued thereafter. The lavas are composed entirely of alkaline or peralkaline rhyolites, and blocks of quartz syenite are present in the associated volcanic breccias.

In the isolated *cumulo-domes* the viscous lava has spread laterally around the point of emission; the best example of this is at Sosso (Fig. 1). The rhyolites are vitreous at the surface and granophyric in depth and always poor in ferromagnesian minerals.

RECENT IGNIMBRITIC VOLCANOES

These are extremely flattened cupolas centred on vast calderas which, for the want of a better word, have been named 'shield sheets' ('boucliers-nappes'). The shield sheet (SCIIIa) of Voôn (Fig. 1) is younger than the erosion of the Hawaiian type volcanoes and pre-dates the period of valley incision which occurred approximately at the beginning of the Quaternary. This sheet covers 2600 km² and has an approximate volume of 130 km². The shield sheet of Yirrigué (SCIIIb) is Quaternary, and extends into all peripheral valleys where it forms 'sillars'. The total area covered is 3200 km² and the approximate volume is 150 km³.

As in the case of older ignimbrites, these rhyolites are characterised by a preponderance of peralkaline varieties such as comendites and pantellerites. The abundance of alkalies is less characteristic than the low alumina and the high iron content, with 3-6% total iron and a clear preponderance of Fe_2O_3 except in the obsidians.

THE FINAL VULCANISM

The final phase of volcanic activity represents only 2% of the whole, and consists mainly of basalts (SN3 and SN4) and hybrid volcanics (SH), largely trachyandesites. Of these, the Recent basalts belong to an alkali association probably similar to the initial association (SN1) but poorer in silica and more potassic. They are represented by numerous small volcanoes of Strombolian type.

The hybrid vulcanism (SH) is spatially connected with the ignimbrite calderas, and has produced several large cones more than 3000 m high; for example, Mousgou, Timi, Tousside (Fig. 1). Of these, Tousside is the only cone in the Tibesti area that can be considered as active, and even that is in a dormant phase. These hybrid volcanoes present several characteristics not previously encountered, for potassic trachyandesites, often with rhombic anorthoclase, were erupted and a reversed evolutionary series consisting of a Pelé dome of sancyite overlain by more fluid doreites, then by andesites, is present at Tousside. Activity at the Trou Natron crater (Fig. 1) has thrown out an initial stage of rhombporphyry syenites and microsyenites. It seems clear that a shallow and strongly differentiated magma chamber emptied itself after crystallisation had started near the roof. A study of this potassic vulcanism suggests that it is a hybrid association, which is genetically comparable to that of Mont-Dore where vulcanism is connected with the collapse of the Bourboule graben (Glangeaud, 1943, 1946).

COMPARISON WITH OTHER PROVINCES

Tibesti has much in common with the Tertiary Scottish Hebridean Province in: (*a*) its sequence of eruptive mechanisms, characterised by plateau basalts followed by basic central volcanoes, then acid vulcanism with caldera formation; and (*b*) the type of volcanic association, namely alkali basalts followed by tholeiites and then by an independent acidic association. The equivalent of those rhyolites of Tibesti, which are independent of the central basic volcanoes, are however unknown in Scotland and are of lesser importance in Iceland.

In Ethiopia and neighbouring regions, there appears to be a similar succession on a larger scale. In the Cameroons two basic associations are known, in the same order as at Tibesti (Gèze, 1943; Jérémine, 1943), and it is interesting to note that the tholeiitic association in the Cameroons has a potassic tendency and is also related to central volcanic activity and caldera formation. Furthermore, in the Cameroons, facies with mainly porphyritic plagioclase are characteristic. Although the connection between these phenomena is not yet understood, it can reasonably be said that this is not a chance association; both the accumulation of plagioclase phenocrysts and the formation of calderas could be connected with the presence of a secondary chamber at shallow depth.

Eruptive processes

The *flood basalts* form two parallel outcrops running north-west to south-east, separated by the Zoumeri Valley in which the Nubian sandstones crop out (Fig. 1). In that region, the northern outcrop is particularly suitable for a study of eruptive processes as it is 20–30 km wide and stretches for over 100 km before disappearing under more recent formations to the south-east. The maximum thickness of more than 300 m of volcanic rocks can be observed along the axis of the trap which corresponds to the drainage divide. Parallel to the axis of the divide, the sections expose a sequence of extensive horizontally-disposed flows typical of plateau basalt accumulations elsewhere. However, across the divide axis, the section is reminiscent of an Hawaiian shield volcano of comparable dimensions, being gently convex upwards. These flows were fed through a system of parallel dykes, which cut across the volcanic sequence and the underlying Nubian sandstones (Fig. 4). Observations made in the extreme north-west, at E. Dilenao, indicate that the dyke trend is parallel to the axis of the divide and that the dykes dip at angles between 70° and 90°. The dykes are usually basaltic or doleritic and between 0·5 m and 2·5 m thick; rare trachyte and phonolite dykes up to 8 m thick are present. There has been no mechanical displacement along the dykes and, conversely, north-north-east- or north-east-trending faults in this region never show any sign of injected volcanic products. It seems therefore that the thickening of the basalt on the trap axis corresponds to the injection zone, the dykes being situated in the tension joints in the convex part of a flexure (Fig. 4). The numerous trachyte and phonolite outcrops observed on the plateau are most densely distributed along the axis and are also fed by dykes.

In the extreme north-west of Ourari the morphology shows that the basalts of the southern flank are more recent, but it is not possible to date

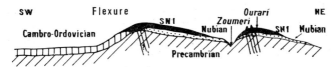

FIG. 4. A section across the traps in the Ourari–Zoumeri region; the distance between the two flexures is 70 km. The Ourari dyke swarm is observed; that adjacent to the flexure affecting the Cambro-Ordovician rocks to the south-west is inferred.

them more definitely. To the east, however, owing to the interstratification of the ignimbrite sheets, accompanied sometimes by fossiliferous volcano-lacustrine deposits, a basalt chronology can be established and the following points are worthy of note (Fig. 5): (a) the disposition of ignimbrites or of

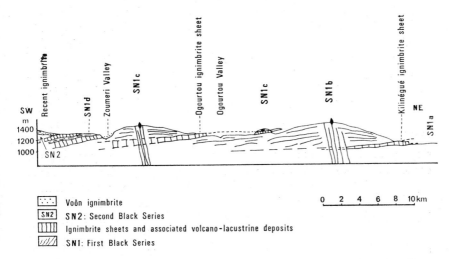

FIG. 5. Schematic section across the Ourari trap to illustrate the south-westward migration of the feeder zone with time.

lacustrine formations shows that at least on the south-west flanks, the convexity or thickening is original and not erosional; (b) two stages of thickening clearly correspond to two feeder zones; and (c) towards the south-west the basalts become more and more recent and there has been a migration of the emission zone in that direction with time. These conclusions can be extrapolated for the whole of Tibesti.

A relationship between the dyke system and the tension joints, which are open along the convex part of the flexured substratum, has been accepted and a comparison has been made with the great dyke system associated with the flexure on the east coast of Greenland, described by Wager and Deer in 1938 (Vincent, 1963). Originally, it was thought that the Tibesti flexure was earlier than the vulcanism; however, more recent data suggest that the flexure was contemporaneous with the vulcanicity and that both have migrated from north-east to south-west as a wave of flexuring and vulcanism.

Although not as volumetrically significant as the plateau basalts, vulcanism of the 'Atakor type' (see p. 306) is also fed by dykes. Generally, this phase of activity produced columnar piston-like protrusions which forced their way through and locally deformed the older strata; occasionally elongate vents overlying dyke lineaments were produced. Vents of the Atakor type are most numerous along the axes of the volcanic traps and in places their abundance can be used to locate the axis where the basalts have been entirely eroded away. However, there is also an alignment of these vents along the direction of the old faults, which suggests that the magma made use of the easiest access routes, at the intersections of faults. The flexures therefore seem to have played an active role in the localisation of lava emission whereas the fault system has had an essentially passive role.

The outpouring of flood basalts was followed by the formation of the *Hawaiian type volcanoes* and, thereafter, volcanic activity continued in two different directions referred to here as the Oyoye type and the Toôn type. In the Oyoye type the summit of the volcano has been replaced by an irregular, star-shaped depression measuring 3×2 km, whose centre is marked by a massif with radiating vertical dykes. These massifs are composed of pneumatolised rhyolites in the form of extrusions which may have been injected along the original vent. On the flanks of the volcano, the dip of the beds increases as the central depression is approached and this seems to reflect uplift which has been limited to the central part of the volcano. There is no ring-fault in the summit area; the depression has been caused by preferential erosion in the uplifted vent. The Tieroko Volcano (Fig. 1) is comparable to Oyoye but is larger, having a central depression that measures 6×9 km.

The Toôn volcano type is characterised by the presence of a caldera. At Toôn itself (Fig. 1) this forms an elliptical depression of 11×9 km. The flanks of the volcano are much dissected, being cut into triangular lava fields that are often flanked by radial dykes. The caldera has been preserved from erosion by the presence of resistant cone-sheets, and is only breached at one point. The interior of the depression is entirely occupied by the products of rhyolitic vulcanism that did not flow out of the caldera.

Observations made on the western flank of this volcano are shown in Fig. 6 and can be summarised as follows: (a) the radial dykes are earlier than the cone-sheets and the collapse of the caldera; (b) the boundary fault is strictly parallel to the cone-sheets and, like them, is inclined inwards; (c) in the interior of the caldera the volcanic products form several sequences of breccias, tuffs and lava flows whose dip increases towards the edge of the caldera, reaching 60° at one point; (d) the last volcanic phase in the caldera, which is terminated by a thick layer of rhyolites, has a much shallower dip and lies unconformably on the earlier rocks; (*e*) the position of the sills and the disposition of the final lava flows seem to indicate that the boundary fault was a feeder channel; and (*f*) no basic vulcanism occurred after the

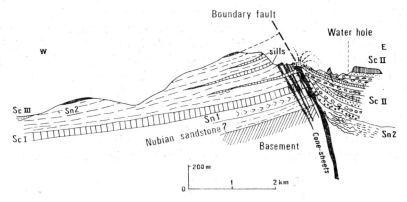

FIG. 6. Section across the western flank of the Toôn Volcano.

collapse of the caldera. Of these features, the feeder role played by the boundary fault is particularly well seen in the Yéga caldera.

This sequence of events suggests that there were two stages in the formation of the caldera (Vincent, 1960, 1963): an initial stage of doming and the formation of a pre-caldera; and a volcanic phase within the caldera with the effusion of rhyolitic products. During the initial stage, radial fractures formed,

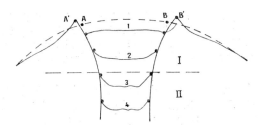

FIG. 7. Circular fault of blunderbuss-form ('en tromblon'). A′ and B′ are the positions of points A and B respectively after the initial upward doming which produces the subsidence of the pre-caldera; I is the level of calderas, and II is the level of ring-complexes.

as at Oyoye, and were followed by the development of concentric circular fractures into which rhyolitic cone-sheets were injected; the boundary fault formed part of the same system of fractures. In the Tibesti calderas the lavas within the caldera are upwarped towards the periphery and this indicates that the diameter of the opening decreases downwards. If the dip of the circular fracture increases until it becomes vertical at depth, the volcanic strata which were deposited in the upper part of the caldera 'funnel' would develop a disposition rather like a 'pile of saucers' on subsidence into the cylinder (Fig. 7). On this interpretation, the space form of the caldera boundary fault is comparable to the shape of a blunderbuss ('*en tromblon*').

Anderson (1936) has shown that at the top of a magma chamber, that has a magma pressure greater than its surroundings, a system of conical fractures develops which is analogous to tensional fractures; the dip of these fractures increases upwards. Whilst retaining Anderson's dynamic interpretation, it is still theoretically possible to obtain the 'blunderbuss' fracture-form if the walls of the magma chamber, and thus the isobaric surfaces, were steeply inclined (see Fig. 8). According to this interpretation, subsidence can be explained, without invoking an eruption, by the distension of the roof following the doming-up of the volcano. Such a mechanism of formation of pre-calderas is thus analogous to Cloos' (1939) experiments on the development of graben by upward pressure on a layer of moist clay. Moreover, the

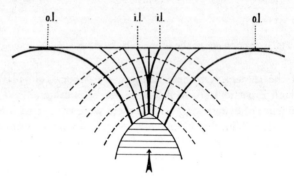

FIG. 8. Swarm of steeply dipping cone-sheets displaying blunderbuss form; o.l, outer limit representing the points at which the cone-sheets are tangential; i.l, inner limits where cone-sheets show a minimum divergence.

interpretation also serves to explain cone-sheets in the Toôn type of caldera, particularly their absence in the central part of the subsided caldera and the clearly-defined outer limit of their occurrence (Fig. 8).

In the classic theory for the formation of a caldera (Williams, 1941), the boundary fault has an outward dip at depth; consequently the magma has no difficulty in rising to the surface. However, if the fault is conical and closing downwards, the central block will seal the vent by its own weight. In the Toôn caldera, there is an alternating sequence of pyroclastic horizons and lava flows; the vent area must have opened and subsided several times and this would explain the observed unconformity (Fig. 6). After a lapse of time, during which gas pressure built up, the vent opened. In contrast to the initial stage of pre-caldera development, a zone of weakness already existed in the form of the boundary fault. No new cone-sheets were therefore formed; the energy was used to raise the 'stopper' and probably to distend the supporting walls which, in turn, facilitated further subsidence after each eruption. Whatever the mechanism, the volcanic activity was rhythmic, and the eruption seems to have been controlled by the rate of build-up of gas pressure and the weight of the 'stopper' which acted as a valve; this system was self-regulating owing to the funnel-form of the boundary fault.

The *large elevated domes* are complex structures which are characterised by

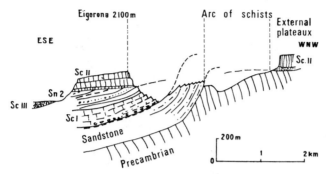

FIG. 9. South-eastern flank of the Abéki Volcano.

an association of extrusive and intrusive events. The structure of Abéki has been studied in most detail; it is 40 km wide with a central block forming positive relief of 1200 m over an area of 11×17 km (Fig. 1). The centre is separated from the flanks by a half circle of uplifted Precambrian schists (Fig. 9).

Only the principal stages in the formation of these domes will be discussed here; for further details and interpretation the reader is referred to

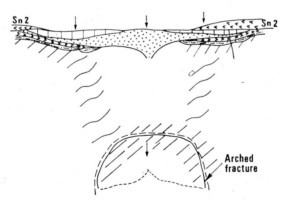

FIG. 10. Subsidence of the early cone at Abéki, and the formation of an arched tensional fracture.

Vincent (1963). The probable stages in the development were: (*a*) formation of a vast pyroclastic cone of debris containing blocks of syenite and peralkaline rhyolites whose roundness suggests fluidisation (c.f. Reynolds, 1954); (*b*) part of the cone collapsed (Fig. 10) and lavas flowed into the basin structure so formed; (*c*) uplift of the central part and asymmetric flexing of the flanks affected all earlier formations including the pre-volcanic sandstones

and Precambrian schists; and (d) emission of the upper series (SCII) produced
the great lava plateaux of the central part, and the unconformable lava
plateaux of the flanks (Fig. 11).

  Ignimbrites represent 700 km³ in the western and central parts of Tibesti

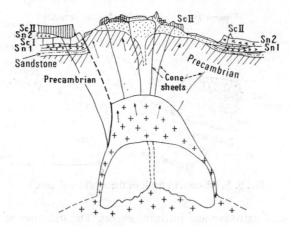

FIG. 11. The present erosional level of the Abéki Volcano.

and a detailed study of these rocks has been given in Vincent (1960, 1963).
Only the essential characteristics are described herein so that the volcanic-
structural relationships can be understood. From the genetic point of view
there are two fundamental types of ignimbrite: type 1 shows a continuous
vitreous or micropumiceous structure and is the product of a froth or foam
lava; and type 2 has a discontinuous clastic fabric suggesting that gas was the

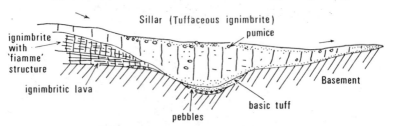

FIG. 12. Semi-schematic section showing the lateral passage of a plateau
ignimbrite sheet into a valley flow of tuffs (the valley runs perpendicular to
the section). To the right the sillar flows up a reverse slope.

supporting medium—this is a vitroclastic ignimbrite or the welded tuff of
English authors. The eruptive sequence shows that type 1 formed first and
evolved towards type 2; this change is progressive and occurs from the
centre of the sheet outwards in both vertical and lateral directions. Both
types may be channelled into a pre-existing valley (Fig. 12) and so produce a
sillar; these 'rootless' bodies can occur at some 10 km from the original point
of eruption.

  A study of the fiamme-matrix relationships indicates that the expansion

of the gases continued in the ignimbrite lavas even after complete immo-
bilisation of the sheet. Conversely, the fragmentation that resulted in the
production of vitroclastic ignimbrites may have been operative from the
time of emission.

The surface of the ignimbrite sheets is usually flat and sub-horizontal.
The two most recent units, SCIIIa and SCIIIb, and particularly the latter,
resemble vast, flattened, slightly convex cones 100 km in diameter and with
slopes ranging between 2° and 3°. At the summit of these cones there is a vast
collapse caldera which is circular at Yirrigué (SCIIIb), measuring 14 × 13 km,
and roughly rectangular at Voôn (SCIIIa) where it is 18 × 11 km.

Morphologically, the ignimbrite sheets, with their summit calderas, are
not unlike the Hawaiian shield volcanoes; however, they are quite distinct in
the following characteristics: (a) their height (about 1000 m) bears no relationship to the thickness of eruptive products which are only 100 m thick at
the centre of the cone—this can only be explained by doming of the basement
prior to the emission, a phenomena verified at Voôn where the ignimbrites
rest unconformably on earlier deformed volcanic products; (b) the collapse of
the caldera terminated the main vulcanism, the caldera bottom remaining flat;
and (c) parasitic hybrid or basic vulcanism developed subsequently along the
caldera boundary fault or outside it.

The injection zone of an ignimbrite is only known for the oldest sheet
(SCIIIa) for it is only there that erosion has been adequate. It consists of a
radiating swarm of some hundred dykes oriented dominantly in two directions which intersect at 75°. These dykes have invaded an ellipsoidal area of
some 11 × 8 km, a surface area less than that of the caldera. The dykes are
between 3 m and 15 m thick and are of alkaline and peralkaline rhyolites and
quartz syenites. Ignimbrites are absent in this injection zone suggesting that
the ignimbrite character developed at the time of emission or much later.
Several plugs occur in the central part of the structure and are probably due
to the injection of degassed epimagma at the end of the volcanic cycle.

The Tibesti ignimbrite vulcanism belongs to a particular central type
which has some unexpected analogies with the Hawaiian vulcanism. All its
products formed from a pyromagma very rich in gas, a 'living lava' which
continued its evolution after eruption. The considerable difference in behaviour compared with other rhyolitic manifestations, shows the importance
of magma storage and evolution at depth. It is suggested that a chamber of
several cubic kilometers existed at moderate depth and that about 40% of
the magma crystallised to produce phenocrysts; high vapour pressure
resulted, and the gases were emulsified resulting in an increase in volume. For
this chamber to remain a closed system it must have had an impermeable
and resistant roof which was sufficiently ductile to allow the elastic deformation, mainly doming, that took place before the eruption.

So far as can be ascertained, the Tibesti ignimbrites never occur along
major faults. They seem to lie along the flexures which previously fed the
flood basalts, and a single centre characterises each flexure: Kilinégué on the
axis of the Ourari trap; Voôn in the axis of central Tibesti; and Yirrigué on

the axis of the Quaternary flexure affecting the Palaeozoic sandstones (Fig. 1). Nevertheless the Voôn and Tousside ignimbrites clearly post-date the plateau basalts and it is suggested that the emplacement or formation of the magma in the chamber took place at the time of flexuring, and that the observed time-lag between flexuring and the emission of these acidic rocks corresponded to the period during which the magma built up sufficient vapour pressure to produce an ignimbritic eruption. The flexures would therefore appear to have played an active role in the control of ignimbrite vulcanism.

Controls of the vulcanism

The association of two basaltic magmas in one area has long been a subject of controversy. Although Tibesti has not yet been studied in sufficient detail, it is known that the alkaline association is the older and has a larger geographical extent than the tholeiite association. Each magma type corresponds to a different mode of extrusion, either fissure or central vent type, with different tectonic controls, either north-west-trending flexures or north-north-east-trending faults respectively.

The existence of two basalt types is most satisfactorily explained by invoking the presence of two independent magmas, and a hypothesis involving differences in the depth of fusion is relevant (Kennedy, 1938). It is suggested that during the course of its upward movement, the alkaline magma has played a thermal role in the generation of tholeiitic magma, which would thus have originated at a shallower depth.

Until the late Quaternary the Tibesti Province was characterised by a coexistence of these basaltic rocks with rhyolitic vulcanism and it has been argued that these two contrasted types resulted from independent sources (Vincent, 1960, 1963). The reasons for this suggestion are: the absence of intermediate types; the volume of rhyolites is too great to have resulted from differentiation of a basaltic magma; the alkali rhyolites are not the differentiation products that would have been produced from the basalts of the area (trachytes or phonolites from basalts of SN1 and calc-alkaline dacites from SN2); and finally, basalts occur without rhyolites outside Tibesti. It is therefore necessary that the fundamental controls of the basic and acid magmas be discussed separately. The alkali basalt will be considered first as this is the earliest eruptive.

Unlike later series, the First Black Series (SN1) extends far beyond Tibesti to Egheï and the Haroudj, where it occupies areas greater than those in Tibesti. Chemical analyses are few, but petrographic descriptions show that these rocks are similar, all being alkaline basalts. The basaltic magma of this initial vulcanism reached the surface via open tension joints in the flexure axes of curved crustal blocks.

Rhyolitic volcanic rocks are present only in the interior of Tibesti and their presence seems to be related to two features: localisation on a zone of maximum uplift; and the presence of an earlier basaltic vulcanism. Whilst it is impossible to assess the relative importance of these two factors in the generation of rhyolitic volcanic rocks, it is clear that neither factor is sufficient by itself. For example, there are no rhyolites associated with the immense basaltic traps of the Libyan Plain and none are known in Tibesti earlier than the first basalts, despite the quasi-permanent presence, since Hercynian movements, of a swell. Therefore, the coupling of these two factors seems to be essential.

Norms of the Tibesti rhyolites show little variation except in the proportion of ferromagnesian constituents, which range from 2% to more than 20% of pyroxene in the pantellerites. If one plots the proportions of salic constituents on Tuttle and Bowen's (1958) diagram for SiO_2-$NaAlSi_3O_8$-$KAlSi_3O_8$-H_2O, established for 2000 atmospheres, the majority of points are grouped near the ternary minimum (Vincent, 1963, p. 212). The salic part of these rhyolites thus corresponds to the most readily melted portion of that system, and therefore to the first liquids produced by differential anatectic fusion (Wyart and Sabatier, 1959; Winkler, 1960). The combination of uplift and basaltic magmatism that appear to control the production of rhyolitic magma must have led to the production of temperatures in the range 700° to 750° which are necessary for anatexis. With respect to the role of upward crustal warping, it has been noted earlier that warping reached its maximum before the Middle Eocene and before the first volcanic effusions, but that uplift had occurred several times at some places since Hercynian times. Successive isostatic readjustments of this type must have been accompanied by a thickening of the crust, the basal part of which must have reached temperatures high enough for anatexis; this is the hypothesis suggested to explain the Mont-Dore vulcanism (Glangeaud, 1943). More recently Bailey (1964), referring to the African Rift Valley vulcanism, has shown the possible relation between the accretion of light material at the base of the crust and the generation of alkaline magma (trachy-phonolite) by partial fusion and by transportation of fugitive constituents from the mantle. Moreover, he supports the view that relief of pressure, connected with upward crustal warping, is an important factor in magma genesis. The sialic magmas of Tibesti are oversaturated and therefore probably of shallower origin, but pressure relief could be a factor which has contributed to their formation.

The initial basaltic magma is not therefore considered to be the parent of the rhyolites. Its role seems to have been entirely thermal, providing the energy necessary for the already reheated sialic rocks to attain the point of anatectic fusion. From the thermodynamic point of view this is quite possible (Smith, 1963; Glangeaud and Letolle, 1965). Geological relationships raise no objections to this idea for the initial basalts and the first ignimbrites are controlled by the same flexures. The absence of hybridisation between the two magmas must be explained by differences in the time of their eruption and by the migration of the feeder channels related to the migration of the flexure.

In addition to these suggested relationships, it is possible that faulting associated with uplift may have been the locus of important thermal anomalies and could, therefore, have contributed to anatexis. At Adamaoua, in the

Cameroons, Cretaceous formations have been affected by thermo-metamorphism of the 'Pyreneen' type in the neighbourhood of faults (Vincent,
1968). In the Basin and Range Province of the western United States numerous
K-Ar ages show that thermo-metamorphism was approximately contemporaneous with uplift (Mauger *et al.*, 1968). Without offering an explanation
it is interesting to note that the nature of faulting in both cases is similar to
that of Tibesti.

Once the basaltic and rhyolitic magmas had been formed, faulting continued to play a very critical role. In particular, in the acid vulcanism, faults
or the absence of them controlled the accumulation and utilisation of the
gases, and the boundary fault exercised an important influence in the
rhythmic vulcanism of the 'blunderbuss' calderas. Moreover, the localisation
of certain emission points of the vulcanism of 'Atakor type' is also related to
the faults. Finally, the generation of tholeiitic magma obeyed a similar
control and all centres of oversaturated basaltic activity are situated along
major faults.

Acknowledgements

Figs. 3, 4, 6, 7 and 9-12 are from Memoir 23 of the Bureau de Recherches
Géologiques et Minières (1963) and are reproduced with the permission
of the B.R.G.M. The Gemini photograph (Plate 1) was provided by the
National Space Science Data Centre of N.A.S.A. and is reproduced with
their kind permission.

REFERENCES

ANDERSON, E. M. 1936. The dynamics of formation of cone-sheets, ring-dykes and
    caldron-subsidences. *Proc. r. Soc. Edinb.*, **56**, 128.
BAILEY, D. K. 1964. Crustal warping—a possible tectonic control of alkaline magmatism.
    J. geophys. Res., **69**, 1103.
BIZARD, CH., BONNET, A., FREULON, J. M., GERARD, G., DE LAPPARENT, A., VINCENT,
    P. M., and WACRENIER, P. 1955. La série géologique entre le Djado et le Tibesti
    (Sahara oriental). *C. R. Acad. Sci. Paris*, **241**, 1320.
BONNET, A., FREULON, J. M., DE LAPPARENT, A., and VINCENT, P. M. 1955. Observations
    géologique sur l'Ennedi, le Mourdi et les Erdis (Territoire du Tchad, A.E.F.). *C.R.
    Acad. Sci. Paris*, **241**, 1403.
CLOOS, H. 1939. Hebung-Spaltung-Vulkanismus. *Geol. Rdsch.*, **30**, 405.
GEZE, B. 1943. Géographie physique et géologie du Cameroun occidental. *Mém. Mus.
    Hist. nat.*, **17**.
—— HUDELEY, H., VINCENT, P. M., and WACRENIER, P. 1959. Les volcans du Tibesti
    (Sahara du Tchad). *Bull. volc.*, **22**, 135.
GLANGEAUD, L. 1943. Evolution des magmas du massif volcanique du Mont-Dore. *Bull.
    Soc. géol. Fr.*, **13**, Ser. 5, 429.
—— 1946. Introduction à l'étude thermodynamique de la pétrogenèse profonde. *Bull.
    Soc. géol. Fr.*, **16**, Ser. 5, 563.
—— and LETOLLE, R. 1965. La théorie des deux magmas fondamentaux. *Geol. Rdsch.*, **55**,
    316.
JEREMINE, E. 1943. Contribution à l'étude pétrographique du Cameroun occidental.
    Mém. Mus. Hist. nat., **17**.

JUNG, J., and BROUSSE, R. 1958. Précisions nouvelles sur la constitution et sur l'origine des associations volcaniques. *Bull. Soc. fr. Minér. Crist.*, **81**, 133.

KENNEDY, W. Q. 1938. Crustal layers and the origin of magmas: petrological aspects of the problem. *Bull. volc.*, **3**, 23.

LELUBRE, M. 1946. Le Tibesti septentrional. Esquisse morphologique et structurale. *C.R. Acad. Sci. colon.*, *Paris*, **6**, 337.

MAUGER, R. L., DAMON, P. E., and LIVINGSTONE, D. E. 1968. Cenozoic argon ages on metamorphic rocks from the Basin and Range Province. *Am. J. Sci.*, **266**, 7.

REYNOLDS, D. L. 1954. Fluidization as a geological process. *Am. J. Sci.*, **252**, 577.

SMITH, R. G. 1963. *Physical geochemistry*. Reading, Mass.

TUTTLE, O. F., and BOWEN, N. L. 1958. Origin of granite in the light of experimental studies in the system $NaAlSi_3O_8$—$KAlSi_3O_8$—SiO_2—H_2O. *Mem. geol. Soc. Am.*, **74**.

VINCENT, P. M. 1960. Dynamisme et structures des volcans rhyolitiques du Tibesti (Sahara du Tchad). *Rev. Géogr. phys. Géol. dynam.*, **3** (2), 229.

—— 1963. Les volcans tertiaires et quaternaires du Tibesti occidental et central (Sahara du Tchad). *Mém. Bur. Rech. géol. min.*, **23**.

—— 1968. Attribution au Crétacé de conglomérats métamorphiques de l'Adamaoua (Cameroun). *Annls. Fac. Sci. Yaoundé*, **1**.

WACRENIER, P. 1956. Aperçu sur l'antécambrien du Tibesti (A.E.F.). *Int. geol. Congr.*, **20**, *Asoc. Serv. geol. Afr.*, 281.

—— HUDELEY, H., and VINCENT, P. M. 1958. Notice explicative de la carte géologique provisoire du Borkou-Ennedi-Tibesti au 1: 1,000,000. *Dir. Mines Géol. A.E.F.*

WAGER, L. R., and DEER, W. A. 1938. A dyke-swarm and crustal flexure in East Greenland. *Geol. Mag.*, **75**, 39.

WILLIAMS, H. 1941. Calderas and their origin. *Univ. Calif. Publ. geol. Sci.*, **25**, 239.

WINKLER, H. G. F. 1960. La genèse du granite et des migmatites par anatexie expérimentale. *Rev. Géogr. phys. Géol. dynam.*, **3** (2) 67.

WYART, J., and SABATIER, G. 1959. Transformation des sédiments pélétiques à 800° C et granitisation. *Bull. Soc. fr. Minér. Crist.*, **82**, 201.

J. B. DAWSON

15 The structural setting of African kimberlite magmatism

ABSTRACT. *The kimberlites of Africa are found in a variety of structural settings: on the stable cratons and their upwarped margins, and also in graben structures in circum-cratonic orogenic belts. The common factors in these settings appear to be epeirogenic uplift and the formation of deep-seated fracture systems. In certain cases these fracture systems appear to be controlled by structures in the basement, in others the fractures transect these structures. The possibility is considered that the kimberlite magmatism may be the result of convection cells in the upper mantle, and the confinement of diamondiferous kimberlites to the ancient shields is discussed.*

Introduction

The ultrabasic rock *kimberlite* was first discovered to be the parent rock of diamond in a volcanic vent at Jagersfontein, South Africa, in 1870, but it was named after the town of Kimberley which grew up around another diamondiferous vent discovered in 1871. It is perhaps appropriate that kimberlite should have its type locality in Africa as it is within this landmass that this rare rock-type is most widespread, being found in a region extending from well north of the equator to almost the southern tip of the continent. The upland interior of South Africa and the central plateau of Tanzania are the areas of the most intense intrusion; in the other countries—Mali, Guinea, Sierra Leone, Ivory Coast, Gabon, Angola, Congo, Kenya, Zambia and Rhodesia—kimberlite occurs on a relatively minor scale.

The environment of kimberlite is discussed here within the tectonic framework of Africa, defined by Kennedy (1964) and enlarged upon by Clifford (1966, 1967), and Cahen and Snelling (1966). These authors have shown that the primary units were older cratons—the West African Craton, the Angola-Kasai Craton, the Tanzania Craton and the Transvaal-Rhodesia Craton—which have not suffered orogeny since at least 1500 m.y. ago. After the Kibaran orogenesis about 1100 ± 200 m.y. ago the consolidated Kibarides linked up the Angola-Kasai and Tanzania shields to form, with the addition of the accreted Irumide Fold-Belt, the Congo Craton; in the south of the continent the Orange River (or Natal–Namaqualand) Belt was fused to the Transvaal-Rhodesias hield to form the Kalahari Craton. These two and the West African Craton acted as stable blocks during the late Precambrian–early Palaeozoic Katangan Orogeny when the Damaride, Mozambique and West Congo Fold-Belts were formed. The distribution of kimberlites in Africa and their relation to the major cratons and fold belts are shown on Fig. 1.

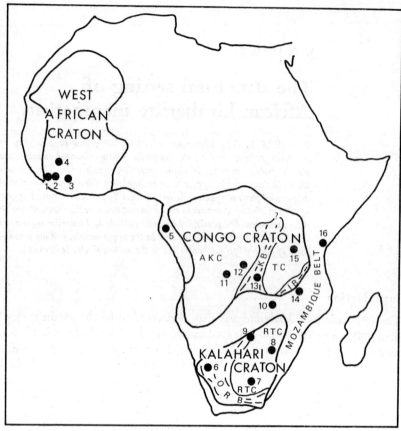

FIG. 1. Distribution of African kimberlites relative to the major structural units:
1, Sierra Leone; 2, Upper Guinea; 3, Ivory Coast; 4, Kenieba area, Mali; 5, Gabon;
6, South-West Africa; 7, South Africa (see Fig. 2); 8, Lochard area, Rhodesia;
9, Letlhekane, Botswana; 10, Luangwa Valley, Zambia; 11, Chicapa graben area,
Angola; 12, Bakwanga–Kalonji area, Congo; 13, Kundulungu Plateau, Congo;
14, Ruhuhu graben, Tanzania; 15, Tanzania Plateau (see Fig. 5); 16, Mrima Hill
area, Kenya.

Abbreviations: AKC, Angola-Kasai Craton; TC, Tanzania Craton; RTC,
Rhodesia-Transvaal Craton; KB, Kibaran Fold Belt; IB, Irumide Belt; ORB,
Orange River Belt. Structural units after Clifford (1966); the four *older cratons*
referred to in this paper are the Angola-Kasai, Tanzania, Rhodesia-Transvaal,
and West African Cratons.

Kimberlites in the stable cratonic areas

Kimberlite is most extensively developed on the major cratons and is intruded
into formations underlain by both the primary shields and the adjacent
Kibaran Fold-Belts (Fig. 1).

SOUTHERN AFRICA

The inland plateau of South Africa is the region of the most concentrated
kimberlite intrusion in the world and is also the most thoroughly investigated

kimberlite province. The kimberlites are spread over a huge roughly oval area north of the Cape Fold Mountains and cover most of the interior plateau except in the south-east where they are conspicuously absent (Fig. 2).

Kimberlites of at least two ages are known. The earlier are those of the Premier Mine, Pretoria District, which have recently been shown to be Precambrian (Allsopp et al., 1967), thus linking them with the chain of alkali plugs and volcanoes known as the Franspoort Line (Shand, 1922). The rocks include syenites, nepheline-syenites, trachytes and carbonatites (Verwoerd, 1966), and were presumably intruded along a major north-north-west to south-south-east fracture system.

Most kimberlites in South Africa are of Cretaceous age and are believed to have been intruded during a period of strong uplift of the continent with attendant downwarping or faulting around the periphery as a result of deepening of the contiguous ocean basins. From Zululand to the Transkei (Fig. 2) successive strips of land were bent down on the seaward side. Along the south coast, the faults run east-west with downthrows of 10,000 ft towards the sea (Du Toit, 1954, p. 571). The faults then swing north-westward following the west coast, and in South-West Africa there has been severe faulting with a north-north-west strike and coastwards downthrows. During this period, the intrusions on the uplifted plateau were mainly of kimberlite. Other rare rock-types such as olivine-melilitite, olivine nephelinite, monchiquite and carbonatite do occur, however, principally between the main kimberlite area and the south and west coasts, with some overlap in South-West Africa, Bushmanland and the Sutherland and Riversdale districts (see Fig. 2). In this overlap zone, which coincides with the Orange River Belt, the kimberlites are not diamondiferous. A similar broad regional zoning has also been observed in the Russian Yakutia Province where the diamondiferous kimberlites in the centre of the Anabarsk shield are surrounded by barren kimberlites and alkalic intrusions (Milashev, 1965).

The kimberlite intrusions of South Africa occur in clusters, as, for example, in the Kimberley, Winburg, Kroonstadt, Gibeon and Bushmanland areas (Fig. 2), and there is a tendency for diatremes to occur in zones or belts, the best example being the Kimberley pipes which occur along a broad north-west to south-east zone. Another conspicuous chain stretches from Preiska to Britstown, coinciding with the strike of the Kheis rocks of the basement (Du Toit, 1954).

Some pipes in these zones (e.g. the Kimberley Mine) have been excavated sufficiently deeply to demonstrate that they contract downwards into dykes or groups of dykes, the intersection of which is the cause of the location of the pipe. For instance, the Kimberley pipe expands above the intersection of two dykes, one running north-east to south-west, the other trending north-west to south-east, the latter coinciding with the direction of the zone of diatremes. With the further evidence of the diatreme-dyke relationships of the kimberlites in the deeply-dissected mountains of Lesotho (Dawson, 1960, 1962), it is safe to infer that zones of kimberlite vents are the surface expressions of dykes at depth, and the apparently random distribution within the zone is the result of near-surface weaknesses. Also in South Africa there are a number of visible kimberlite dyke swarms which, according to Hallam (in preparation), have dominant strike trends of north-west to south-east, north-east to south-west, west-north-west to east-south-east and east-west. In some cases the trend of the diatreme zones and dyke swarms coincides with deeper structures; for instance, the broad north-north-west-trending zone in the Gibeon–Keetmanshoop area of South-West Africa (Wagner, 1916; Janse, 1964) is parallel to the strike of the basement gneisses (Martin, 1965), and to a zone of Mesozoic faults. Similarly the east-west kimberlite dykes in the Swartruggens area are parallel to the regional strike of quartzites and shales of

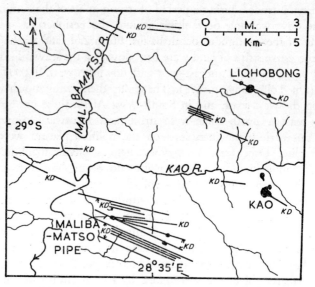

Fig. 3. The trend of kimberlite dykes (KD) in the Kao area of Lesotho. Kimberlite diatremes are shown in solid black (modified after Dawson, 1960).

the Transvaal System. Crockett and Mason (1968) recognise a zone of kimberlite intrusion 600 km long and 280 km wide extending from the Postmasburg–Britstown area eastwards to Lesotho and including the Kimberley, Koffiefontein, Jagersfontein and Winburg areas (Fig. 2). The axis of this zone lies in a direction between east-west and west-north-west to east-south-east, and has been termed the 'Lesotho Trend' by these authors, who also indicate a certain degree of clustering of kimberlite intrusions where the Lesotho Trend intersects north-south-trending basement rocks in the Postmasburg–Prieska area. The concept of the Lesotho Trend (which is supported by evidence that the zone of kimberlite intrusion is one of seismic activity), requires a certain amount of modification as it is apparent that not all the kimberlite dykes or diatreme zones follow that trend (see Fig. 2). For instance, in the Kimberley–Boshof area, there are three dominant dyke directions which are north-west to south-east, north-east to south-west and east-west; Hallam (in preparation) has shown that the east-west direction (which is also seen in the Jagersfontein–Koffiefontein area) coincides with the foliation in the basement granite-gneiss encountered in the deepest mines. The north-west and north-east-trending dykes, however, cut across this trend and are probably due to kimberlite infilling deep-seated fractures that transect and are not deflected by the crustal rocks. The same must be true of the dykes of the Postmasburg area, where the east-west strike (Fig. 2) is not that of the basement Kheis rocks, and of the north-east to south-west diamondiferous dykes of the Belsbank area north of Kimberley. It is apparent then that this area of the most concentrated kimberlite intrusion embracing

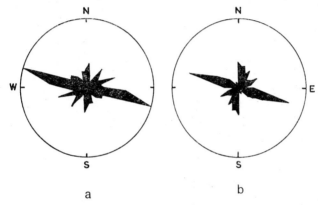

Fig. 4. Strike frequency diagrams for: (*a*) Karroo basalt and dolerite dykes in the Maseru–Butha Buthe and Mokhotlong areas, Lesotho; and (*b*) joints in Karroo sediments in the Maseru–Butha Buthe area and in Drakensberg basalts in the vicinity of the Kao kimberlite pipes. There is a strong coincidence between the trend of the kimberlite dykes (see Fig. 3) and the main trends of both the Karroo dykes and the joints.

most of the highly diamondiferous diatremes in South Africa, lies at the intersection of three fundamental fracture trends. It is significant that this has a parallel in the Yakutia Province where the diatremes of the Olenek–Markha area are sited at the intersection of two zones of north-west and north-east strike (Arsenyev, 1962).

In Lesotho and the neighbouring parts of the Orange Free State and Natal, the kimberlite dykes strike west-north-west to east-south-east (Fig. 3) and are parallel to the dominant direction of Karroo dolerite dykes of the area; in the upper Orange River Valley, north of Mokhotlong (Fig. 2), there is an example of kimberlite forming a composite dyke with dolerite (Dawson, 1960). This trend also coincides with the main joint direction in the bedded Karroo rocks (Fig. 4). Although nothing is known of the deeply-buried basement rocks, it is evident that this direction does not coincide with the strike of the Orange River Belt beneath Lesotho (Fig. 2) as inferred by

Nicolaysen and Burger (1965), and again it must be assumed that the kimberlites were intruded along fundamental fractures, already exploited by the earlier dolerite dykes.

The newly discovered group of kimberlite pipes at Letlhekane in north-central Botswana (not shown on Fig. 2) is apparently close to the edge of the craton, being situated where rocks of the Limpopo Belt (a 2100-1950 m.y. orogenic zone which lies between the Rhodesian and Transvaal shields and forms part of the Kalahari Craton) intersect the circum-cratonic Damaran Belt. The kimberlites 'are associated with the junction between the Limpopo and Damara–Zambezi Trends. The pipes are aligned along the Limpopo Trend but the alignments appear to be displaced by faults related to the Damara–Zambesi Trend' (G. T. Lamont, in Crockett and Mason, 1968).

ANGOLA—CONGO

The diamondiferous kimberlites of north-eastern Angola intrude Proterozoic and Mesozoic sediments in the valleys of the Luachimo and Chicapa Rivers (Fig. 1) and are believed to be Upper Jurassic in age (Real, 1968). The intruded sediments infill the Lucapa graben, the north-east to south-west trend of which coincides with that of the rocks of the basement complex. Within the graben, some kimberlites intrude the Cangoa Fault which is parallel to the edge of the graben, and which was initiated or reactivated in Jurassic times. A line extending south-westwards from the Lucapa graben links up with a north-east-trending chain of Karroo ring-complexes, which include alkali and carbonatite complexes (De Sousa Machado, 1958). Similarly, a line projected north-eastwards from the graben trends directly towards the Mesozoic kimberlites of the Congo, which are found along a zone of similar trend in the Bakwanga–Kalonji area (see Fig. 1) (Fieremans, 1966). It appears that the Angola–Congo kimberlites, and the ring-complexes of central Angola, were intruded along a major north-east to south-west zone that fractured the Angola-Kasai shield in Mesozoic times, this fracture zone possibly being due to reactivation of older lines of weakness.

In the south-eastern Congo, non-diamondiferous kimberlites intrude the Precambrian rocks of the Kundelungu Plateau (Verhoogen, 1938) which is underlain at depth by rocks of the Kibaran Orogenic Belt (Fig. 1).

TANZANIA

The central plateau of Tanzania is second only to South Africa in the abundance of its kimberlite intrusions. The kimberlites are intruded mainly into the ancient granitic shield south of Lake Victoria (Fig. 5). More than 120 are known, largely concentrated in the Mabuki, Shinyanga, Nzega, Kahama, and Kimali areas and on the Iramba Plateau west of Singida; only the Mwadui pipe near Shinyanga, and one near Kahama, are diamondiferous (Edwards and Howkins, 1966). Individual kimberlites tend to be related to lines of weakness, such as contacts between the basement granite and Nyanzian metasediment pendants; those on the Iramba and Singida Plateaus are intruded into rocks cut by numerous north-east to south-west faults whose latest movements were Tertiary, but which were probably in existence at the presumed Cretaceous time of the kimberlite activity. There is no alignment within each group of kimberlite intrusions, but the groups themselves lie on a very broad north-north-west to south-south-east zone running some 200 miles from the southern shores of Lake Victoria (Fig. 5). This pattern may indicate a zone of fundamental fractures at depth, the kimberlites being erupted where this zone is transected either by the Nyanzian pendants or by the north-east-trending fractures of the Singida–Kimali area.

The Igwisi Hills, west of Tabora, are two small olivine-rich tuff-cones, from one of which an olivine-rich, highly carbonated lava has been extruded; they have been interpreted as the surface expression of Recent kimberlite vents (Sampson, 1953; Fozzard, 1959). By contrast with volcanics of similar age which erupted along the late Tertiary and Recent rift valleys encircling the Tanzania shield (see King, this volume, p. 263), the Igwisi volcanoes lie in the centre of the craton. Owing to the extensive superficial deposits it is not known whether major fracture zones exist in the area although, like some isolated kimberlites to the south of Dodoma and south-west of Tabora, the Igwisi extrusives lie close to the inferred contact between the Dodoman Belt and the granitic shield (Fig. 5).

WEST AFRICA

Diamondiferous kimberlites are known in the Kenieba area of Mali; in Upper Guinea; in the Sefada–Koidu area of Sierra Leone; and in the Seguela area of the Ivory Coast. All these areas are in the West African Craton, though the last is close to its western margin (Fig. 1).

In Sierra Leone the country rocks are granites and north-north-west to south-south-east-trending schist belts, intruded by two sets of dolerite dykes striking roughly east-west and north-north-east to south-south-east. The kimberlites are dykes that strike N. 55° E. (Grantham and Allen, 1960), and do not follow the planes of weakness exploited by the earlier dolerites. Their intrusion was controlled by shear-belts (Andrews-Jones, 1966) that, like the fundamental fractures in South Africa, cut across all the earlier regional structures.

The same N. 55° E. trend of kimberlite dykes has also been found in Upper Guinea, where in addition there are trends of N. 85° E. and E. 40° S. linking up groups of kimberlite intrusions (Bardet, 1963). As in Sierra Leone, these kimberlite trend-lines do not coincide with regional schist belts or dolerite dyke trends.

Little has been published on the Mali or Ivory Coast kimberlites. The former occur in small vents cutting Precambrian rocks exposed in a small dome surrounded by Palaeozoic sediments (ibid.), whereas the latter form part of a dyke-swarm containing fitzroyite and leucite lamproite (Dawson, 1967a). Like the Sierra Leone and Guinea intrusives, they occur on the upwarped western and southern rim of the craton.

Kimberlites in the circum-cratonic belts

Compared with their sporadic concentration in the older cratons (Fig. 1), kimberlites are sparse in the circum-cratonic orogenic belts.

In South Africa kimberlites and associated olivine melilitites occur in the Upper Palaeozoic Cape Fold Belt in the Riversdale district (Fig. 2); like the kimberlites in the orogenic belts of Kibaran age, they are non-diamondiferous.

In East Africa, kimberlites intrude the Mozambique Belt in or near areas that have suffered major fracturing at Mesozoic or later dates. Kimberlites intrude Karroo sediments overlying the Mozambique Belt in the Ruhuhu graben of south-west Tanzania (McKinlay, 1958) and also in the lowlands of Kenya, where they are associated with the Mrima Hill alkalic carbonatite complex (Fig. 1) (L. G. Murray, personal communication). At Mrima Hill the kimberlite dykes strike between S. 66° W. and N. 8° W.; some are parallel to the inferred north-south trend of the underlying Mozambique Belt, but others cut across this trend. A similar feature is found in Zambia where a group of 14 vents and 4 dykes of barren kimberlite intrude Karroo sediments in the Luangwa graben (Fig. 1), 50 miles east of Kanoma (L. G. Murray, personal communication). The basement rocks are obscured by the sediments of the graben, but the area of intrusion is probably close to the boundary between the Irumide Fold Belt and the Mozambique Belt. The trend of the kimberlite vent zone and the strike of the dykes is north-west to south-east, in contrast to the north-north-east to south-south-west direction of the graben, and to the known north-east to south-west or east-south-east to west-south-west trend of the basement rocks outside the graben.

The late Tertiary–Recent rift valleys, which are bounded by very deep-seated fractures following the trend of earlier fold-belts (Brock, 1966) contain few kimberlites. In northern Tanzania the Lashaine tuff-cone is one of many similar cones erupted on the floor of the Eastern Rift Valley following major late-Pleistocene faulting. On account of its suite of ejected blocks of garnetiferous and garnet-free peridotites, the volcano has been described as the possible surface expression of a modern kimberlite diatreme (Dawson, 1964).

From this summary, three main conclusions emerge:

1. Kimberlites are abundant only on the post-Kibaran cratons.

2. They were intruded for the most part into major fundamental fractures that cut both the cratonic areas and the circum-cratonic belts during periods of epeirogenic uplift. Some of these fractures were controlled by existing planes of weakness, but others transect the deepest known structures and apparently were independent of any earlier structural weaknesses.

3. As pointed out by Clifford (1966), diamondiferous kimberlites are confined to the older cratons (Fig. 1), whereas those in the more youthful Kibaran, Katangan or Palaeozoic orogenic belts are barren.

Convection cells in the mantle and the formation of kimberlite

The fractures occupied by the kimberlite dykes are tensional and in South and West Africa commonly form a simple geometrical pattern. The magmatism is also seen to be linked with large-scale tectonics—the Mesozoic uplift of the continent, the marginal downwarping and the deepening of the surrounding ocean basins (Kennedy, 1965). It would seem, therefore, that the whole tectonic/magmatic cycle can best be explained by a series of convection cells rising beneath the continent and descending either at its margins or beneath areas now occupied by major sedimentary basins. On this view the uprising cells lay beneath the areas of maximum vertical uplift, and the fundamental fractures resulted from dilation caused by sub-crustal flow. The rising convection cells were accompanied by melting in the upper mantle, kimberlite being one of the products.

Although the origin of kimberlite in the mantle appears to be beyond dispute (in view of its high-pressure mineralogy), nonetheless, in certain aspects of its chemistry, kimberlite is unlike other peridotitic rocks (Dawson, 1960, 1967b) and must be regarded as highly modified mantle material. Furthermore it is difficult to envisage how such a potassium-rich and volatile-rich rock could be derived from the 'barren' peridotites (pyrolite and garnet pyrolite) regarded by many authors (Clark and Ringwood, 1964; O'Hara, 1965; Harris et al., 1967; Ito and Kennedy, 1967) as the principal material in the upper mantle. In view of the high potassium content and the apparent confinement of kimberlite to areas with great crustal thickness it has previously been suggested that assimilation of granitic crustal material played a major role in the formation of kimberlite (Dawson, 1967b). Recently, however, serious consideration has been given to phlogopite as a primary mineral in the mantle (Boyd, 1966; Kushiro et al., 1967). This view is reinforced by: (i) the presence in kimberlite of brown xenocrystal mica—the 'first-generation' mica (Dawson, 1962); (ii) the presence of both biotite (Williams, 1932, p. 421) and muscovite (Meyer, 1968) as inclusions in diamond; and (iii) the discovery of primary phlogopite in the metamorphosed alpine peridotite of Totalp, Switzerland (Peters, 1968). In addition, I have found brown primary mica in one specimen of garnet lherzolite, in several specimens of spinel lherzolite, and in a mica dunite which, together with other ultramafic and granulite blocks, have been ejected from the Lashaine tuff-cone in north Tanzania (Dawson, 1964; Dawson et al., in press). It is suggested therefore that restricted or partial melting of a mica-bearing peridotite may give rise to a relatively potash- and water-rich fluid which, when mixed with unmelted mantle material, forms kimberlite. This suggestion is slightly different from the inference of Kushiro et al. (1967) that kimberlite may be a melt in its own right; it also carries the corollary that the mantle is chemically more heterogeneous than is generally supposed and is thus in keeping with the views of Sobolev and Sobolev (1965) and Davidson (1967).

More complete melting, involving the garnet peridotite of the mantle, may give rise to the continental flood-basalts and associated intrusives which, on many continents, are found in the same areas as the kimberlites. Various authors (see Verschure, 1967) have postulated a genetic connection between earlier flood-basalt activity and the subsequent intrusion of kimberlite. Whilst this relationship appears to be valid in many kimberlite provinces there are certain notable exceptions, such as Tanzania and Arkansas, where no such connection is known. Moreover, strong vertical uplift and extensive magmatism are not necessarily correlative since, in South Africa at least, the vast volumes of the Karroo plateau basalts were accompanied by much smaller vertical movements than those preceding the areally insignificant kimberlite activity. It is suggested that the common factor is the tectonic setting on cratons, and whether kimberlite or basaltic activity results would depend upon the degree and rate of melting of the mantle (see O'Hara, 1965). These factors would in turn be dependent upon the magnitude of convection and associated heat-flow in the mantle, and the extent to which the energy released was translated into vertical uplift or melting.

Diamond-bearing kimberlites

As noted by Clifford (1966), diamond-bearing kimberlites are confined to the older cratons (Fig. 1) (although the non-diamondiferous type are found there also) and this relationship is so consistent that it cannot be fortuitous. Diamonds are only formed and preserved in very exceptional conditions, the requisite conditions being the correct chemical environment, the necessary physical conditions of high pressure and temperature, and some means of preserving the diamond, once formed, both at depth and during its transport to the surface. It appears highly probable that the necessary combination of chemical and physical conditions occurs only rarely; and the highly critical conditions for diamond crystallisation are evidenced by the carbon inter-growths and coatings on some diamonds which are the reflection of delicate oscillations across the diamond/graphite equilibrium boundary. Just how far are these conditions peculiar to the ancient shield areas?

Diamonds result from the crystallisation of carbon in an ultrabasic (peridotitic) environment, under reducing conditions, and experimental work (Bovenkerk et al., 1959) suggests that small amounts of nickel, chromium, cobalt and titanium (all abundant trace elements in kimberlite) may act as catalysts. The presence of dissolved carbon in the silicate melt is a pre-requisite, and according to Kennedy and Nordlie (1968) 'the concentration of diamond in peridotite from the diamond mines in roughly one part of carbon per 20 million parts of silicate. This is probably close to the limit of solubility of carbon in molten silicate'. The ultrabasic chemical environment will apply to much of the mantle, and there appears to be no reason to assume that it is confined to those parts beneath the ancient shields.

Very high pressures are necessary for the formation of diamond, such pressures only being attained at depths well within the upper mantle. As in the case of the chemical environment, there is no reason to assume that high pressures should not be found in sub-oceanic mantle as well as in the sub-crustal mantle; nor would large differences be expected in the mantle beneath the older cratons and their surrounding orogenic belts. Nonetheless, the fact that diamond has only rarely been found in peridotite nodules in kimberlite compared with the host kimberlite and eclogite nodules (the latter nowadays being regarded by many petrologists as basaltic magma, resulting from partial fusion of mantle peridotite, which subsequently crystallised under high pressures in the mantle) implies that modified mantle material is a more favourable environment for diamond formation. This poses the question as to the cause and nature of the modification, and its apparent confinement to the older cratons. One way in which the latter are different from other areas of the earth is that they have a very low heat-flow compared with other continental areas and, by inference, the mantle beneath the shields must be relatively cool (Clark and Ringwood, 1964). Moreover, the evidence that kimberlite contains high amounts of K_2O, FeO, H_2O, CO_2, Ba, Sr, Nb, Zr, and rare earths (Dawson, 1967b) suggests incipient, rather than widespread, melting with the accumulation of large amounts of volatiles, mainly H_2O and CO_2. It is possible that the relatively cool mantle beneath the shields may be a more conducive environment to volatile accumulation than hotter mantle areas. The importance of high concentrations of CO_2 has been emphasised by Kennedy and Nordlie (1968) whose thermodynamic calculations show that a partial pressure of CO_2 essentially equal to that of the confining pressure is necessary if diamond is to be stable in an environment containing substantial amounts of ferrous and ferric iron. From this it appears that the necessary conditions for diamond formation are an accumulation of volatiles under mantle conditions. The low heat-flow under the shields may be one reason why the accumulations take place in this part of the mantle.

The specific conditions for the formation of diamond in nature is still a matter for further investigation. Very high pressures are undoubtedly necessary for the formation of diamond and its associated pyrope-rich garnet, and also for the formation of the rare inclusions of coesite that have been found as inclusions within synthetic diamonds (Milledge, 1961; Harris, 1968). On the basis of the graphite/diamond equilibrium curve and its intersection with their garnet peridotite solidus, Kennedy and Nordlie (1968) state: 'The diamond stability field where peridotite is also partially liquid represents the only conditions of temperature and pressure under which diamond can form in the earth. Thus the minimum pressure-temperature condition for the natural origin of diamond is c. 1800°C and 68 kilobars. This is equivalent to conditions probably not encountered at depths shallower than 200 km.' However, it is possible that the peridotite solidus (and hence the conditions of diamond formation) may be considerably modified if mica is a primary mineral in the mantle, since the experiments of Seifert and Schreyer (1968) have shown that mixtures of forsterite and enstatite show partial melting in the low pressure range 1-5 kb at approximately 550-700°C if small amounts of K_2O are added to the system in the presence of excess water.

After they have been formed at depth, diamonds are transported upwards to the surface, and the physical conditions under which diamonds are now found in kimberlite are vastly different from those under which they crystallised in the mantle. To attain the present position one of two things must have taken place: either the kimberlite with its diamond was emplaced from depth sufficiently rapidly to prevent resorption of the diamond or its inversion to a low-pressure form; or, during the emplacement of kimberlite, diamond facies conditions were maintained up to a relatively shallow depth in the crust. The rapid emplacement hypothesis has been advocated by Davidson (1967) and Kennedy and Nordlie (1968). A major part of the latter two authors' arguments hinge on the absence of thermal metamorphic effects in the kimberlites occupying diatremes and that 'the single character to be emphasised is that diamond pipes are filled with fragmental debris' (*ibid.*, p. 497). These explosion-intrusion hypotheses infer that kimberlite has never been hot or liquid. To be sure, the absence of thermal metamorphic effects in kimberlite pipes has excited comment since the days of Wagner (1916), and the numerous features of fluidisation intrusion seen in many kimberlite bodies (Dawson, 1962) testify to the rapid emplacement of diatreme kimberlite. However, the fragmentation of the kimberlite is an integral part of the formation of the diatremes themselves, and is a near-surface phenomenon, taking place at depths of probably no more than 2-3 km (Dawson, 1960, 1967a; Sobolev 1960). Fragmented kimberlite is characteristic only of diatremes, and the massive kimberlites occurring in the sub-diatreme dyke-swarms or in the deepest parts of diatremes show fluidal textures, such as alignment of the long axes of inequidimensional minerals or xenoliths. These features have been accepted by workers on the Yakutian kimberlite province (Milashev *et al.*, 1963; Bobrievich *et al.*, 1964; Milashev, 1965) as indicating that the kimberlite was liquid, and the term 'magmatic kimberlite' is recognised terminology in the Russian literature (Rhabkin *et al.*, 1962). In addition I have observed in specimens from South Africa a variety of textures, including trachytic texture, vortex or eddy structures, layering and flowage differentiation, that could only have formed in a very fluid medium. It should be stressed that in some of these cases there is still abundant fresh olivine which would seem to exclude the possibility that these textures are due to the type of plastic flowage that is often observed in Alpine-type serpentinites. The evidence from both Russia and South Africa indicates that kimberlites were fluid at depths of not more than 2-3 km below the land-surface at the time of their emplacement.

It is also apparent that the most diamondiferous kimberlite diatremes are also the largest (e.g. Mwadui, Premier, Kimberley, Wesselton) and it is not unreasonable to infer that the high diamond content and the large size of the diatremes are both a reflection of the extremely high pressures in the sub-diatreme dykes. In addition the crystallographic perfection of many diamonds from these large diatremes suggests that they have suffered neither resorption nor inversion; nor have they travelled far from a locus in which they were in complete equilibrium. The suggestion is that very high pressures, perpetuating diamond facies conditions, were maintained in the dykes up to the relatively high levels in the crust where, with the reduced lithostatic pressure, the high pressure gases would exsolve from the magma to begin fluidisation and effect the final upward emplacement of the kimberlite with its diamonds into the diatremes. For these very high pressures to have been retained up to high levels in the crust, the wall rocks of the fractures now occupied by the kimberlite dykes must have had sufficient strength to confine the pressures. The possible differences between the diamondiferous kimberlites on the older cratons and the non-diamondiferous kimberlites in the circum-cratonic areas may hinge on this point. A diamondiferous kimberlite under very high pressure infilling a fracture in one of the orogenic belts would come into contact with faults, fractures, foliations and other planes of weakness along which gases from the magma could escape, thereby lowering the pressures during the ascent, and promoting inversion or resorption of the diamonds. On the other hand, if a kimberlite was intruding a fracture in an ancient shield area, the rigidity of the granitised, more homogeneous wallrocks may well prevent lateral dispersion of the high pressure gases, thereby retaining the high pressure regime up to the relatively high level where fluidisation intrusion would take over.

In summary, the ancient shield areas differ from other parts of the earth's crust in having a higher degree of rigidity, and the mantle below them is cooler than in other parts of the mantle. It is tentatively suggested that these two features may have some bearing on the confinement of diamondiferous kimberlites to the older cratons. The relatively cool mantle below the shields may be the most favourable part for accumulations of the high amounts of volatile matter that appear to be necessary for diamond formation. The rigid masses of the ancient shields will provide wall-rocks of sufficient strength to contain the high pressures that perpetuate diamond-facies physical conditions up to high levels in the crust.

Acknowledgements

I wish to thank Dr L. G. Murray of the Anglo-American Corporation of South Africa for information on the kimberlites of the Luangwa Valley and the Mrima Hill area, and Mr C. D. Hallam of De Beers Consolidated Mines, Kimberley, for the loan of the pre-publication manuscript of his paper on the Kimberley mines.

REFERENCES

ALLSOPP, H. L., BURGER, A. J., and VAN ZYL, C. 1967. A minimum-age for the Premier kimberlite pipe yielded by biotite Rb-Sr measurements, with related galena isotope data. *Earth Planetary Sci. Let.*, **3**, 161.

ANDREWS-JONES, D. A. 1966. Geology and mineral resources of the northern Kambui schist belt and adjacent granulites. *Bull. geol. Surv. Sierra Leone*, **6**.

ARSENYEV, A. A. 1962. Laws of the distribution of kimberlites in the eastern part of the Siberian platform. *Dokl. Akad. Nauk. S.S.S.R., Earth Sci. Sect.*, **137**, 355.

BARDET, M. G. 1963. Controle géotectonique de la repartition des venues diamantifères dans la monde. *Chron. Mines. Rech. min.*, **328-9**, 67.

BOBRIEVICH, A. P., *et al.* 1964. *Petrography and mineralogy of the kimberlite rocks of Yakutia,* Moscow (in Russian).

BOVENKERK, H. P., BUNDY, F. P., HALL, H. T., STRONG, H. M., and WENTORF, R. H. 1959. Preparation of diamond. *Nature, Lond.*, **184**, 1094.

BOYD, F. R. 1966. Electron-probe study of diopside pyroxenes from kimberlites. *Yb. Carneg. Instn.*, **65**, 252.

BROCK, B. B. 1966. The rift valley craton. *Pap. geol. Surv. Can.*, **66-14**, 99.

CAHEN, L., and SNELLING, N. J. 1966. *The geochronology of equatorial Africa*, Amsterdam.

CLARK, S. P., and RINGWOOD, A. E. 1964. Density distribution and constitution of the mantle. *Rev. Geophys.*, **2**, 35.

CLIFFORD, T. N. 1966. Tectono-metallogenetic units and metallogenic provinces of Africa. *Earth Planetary Sci. Let.*, **1**, 421.

—— 1967. The Damaran episode in the Upper Proterozoic-Lower Paleozoic structural history of southern Africa. *Spec. Pap. geol. Soc. Am.*, **92**.

CROCKETT, R. N., and MASON, R. 1968. Foci of mantle disturbance in southern Africa and their economic significance. *Econ. Geol.*, **63**, 532.

DAVIDSON, C. F. 1964. On diamantiferous diatremes. *Econ. Geol.*, **59**, 136.

—— 1967. The so-called 'cognate xenoliths' of kimberlites. In *Ultramafic and related rocks*, p. 342. (Ed. P. J. Wyllie), New York.

DAWSON, J. B. 1960. A comparative study of the geology and petrography of the kimberlites of the Basutoland province. Unpublished Ph.D. thesis, Univ. Leeds.

—— 1962. Basutoland kimberlites. *Bull. geol. Soc. Am.*, **73**, 545.

—— 1964. Carbonate tuff cones in northern Tanganyika. *Geol. Mag.*, **101**, 129.

—— 1967a. A review of the geology of kimberlite. In *Ultramafic and related rocks*, p. 241. (Ed. P. J. Wyllie), New York.

—— 1967b. Geochemistry and origin of kimberlite. In *Ultramafic and related rocks*, p. 269. (Ed. P. J. Wyllie), New York.

—— POWELL, D. G., and REID, A. M. (in press). Ultrabasic lava and xenoliths from the Lashaine volcano, northern Tanzania. *J. Petrology*.

DE SOUZA MACHADO, F. J. 1958. The volcanic belt of Angola and its carbonatites *C.C.T.A., Geol. Publ.*, **44**, 309.

DU TOIT, A. L. 1954. *The geology of South Africa*, 3rd ed. Edinburgh.

EDWARDS, C. B., and HOWKINS, J. B. 1966. Kimberlites in Tanganyika, with special reference to the Mwadui occurrence. *Econ. Geol.*, **61**, 537.

FIEREMANS, C. 1966. Contribution a l'étude petrographique de la brêche kimberlitique de Bakwanga. *Mem. Inst. Geol., Univ. Louvain*, **24**.

FOZZARD, P. M. H. 1959. Further notes on the volcanic hills at Igwisi. *Rec. geol. Surv. Tanganyika*, **6**, 69.

GRANTHAM, D. R., and ALLEN, J. B. 1960. Kimberlites in Sierra Leone. *Overseas Geol. Mineral Resources*, **8**, 5.

HARRIS, J. W. 1968. The recognition of diamond inclusions. Pt. 1: Syngenetic mineral inclusions. *Ind. Diamond Rev.*, **28**, 402.

HARRIS, P. G., REAY, A., and WHITE, I. G. 1967. Chemical composition of the upper mantle. *J. geophys. Res.*, **72**, 6359.

ITO, K., and KENNEDY, G. C. 1967. Melting and phase relations in a natural peridotite to 40 kilobars. *Am. J. Sci.*, **265**, 519.

JANSE, A. J. A. 1964. Kimberlites and the related rocks of the Nama Plateau, South-West Africa. Unpublished Ph.D. thesis, Univ. Leeds.

KENNEDY, G. C., and NORDLIE, B. E. 1968. The genesis of diamond deposits. *Econ. Geol.*, **63**, 495.

KENNEDY, W. Q. 1964. The structural differentiation of Africa in the Pan-African ($\pm$500 m.y.) tectonic episode. *8th Ann. Rep. Res. Inst. African Geol., Univ. Leeds*, 48.

—— 1965. The influence of basement structure on the evolution of the coastal (Mesozoic and Tertiary) basins of Africa. In *Salt basins around Africa*, p. 7. Inst. Petrol., London.

KUSHIRO, I., SYONO, Y., and AKIMOTO, S. 1967. Stability of phlogopite at high pressures and the possible presence of phlogopite in the earth's upper mantle. *Earth Planetary Sci. Let.*, **3**, 197.

MARTIN, H. 1965. *The Precambrian geology of South-West Africa and Namaqualand.* Precambrian Res. Unit, Univ. Cape Town.

McKINLAY, A. C. M. 1958. Kimberlite intrusions cutting Karroo sediments in the Ruhuhu depression of south-west Tanganyika. *Rec. geol. Surv. Tanganyika,* 5, 63.

MEYER, H. O. A. 1968. Mineral inclusions in diamonds. *Yb. Carneg. Instn.,* 66, 446.

MILASHEV, V. A. 1965. *Petrochemistry of the kimberlites of Yakutia and some factors in diamond formation,* Moscow (in Russian).

—— KRUTOYARSKI, M. A., RABHKIN, M. I., and ZIRLICH, Z. N. 1963. *Kimberlitic rocks and picritic porphyries of the north-eastern part of the Siberian platform,* Moscow (in Russian).

MILLEDGE, H. J. 1961. Coesite as an inclusion in G.E.C. synthetic diamonds. *Nature, Lond.,* 190, 1181.

NICOLAYSEN, L. O., and BURGER, A. J. 1965. Note on an extensive zone of 1000 million-year old metamorphic and igneous rocks in southern Africa. *Sci. Terre.,* 10, 497.

O'HARA, M. J. 1965. Primary magmas and the origin of basalts. *Scott. J. Geol.,* 1, 19.

PETERS, T. 1968. Distribution of Mg, Fe, Al, Ca and Na in coexisting olivine, ortho-pyroxene and clinopyroxene in the Totalp serpentinite (Davos, Switzerland) and in the Alpine metamorphosed Malenco serpentinite (N. Italy). *Contr. Miner. Pet.,* 18, 65.

RABKHIN, M. I., KRUTOYARSKI, M. A., and MILASHEV, V. A. 1962. Classification and nomenclature of the Yakutian kimberlites. *Trudy Issled. Inst. Geol. Arktiki,* 121, 154 (in Russian).

REAL, F. 1958. Sur les roches kimberlitiques de la Lunda (Angola). *Bull. Fac. Sci. Univ. Lisbon,* 26.

SAMPSON, D. N. 1953. The volcanic hills at Igwisi. *Rec. geol. Surv. Tanganyika,* 3, 48.

SEIFERT, F., and SCHREYER, W. 1968. Die Möglichkeit der Entstehung ultrabasischer Magmen bei Gegenwart geringer Alkalimengen. *Geol. Rdsch.,* 57, 349.

SHAND, S. J. 1922. The alkaline rocks of the Franspoort line, Pretoria District. *Trans. geol. Soc. S. Africa,* 25, 81.

SOBOLEV, V. S. 1960. Conditions of formation of diamond. *Geol. i Geofiz.,* 1 (in Russian).

—— and SOBOLEV, N. V. 1965. Xenoliths in kimberlites of northern Yakutia and the structure of the mantle. *Dokl. Akad. Nauk. S.S.S.R., Earth Sci. Sect.,* 158, 22.

VERHOOGEN, J. 1938. Les pipes de kimberlite de Katanga. *Annls. Serv. Mines. Comité Spécial Katanga,* 9.

VERSCHURE, R. H. 1966. Possible relationships between continental and oceanic basalts and kimberlites. *Nature, Lond.,* 211, 1387.

VERWOERD, W. J. 1966. South African carbonatites and their possible mode of origin. *Annls Univ. Stellenbosch,* 41, Ser. A.

WAGNER, P. A. 1916. The geology and mineral industry of South-West Africa. *Mem. S. Afr. Dept. Mines,* 7.

WILLIAMS, A. F., 1932. *The genesis of the diamond,* London.

J. R. VAIL

16   **Tectonic control of dykes and related irruptive rocks in eastern Africa**

ABSTRACT. *Four major provinces of dyke intrusion and related igneous activity in the eastern half of Africa are considered and regional tectonic controls sought for their present distribution. Shortage of detailed geological, geochronological and palaeomagnetic data necessarily limits the interpretation of the relationships between igneous rocks and the structural environment during emplacement.*

Geological and palaeomagnetic studies in Rhodesia, South Africa and Botswana have distinguished Umkondo, Mashonaland and Waterberg dolerites emplaced about 1750-1950 m.y. ago and a Pilansberg suite about 1450 m.y. ago. In Tanzania, so-called Bukoban dolerites may have been intruded at around 900, 1900 and 2500 m.y. ago whereas extensive dykeswarms in Egypt, Sudan and Saudi Arabia include dykes that preceded and post-dated c. 600 m.y. old granites. In south-eastern Africa, Mesozoic igneous activity was widespread and includes Karroo dolerites, basalts and rhyolites, ring-complexes and plugs of granite, gabbro, syenite, carbonatite and some kimberlites.

Rift valley tectonism controlled the localisation of some of the Mesozoic rocks but evidence of a similar control for the Precambrian irruptives is lacking and their distribution appears to be fundamentally influenced by the orogenic and cratonic pattern which existed at the time of their emplacement.

Introduction

Dyke intrusions of many different ages and types are widespread throughout the eastern half of Africa. In most territories the dykes have been given group names, such as Karroo dolerites in South Africa or Bukoban dykes in Tanzania, and in many places well-developed dyke-swarms have been mapped. Nevertheless, the majority of these irruptive provinces have yet to be carefully studied. The only comprehensive work on African dykes and sills is the outstanding investigation by Walker and Poldervaart (1949) on the Karroo dolerites. However, in that study it was the petrological rather than the structural aspect that was examined.

In the present review, the tectonic control of magmatism as expressed by the distribution pattern of irruptives will be sought. The rocks to be considered include dykes, sills and irregular bodies and associated sub-volcanic ring-complexes, pipes, and to a lesser degree the contemporaneous volcanic outpourings, which are usually of plateau-type lavas. When these igneous masses are plotted on a suitable scale a pattern emerges, showing the temporal and spacial relationships of igneous activity in eastern Africa.

Incomplete mapping of both the dykes and the surrounding basement structure imposes serious limits on such a continent-wide survey. Because few reliable isotopic analyses have been made either on the igneous rocks or on the associated country-rocks, the ages of many of the Precambrian igneous and tectonic events must remain speculative. Palaeomagnetic studies have provided a useful additional source of information in distinguishing igneous groups, but again the lack of complete data detracts from the usefulness of this method when applied to continental studies. The sources for the present study have been mainly the various territorial geological surveys, publications in scientific journals and my own observations. As a consequence of the uneven distribution of data the apparent lack of dykes in some areas (Figs. 1-4) is probably due to incomplete investigation rather than an absence of igneous activity. Furthermore, some extensive dyke-swarms of petrologically similar rocks could be of very different ages and have been emplaced under different tectonic regimes; these are, as yet, indistinguishable. However, by extrapolation and simplification, the study of dykes and related irruptive rocks and the development of magmatic activity in relation to the tectonism prevailing in eastern Africa can be illustrated.

Major periods of dyke intrusion and related activity along the eastern side of Africa are considered in four principal areas:

(i) Precambrian basic intrusives of the Umkondo–Mashonaland, Waterberg and Pilansberg igneous provinces of the Transvaal, Botswana and Rhodesia.

(ii) Precambrian basic rocks of the so-called Bukoban dyke-swarms and lavas in Tanzania, Kenya, Uganda, Congo, Burundi, Zambia and Malawi.

(iii) Late Precambrian igneous activity adjacent to the Red Sea area in Egypt, Sudan and Saudi Arabia.

(iv) Mesozoic igneous activity of southern Africa, including the Karroo dolerites and lavas and the basic and acid ring-complexes, alkali plugs and some kimberlites of South Africa, Lesotho, Botswana, Rhodesia, Malawi, Mozambique and Swaziland.

Precambrian igneous activity in the Transvaal, Botswana and Rhodesia in south-eastern Africa

TRANSVAAL AND BOTSWANA

In the north-western Transvaal and extending across into eastern Botswana are a group of arenaceous and argillaceous sediments which are collectively referred to as the Waterberg System (Fig. 1). They are generally undisturbed and lie unconformably on older sedimentary sequences of the Transvaal System and contain fragments of igneous rocks of the Bushveld Complex. The lower part of the succession has been referred by Truter (1949) to the Loskop System, which he considered to be separate from the Waterberg System. Recent geochronological work (Oosthuyzen and Burger, 1964), however, indicates that the two units probably belong to the same period

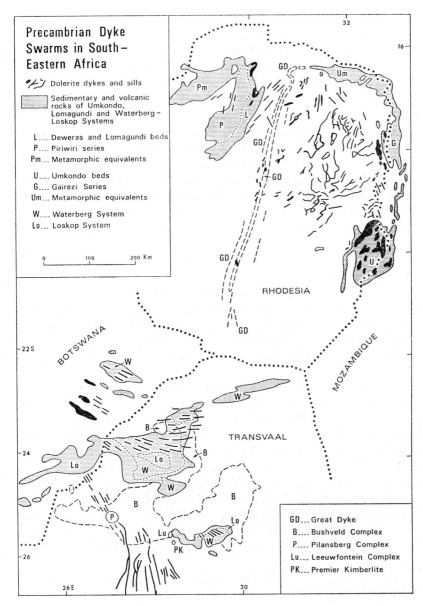

Precambrian Dyke
Swarms in South–
Eastern Africa

Dolerite dykes and sills

Sedimentary and volcanic
rocks of Umkondo,
Lomagundi and Waterberg–
Loskop Systems

L.... Deweras and Lomagundi beds
P.... Piriwiri series
Pm... Metamorphic equivalents

U.... Umkondo beds
G.... Gairezi Series
Um... Metamorphic equivalents

W.... Waterberg System
Lo... Loskop System

0   100   200 Km

GD... Great Dyke
B.... Bushveld Complex
P.... Pilansberg Complex
Lu... Leeuwfontein Complex
PK... Premier Kimberlite

FIG. 1.

and thus should be regarded as constituting a single *Waterberg System*. The
north-easternmost outcrop of the Waterberg beds in the Transvaal is underlain by thick lavas with intercalated arenaceous sediments. Both of these are
considered by the Geological Survey of South Africa to belong to a much
earlier sequence of rocks although at one time the volcanics were included in
the Waterberg System.

In Botswana and the Transvaal numerous pre-Karroo (i.e. pre-Upper

Palaeozoic) dolerites are intruded into the Waterberg System sediments and older rocks. In the past they were referred to as post-Waterberg diabase. In general the dolerites are in the form of long, narrow dykes which strike south-east or are elongated large sills, such as the Shushong sill in Botswana. A large swarm of basic dykes with a general east-west trend cuts Waterberg and older beds in the northern Transvaal (Fig. 1). Karroo dolerite dykes are also present.

A swarm of post-Waterberg composite dolerite-syenite dykes is centred on the Pilansberg alkaline complex (Fig. 1), which is intruded into the western part of the Bushveld Igneous Complex. The Pilansberg dyke swarm has a north-north-west to south-south-west trend and is genetically related to the Pilansberg volcanic centre from which they fan out over a strike distance of about 275 km. In places some of the dykes follow post-Bushveld granite faults.

RHODESIA

Around the eastern, north-eastern and north-western margins of Rhodesia a sequence of arenaceous, calcareous and argillaceous rocks unconformably overlies 'Basement Complex' granites and gneisses and Archaean formations (Fig. 1). Towards the outer margin of Rhodesia they become progressively folded and metamorphosed, but towards the centre they are only slightly disturbed.

In the eastern districts the sediments have been referred to the Umkondo System, the Gairezi Series and the Frontier System (Swift, 1961) but these are now all considered to be facies variations of the *Umkondo System* (Vail, 1968a; Slater, 1967). In the southernmost outcrop of the Umkondo System, andesitic basalts up to 190 m thick occur at the highest stratigraphic horizon (Swift, 1961). In the north-western districts, lithologically similar sequences have been termed the Deweras Series, the Lomagundi System and the Piri-wiri Series, and include highly metamorphosed gneisses in the Urungwe and Kariba districts. These various units are all considered here to be sedimentary and metamorphic facies of the *Lomagundi System*. Volcanic rocks are developed near the base of the succession where thick amygdaloidal basaltic lavas are present.

Basic intrusive rocks cut the Umkondo and Lomagundi Systems and are also well-developed in the north-central parts of Rhodesia (Fig. 1). They are generally known as Umkondo dolerites, although those in the west have not been included due partly to the uncertainty in correlating across the country and to metamorphic alteration. The basic intrusive rocks are quartz dolerites and take the form of irregular sill-like masses and irregular dykes up to tens of metres thick, usually without any preferred orientation. However, where they are intruded into the relatively flat-lying sediments they form horizontal sheets up to about 410 m thick (Tyndale–Biscoe, 1958). Where the igneous rocks have been metamorphosed they are contorted and altered to amphibolites.

AGE OF IGNEOUS ACTIVITY

Before the combination of palaeomagnetic measurements and geochronological analyses was applied to the basic igneous intrusives of the Transvaal and Rhodesia no meaningful correlation was possible and no subdivisions had been suggested. In recent years new data have substantially changed the situation and provide an example of what can be achieved by using these techniques.

The age of the Lomagundi and Umkondo Systems has long been controversial and is still unresolved. However, geochronology has provided a time-range for the systems. Minimum K-Ar ages for phyllites and slates of the Lomagundi System (as here used) are of the order of 1975 m.y. (Vail *et al.*, 1968); supporting evidence from whole rock Rb-Sr ages of intrusive granites at 2080 m.y. (Clifford *et al.*, 1967) and from galena veins at 1980 m.y. (see Vail *et al.*, 1968) confirms an upper age of about 2000 m.y. Likewise, the Umkondo System has yielded K-Ar ages from hornfels of 1785 m.y. (Snelling, 1966) although K-Ar age measurements on the dolerites themselves range from about 640 m.y. to 1640 m.y. (Jones and McElhinny, 1966).

Palaeomagnetic work has indicated two interesting features about the igneous rocks. First, the Umkondo dolerites are very closely related to the Umkondo volcanic rocks (McElhinny, 1966); and secondly, there are two distinct palaeomagnetic groups of dolerites, termed by McElhinny and Opdyke (1964) the Umkondo eastern group and the Mashonaland western group. It was initially thought that these two groups were very different in age, but more recent work (Jones and McElhinny, 1967) indicates that the Mashonaland group is probably only slightly older than the Umkondo dolerites, and that both can be considered as being within the limits of the Umkondo System.

Age determinations on the Transvaal and Botswana dolerites have also not been entirely satisfactory. Dolerite sills in Botswana have yielded mineral ages of 590-940 m.y. and a whole rock age of 1050 m.y. (Jones and McElhinny, 1966), whereas the Pilansberg Complex and related dykes have been dated at 1250 m.y. (Snelling, 1963) and 1330 m.y. (Van Niekerk, 1962) respectively. The cogenetic Leeuwfontein Complex provided U-Th-Pb ages of 1420 m.y. (Oosthuyzen and Burger, 1964), clearly suggesting an age of about 1420 m.y. for the post-Waterberg Pilansberg activity. A minimum age limit for the Waterberg sediments is provided by an intrusive granophyric granite dated by U-Th-Pb analysis at 1790 ± 70 m.y. (Oosthuyzen and Burger, *ibid.*) and by the kimberlite pipe at Premier Mine which contains xenoliths of Waterberg rocks; galena from the pipe indicates an age of 1750 ± 100 m.y. (Allsopp *et al.*, 1967). A maximum age is provided by the Bushveld Igneous Complex at about 1950 ± 150 m.y. (Nicolaysen *et al.*, 1958).

Palaeomagnetic work on the Waterberg red beds has indicated three distinct poles upwards through the succession which Jones and McElhinny (1967) interpret as being due to polar wander during Waterberg deposition and consolidation. It was noted that the palaeomagnetic pole position for the so-called 'post-Waterberg diabase' corresponds with the upper pole position

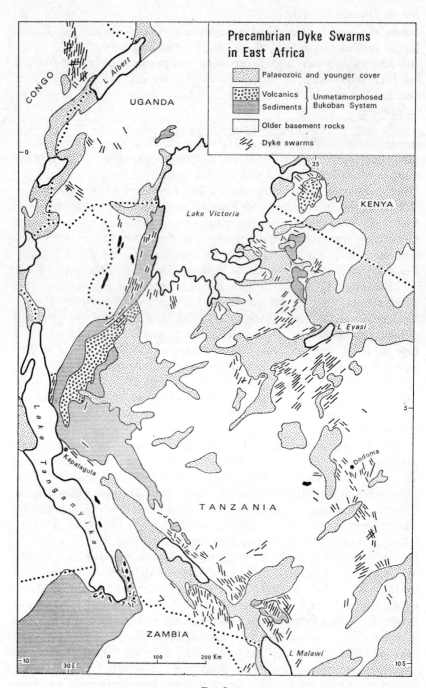

FIG. 2.

from the sediments thus suggesting the diabase belongs to the Waterberg System and should be referred to as Waterberg diabase. It had already been shown (Jones and McElhinny, 1966) that the Waterberg basic rocks were probably intruded contemporaneously with the Umkondo dolerites in Rhodesia, so the Umkondo and Waterberg Systems probably belonged to the same general period of deposition.

The doleritic intrusive rocks (Umkondo, Mashonaland and Waterberg groups) all can be considered to belong to the same irruptive period, i.e. between about 1750 and 1950 m.y. ago, whereas the Pilansberg Complex and dyke-swarm were emplaced 300-500 m.y. later.

Precambrian basic igneous activity in Tanzania and adjacent areas of East Africa

A series of almost unmetamorphosed arenaceous, calcareous and argillaceous beds with basaltic lavas near the top occurs along the western borders of Tanzania from the Uganda shore of Lake Victoria, across into Burundi and southwards almost to Lake Malawi (Fig. 2). These rocks, although given local names, constitute the Bukoban System (Quennell, 1956). Apart from their main occurrence they are correlated with smaller outliers mainly east of Lake Victoria in Kenya (Kisii Series) and Tanzania (Ikorongo Series). An outlier near Lake Tanganyika, known as the Abercorn Sandstone, is generally regarded as being continuous with Kundelungu (Upper Katanga System) beds in Zambia (see Cahen, this volume, p. 103). The Bukoban System is generally terrestrial in origin, with minor volcanics, and has suffered only slight folding. Possible metamorphosed equivalents cover extensive areas in the east.

Throughout much of Tanzania considerable numbers of dolerite dykes have been encountered and because similar dolerites are to be found associated with the Bukoban sedimentary and volcanic rocks they have all been given the general name of Bukoban dykes despite the fact that many of them occur hundreds of kilometres from the nearest Bukoban type rocks. Characteristically the dykes occur in swarms in which individuals are usually a few metres wide and strike parallel to one another. One swarm near Lake Eyasi extends for a strike length of at least 200 km before disappearing beneath younger rocks, and is up to 200 km in width (Fig. 2). Another swarm strikes roughly parallel to the Lake Malawi–Tanganyika Rift trough in the south-west and is about 300 km in length and some 150 km wide (Fig. 2).

Harpum (1955) established the presence of a basic igneous province termed the Bukoban Eruptive Province in Tanzania. From detailed work in the south-west of the country Harpum described dolerite dykes up to 9 km long which are mineralogically similar to nearby amygdaloidal basalts. These rocks were affected by folding; subsequently, a gabbro sheet was intruded along the plane of a thrust for nearly 160 km and pigeonitic dolerite dykes were emplaced. Harpum considered these various phases to constitute an igneous sub-province formed entirely during the Bukoban period.

The Bukoban basalts east of Lake Tanganyika extend for some 500 km (Fig. 2). They are believed to be made up of more than 900 m of amygdaloidal basalt in which olivine is rare and quartz and pigeonite common. No feeders have been found but occasional dolerite dykes appear in the volcanic rocks near the northern end of the basalt outcrop.

Most of the remaining dolerite dykes in Tanzania have been referred to as Younger Dolerites (*ibid.*) and it was previously thought that many were of Karroo (Mesozoic) age. However, Harpum gives reasons for regarding them as Precambrian and although not exactly the same age he considered them to be related to the Bukoban Eruptive Province. Following Teale (1930) several tholeiitic types were distinguished including: olivine-, augite-, quartz- or granophyric-quartz-dolerites. Noritic rocks at Kapalagula (Fig. 2) were excluded by Harpum on the grounds that they are pre-Bukoban. Although Harpum considered all olivine dolerites to lie south of a line between the southern end of Lake Tanganyika and just north of Dodoma (Fig. 2), some olivine dolerites do occur near Lake Eyasi. Harpum included the Kisii Series basalts, rhyolites and andesites in the Bukoban, and Huddleston (1951) showed that some unmetamorphosed quartz and pigeonitic dolerites cut the Kisii Series rocks.

On the north-west side of Lake Albert (Fig. 2), in the district of Kilo in the Congo Republic, numerous dykes have been intruded into the granitoid Precambrian basement rocks (Woodtli, 1959). The intrusives are ophitic pigeonitic dolerites, olivine being typically absent. They occur as individual dykes up to several tens of kilometres in length, but only a few metres wide, which make up a northerly-trending swarm about 150 km in length. Some of the dykes strike east-west and at least two ages are indicated. Woodtli tentatively suggested from geomorphological evidence, and their petrographic resemblance to the Karroo dolerites, that the dykes were Mesozoic and Quaternary and that they were emplaced during a period of tension related to the rift structures. The age of the dykes is quite unknown and they could easily be related to the Bukoban eruptive events of Tanzania.

The age of the Bukoban rocks has not yet been established and such age determinations as are available provide an ambiguous answer. It has long been considered (Quennell, 1956; Haughton, 1963) that the Bukoban System was a correlative of the Katanga System of Zambia and the Congo (and therefore about 600-900 m.y. old). Recent age determination work (Snelling, 1967; Cahen and Snelling, 1966) has indicated that the Bukoban sediments are younger than about 1300 m.y. while the Abercorn Sandstone is intruded by a dolerite which yielded a pyroxene K-Ar age of 940 ± 40 m.y. This suggests the Bukoban System may be equivalent to the Karagwe-Ankolean in age. Basic lavas from the Kisii Series have yielded whole rock K-Ar ages of 675 m.y. (*ibid.*) presumably reflecting the effects of the nearby Mozambique Belt.

Preliminary K-Ar dating of so-called Bukoban dolerites from other parts of Tanzania has provided a wide range of ages (Snelling, 1967; Hepworth, personal communication). Dolerites in the vicinity of Lake Eyasi and elsewhere have given ages of *c.* 2500 m.y., 1900 m.y. and 900 m.y. How far these results reflect ages of intrusion or subsequent metamorphic events is not yet known, but it seems that the basic dyke-swarms of East Africa belong to several different periods and perhaps to different tectonic regimes, although subsequent activity has in places followed old lines of weakness. Until reliable age determinations and systematic palaeomagnetic work has been done the dykes must remain grouped with the Bukoban.

Late Precambrian igneous activity adjacent to the Red Sea in North Africa and Arabia

The Eastern Desert of Egypt, the Sinai Peninsula, the Red Sea Hills of Sudan and western Saudi Arabia are all Precambrian terrains in which a particularly striking feature is the unusual abundance of dyke intrusions (Fig. 3).

The dykes are usually narrow, sub-vertical bodies of basic and acid rocks which in places occupy more terrain than the invaded country-rocks. Along the western side of the Red Sea the dykes have north, north-east, east-west, or south-east trends and they cover the entire Precambrian outcrop area which is about 200 km wide and over 1400 km in length. Lack of detailed mapping in Sudan and Ethiopia limits the knowledge of their southwards extent. In Saudi Arabia the dykes are not so continuous over wide areas (see Brown *et al.*, 1962) but this may be partly accounted for by lack of detailed mapping. Nevertheless, numerous dykes are present in an area of 1500×500 km.

The Precambrian terrains adjacent to the Red Sea must be amongst the most intensely dyke-intruded areas in the world and it is regrettable that so little information is as yet available from this remarkable province. Schürmann (1966) has summarised the data for the northern Red Sea area. He noticed that the dominant trend directions are perpendicular to the coast, occasionally parallel to it, or at such an angle as to form a conjugate pattern. Schürmann, however, warns that younger block faulting might be partly responsible for the differences in strike. Many dykes were intruded parallel to the cleavage directions in older plutonic granites. In the Eastern Desert of Egypt an early north-south cleavage was followed by acid dykes and the secondary east-west cleavage by basic dykes. A rough sequence of intrusion has been noted in which the oldest diabase and acid porphyries strike dominantly north-west, followed by microdiorites striking north-east, and the youngest acid alkaline and dolerite dykes striking north-south. Composite dykes are also present.

The dykes adjacent to the Red Sea belong to more than one period of igneous activity. Without direct geochronological or palaeomagnetic data it is not possible to determine these periods. However, many of the dykes are associated with various granite masses, and some of these have been isotopically dated.

The Aswan Granite in Egypt yields a Rb-Sr whole rock isochron age of 600±20 m.y. (Leggo, 1968). This supports other apparent ages determined by K-Ar, Rb-Sr and Pb methods from similar rocks in Egypt and Arabia. Gattarian granites from the northern part of the Eastern Desert gave mineral K-Ar and Rb-Sr ages of about 500 m.y. (Schürmann, 1966, p. 207). In Saudi Arabia older Rb-Sr ages have been encountered in the range 800–1050 m.y.

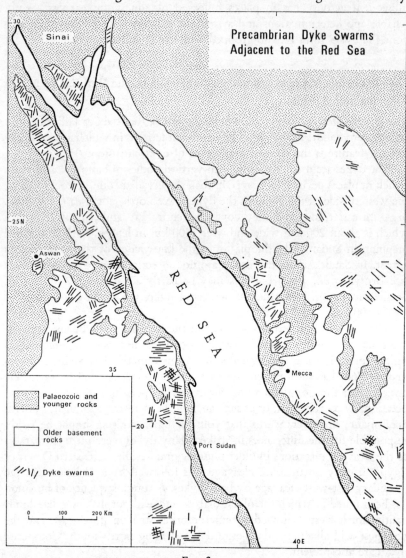

Fig. 3.

A single K-Ar age determination is available for one of the dykes in north-eastern Sudan which provided an apparent age of 740 ± 80 m.y. (Whiteman, 1968). Schürmann discusses the difficulties in interpreting the geochronological results from Precambrian terrains which have yielded Cambrian or Ordovician apparent ages, and which are overlain by fossiliferous Cambrian sediments.

In Arabia, dykes of acid and basic composition occur in Precambrian gneisses and in the so-called younger granites. The latter have yielded Rb-Sr ages of between 500 and 600 m.y. (Brown and Jackson, 1960). Some of the dykes pre-date the granites, indicating that there must be at least two quite different periods of dyke emplacement. In certain areas swarms of dykes have uniform strike directions, but the orientation of these swarms varies and in some places older swarms can be seen to strike in different directions to the younger sets.

The Red Sea igneous province can thus be seen to include extensive dyke intrusions over an area nearly 2000 km in length. They vary in composition from acid to basic; and, in age, they extend from before the Gattarain granites (late Precambrian to early Palaeozoic), to post-granite and pre-Mesozoic and younger cover. Although the dykes have a variety of strike directions, in general they trend across the axis of the Red Sea, either perpendicular to it or inclined in a conjugate system.

Mesozoic igneous activity in southern Africa

Most of South Africa and northwards across Rhodesia and Mozambique to southern Malawi belongs to the huge igneous province of the Karroo (see Cox, this volume, p. 212). Igneous rocks of Mesozoic age are found, in places in great abundance, within an area extending for nearly 3000 km north from the Cape and for 1000 km across the width of the southern extremity of Africa (Fig. 4).

Sedimentary strata, mainly of continental facies of the Karroo System (Upper Carboniferous to Lower Jurassic), cover much of South Africa, and extend intermittently northwards, in places beneath younger sediments, into Tanzania and the Congo. At the top of the sequence the Stormberg Series is terminated by Drakensberg volcanics. The type area for the extrusive rocks is in the highlands of Lesotho (formerly Basutoland) where 1500 m of flat-lying amygdaloidal basalts occur. Intruded into the volcanic pile there are numerous sub-vertical tholeiitic dolerite dykes that acted as feeders. Some of the dykes are of considerable lateral extent: one can be traced for 85 km and has a width of 300 m (Walker and Poldervaart, 1949). In the Transkei the so-called Gap Dykes extend for 150 km (Mountain, 1943). In the eastern Transvaal, the Lebombo Monocline (Du Toit, 1929) affects basalts and rhyolites along a narrow zone in which, at least in parts, dykes are associated and strike parallel to the trend of the monocline. In eastern Botswana and southern Rhodesia (Fig. 4) numerous dykes, mainly of dolerite but also granitic, as well as ring-complexes are associated with the Karroo extrusive rocks (Cox et al., 1965; Vail et al., in press). The irruptive rocks continue north-eastwards through central Mozambique (Vail, 1968b) to central Malawi where they pre-date the Lupata volcanic episodes (Vail, 1967b) and the Chilwa Alkaline Province intrusive suite (Dixey et al., 1937). Locally, some of the dykes are associated with central intrusive complexes and take the form of radiating swarms, such as Marangudzi Complex in Rhodesia (Cox et al., 1965) and Gorongosa in Mozambique (Vail, 1963).

There is a conspicuous absence of the intense dyke intrusions from the western regions. For example, the Precambrian Rhodesia craton is almost devoid of dykes of this age although small areas of Karroo basalts occur in the centre of the country and more extensively in the west (Cox *et al.*, 1967).

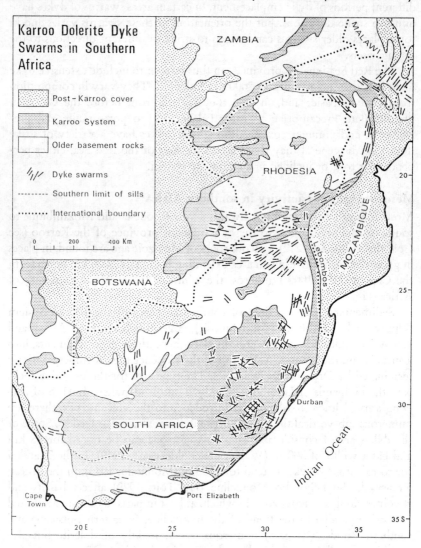

FIG. 4.

Despite the difficulty of recognising Mesozoic rocks in Precambrian terrains there is a real absence of dykes from these regions. The northernmost known Karroo dolerite dykes are in northern Malawi (Fitches, 1968) but rocks of this age may occur in Tanzania.

Although South Africa has been more fully mapped geologically than most of Africa, and the dolerite dykes and sills in particular have been the subject of careful study (Walker and Poldervaart, 1949), the Karroo igneous rocks of the type area in the southern Cape have not yet been systematically mapped. In Fig. 4 an attempt is made to show the distribution of Mesozoic dykes; the sills are omitted for clarity, and Walker and Poldervaart's line indicating the southern limit of Karroo dolerites is amended to show the limit of sills, since dolerite dykes near Cape Town and in pre-Karroo rocks further east are possibly also Karroo in age. North of Port Elizabeth pre-Cretaceous pyroclastics and lavas also are considered to belong to the Stormberg vulcanism (Haughton, 1963).

Within the Mesozoic Igneous Province intrusive rocks younger than the Karroo irruptives also occur (Woolley and Garson, this volume, p. 237). They are usually more alkaline, in contrast to the tholeiites, and include the Chilwa Alkaline Province syenites and the widespread carbonatites of southern Africa (Tuttle and Gittins, 1966). In north-eastern Rhodesia large quartz porphyry dykes and composite dolerites form north-west-trending swarms (Fig. 4). The later kimberlites can also be included in the province although they tend to occupy stable shield areas and to avoid the zones of recent movement and magmatism where the Karroo igneous rocks are commonly to be found; this is not to say they do not occur associated with Karroo rocks, for in Lesotho, for example, kimberlite pipes occur closely related to Karroo dolerite rocks and lava flows (Dawson, this volume, p. 323).

The igneous activity associated with the Karroo volcanic episode is confined to the Mesozoic. Vail (1968b) and Woolley and Garson (this volume, p. 243) have summarised the available geochronological data and have shown that age determinations indicate a duration of igneous activity of about 100 m.y. lasting from about 200 m.y. ago till the end of the Cretaceous. Palaeomagnetic information is limited to half a dozen sites, with a polarity reversal shown at several localities (Brock, 1968). Nevertheless, the Karroo igneous rocks show distinctive pole positions which enable them to be distinguished from other basic dykes where other means of identification are not available.

Regional tectonic control of igneous activity in eastern Africa

Igneous intrusives such as linear dykes and lines of igneous complexes usually gain access to the upper parts of the crust along zones of weakness. Locally, these planes of easy access are provided by such features as joint intersections, faults, foliations, cleavage planes, or lithological contacts. On a broader scale there appears to be a more fundamental overall control which is probably deep-seated and may reflect upper mantle and lower crustal conditions. Indirect evidence for such regional controls of irruptive activity may be deduced from the four areas of dyke emplacement considered above.

The fundamental geological structure of the southern African continent can be broadly identified as a mosaic of overlapping orogenic zones about their respective stable cratonic shields (Cahen and Snelling, 1966; Vail, 1967a); some of these are outlined in Fig. 5.

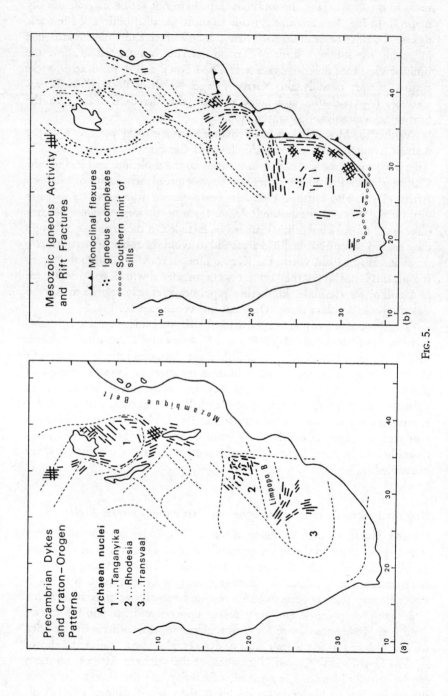

FIG. 5.

In Rhodesia and Transvaal there are indications that the Archaean cratons were bordered in early Precambrian times by once active orogenic gneiss belts (Johnson and Vail, 1965; Vail, 1968c). These structures were then penetrated by basic magma along a north-north-east axis to form the Great Dyke, Bushveld Complex and other intrusive bodies (Cousins, 1959; Bichan, this volume, p. 51). In Rhodesia, sedimentary formations accompanied by volcanic extrusions were deposited around the margins of the craton. Dolerites, probably closely associated with the vulcanism, were intruded as flat sills and sheets in the sedimentary rocks and as irregular bodies randomly oriented and only locally following lines of weakness in the granitoid Archaean basement rocks. Basic igneous rocks are notably absent from the high-grade gneisses constituting the Mozambique Belt in the east (Fig. 5a). In northern Transvaal, pre-Waterberg volcanic rocks border the Archaean craton and Waterberg diabase dykes occur along the edge of the Limpopo Belt or diagonally across it in the west (Fig. 5a). Four hundred million years later new lines of weakness were followed by the Pilansberg dyke-swarms across the stable cratonic platform of the Transvaal.

A somewhat similar situation occurs in Tanzania where Bukoban-type sediments and extrusive rocks occur around the margins of the Archaean craton. Dyke-swarms of various ages follow the margins of the active gneiss belts. Some swarms, for example those near Lake Eyasi (Fig. 2), cut diagonally across the old craton and in turn are followed by later rift movements (Tertiary to Recent). As in the case of the Precambrian dykes in Rhodesia, basic dykes are notably absent from the Mozambique Belt in Tanzania. This is different from north-east Africa and Arabia where Pan-African (Mozambique belt-type) orogenesis and isotopic events also prevailed. In the north, in strong contrast to Kenya, Tanzania and Mozambique, the Precambrian terrain is intensely invaded by basic dyke-swarms and granites. Two fundamentally different geological environments exist south and north of the Kenya–Ethiopia–Gulf of Aden zone of rift faulting and Tertiary vulcanism.

Major continental structures during late Precambrian times in North Africa are not clearly understood, but the tectonic pattern seems to have been less dependent on craton-orogen relations than in southern and eastern Africa. The dominant structure in this area is the geologically recent feature of the Red Sea Rift. The numerous dyke-swarms in the adjacent Arabian and North African cratons generally strike normal and parallel to the rift, or more usually they form an oblique conjugate system on either coast. The diagonal pattern suggests tension across the axis of the present Red Sea as the dykes cut freely across basement structural trends.

Mesozoic igneous activity in southern Africa occurred over a stable Palaeozoic cratonic area, the margins of which have suffered folding along the Cape Fold Belt in the south and monoclinal warping and fracturing along the eastern side. The craton was the site of the Karroo depositional basin but the Karroo dolerites are not confined to the basin. In regions of sediments the igneous bodies tend to take the form of flat sills as well as dykes, whereas in the older basement rocks dyke-swarms or irregular masses are more common.

Some dykes radiate from ring-complexes but they are of local importance, and the dykes parallel to the monoclinal flexures are more numerous and widespread. Around the south-east margins a conjugate system of dyke intrusion is developed (Fig. 5b).

In Rhodesia, the Karroo dolerites tend to avoid the central Archaean craton and preferentially follow the old orogenic belts. The Cape Fold Belt, on the other hand, has a remarkable absence of igneous intrusions. The distribution of Mesozoic irruptive rocks outside the Karroo basin is influenced by the rift valley fracture pattern (Vail, 1968b). Younger ring-complexes and associated dyke-swarms occur in, along, and at the inter-sections of the rift valleys (Fig. 5b). Kimberlite pipes, on the other hand, are widely scattered and only in a few regions are they located near rift fractures. Vail (1967a) has pointed out that the diamondiferous kimberlites are usually found cutting the Archaean cratonic nuclei (2650 m.y. or more in age) and those in the more youthful orogenic belts are not normally diamondiferous; this subject is dealt with by Dawson elsewhere in this volume (p. 330).

A consideration of selected irruptive regions in eastern Africa clearly shows that there have been several episodes of emplacement of dykes and related rocks. The magmatic activity has been subjected to an overall tectonic control by major crustal structures. Locally sills, rather than dykes, develop where confining pressures allow the igneous rocks to be emplaced hori-zontally rather than vertically. If the dykes represent tensional fractures then considerable expansion of the crust could have occurred. Du Toit (1954) considered that the Karroo dykes along the Lebombo Monocline are associated with warping movements, and Walker and Poldervaart (1949) considered the Karroo igneous activity to be related to the break-up of Gondwanaland by continental drift (see Cox, this volume, p. 228). Many of the African dykes strike perpendicular to the long axis of the continent and might indicate expansion in that direction but no evidence of continental disruption during earlier times is available from the configuration of the African dyke-swarms. Further geochronological and palaeomagnetic studies may throw light on these problems.

REFERENCES

Allsopp, H. L., Burger, A. J., and Van Zyl, C. 1967. A minimum age for the Premier kimberlite pipe yielded by biotite Rb-Sr measurements, with related galena iso-topic data. *Earth Planetary Sci. Let.*, **3**, 161.

Brock, A. 1968. Palaeomagnetism of the Nuanetsi Igneous Province and its bearing upon the sequence of Karroo igneous activity in southern Africa. *J. geophys. Res.*, **73**, 1389.

Brown, G. F., and Jackson, R. O. 1960. The Arabian Shield. *Int. geol. Congr.*, **21(9)**, 69.

—— Jackson, R. O., Bogue, R. G., and Maclean, W. H. 1962. Geology of southern Hijaz quadrangle, Kingdom of Saudi Arabia. *Misc. geol. Invest. U.S. geol. Surv.*

Cahen, L., and Snelling, N. J. 1966. *The geochronology of equatorial Africa*, Amsterdam.

CLIFFORD, T. N., REX, D. C., and SNELLING, N. J. 1967. Radiometric age data for the Urungwe and Miami Granites of Rhodesia. *Earth Planetary Sci. Let.*, **2**, 5.

COUSINS, C. A. 1959. The structure of the mafic portion of the Bushveld Igneous Complex. *Trans. geol. Soc. S. Afr.*, **62**, 179.

COX, K. G., JOHNSON R. L., MONKMAN, L. J., STILLMAN, C. J., VAIL, J. R., and WOOD, D. N. 1965. The geology of the Nuanetsi Igneous Province. *Phil. Trans. r. Soc. Lond.*, **257**, Ser. A, 71.

—— MACDONALD, R., and HORNUNG, G. 1967. Chemical and petrographic provinces in the Karroo basalts of southern Africa. *Am. Miner.*, **52**, 1451.

DIXEY, F., CAMPBELL SMITH, W., and BISSET, C. B. 1937. The Chilwa Series of southern Nyasaland. *Bull. geol. Surv. Nyasaland*, **5**.

DU TOIT, A. L. 1929. The volcanic belt of the Lebombo—a region of tension. *Trans. r. Soc. S. Afr.*, **18**, 189.

—— 1954. *The geology of South Africa*, 3rd ed., Edinburgh.

FITCHES, W. R. 1968. New K/Ar age determinations from the Precambrian Mafingi Hills area of Zambia and Malawi. *12th Ann. Rep. Res. Inst. African Geol., Univ. Leeds*, 12.

HARPUM, J. R. 1955. Recent investigations in pre-Karroo geology in Tanganyika. *C.R. Ass. Serv. géol. Afr., Nairobi Meeting* (1954), 165.

HAUGHTON, S. H. 1963. *The stratigraphic history of Africa south of the Sahara*, Edinburgh.

HUDDLESTON, A. 1951. Geology of the Kisii District. *Rep. geol. Surv. Kenya*, **18**.

JOHNSON, R. L., and VAIL, J. R. 1965. The junction between the Mozambique and Zambesi Orogenic belts; north-east Southern Rhodesia. *Geol. Mag.*, **102**, 489.

JONES, D. L., and McELHINNY, M. W. 1966. Palaeomagnetic correlation of basic intrusions in the Precambrian of southern Africa. *J. geophys. Res.*, **71**, 543.

—— and McELHINNY, M. W. 1967. Stratigraphic interpretation of palaeomagnetic measurements on the Waterberg Red Beds of South Africa. *J. geophys. Res.*, **72**, 4171.

LEGGO, P. J. 1968. Some recent isotope investigations. *12th Ann. Rep. Res. Inst. African Geol., Univ. Leeds*, 45.

McELHINNY, M. W. 1966. The palaeomagnetism of the Umkondo lavas, eastern Southern Rhodesia. *Geophys. J. r. astr. Soc.*, **10**, 375.

—— and OPDYKE, N. D. 1964. The palaeomagnetism of the Precambrian dolerites of eastern Southern Rhodesia, an example of geologic correlation by rock magnetism. *J. geophys. Res.*, **69**, 2465.

MOUNTAIN, E. D. 1943. The dikes of the Transkei Gaps. *Trans. geol. Soc. S. Afr.*, **46**, 55.

NICOLAYSEN, L. O., DE VILLIERS, J. W. L., BURGER, A. J., and STRELOW, F. W. E. 1958. New measurements relating to the absolute age of the Transvaal System and of the Bushveld Igneous Complex. *Trans. geol. Soc. S. Afr.*, **61**, 137.

OOSTHUYZEN, E. J., and BURGER, A. J. 1964. Radiometric dating of intrusives associated with the Waterberg System. *Annls. geol. Surv. S. Afr.*, **3**, 87.

QUENNELL, A. M. 1956. The Bukoban System of East Africa. *Int. geol. Congr.*, **20**, El Sistema Cámbrico, 281.

SCHÜRMANN, H. M. E. 1966. *The Precambrian along the Gulf of Suez and the northern part of the Red Sea*, Leiden.

SLATER, D. 1967. A correlation of the Gairezi Series with the Umkondo System, Rhodesia. *Trans. geol. Soc. S. Afr.*, **70**—in press.

SNELLING, N. J. 1963. Age determination unit. In *Ann. Rep. Overseas geol. Survs.* (for 1962), p. 30.

—— 1966. Age determination unit. In *Ann. Rep. Overseas geol. Survs.* (for 1965), p. 44.

—— 1967. Age determination unit. In *Ann. Rep. Inst. geol. Sci.* (for 1966), p. 142.

SWIFT, W. H. 1961. An outline of the geology of Southern Rhodesia. *Bull. geol. Surv. S. Rhodesia*, **50**.

TEALE, E. O. 1930. Dolerites. *Ann. Rep. geol. Surv. Tanganyika*, 9.

TRUTER, F. C. 1949. A review of volcanism in the geological history of South Africa. *Proc. geol. Soc. S. Afr.*, **52**, xxix.

TUTTLE, O. F., and GITTINS, J. 1966. *Carbonatites*, London.

TYNDALE-BISCOE, R. 1958. The geology of a portion of the Inyanga District. *Short Rep. geol. Surv. S. Rhodesia*, **37**.

VAIL, J. R. 1963. Notes on the Inhamangombe and Gorongosa dyke swarms, Mozambique. *7th Ann. Rep. Res. Inst. African Geol., Univ. Leeds*, 28.

—— 1967a. Distribution of non-orogenic igneous complexes in southern Africa and their tectonic setting. *11th Ann. Rep. Res. Inst. African Geol., Univ. Leeds*, 33.

—— 1967b. Contribution to the discussion on the age of the Lupata rocks, lower Zambesi Valley, Mozambique by G. Flores. *Trans. geol. Soc. S. Afr.* **67**, 283.

—— 1968a. Aspects of the stratigraphy and structure of the Umkondo System in the Manica belt of Southern Rhodesia and Mozambique and an outline of the regional geology. *Trans. geol. Soc. S. Afr.*, **68**, 13 and 248.

—— 1968b. The southern extension of the East African Rift System and related igneous activity. *Geol. Rdsch.*, **57**, 601.

—— 1968c. Significance of the tectonic pattern of southern Africa. *Tectonophys.*, **6**, 403

—— HORNUNG, G., and COX, K. G. (in press). Karroo basalts of the Tuli Syncline, Rhodesia. *Bull. volc.*

—— SNELLING, N. J., and REX, D. C. 1968. Pre-Katangan geochronology of Zambia and adjacent parts of central Africa. *Can. J. Earth Sci.*, **5**, 621.

VAN NIEKERK, C. B. 1962. The age of the Gemsbok Dyke from the Venterspost Gold Mine. *Trans. geol. Soc. S. Afr.*, **65**, 105.

WALKER, F., and POLDERVAART, A. 1949. Karroo dolerites of the Union of South Africa. *Bull. geol. Soc. Am.*, **60**, 591.

WHITEMAN, A. J. 1968. Formation of the Red Sea Depression. *Geol. Mag.*, **105**, 231.

WOODTLI, R. 1959. Les intrusions doleritiques de Kilo (NE du Congo Belge) et leurs relations avec la tectonique. *Int. geol. Congr.*, **20**, *As. Serv. geol. Afr.*, 485.

J. M. ROOKE

17   Geochemical variations in African granitic rocks, and their structural implications

ABSTRACT. *On the basis of spectrographic, chemical and X-ray fluorescence analyses of approximately 200 samples of granites and kindred rock types from many parts of Africa, non-orogenic complexes in Nigeria, Nuanetsi (Rhodesia), Botswana, the Chilwa Province of Malawi, northeast Sudan and South Arabia are contrasted with geochemical data from orogenic occurrences in other regions, including Egypt, Somalia, Uganda, Zambia, Malawi, Rhodesia and South-West Africa.*

Introduction

In the past decade many research geologists have been privileged to work in Africa under the aegis of the Research Institute of African Geology at Leeds University, and a large proportion of their collections has passed through the analytical laboratories of the Earth Sciences Department. Whereas individual workers in the field are mainly concerned with geochemical detail in their own areas, the analyst is in an ideal position to co-ordinate and take a broader view. Recognising this opportunity, Professor W. Q. Kennedy initiated a project aimed at the identification and delimitation of geochemical provinces in Africa; the work has been carried out in the spectrographic section, and every sample in this study has been analysed for trace constituents by the optical emission spectrographic technique at Leeds (Rooke and Fisher, 1962). Miss J. Brown, Mrs A. M. Fisher and Miss P. W. Roe have given valuable assistance.

Most of the specimens have been chemically analysed for major elements by classical methods in the Leeds laboratories by Dr O. Von Knorring, Mrs M. H. Kerr and Miss J. R. Baldwin; two sets of results are by Dr G. Hornung using X-ray fluorescence; and many of the samples from Malawi have been chemically analysed in Glasgow by Mr W. H. Herdsman. Other smaller groups of analyses for major constituents are acknowledged in the text, and I should like to express my gratitude to all these analysts, and to the Directors and staff of Geological Surveys throughout Africa, past and present members of the Research Institute of African Geology, and members of the Department of Earth Sciences at Leeds University, for their generous help and ready co-operation in supplying samples and information for this project. I am indebted to staff of the Electronic Computing Laboratory at Leeds University for advice and assistance. Grateful thanks are due to the secretaries in the Department of Earth Sciences who have had the onerous task of typing the tables of data, and also to the technical staff for their skilful practical support on numerous occasions.

This chapter presents an interim report; acidic rock groups of widely varying ages, tectonic environment and geographical distribution in Africa are located (Fig. 1) and described. It is fully recognised that the sampling is far from random, and that further consolidation of the analytical data is essential; the need to assess these results in relation to geochemical variations in basaltic associations has not been overlooked.

Location of samples

NIGERIA

Nineteen samples from the Younger Granite Province of northern Nigeria (Jacobson et al., 1958; Butler et al., 1962) are included in this work (see Fig. 1, no. 1). The group comprises nine specimens from the Liruei Hills, three from Amo, one from Jos (9° 50′ N., 8° 50′ E.), three from Ropp and three from the Sha-Kaleri Complex. An early volcanic phase is represented, but the majority of samples are from high-level granitic intrusions controlled by ring-fracturing and major block subsidence. They are characterised by tin mineralisation and show sharp discordant contacts with the country rocks; their emplacement was not related in time to any cycle of orogeny, although the area (Fig. 1) lies in terrain affected by the Pan-African thermo-tectonic episode ± 500 m.y. ago (Kennedy, 1964; see Black and Girod, this volume, p. 193).

Isotopic measurements show that the ring-complexes of northern Nigeria are mid-Jurassic in age. Determinations by the K-Ar method on biotites from two granites (Rayfield, Jos Plateau Province; Banke, Zaria Province) have given 162 ± 5 m.y. and 164 ± 5 m.y. respectively; and by the Rb-Sr method, 159 ± 6 m.y. and 162 ± 5 m.y. respectively (Jacobson et al., 1964). Thirteen samples were chemically analysed in Leeds, and six chemical analyses were supplied by the Geological Survey Department of Nigeria.

ASWAN, UNITED ARAB REPUBLIC

Seven porphyritic granites and granitic gneisses from the Aswan Granite Complex (24° 3′ N., 32° 53′ E.; see Fig. 1, no. 2) have been described by Shackleton (in preparation). The intrusion measures approximately 3 miles in diameter and is situated within a broad belt of gneisses and schists on the eastern side of the Nile (Fig. 1). This area lies in an orogenic zone which may be an extension of the main north-south Mozambique Belt (Shackleton, 1964).

Gindy (1954) postulated a migmatitic origin for the Aswan Granite; in his view, it developed in situ from the country rocks by intrusion of acid pegmatitic material, granitisation and orogenic deformation. Isotopic studies on whole rock samples gave an age of 600 ± 20 m.y. and an initial Sr^{87}/Sr^{86} ratio of 0.704 ± 0.001 (Leggo, 1968).

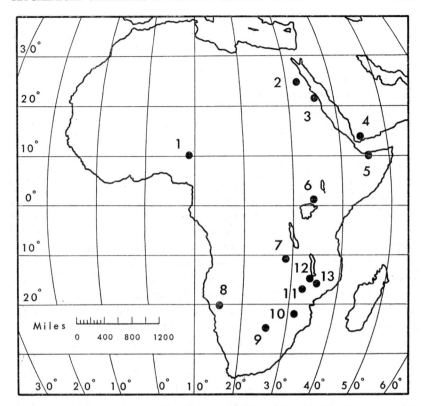

FIG. 1. Outline map of Africa showing locations of samples in this study: 1, Northern Nigeria; 2, Aswan, United Arab Republic; 3, north-east Sudan; 4, Aden, South Arabian Federation; 5, Somalia; 6, south-east Uganda; 7, Luapula, Zambia; 8, Franzfontein, South-West Africa; 9, Gaberones, Botswana; 10, Nuanetsi, south-east Rhodesia; 11, Chirwa, north-east Rhodesia; 12, Malawi Basement Complex; and 13, Chilwa Alkaline Province, Malawi.

All the Aswan samples were analysed in Leeds by Dr G. Hornung using X-ray fluorescence.

NORTH-EAST SUDAN

Delany (1955) drew attention to ring-structures in the northern parts of Sudan. Riebeckite granites of similar occurrence to those in northern Nigeria are found in the region bordering the west coast of the Red Sea Rift (Fig. 1); they are included here with a suite of post or non-orogenic granites, syenites and Tertiary volcanics from north-east Sudan (see Fig. 1, no. 3; and Gass, 1955).

The acid rocks, located at approximately 21° 45′ N., 36° 30′ E., comprise three rhyolites (Asotriba volcanics) and two granites. An intermediate rock grouping includes four syenites (two from the Salala Ring-Complex) and two Tertiary trachytes (21° 20′ N., 36° 10′ E.). These localities lie in a region affected by the Mozambiquian Orogeny. Field evidence suggests close affinities between the plutonic and volcanic phases and also indicates a post-Jurassic age, but the latter is contradicted by recent tentative isotopic data which give the age as very approximately 600 m.y. (Gass, personal communication).

All eleven samples have been chemically analysed in Leeds; it is recognised that the two sub-groups are rather small and any conclusions using these geochemical data must be cautious. An intensive study of a wider selection of samples from the area is presently in hand (Gass and Neary, in preparation).

ADEN, SOUTH ARABIAN FEDERATION

A suite of peralkaline rhyolites and trachytes from an extensive Tertiary volcanic field to the west of Aden, on the south coast of Arabia (12° 45′ N., 44° 55′ E.; see Fig. 1, no. 4) has been described (Gass et al., 1965; Cox et al., in preparation). Two volcanoes sampled in the present study—those at Aden and Little Aden—are closely related in time and space and therefore are combined as one locality. Original composite cones of basic lavas with rhyolitic effusives and agglomerate were later modified, and the caldera which were formed were infilled with horizontal extrusives of intermediate composition. The acidic group of samples consists of nine rhyolites and four comendites; in the intermediate grouping there are four trachytes and two trachyandesites. Both volcanoes are approximately equally represented in the two sets of data.

These occurrences lie close to the Red Sea–Gulf of Aden Rift System, where volcanic activity is probably associated with rift development. The rocks have given radiometric ages varying from 5 to 10 m.y. (Dickinson et al., 1969), hence their eruption took place 10–20 m.y. after the rift from the mainland of Africa had occurred. A large proportion of samples have initial Sr^{87}/Sr^{86} ratios close to 0·704, which points to a mantle origin (ibid.). The authors postulate that much of the observed chemical differentiation occurred at depth, although there is strong evidence that a high-level magma chamber existed immediately below the volcanoes. All the samples have been analysed chemically in Leeds.

SOMALIA

A representative collection of granitic rocks from the basement complex of northern Somalia (Fig. 1, no. 5) has been made in the course of an investigation into the geology of the Darkainle nepheline syenite complex (Gellatly, 1963). Twenty-four samples, including five of syenitic composition, have been taken from an area extending from 10° 0′ N. to the Gulf of Aden (Fig. 1) and from 43° 0′ E. to 49° 0′ E., over the Hargeisa, Berbera, Burao and Erigavo Districts from Darkainle in the west to Las Khoreh in the east (Mason and Warden, 1956; Hunt, 1958, 1960; Gellatly, 1960; Greenwood, 1960).

The group consists mainly of metasomatic and migmatised granites and granitic gneisses, with some discrete intrusive masses and older plutons; a few younger discordant plutons are included, but none showing evidence of ring-faulting or cauldron subsidence. Granitisation and migmatisation occurred at various stages in the Basement System, and three periods of folding and metamorphism have been noted (Gellatly, 1963). Northern Somalia is remote from the stable cratonic areas of the continent, and may well be part of a northern extension of the Mozambique Belt (Clifford, 1968, p. 320).

Although the bulk of the samples are thought to be Precambrian, age determinations by K-Ar on biotites from an adamellite (north of Mait Pass) and the Las Bar Granite give 500 ± 20 m.y. and 515 ± 25 m.y. respectively (Snelling, 1963). All the specimens were analysed chemically in Leeds.

SOUTH-EAST UGANDA

Gneisses and volcanics of the Nyanzian System in the south-east corner of Uganda (Fig. 1, no. 6) provide a setting for a number of acid intrusions, some of which have been chemically analysed and described by Von Knorring (1960) in a study of the Lunyo area. A group of seven granites, granodiorites and porphyries (excluding the Lunyo Granite) from the Bukedi and Busoga Districts (0° 30' N., 34° 0' E.) is incorporated in this work, together with three samples from other plutons in the region, supplied and chemically analysed by the Uganda Geological Survey.

Old (1968), discussing the Masaba and Buteba Granites, suggests that they were emplaced approximately 2600 m.y. ago, and remobilised at around 1800 m.y.

LUAPULA, ZAMBIA

The stratigraphical succession in the mid-Luapula valley along the Zambia–Congo Republic frontier (Fig. 1) has been examined by Abraham (1959). Acid volcanics and granitic rocks form the basement complex in the region to the north-west of Fort Rosebery (10° 50' S., 28° 40' E.); the Luapula volcanics extend in a belt for at least 300 miles from the Congo pedicle in the south to the Marungu Plateau in the north, and five samples of rhyolites, rhyodacites or dacites have been studied (Fig. 1, no. 7). Two granites from the Luongo intrusion are also included; this body was emplaced after the volcanics but prior to the deposition of the Plateau Series of the Roan System, and Abraham (*ibid.*) considers that there is a genetic relationship between the two groups of acidic rocks.

The age of the rocks is certainly pre-Katanga (Abraham, personal communication), but the chronology of pre-Katangan events is somewhat ambiguous (Snelling *et al.*, 1964a). Age determinations have shown the overlying Plateau Series and Abercorn Sandstone to be in the time range 940–1800 m.y. (Snelling, 1967; see also Cahen, this volume, p. 103), and the neighbouring and possibly correlative Kate Granite has an age of 1800 m.y. (Page, 1960). The age of the Luapula volcanics is therefore thought to be greater than 1800 m.y.

FRANZFONTEIN, SOUTH–WEST AFRICA

Basement granites of South–West Africa (Fig. 1, no. 8) have been studied in detail by Clifford et al. (1962). Eleven samples in the present investigation were taken from the core of the Huab Anticline in the Franzfontein region (20° 10′ S., 15° 15′ E.), where granitic rocks occupy an area of over 3000 square miles (Fig. 1). Crystallisation formed part of a major period (the Huabian episode) of batholithic granite emplacement, and an intimate association with regionally developed pre-Otavi schists and gneisses indicates that this episode was a plutonic-tectonic event. Isotopic work on constituent minerals and whole-rock samples has given an age of 1650 ± 80 m.y. for the emplacement of these granitic rocks (Clifford et al., 1969), with a Sr^{87}/Sr^{86} initial ratio of 0.708 ± 0.002. The localities lie within a belt subsequently affected by the Damaran Orogeny (450-550 m.y.) which extends across the African continent in a north-easterly direction, separating the Congo and Kalahari cratonic areas. All the specimens in this acidic rock group were chemically analysed in Leeds.

GABERONES, BOTSWANA

An area of approximately 1000 square miles in south-east Botswana (24° 40′ S., 25° 50′ E.) is occupied by the discordant Gaberones Granite (Fig. 1, no. 9) which has been mapped by Wright (1958, 1961). The complex is anorogenic, and takes the form of a differentiated laccolith, the result of slow updoming of a viscous magma under a shallow cover of its own extrusive phase. Seven samples in this present investigation represent a succession of felsite, granophyre and granitic rocks including two central rapakivi granites. All were chemically analysed in Leeds.

Snelling (1966) reports a K-Ar age of 2305 ± 115 m.y. for hornblende from the central granite. Whole-rock measurements by the Rb-Sr method on three samples of the Gaberones Granite have given an age of 2340 ± 50 m.y. (McElhinny, 1966), and an initial ratio for Sr^{87}/Sr^{86} at 0.714 ± 0.031.

NUANETSI, SOUTH–EAST RHODESIA

The Nuanetsi Igneous Province is situated in south-east Rhodesia (Fig. 1, no. 10) immediately to the north of the Limpopo River, and close to the border with Mozambique (approximately 22° 0′ S., 30° 30′ E.). It lies at the intersection of the Lebombo Monocline (a boundary between the uplifted central portion of southern Africa and the depressed Mozambique 'geosynclinal' area to the east) and the east-north-east-trending Limpopo Orogenic Belt ($\sim$2000 m.y.) (see Cox, this volume, p. 212). This ancient belt apparently controlled the location of intrusive activity during the late-Karroo period (Cox et al., 1965).

Twenty samples are included, three of them being rhyolites from an early extrusive phase, the remainder drawn from later intrusive phases including granophyres from Masukwe, Dembe-Divula and the Maose-Malibangwe area, and granites from six non-orogenic ring-complexes.

Manton (1968) gives the age of the extrusives at 206 ± 13 m.y. by the

Rb-Sr whole-rock method. The age of the intrusives is also early Jurassic, with some variation amongst the complexes; Marangudzi approximately 200 m.y.; Masukwe and Dembe-Divula, 177 ± 7 m.y.; and the Main Granophyre approximately 195 m.y. Measured initial Sr^{87}/Sr^{86} ratios for the rhyolites and for the granites from the ring-complexes are $0·7081 \pm 0·0008$ and $0·7085 \pm 0·0007$ respectively. These low initial ratios agree with those for acid rocks at the southern end of the Lebombo Monocline.

Major element concentrations are not complete for the whole group, but thirteen samples have been chemically analysed in Leeds.

CHIRWA, NORTH–EAST RHODESIA

Mobilised Archaean granites occurring in the north-east corner of Rhodesia (17° 30′ S., 32° 45′ E.; Fig. 1, no. 11) have been described by Johnson (1968). Fourteen analysed samples are included in the present investigation; they comprise nine granites and granodiorites from the Chirwa Granite and its vicinity, two granites from the Nyalugo sheet and three from Matisi Bridge. The latter represent original cratonic material of an age similar to that of the neighbouring Mtoko batholith (> 2300 m.y.) which was studied isotopically by Snelling et al. (1964b). The rest of this group of samples are thought to have been remobilised during the Mozambiquian Orogeny, c. 550 ± 100 m.y. ago; a Rb-Sr age determination on biotite from the Chirwa Granite gave 460 m.y. but the whole-rock result is much older (Johnson, 1968). All the samples have been analysed in Leeds by X-ray fluorescence by Dr G. Hornung.

MALAWI BASEMENT COMPLEX

Twelve granitic rocks of basement or orogenic type have been collected in central and southern Malawi (Fig. 1, no. 12; see Bloomfield, this volume, p. 119). They include three magmatic occurrences of probable Lower Palaeozoic age from the southern and eastern shores of Lake Malawi. In addition there are specimens from the Precambrian Basement Complex in the vicinity of Tundulu (Garson, 1962), Fort Manning area, Fort Johnston and west of Lake Malombe (Holt, 1961). Two of the specimens are from the Ntonya infracrustal ring-complex south of Zomba. Bloomfield (1965b) believes that these have been formed by diapiric movements during intense lateral compression. Ages are still in doubt, but the majority of samples are thought to be Precambrian. All but two of the chemical analyses were by W. H. Herdsman.

Basement syenites and perthosites have been gathered from widely scattered localities throughout Malawi, ranging from Port Herald area in the south (17° 0′ S., 35° 10′ E.) to Rumpi in the northern region (11° 0′ S., 33° 50′ E.) and the geology has been described by Bloomfield (1968). Most of the 17 samples are of Precambrian age but some are thought to be Lower Palaeozoic; the infracrustal ring-complexes are represented. One sample was chemically analysed in Leeds, four by J. H. Scoon at Cambridge, and the remainder by W. H. Herdsman.

CHILWA ALKALINE PROVINCE OF MALAWI

In the region of Lake Chilwa in southern Malawi (Fig. 1, no. 13) there occurs a group of syenogranitic and nepheline syenite plutons associated with volcanic vents infilled with carbonatite, and cut by various minor intrusions (Dixey et al., 1955; Woolley and Garson, this volume, p. 237). Generally in the form of ring-complexes less than 10 miles in diameter, and in sharp contact with the country rocks, they are thought to have been emplaced by cauldron subsidence under essentially epeirogenic (or anorogenic) conditions (Bloomfield, 1965a). The main plutonic phase of the Chilwa Alkaline Province occurred in late-Jurassic to early-Cretaceous times; the ring-complexes lie within the Mozambique Orogenic Belt but are significantly younger than Mozambiquian orogenic activity, and the eastern escarpment of the main north-east trending Rift Valley Fault cuts through the group. Vail (1967) emphasises that the Lake Malawi–Shire Valley Fault is a region of overlap of earlier and later phases of post-Karroo dislocation.

Thirty-five samples from the province have been included, 13 acidic and 22 of intermediate composition. The acid rocks are predominantly from Mlanje Mountain (15° 55′ S., 35° 40′ E.), and are mainly granites of riebeckite type (Walshaw, in preparation). A radiometric age of 128 ± 6 m.y. by K-Ar on biotite from quartz syenite has been quoted (Snelling, 1966). Dyke samples from Zomba Mountain and East Mongolowe are also included (Bloomfield, 1965a; Vail and Mallick, 1965). In the acidic group as a whole six samples were chemically analysed in Leeds, and seven by W. H. Herdsman.

Syenitic rocks from Mlanje Mountain (Walshaw, 1966) form half the total of intermediate samples from the Chilwa Alkaline Province. Snelling (1966) gives an age of 116 ± 6 m.y. by K-Ar on biotite from a perthosite in the outer ring of this complex. Also included are three syenites from the Zomba Mountain granite/syenite pluton (Bloomfield, 1965a), and eight alkaline rocks from the Fort Johnston District (Holt, 1961), Chikala Hill (Stillman and Cox, 1960), Chaone (Vail and Monkman, 1960), West Mongolowe (Vail and Mallick, 1965) and Chinduzi Hill (Bloomfield, 1965a). Age determinations on biotite from pulaskite at Chaone by the K-Ar method gave 116 ± 6 m.y. (Snelling, 1966), and on zircon by the Pb/α method 138 ± 14 m.y. (Bloomfield, 1961). Of the 22 samples in the intermediate group nine were analysed in Leeds, and the remainder by W. H. Herdsman, all by the wet chemical method.

Data processing and presentation

This study of geochemical provinces requires that comparisons be made between groups of analyses of similar rock types from different localities in order to assess overall agreement or disparity. A simple method is available in the statistical technique of analysis of variance, whereby mean values of several groups of observations may be tested for the significance of their differences (Miller and Kahn, 1962, p. 134). The method assumes that the data are normally distributed although it has been stated (Quenouille, 1959) that the procedure is insensitive to non-normality provided that all observations are similarly affected. However, it would be wrong to assume normal distribution for all elements in igneous rocks, particularly in the light of accumulated evidence (Ahrens, 1963, 1966). The prevalence of positively skewed lognormal-type distribution suggests that a logarithmic transformation might be applicable over a fairly wide field (Rodionov, 1962) but it must be borne in mind that, where skewness is not pronounced, such a transformation could give over-correction.

Frequency distribution diagrams for major and trace constituents in African granites (Rooke, 1964) show that no single rule can be adopted to cover all cases, since abnormal patterns may arise from the mixing of results from several distinct populations. Somewhat daunted by the prospect of determining the distribution of numerous groups of data in the present work, I decided at the outset to make two alternative assumptions—normal and lognormal—and conduct all stages of computation in duplicate. Tables of variance have been drawn up to give a direct comparison of results from these two possible forms of distribution. There may be occasions when neither condition is strictly true, and it is questionable whether the statistical technique is sufficiently robust in these circumstances; but a discussion of the problem in general terms (Cochran, 1947) indicates that no serious error is introduced and efficiency remains high for moderate departures from normality. Furthermore, the strain imposed on a second basic assumption of the statistical method—that all variances are homogeneous—can be reduced considerably by logarithmic transformation of raw geochemical data. In practice the conclusions reached by the two-fold approach show remarkable consistency for the great majority of elements examined.

A survey of this size could only be undertaken with the aid of electronic computing in the Department of Computational Science at Leeds University, and a program written by Dr C. J. Bell was used in the early stages of the project. More recently Dr D. Wood, presently at the Courant Institute of Mathematical Sciences in New York, developed a series of three computer programs specifically designed to accommodate a growing file of geochemical data, and to deal with the calculations necessary in a search for significant variations. Details are to be published elsewhere, but in brief, it is possible to select groups of samples by a variety of criteria, apply logarithmic conversion to the data if required, and calculate mean value, range, sums of squares, variance and standard deviation, and finally the variance ratio F with appropriate degrees of freedom.

Appendix Tables 1(a)-(j)[1] collate the arithmetic mean values for the various localities in Africa, and for other wider groupings. Geometric mean values are not quoted, although it will be appreciated that significance tests using logarithmic conversion refer to geometric means. It has been necessary to show the number of samples in three categories: those less than the limit of detection (<); the determined values (n); and those greater than the

[1] For Appendix Tables, see pp. 381-414.

maximum working range of the analytical method (>). Partial analysis of some specimens explains any lack of consistency amongst the sample totals for any one group.

Since the limits of detection vary considerably it is thought undesirable to assume zero level for the elements quoted as 'less than' the limit, hence all averages are calculated from determined values only, and subsequent tests for significance also ignore the indeterminate groups of results. Consequently the analysis of variance tables (Appendix Tables 2(a)-(s)) omit those elements having more than about 10% of samples in the 'less than' or 'greater than' combined categories. An attempt is made to take the latter into account by visual assessment, and a question-mark under the heading of significance (e.g. for Mg in Appendix Table 2(a)) indicates clear evidence of a difference in analytical results by inspection.

Analytical results

Analyses for a total of 212 samples are available, of which 156 are acidic and 56 of intermediate composition. This subdivision is based on mineralogical considerations and proportion of normative quartz, using 10% as a guiding factor; the lower limit for syenitic rocks has been chosen, somewhat arbitrarily, as 55% SiO_2. Whilst it is debatable whether the two types should be separated, since they may well be cogenetic, it is necessary to set limits.

'Grand average' compositions of the acid and intermediate rock groups are presented in Appendix Tables 1(a) and (b), together with (in Appendix Table 1(a)) ranges of determined values and standard deviations for all elements. In normal practice the trace analyses would be rounded-off to accord with known experimental error, but here the statistical mean is followed more closely. However, the arithmetic mean for African rocks may be the expression of an unjustified amalgamation of several distinct populations, each having a unique geochemical association. It is the aim of this work to seek out such provinces using the technique of analysis of variance, whereby selected groups of samples may be inter-compared for significant variations in their average concentrations. An essential pre-requisite is that the subdivisions under test should be chosen by *prima facie* criteria totally unconnected with the analytical data. Differences shown to be 'highly significant' or 'significant' (Appendix Tables 2(a)-(s)) are discussed in the following sections.

ANOROGENIC AND OROGENIC ROCKS

Examples of a predominantly thermal anorogenic type of igneous intrusion are to be found amongst the discordant 'Younger Granites' such as those of Nigeria, but anorogenic events are not essentially tied to any particular geological age or form of complex. Amongst the *acid rock* localities under consideration, those of Aden, north-east Sudan, Nigeria, Nuanetsi (south-east Rhodesia), the Chilwa Province of Malawi and Gaberones (Botswana) are allocated to the 'anorogenic' group; they are compared with rocks whose formation was intimately associated with orogeny at seven localities: Aswan (U.A.R.), Somalia, south-east Uganda, Luapula (Zambia), Chirwa (north-east Rhodesia), Malawi and South-West Africa. Arithmetic mean values are shown in Appendix Table 1(a), and significant differences in Appendix Table 2(a).

The average values of Si, Fe^{2+}, K and H_2O+ do not differ significantly between the two groups. However, five major elements and several trace constituents show highly significant variations: Al, Ca, Mg and Sr are low in the anorogenic rocks, whilst Fe^{3+}, Na, Mn, Ga, Li, Nb, Y and Zr are distinctly high; and these conclusions are reached also after logarithmic conversion of the data. It may be that the observed differences reflect early and late stages in fractional crystallisation, and it seems likely that there are two distinct populations. The possibility arises that the two groups represent a real discontinuity in either mode of origin or subsequent development.

INTRUSIVE AND EXTRUSIVE ANOROGENIC ROCKS

In this comparison of intrusive and extrusive anorogenic rocks there is some tendency towards regional bias as the Aden representatives are entirely volcanic, but extrusive phases from Nigeria, Nuanetsi, Malawi, Botswana and north-east Sudan make up about 50% of the total (Appendix Table 1(c)).

Significant differences are observed in only two major constituents, the most noticeable being Fe^{3+}, which is enriched in the rhyolites; associated with this element there are increases in Mn and perhaps in Nb (provided the distribution is normal). The other main constituent of note is K, which is lower in the rhyolites, together with diminished amounts of Rb and Li (Appendix Table 2(b)).

REGIONAL VARIATIONS IN THE ANOROGENIC ACID ROCKS OF AFRICA

Subdivision of the anorogenic acid rock group into six localities at Aden, Chilwa Province (Malawi), northern Nigeria, Nuanetsi Province (Rhodesia), Gaberones (Botswana) and north-east Sudan (Appendix Tables 1(d) and (e)) yields an impressive number of highly significant variations (Appendix Table 2(c)).

It is remarkable that, despite the numerous highly significant differences in both major and trace constituents, the variations in Al are only 'probably' significant, and visual assessment of Sr indicates reasonable consistency. Aden and Sudan mark the lower and higher limits in the average values for Si; Sudan has low Ti, whereas Nuanetsi and Gaberones carry higher concentrations of this element (Appendix Table 1(e)). Total iron contents are unusually high in Aden (3·62%) and Chilwa (3·45%) but very low in Sudan (1·01%); Mn reaches very high levels in Aden and Chilwa (Appendix Table 1(d)). Mg is difficult to assess because of the indeterminate groups, but it has its highest concentration in Nuanetsi and Gaberones. Ca is present in relatively large amounts in Aden (0·75%), Nuanetsi (0·97%) and Gaberones (0·83%), and it may be noted that there is similarity with some orogenic values (Appendix Tables 1(f) and (h)). However, Aden has exceptionally high Na concentrations with rather low K, whilst in Nuanetsi and Gaberones there is a marked drop in Na, coupled with high K. Chilwa carries a similar 'enrichment' in K, but Sudan has the lowest average content of this element amongst the anorogenic locations.

Considerable local diversity is apparent (Appendix Tables 1(d) and (e)) in the trace elements: Ba is high in Aden, Nuanetsi, Gaberones and Sudan; Ga is unusually low in Nuanetsi and Sudan, and relatively enriched in Chilwa and Nigeria; Li has its greatest values in Chilwa, Nigeria and Gaberones. The analyses of Mn by spectrographic and by chemical methods are in good agreement, showing peak levels in Aden and Chilwa. There is a lack of Nb at Gaberones and Sudan, and the latter area is undoubtedly depleted in Rb, Y and Zr (Appendix Table 1(e)). Chilwa, Nigeria and Aden carry high amounts of Nb and Zr, whilst Rb and Y are also at high levels in Chilwa and Nigeria. Gaberones shows an excessive amount of Rb, and fairly high Zr.

At this stage it is evident that Chilwa and Nigeria show similarities in their high levels of Ga, Li, Nb, Rb, Y and Zr, coupled with unusually low Ba (Appendix Table 1(d)). In Aden, on the other hand, relatively high levels of Ga, Nb and Zr are associated with low Li and Rb, and high Ba. Sudan averages are remarkably low in Ga, Nb, Rb, Y and Zr, and high in Ba (Appendix Table 1(e)). Nuanetsi is similar to Sudan in Ba, Ga, Li and Y; Gaberones also has high Ba and low Nb, but fairly high Zr.

Averages for the anorogenic localities may be compared with those obtained in the orogenic and anorogenic associations (Appendix Table 1(a)), and it is clear that the range of average values in rocks of anorogenic type includes levels of concentration which are typical of orogenic occurrences (Appendix Tables 1(f)-(h)): in other words, certain anorogenic groups resemble basement or orogenic rocks rather closely in composition. This is particularly true of the trace elements in Sudan, but Nuanetsi and Gaberones also show signs of this affiliation.

REGIONAL VARIATIONS IN THE OROGENIC ACID ROCKS OF AFRICA

The orogenic acid association is made up of rocks from seven localities: Aswan (U.A.R.), Somalia, south-east Uganda, Luapula (Zambia), Franzfontein (South-West Africa), Chirwa (north-east Rhodesia) and Malawi (Appendix Tables 1(f)-(h)). Significant differences amongst the averages for the various elements are set out in Appendix Table 2(d), and those of most importance (i.e. 'highly significant' and 'significant') are restated and discussed below: Ti and Fe^{2+} are included as logarithmic conversion shows them to be of interest, although the values quoted are in fact arithmetic means. Total iron is included although it has not been rigorously tested.

Variations in the orogenic locations are fewer than amongst the anorogenic complexes, and in particular it is noted that the differences in Mg and Ca are not significant; furthermore the changes in Si and Al are rated only at the

	Ti	Fe^{3+}	Fe^{2+}	Total Fe	Na	K	H_2O+
			Percentage				
Aswan	0·13	1·57	1·00	2·57	3·30	4·06	tr-0·46
Somalia	0·16	0·99	0·70	1·69	3·05	3·42	tr-0·31
Uganda	tr-0·20	0·69	1·37	2·06	3·29	3·22	0·70
Zambia	0·27	0·98	1·07	2·05	2·83	3·18	0·81
S.W.A.	0·26	1·15	0·72	1·87	2·70	4·24	0·45
Rhodesia	0·25	0·89	1·18	2·07	2·35	3·84	0·45
Malawi	0·25	1·55	1·36	2·91	2·71	5·23	0·45
Orogenic average	tr-0·21	1·09	1·03	2·12	2·86	3·90	tr-0·48

	Cr	Ga	Pb	Rb	Zr
		ppm			
Aswan	<10	26	23	195	235
Somalia	37	25	<5-24	90	205
Uganda	91	27	19	<50-390	165
Zambia	57	14	20	<50-130	130
S.W.A.	19	22	21	125	400
Rhodesia	44	22	35	205	310
Malawi	<10-26	31	36	300	490->1000
Orogenic average	<10-43	24	<5-26	<50-170	285->1000

level of 'probable' significance (Appendix Table 2(*d*)). These results have a bearing on the plotting of graphs and diagrams to illustrate observed differences; clearly there is no justification in using the averages for such elements, since the spread of points has no significance within the specified group.

Special items of interest in the summary of data for orogenic localities are the low concentrations of Ti in Aswan and Somalia, and also the unusually high total iron contents in Aswan and Malawi. There is an 'enrichment' of Na in Aswan, Somalia and Uganda compared with orogenic regions to the south. Low values for K have been found in Uganda and Zambia and an exceptionally high concentration in Malawi, and although the anorogenic rocks of Chilwa in Malawi have relatively high K on average (Appendix Table 1(*d*)), the concentration is not comparable with that of the Malawi Basement, which is unique in this survey. H_2O+ has appeared amongst the highly significant differences for the first time, and the results illustrate that this is due to the analyses for Uganda and Zambia (Appendix Table 1(*g*)).

Trace elements indicate an enrichment of Cr in Uganda and Zambia, and of Pb in Rhodesia and Malawi. Ga shows fair consistency except for a low figure in Zambia and a peak in Malawi; Rb is unusually low in Somalia. With regard to Zr, Uganda and Zambia are depleted, but South-West Africa and Malawi have high concentrations—and indeed the latter corresponds well with the levels found in anorogenic rocks; whereas it is worth noting that Aswan and Somalia are anomalously low.

Transcontinental geochemical variations

Despite an earlier conclusion that thermo-tectonic history causes significant geochemical variations, suspicion grows that geographical distribution may well be a deciding factor in the average abundances of certain elements in the acid rocks of Africa. To test this theory a more detailed survey of the similarities and differences between neighbouring complexes is now presented, in a progression from north to south of the continent. This journey is interspersed with occasional long-distance comparisons to maintain a broader view, and attempts to link with the somewhat isolated group of samples in Nigeria (Fig. 1). Emphasis is placed on the larger sets of data.

NORTH–EAST AFRICA AND THE SOUTH ARABIAN FEDERATION

Aden–Somalia

In the north-eastern region Aden and Somalia form the largest groups of samples and the localities are separated only by the Gulf of Aden—a distance of about 250 miles; in terms of age, environment and method of emplacement they are however very different, for the volcanics of Aden are thought to have been derived from the mantle and have been dated at ~ 6 m.y., whilst the Precambrian basement rocks of Somalia are considered to be of metasomatic and migmatitic origin. Appendix Table 2(e) gives eight highly significant differences between the two groups, and to these are added two by visual inspection taking account of indeterminate results.

Analyses shown to be significantly different are taken from Appendix Tables 1(d) and (f) as follows:

	Si	Fe^{3+}	Total Fe	Mn	Mg	Na
			Percentage			
Aden	31·94	2·72	3·62	0·13	tr–0·18	4·25
Somalia	34·24	0·99	1·69	0·03	tr–0·34	3·05

	Ga	Mn	Nb	Zr
		ppm		
Aden	37	1300	195	750–>1000
Somalia	25	300	<10–38	205

Briefly, Aden is on average low in Si and Mg, but enriched in Fe^{3+} and in total iron, and also in Mn, Na, Ga, Nb and Zr. Amongst the major constituents there are no significant variations in the values of Ti, Al, Fe^{2+}, Ca and K between the two occurrences. Logarithmic conversion of the data suggests that the slightly higher level of Rb and lower P at Aden may be significant (Appendix Table 2(*e*)).

Aden–north-east Sudan

Looking further afield we find that the next group in reasonable proximity to Aden is that of north-east Sudan, approximately 900 miles to the north-west. Since this area is represented by few acidic samples the results (Appendix Table 2(*f*)) are not quoted here in full. There are seven highly significant and three significant differences using normal data, and others to be considered with logarithmic conversion and after visual inspection of indeterminate results. In summary, Aden lacks Si relative to Sudan, but carries higher amounts of Ti, Al, Fe^{3+}, Fe^{2+}, Mn, Ca, Na and H_2O+; and in trace constituents the Aden rocks show enhancement of Ga, Mn, Nb, Rb, Y and Zr compared with those of north-east Sudan (Appendix Tables 1(*d*) and (*e*)).

Aden–Nigeria

It is of interest to compare the peralkaline rocks of Aden with those of Northern Nigeria (Appendix Table 2(*g*)), and perhaps a little surprising to find a large number of highly significant variations between the groups of data; these are given in the following analyses abstracted from Appendix Table 1(*d*):

	Si	Ti	Al	Fe^{3+}	Total Fe	Mn	Ca	Na	K
				Percentage					
Aden	31·94	0·20	7·09	2·72	3·62	0·13	0·75	4·25	3·62
Nigeria	34·66	0·12	6·40	0·90	2·10	tr-0·04	0·45	3·27	3·95

	Ba	Ga	Mn	Nb	Pb	Y
			ppm			
Aden	<30-790	37	1300	195	<5-9	93
Nigeria	<30-170	52	480	115->1000	56	220

Again the Aden rocks prove to be low in Si and high in Ti, Al, Fe^{3+}, total iron, Mn, Ca and Na, and they are rather low in K compared with Nigeria. However, Rb shows no significant difference, and the suite of trace elements varies from that previously listed in other tests involving Aden: with respect to Nigeria the Aden volcanics are high in Ba and Mn, but low in Ga, Pb and Y.

N

North-east African Group–Nigeria

Average analyses for Nigeria are given in Appendix Table 1(*d*); values for the combined localities of Aswan, north-east Sudan and Somalia have not been tabulated here, but the tests show that they differ from Nigeria at a high level of significance in Al and Ga, if the data are assumed to have normal distribution, and in Al, Fe^{2+}, P and Ga if lognormal. Visual inspection suggests that Mg, Ba, Li, Nb, Sr, Y and Zr should also be considered. Compared with the test of Nigeria against Aden (given above), far fewer major constituents (listed below) show significant results, but more trace elements are in the present list (Appendix Table 2(*h*)).

	Al	Fe^{2+}	Mg	P
		Percentage		
Nigeria	6·40	1·20	tr–0·10	tr–0·02
N.E. Africa	7·12	0·71	tr–0·31	tr–0·06

	Ba	Ga	Li	Nb
		ppm		
Nigeria	<30–170	52	82–>600	115–>1000
N.E. Africa	560–>2000	25	<5–24	<10–40

	Sr	Y	Zr
	ppm		
Nigeria	<30– 49	220	510–>1000
N.E. Africa	<30–160	<10–49	215

Generally accepted trends of magmatic crystallisation could account for the observed differences in the trace elements, assuming Nigeria to represent a later stage of formation; a similar argument might be applied to the few major constituents listed, but it is curious that there are no significant differences in the average levels of Si, Ti, Fe^{3+}, Na or K between the two regional groups.

CENTRAL AND EAST AFRICA

Somalia–south-east Uganda

We may now draw comparisons southwards between Somalia and the group of specimens from south-east Uganda (Appendix Table 2(*i*)); the average values for each locality are given in Appendix Tables 1(*f*) and (*g*). There are very few differences of significance:

	Fe^{2+}	H$_2$O+	Ni
	Percentage		ppm
Somalia	0·70	tr-0·31	<10
S.E. Uganda	1·37	0·70	<10-27

Total iron in the two areas does not differ much (Somalia, 1·69%; Uganda, 2·06%), although the Fe^{3+}/Fe^{2+} ratio is higher on average in Somalia. No significant difference is found for Si, Ti, Al, Fe^{3+}, Mg, Ca, Na, K and P, but the two levels of Mn may be worth future consideration (Somalia, 300 ppm; Uganda, 590 ppm).

South-east Uganda–Zambia

For the first time complete agreement between all the major constituents is recorded for the two areas of south-east Uganda and Luapula (Zambia), and amongst the trace elements only Ga is put forward as highly significant, with 'possible significance' for Ba.

	Ba	Ga
	ppm	
S.E. Uganda	800->2000	27
Zambia	1400	14

The result for Ba is only true if one sample greater than 2000 ppm in the Uganda group is ignored. It must be noted that these are rather small sets of data, but nevertheless the similarity overall is noteworthy (see Appendix Tables 1(*g*) and 2(*j*)).

Zambia–north-east Rhodesia

Passing on to the next group, the remobilised Basement at Chirwa in north-east Rhodesia, several changes are observed (Appendix Table 2(*k*)). The significant analyses here are abstracted from Appendix Tables 1(*g*) and (*h*).

	Na	H$_2$O+	Ga	Pb	Zr
	Percentage			ppm	
Zambia	2·83	0·81	14	20	130
Chirwa	2·35	0·45	22	35	310

Apart from these constituents there are no significant differences in the concentrations of Si, Ti, Al, Fe^{3+}, Fe^{2+}, Mn, Mg, Ca and K, but a definite depletion in Na is observed southwards.

SOUTHERN AFRICA

North-east Rhodesia–South-West Africa

Comparison of the remobilised Basement at Chirwa in Rhodesia with the basement complex in South-West Africa at Franzfontein 1200 miles to the west (Appendix Table 1(*h*)) yields very few real differences in the averages for the two areas (Appendix Table 2(*l*)):

	Fe²⁺ Percentage	Cr	Pb ppm	Rb
Chirwa	1·18	44	35	205
South-West Africa	0·72	19	21	125

There are no significant differences in Si, Ti, Al, Fe³⁺, Mn, K and perhaps Mg, although the latter is rather difficult to assess. Total iron contents are similar: Chirwa, 2·07%; S.W.A., 1·87%. A probable significant difference is recorded in the case of Ca if logarithmic conversion is used, and under normal circumstances, for Na, Mn and Y. Once again there is confirmation of relatively high concentrations of Pb in north-east Rhodesia, and it may be noted that Aswan, Sudan, Somalia, south-east Uganda, Zambia and South-West Africa are all fairly consistent in having rather lower amounts around 20 ppm Pb.

Gaberones (Botswana) Nuanetsi (Rhodesia)

Breaking the sequence for a moment it is instructive to draw comparisons between two anorogenic groups of widely differing ages, namely those of Gaberones (Botswana) and Nuanetsi (Rhodesia). Averages are presented in Appendix Table 1(*e*), and the analysis of variance in Appendix Table 2(*m*).

	Ga	Nb ppm	Pb	Rb
Gaberones	33	47	58	370
Nuanetsi	22	100	22	<50-215

The tests show that all the major constituents Si, Ti, Al, Fe³⁺, Fe²⁺, Mn, Mg, Ca, Na, K, P and H₂O+, might well belong to a single population, since there are no significant differences. Ga is unusually low at Nuanetsi, having a value akin to the 'orogenic' level of 24 ppm, and in fact matching exactly with concentrations in the South-West African and north-east Rhodesian basements. Pb in the Nuanetsi rocks is in agreement with a widely observed basement level of approximately 20 ppm.

Gaberones has rather high amounts of Rb and Y which may be significant, and perhaps indicative of a later stage of magmatic differentiation, but the relative lack of Nb is rather perplexing in this context.

Gaberones–Nuanetsi–South-West Africa

It is appropriate to return now to the main theme, and having seen few deviations from a regular pattern in the traverse from Somalia through Uganda, Zambia and north-east Rhodesia to South-West Africa, the possible links amongst three occurrences in the southern part of Africa are examined.

Average analyses for South-West Africa, Gaberones (Botswana) and Nuanetsi (Rhodesia) are to be found in Appendix Tables 1(e) and (h), and significant variations amongst the three groups of data are given in Appendix Table 2(n) and are as follows:

	Al	Mg	Ga	Nb	Pb	Y
	Percentage			ppm		
South-West Africa	7·29	tr-0·46	22	<10-23	21	52
Gaberones	6·47	0·27	33	47	58	96
Nuanetsi	6·72	tr-0·22	22	100	22	65

Si, Ti, Fe^{3+}, Fe^{2+} Mn, Ca, Na, K, P and H_2O+ show no significant differences amongst the three localities despite their widely differing ages and modes of emplacement, and the fact that, whilst two complexes are of anorogenic origin, the batholithic granites of South-West Africa are obviously tectonised. Environment may be a factor in the control of Al and Mg, and in an earlier test both these elements were found to be low in anorogenic rocks as compared with basement or orogenic specimens (p. 365); but the extensive suite of element variations mentioned in that test has not been repeated here.

In addition to the trace elements listed above, Ba and Sr are relatively high in the rocks of South-West Africa, a 'probable' significance having been recorded for Sr (Appendix Table 2(n)).

South-West Africa–Gaberones–Nuanetsi–north-east Rhodesia

A high degree of conformity has shown itself amongst three localities in southern Africa (see above), and now an attempt is made to add a fourth to the province—that of Chirwa in north-east Rhodesia. Analyses are to be found in Appendix Tables 1(e) and (h), and the elements of importance are abstracted from Appendix Table 2(o) as follows:

	Al	Fe^{3+}	Total Fe	Cu	Ga	Nb	Pb	Rb	Y
	Percentage					ppm			
South-West Africa	7·29	1·15	1·87	4	22	<10-23	21	125	52
Gaberones	6·47	1·36	2·66	19	33	47	58	370	96
Nuanetsi	6·72	1·93	2·80	31	22	100	22	<50-215	65
Chirwa	7·42	0·89	2·07	35	22	<10-17	35	205	29

Only two major constituents (Al, Fe^{3+}) differ significantly amongst the four groups of rocks, which may be regarded as a single population in respect of Si, Ti, Mn, Mg, Ca, Na, K and H_2O+. The use of logarithmic conversion gives 'probable' significance to Fe^{2+}, and the ratio of ferric to ferrous iron is seen to be lower in Chirwa. The anorogenic localities of Gaberones and Nuanetsi are high in total iron (though these figures have not been tested statistically) and low in Al, and it is interesting to observe that the trace elements do not reflect this clear division. Cu increases in an easterly direction; Ga, Pb, Rb and Y are concentrated at Gaberones; and Nb is high in the Nuanetsi rocks. However, Sr has been recorded as 'probably' significant, and this element is higher in the two orogenic groups.

Nuanetsi–north-east Rhodesia–Malawi basement

In order to relate the Malawi Basement to other regions of southern Africa it is necessary to take account of the two slightly differing facets of an otherwise homogeneous southern province—and therefore to compare the Malawi Basement with Chirwa (north-east Rhodesia), which is a high alumina-low iron orogenic complex, and with Nuanetsi, typifying the low alumina-high iron anorogenic rocks. Averages for the elements are shown in Appendix Tables 1(e) and (h); the variance tests (Appendix Table 2(p)) provide the following points of interest:

	Si	Al	Fe^{3+}	Total Fe	K
		Percentage			
Nuanetsi	33·41	6·72	1·93	2·80	4·40
Chirwa	33·46	7·42	0·89	2·07	3·84
Malawi	31·99	7·66	1·55	2·91	5·23

	Cu	Ga	Li	Nb	Pb	Rb	Sr	Y
				ppm				
Nuanetsi	31	22	<5-13	100	22	<50-215	<30-91	65
Chirwa	35	22	28	<10-17	35	205	190	29
Malawi	<5-13	31	32	<10-58	36	300	290	<10-46

Malawi basement rocks are, on average, low in Si and high in K compared with the other two areas. The trace element Cu is relatively low, whereas Ga, Rb and Sr are unusually enriched in Malawi.

Al, Li and Pb are high in both Chirwa and Malawi, and there is agreement between Nuanetsi and Malawi in high iron contents; Nb and Y vary throughout the group. The number of major constituents showing no significant differences amongst the localities is rather limited compared with the earlier traverse from South-West Africa to Chirwa, and only Ti, Mn, Mg, Ca and H_2O+ can be assumed to represent a single population.

Malawi basement–South-West Africa

The anomalous geochemistry of Malawi may perhaps be confirmed by comparison with Franzfontein, South-West Africa (Appendix Table 1(h)). Differences are shown in Appendix Table 2(q) and quoted in detail below.

	Si	K	Ga	Nb	Pb	Rb
	Percentage		ppm			
South-West Africa	33·61	4·24	22	<10-23	21	125
Malawi	31·99	5·23	31	<10-58	36	300

In this test a deficiency in Si at Malawi is reiterated, and an enrichment in K verified, but there are no significant differences in Ti, Al, Fe^{3+}, Mn, Ca, Na and H_2O+. There is some difficulty in assessing Mg, and it may be lower in Malawi. An increase in Fe^{2+} is 'probably' significant, and in fact the total iron content (2.91%) is a good deal higher than that of South-West Africa (1.87%). Ga, Nb, Pb and Rb are relatively high in the Malawi rocks, but it is evident that the group as a whole is in rather better agreement with South-West Africa than with its more immediate neighbours at Chirwa and Nuanetsi. Nevertheless there are still some important differences, particularly in Si and K.

ABERRANT LOCALITIES

Aden–Chilwa (Malawi)–Nigeria

Taking a broad view the acidic rocks of Africa present a relatively harmonious picture, but several discordant areas become the focus of attention. These are the anorogenic complexes of Nigeria, Aden, and Chilwa, all of which display individuality (Appendix Table 1(d)). Contrasting element abundances are emphasised in Appendix Table 2(r) and are as follows:

	Si	Ti	Fe^{3+}	Fe^{2+}	Mn	Ca	Na	K
				Percentage				
Aden	31·94	0·20	2·72	0·90	0·13	0·75	4·25	3·62
Chilwa	33·15	0·14	1·58	1·87	0·10	tr-0·44	3·61	4·28
Nigeria	34·66	0·12	0·90	1·20	tr-0·04	0·45	3·27	3·95

	Ba	Ga	Mn	Pb
		ppm		
Aden	<30-790	37	1300	<5-9
Chilwa	<30-180	45	1050->2000	<5-46
Nigeria	<30-170	52	480	56

Mg is reasonably comparable in the three groups, and Al shows only 'probable' significance (7.09%, 6.70%, 6.40%). Total iron (3.62%, 3.45%, 2.10%) is unusually low in Nigeria, where Mn is also depleted, and Si is very high in this group on average. Chilwa and Nigeria show agreement in having low Ti, Ca, Na and Ba, and high K, Ga and Pb, in relation to Aden. Some loss of volatile components and an increase in the Fe^{3+}/Fe^{2+} ratio may be expected in the volcanic rocks, but there are nevertheless, some important differences which are not explicable by mode of emplacement.

Aden–Nuanetsi

Finally it is of interest to compare the volcanics of Aden with the anorogenic granites of Nuanetsi (Appendix Tables 1(d) and (e)) since both complexes have produced low initial Sr^{87}/Sr^{86} ratios during the course of isotopic studies.

The elements which differ significantly are listed in Appendix Table 2(s) and are collated here:

	Si	Mn	Na	K
		Percentage		
Aden	31·94	0·13	4·25	3·62
Nuanetsi	33·41	0·05	2·48	4·40

	Cu	Ga	Mn	Nb	Pb	Y	Zr
			ppm				
Aden	5	37	1300	195	<5-9	93	750->1000
Nuanetsi	31 ·	22	390	100	22	65	400

Attention is directed to fewer major constituents compared with the previous section, but more trace elements are in this list. Si, K, Cu and Pb are unusually low in the Aden rocks whereas Mn, Na, Ga, Nb, Y and Zr are relatively enriched. No significant differences are observed in Ti, Fe^{2+}, Mg, Ca and H_2O+ ; total iron is high at Aden (3·62%) compared with Nuanetsi (2·80%), and a slightly higher level of Al at Aden is classified as 'probably' significant.

Petrogenesis

C.I.P.W. norms have been calculated from the average chemical analyses of locality groups (Appendix Table 3) assuming 'trace' levels, as for example in MgO, to be zero. Fig. 2 illustrates the plot of salic normative constituents (except An) on an enlarged part of the quartz-nepheline-kaliophilite diagram (after Schairer, in Turner and Verhoogen, 1960). Concentration of the majority of samples in the low-temperature troughs of this system indicates that they are the end-products of fractional crystallisation, or the first products of anatectic fusion.

Bailey and Schairer (1966) emphasise that the quaternary system $Na_2O-Al_2O_3-Fe_2O_3-SiO_2$ is essentially the peralkaline residua system, and therefore discussion of magmatic origins must be restrained. However, a few general features may be noted. First, there is indiscriminate mingling of anorogenic and orogenic points (see p. 364). Secondly, anomalous geo-chemical features already noted for Aden (no. 4), north-east Sudan (no. 3) and Malawi (nos. 12 and 13) are reflected in the diagram, but the position of Nigeria (no. 1) is in no way unusual. Thirdly, there may be a tendency for rocks from low-numbered (or northern) localities to lie closer to the Q-Ne sideline than high-numbered (southern) groups which are apparently enriched in the potassic component; a close grouping of the southern African complexes (nos. 8, 9, 10 and 11) is evident. In this connection it is pertinent to record that average K/Na ratios for acidic rocks of localities nos. 8-12 inclusive (the 'southern African province' plus Malawi basement) lie within the range 1·56–1·93; all the remaining localities, of both acidic and intermediate type, have average K/Na ratios in the range 0·65–1·23.

All groups having total (Q+ Or+ Ab) greater than 80% are plotted on the $NaAlSi_3O_8$-$KAlSi_3O_8$-SiO_2-H_2O experimental system (Tuttle and

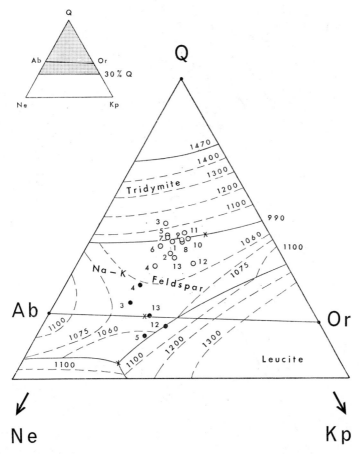

FIG. 2. Enlarged portion (stippled area in small triangle) of the anhydrous system Q–Ne–Kp at atmospheric pressure, showing averages of the salic normative constituents (except An) for acid rock localities (open circles) and for intermediate rock localities (solid circles). Points numbered as in Fig. 1. Axis of low–temperature trough is marked by crosses. (After Schairer, in Turner and Verhoogen, 1960.)

Bowen, 1958) in Fig. 3(a) and (b). Locality points are seen in relation to isobaric fractionation curves for a water vapour pressure of 1 kilobar (Fig. 3(a)), and also (Fig. 3(b)) in comparison with isobaric minima and isobaric eutectics (Luth et al., 1964). A rigorous treatment such as that by Brown (1963) is unwarranted here, but it may perhaps be inferred that the rocks of the southern African province (nos. 8, 9, 10 and 11) were formed under relatively low

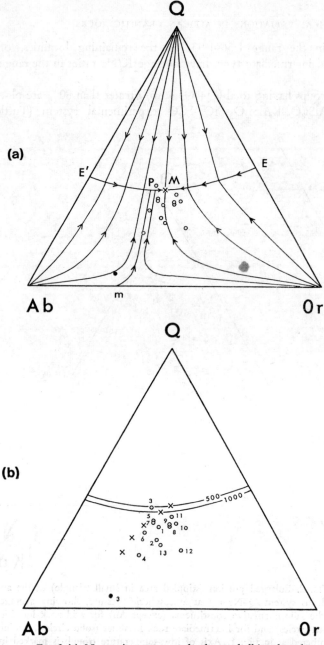

FIG. 3 (a). Normative quartz, orthoclase and albite plotted on the $NaAlSi_3O_8$-$KAlSi_3O_8$-SiO_2-H_2O experimental system (Tuttle and Bowen, 1958). Projection on to the anhydrous base of the tetrahedron, with isobaric fractionation curves shown for a water vapour pressure of 1 kilobar. Average acid rock localities, open circles; average intermediate rock locality, solid circle.

(b) As for Fig. 3 (a). Average compositions shown in relation to the quartz–feldspar boundaries at 0·5 and 1·0 kilobar; crosses mark the isobaric minima at 0·5, 1 and 3 kilobars, and isobaric eutectics at 5 and 10 kilobars (Luth et al., 1964). Points numbered as in Fig. 1.

pressures of water vapour. Malawi basement rocks (no. 12) may be associated with this group, whereas Chilwa Province (no. 13) shows possible affinities with Nigeria (no. 1).

On the other hand the intermediate rocks of north-east Sudan (no. 3) lie on the plagioclase feldspar side of a 'thermal valley' and the acidic rocks of Aden (no. 4) within the valley, which may be evidence in favour of basaltic parent material (Wyllie and Tuttle, 1961); indeed, isotopic studies support a mantle derivation for the Aden volcanics (Dickinson et al., 1969). North-east Sudan acid rocks (no. 3), in contrast, plot on the silica-rich side of the quartz-feldspar boundary indicating an entirely different magmatic origin.

Conclusions

Initially in this study a clear distinction is made between the average analyses of anorogenic acid igneous rocks on the one hand, and orogenic rocks on the other. Elements differing in average levels between the two associations include: (a) Al, Ca, Mg and Sr; and (b) Fe^{3+}, Na, Mn, Ga, Li, Nb, Y and Zr. The former group are relatively high in orogenic acid rocks, and the latter are concentrated to a greater extent in the anorogenic acid rocks.

Orogenic complexes in the basement of Somalia, south-east Uganda, Zambia, north-east Rhodesia and South-West Africa differ only slightly in their overall compositions, except for occasional significant variations in Fe^{2+}, Na, H_2O+, Ba, Cr, Ga, Ni, Pb, Rb and Zr. The acidic basement rocks of Malawi are relatively high in K and low in Si.

Two anorogenic groups situated in the Kalahari Craton are closely similar in composition to two orogenic complexes in southern Africa, whilst retaining the characteristically low Al and Sr of the anorogenic rock association. These four complexes at Franzfontein, Gaberones, Nuanetsi and Chirwa, of diverse formations, tectonic environments and ages (ranging from approx. 2300 m.y. to 180 m.y.), constitute a geochemical province in southern Africa. The region is undoubtedly enriched in potassium and deficient in sodium relative to more northerly locations. This may be the result of large-scale chemical changes, or the variation from north to south of the continent may reflect deep-seated differences in the mantle.

There is a considerable measure of agreement between the anorogenic acid rocks of Nigeria and Chilwa Province, and to some extent with Aden. The latter are rich in Fe^{3+}, Ti, Ca, Na and Ba, and lack Si, Fe^{2+}, K, Ga and Pb in comparison with both Nigeria and Chilwa; analyses of north-east Sudan acidic rocks are distinctly anomalous.

Various suites of associated major and trace constituents have been identified; their interpretation in terms of three simplified models of granite formation (Mehnert, 1968) remains open to discussion.

APPENDIX
TO CHAPTER 17

TABLE 1(*a*)-(*j*)

Average analyses of selected groups of samples.

TABLE 1(b)

Intermediate rocks of Africa of orogenic and anorogenic type

Element	INTERMEDIATE ROCKS OF AFRICA				OROGENIC INTERMEDIATE ROCKS				ANOROGENIC INTERMEDIATE ROCKS			
	Number of samples			Arithmetic mean and indeterminate groups	Number of samples			Arithmetic mean and indeterminate groups	Number of samples			Arithmetic mean and indeterminate groups
	<	n	>		<	n	>		<	n	>	
				%				%				%
Si	—	56	—	28·40	—	22	—	27·55	—	34	—	28·94
Ti	—	56	—	0·38	—	22	—	0·32	—	34	—	0·42
Al	—	56	—	9·29	—	22	—	10·22	—	34	—	8·69
Fe^{3+}	—	56	—	1·90	—	22	—	1·51	—	34	—	2·16
Fe^{2+}	1	55	—	tr-1·88	—	22	—	1·70	1	33	—	tr-2·00
Mn	—	56	—	0·13	—	22	—	0·07	—	34	—	0·16
Mg	3	53	—	tr-0·50	1	21	—	tr-0·60	2	32	—	tr-0·44
Ca	—	56	—	1·46	—	22	—	1·68	—	34	—	1·32
Na	—	56	—	4·65	—	22	—	4·48	—	34	—	4·76
K	—	56	—	4·49	—	22	—	5·11	—	34	—	4·08
H_2O+	—	56	—	0·50	—	22	—	0·42	—	34	—	0·55
H_2O-	1	55	—	tr-0·23	1	21	—	tr-0·12	—	34	—	0·30
P	—	56	—	0·13	—	22	—	0·16	—	34	—	0·11
F	—	7	—	0·21	—	1	—	0·20	—	6	—	0·21
				ppm				ppm				ppm
Ba	—	45	11	490->2000	—	13	9	370->2000	—	32	2	550->2000
Be	22	34	—	<3- 7	8	14	—	<3- 7	14	20	—	<3- 6
Co	51	5	—	<10- 11	19	3	—	<10- 11	32	2	—	<10- 10
Cr	26	30	—	<10- 21	11	11	—	<10- 19	15	19	—	<10- 22
Cu	12	20	—	<5- 12	6	4	—	<5- 10	6	16	—	<5- 13
Ga	—	56	—	33	—	22	—	31	—	34	—	35
La	9	47	—	<100-170	6	16	—	<100-230	3	31	—	<100-140
Li	9	47	—	<5- 21	5	17	—	<5- 15	4	30	—	<5- 24
Mn	—	49	7	1200->2000	—	22	—	860	—	27	7	1500->2000
Mo	24	32	—	<3- 6	13	9	—	<3- 5	11	23	—	<3- 6
Nb	1	55	—	<10-105	1	21	—	<10- 75	—	34	—	125
Ni	49	7	—	<10- 18	16	6	—	<10- 19	33	1	—	<10- 10
Pb	12	43	1	<5- 28->1000	3	19	—	<5- 31	9	24	1	<5- 26->1000
Rb	1	53	—	<50-155	—	20	—	175	1	33	—	<50-140
Sc	39	17	—	<10- 13	20	2	—	<10- 13	19	15	—	<10- 13
Sr	3	49	4	<30-290->1000	—	18	4	470->1000	3	31	—	<30-190
V	14	42	—	<3- 32	4	18	—	<3- 40	10	24	—	<3- 25
Y	4	52	—	<10- 55	4	18	—	<10- 46	—	34	—	59
Zr	—	48	8	480->1000	—	17	5	490->1000	—	31	3	480->1000

383

TABLE 1(c)

Anorogenic acid rocks of intrusive and extrusive occurrence

Element	ACID ROCKS				ACID INTRUSIVES				ACID EXTRUSIVES			
	Number of samples			Arithmetic mean and indeterminate groups	Number of samples			Arithmetic mean and indeterminate groups	Number of samples			Arithmetic mean and indeterminate groups
	<	n	>		<	n	>		<	n	>	
				%				%				%
Si	—	70	—	33–67	—	45	—	33–98	—	25	—	33·10
Ti	—	70	—	0·17	—	45	—	0·16	—	25	—	0·18
Al	—	70	—	6·66	—	45	—	6·62	—	25	—	6·72
Fe^{3+}	—	70	—	1·59	—	45	—	1·19	—	25	—	2·31
Fe^{2+}	—	70	—	1·15	—	45	—	1·33	—	25	—	0·83
Mn	1	69	—	tr–0·07	—	45	—	0·06	1	24	—	tr–0·10
Mg	20	50	—	tr–0·16	10	35	—	tr–0·16	10	15	—	tr–0·16
Ca	4	66	—	tr–0·64	3	42	—	tr–0·61	1	24	—	tr–0·69
Na	—	70	—	3·30	—	45	—	3·16	—	25	—	3·56
K	—	70	—	4·01	—	45	—	4·15	—	25	—	3·75
H$_2$O+	—	70	—	0·48	—	45	—	0·40	—	25	—	0·64
H$_2$O−	2	68	—	tr–0·22	2	43	—	tr–0·16	—	25	—	0·32
P	9	61	—	tr–0·04	7	38	—	tr–0·04	2	23	—	tr–0·03
F	1	33	—	tr–0·18	—	17	—	0·21	1	16	—	tr–0·14
				ppm				ppm				ppm
Ba	6	67	4	<30–470–>2000	4	46	2	<30–420–>2000	2	21	2	<30–590–>2000
Be	21	55	—	<3– 9	10	41	—	<3– 9	11	14	—	<3– 7
Co	73	3	—	<10– 15	48	3	—	<10– 15	25	—	—	<10
Cr	19	58	—	<10– 33	8	44	—	<10– 30	11	14	—	<10– 43
Cu	7	42	—	<5– 14	4	27	—	<5– 18	3	15	—	<5– 9
Ga	—	77	—	37	—	52	—	38	—	25	—	35
La	17	59	1	<100–155–>1000	11	40	1	<100–160–>1000	6	19	—	<100–155
Li	13	62	2	<5– 51–>600	8	42	2	<5– 66–>600	5	20	—	<5– 20
Mn	—	64	1	730–>2000	—	42	1	590–>2000	—	22	—	980
Mo	51	26	—	<3– 6	34	18	—	<3– 6	17	8	—	<3– 6
Nb	5	69	3	<10–125–>1000	3	46	3	<10–100–>1000	2	23	—	<10–175
Ni	64	13	—	<10– 14	42	10	—	<10– 12	22	3	—	<10– 22
Pb	15	62	—	<5– 38	5	47	—	<5– 42	10	15	—	<5– 27
Rb	3	64	—	50–215	2	46	—	50–255	1	18	—	50–120
Sc	47	3	—	<10– 13	28	1	—	<10– 10	19	2	—	<10– 15
Sr	23	54	—	<30– 70	17	35	—	<30– 71	6	19	—	<30– 69
V	53	24	—	<3– 15	34	18	—	<3– 16	19	6	—	<3– 12
Y	1	76	—	<10–120	1	51	—	<10–130	—	25	—	99
Zr	—	65	12	510–>1000	—	45	7	460–>1000	—	20	5	610–>1000

TABLE 1(d)

Anorogenic acid rocks of Aden, Malawi and northern Nigeria

Element	ADEN, SOUTH ARABIAN FEDERATION				CHILWA ALKALINE PROVINCE, MALAWI				NORTHERN NIGERIA			
	\<	n	\>	Arithmetic mean and indeterminate groups	\<	n	\>	Arithmetic mean and indeterminate groups	\<	n	\>	Arithmetic mean and indeterminate groups
				%				%				%
Si	—	13	—	31·94	—	13	—	33·15	—	19	—	34·66
Ti	—	13	—	0·20	—	13	—	0·14	—	19	—	0·12
Al	—	13	—	7·09	—	13	—	6·70	—	19	—	6·40
Fe³⁺	—	13	—	2·72	—	13	—	1·58	—	19	—	0·90
Fe²⁺	—	13	—	0·90	—	13	—	1·87	—	19	—	1·20
Mn	—	13	—	0·13	—	13	—	0·10	1	18	—	tr-0·04
Mg	7	6	—	tr-0·18	4	9	—	tr-0·10	4	15	—	tr-0·10
Ca	—	13	—	0·75	4	9	—	tr-0·44	—	19	—	0·45
Na	—	13	—	4·25	—	13	—	3·61	—	19	—	3·27
K	—	13	—	3·62	—	13	—	4·28	—	19	—	3·95
H₂O+	—	13	—	0·79	—	13	—	0·49	—	19	—	0·29
H₂O-	—	13	—	0·43	1	12	—	tr-0·15	1	18	—	tr-0·15
P	2	13	—	tr-0·03	5	8	—	tr-0·04	2	17	—	tr-0·02
F	1	12	—	tr-0·16	—	—	—	—	—	17	—	0·20
				ppm				ppm				ppm
Ba	2	11	—	\<30-790	2	11	—	\<30- 180	2	17	—	\<30-170
Be	5	8	—	\<3- 7	—	13	—	12	—	19	—	10
Co	13	—	—	\<10	13	—	—	\<10	19	—	—	\<10
Cr	10	3	—	\<10- 52	8	5	—	\<10- 25	—	19	—	20
Cu	—	10	—	5	2	10	—	\<5- 12	—	9	—	12
Ga	—	13	—	37	—	13	—	45	—	19	—	52
La	—	13	—	140	2	11	—	\<100- 185	4	14	1	\<100-210->1000
Li	—	13	—	21	1	12	1	\<5- 80	—	17	2	82->600
Mn	—	13	—	1300	7	12	1	1050->2000	—	19	—	480
Mo	10	3	—	\<3- 6	7	6	—	\<3- 4	14	5	—	\<3- 13
Nb	—	13	—	195	1	11	1	\<10- 155->1000	—	17	2	115->1000
Ni	10	3	—	\<10- 22	13	—	—	\<10	19	—	—	\<10
Pb	10	3	—	\<5- 9	5	8	—	\<5- 46	—	19	—	56
Rb	—	10	—	125	—	11	—	230	—	19	—	235
Sc	8	10	—	\<10- 15	13	—	—	\<10	9	—	—	\<10
Sr	3	2	—	\<30- 77	6	7	—	\<30- 48	11	8	—	\<30- 49
V	11	13	—	\<3- 10	9	4	—	\<3- 6	16	3	—	\<3- 12
Y	—	13	—	93	1	12	—	\<10- 125	—	19	—	220
Zr	—	10	3	750->1000	—	10	3	570->1000	—	13	6	510->1000

TABLE 1(e)

Anorogenic acid rocks of Rhodesia, Botswana and north-east Sudan

Element	NUANETSI PROVINCE, RHODESIA — Number of samples <	>	NUANETSI — Arithmetic mean and indeterminate groups	GABERONES COMPLEX, BOTSWANA — Number of samples <	>	GABERONES — Arithmetic mean and indeterminate groups	NORTH-EAST SUDAN — Number of samples <	>	NORTH-EAST SUDAN — Arithmetic mean and indeterminate groups
			%			%			%
Si	13	—	33-41	7	—	33.98	5	—	35.99
Ti	13	—	0.24	7	—	0.25	5	—	0.08
Al	13	—	6.72	7	—	6.47	5	—	6.48
Fe^{3+}	13	—	1.93	7	—	1.36	5	—	0.71
Fe^{2+}	13	—	0.87	7	—	1.30	5	—	0.30
Mn	13	—	0.05	7	—	0.04	5	—	0.04
Mg	12	1	tr-0.22	7	—	0.27	4	1	tr-0.13
Ca	13	—	0.97	7	—	0.83	5	—	0.29
Na	13	—	2.48	7	—	2.72	5	—	3.10
K	13	—	4.40	7	—	4.25	5	—	3.22
H_2O+	13	—	0.52	7	—	0.39	5	—	0.42
H_2O-	13	—	0.27	7	—	0.12	5	—	0.08
P	13	—	0.06	7	—	0.06	5	—	0.03
F	4	—	0.13	—	—	—	5	—	0.03
			ppm			ppm			ppm
Ba	16	4	650->2000	—	7	650	—	5	640
Be	12	7	<3- 4	—	7	8	4	1	<3- 3
Co	16	3	<10- 15	7	—	<10	5	—	<10
Cr	1	19	<10- 32	—	7	29	—	5	84
Cu	—	7	31	—	6	19	5	—	<5
Ga	—	20	22	—	7	33	—	5	20
La	6	14	<100-115	—	7	125	5	—	<100
Li	7	13	<5- 13	4	3	<5- 74	2	3	<5- 22
Mn	—	8	390	—	7	420	—	5	330
Mo	12	8	<3- 5	4	3	<3- 3	5	—	<3
Nb	—	20	100	—	7	47	4	1	<10- 10
Ni	10	10	<10- 12	7	—	<10	5	—	<10
Pb	—	20	22	—	7	58	—	5	16
Rb	2	13	<50-215	—	7	370	4	1	<50- 61
Sc	6	—	<10	7	—	<10	4	1	<10- 10
Sr	2	18	<30- 91	—	7	65	—	5	<30- 55
V	10	10	<3- 18	5	2	<3- 20	5	—	<3
Y	—	20	65	—	5	96	—	5	40
Zr	—	20	400	—	7	570	—	5	220

TABLE 1(f)

Acid rocks of the United Arab Republic and Somalia

	ASWAN, UNITED ARAB REPUBLIC			SOMALIA			
Element	Number of samples			Arithmetic mean and indeterminate groups	Number of samples			Arithmetic mean and indeterminate groups
	<	n	>		<	n	>	
				%				%
Si	—	7	—	33·09	—	19	—	34·24
Ti	—	7	—	0·13	—	19	—	0·16
Al	—	7	—	7·83	—	19	—	7·03
Fe^{3+}	—	7	—	1·57	—	19	—	0·99
Fe^{2+}	—	7	—	1·00	—	19	—	0·70
Mn	—	7	—	0·03	—	19	—	0·03
Mg	—	7	—	0·25	1	18	—	tr-0·34
Ca	—	7	—	0·68	—	19	—	1·09
Na	—	7	—	3·30	—	19	—	3·05
K	—	7	—	4·06	—	19	—	3·42
H_2O+	1	6	—	tr-0·46	2	17	—	tr-0·31
H_2O-	7	—	—	tr	—	19	—	0·09
P	3	4	—	tr-0·07	—	19	—	0·07
F	—	—	—	—	—	—	—	—
				ppm				ppm
Ba	—	7	—	540	—	14	5	550->2000
Be	4	3	—	<3- 5	12	7	—	<3- 6
Co	7	—	—	<10	19	—	—	<10
Cr	7	—	—	<10	—	19	—	37
Cu	—	7	—	4	9	8	—	<5- 14
Ga	—	7	—	26	—	19	—	25
La	6	1	—	<100-200	14	5	—	<100-205
Li	2	5	—	<5- 22	8	11	—	<5- 26
Mn	1	6	—	<30-300	—	19	—	300
Mo	6	1	—	<3- 6	15	4	—	<3- 7
Nb	1	6	—	<10- 48	11	8	—	<10- 38
Ni	7	—	—	<10	19	—	—	<10
Pb	—	7	—	23	1	18	—	<5- 24
Rb	—	7	—	195	—	19	—	90
Sc	6	1	—	<10- 10	15	4	—	<10- 26
Sr	—	7	—	130	2	17	—	<30-195
V	2	5	—	<3- 14	6	13	—	<3- 24
Y	1	6	—	<10- 55	7	12	—	<10- 50
Zr	—	7	—	235	—	19	—	205

TABLE 1(*g*)

Acid rocks of south-east Uganda and Zambia

Element	SOUTH-EAST UGANDA Number of samples < n >			Arithmetic mean and indeterminate groups	LUAPULA, ZAMBIA Number of samples < n >			Arithmetic mean and indeterminate groups
				%				%
Si	—	10	—	33·06	—	6	—	32·79
Ti	1	9	—	tr-0·20	—	6	—	0·27
Al	—	10	—	7·59	—	6	—	7·53
Fe^{3+}	—	10	—	0·69	—	6	—	0·98
Fe^{2+}	—	10	—	1·37	—	6	—	1·07
Mn	—	10	—	0·05	—	6	—	0·05
Mg	—	10	—	0·49	—	6	—	0·70
Ca	1	9	—	tr-1·35	—	6	—	1·55
Na	—	10	—	3·29	—	6	—	2·83
K	—	10	—	3·22	—	6	—	3·18
H_2O+	—	10	—	0·70	—	6	—	0·81
H_2O-	—	10	—	0·07	—	6	—	0·10
P	—	10	—	0·06	—	6	—	0·05
F	—	5	—	0·11	—	—	—	—
				ppm				ppm
Ba	—	9	1	800->2000	—	6	—	1400
Be	7	3	—	<3- 22	6	—	—	<3
Co	10	—	—	<10	4	2	—	<10- 30
Cr	—	10	—	91	—	6	—	57
Cu	—	3	—	17	—	—	—	—
Ga	—	10	—	27	—	6	—	14
La	8	2	—	<100-115	4	2	—	<100- 135
Li	2	8	—	<5- 38	1	5	—	<5- 17
Mn	—	9	—	590	—	—	—	—
Mo	9	1	—	<3- 3	6	—	—	<3
Nb	9	1	—	<10- 30	6	—	—	<10
Ni	2	7	—	<10- 27	1	5	—	<10- 17
Pb	—	10	—	19	—	6	—	20
Rb	8	2	—	<50-390	2	4	—	<50- 130
Sc	4	—	—	<10	—	—	—	—
Sr	—	9	1	210->1000	—	6	—	540
V	4	6	—	<3- 52	—	6	—	56
Y	7	3	—	<10- 33	3	3	—	<10- 32
Zr	—	10	—	165	—	6	—	130

Orogenic acid rocks of South-West Africa, Rhodesia and Malawi

Element	FRANZFONTEIN, SOUTH-WEST AFRICA			CHIRWA, NORTH-EAST RHODESIA			MALAWI		
	Number of samples		Arithmetic mean and indeterminate groups	Number of samples		Arithmetic mean and indeterminate groups	Number of samples		Arithmetic mean and indeterminate groups
	n <	n >		n <	n >		n <	n >	
			%			%			%
Si	11	—	33·61	14	—	33·46	12	—	31·99
Ti	11	—	0·26	14	—	0·25	12	—	0·25
Al	11	—	7·29	14	—	7·42	12	—	7·66
Fe^{3+}	11	—	1·15	14	—	0·89	12	—	1·55
Fe^{2+}	11	—	0·72	14	—	1·18	12	—	1·36
Mn	11	—	0·04	14	—	0·03	12	—	0·05
Mg	3	8	tr-0·46	14	—	0·54	12	—	0·26
Ca	11	—	0·89	14	—	1·38	12	—	1·05
Na	11	—	2·70	14	—	2·35	12	—	2·71
K	11	—	4·24	14	—	3·84	12	—	5·23
H_2O+	11	—	0·45	14	—	0·45	12	—	0·45
H_2O-	10	1	tr-0·06	—	—	—	12	—	0·14
P	11	—	0·06	—	—	—	12	—	0·07
F	—	—	—	—	—	—	—	—	—
			ppm			ppm			ppm
Ba	9	2	1250->2000	—	7	880->2000	9	3	790->2000
Be	7	4	<3- 3	6	8	<3- 4	9	3	<3- 6
Co	11	—	<10	14	—	<10	12	—	<10
Cr	—	11	19	—	14	44	1	11	<10- 26
Cu	1	—	4	—	14	35	2	9	<5- 13
Ga	—	11	22	—	14	22	—	12	31
La	—	11	85	7	7	<100-125	2	10	<100-140
Li	2	9	<5- 18	—	14	28	—	12	32
Mn	—	11	450	—	14	245	—	10	430
Mo	11	—	<3	7	7	<3- 3	4	8	<3- 5
Nb	2	9	<10- 23	5	9	<10- 17	2	10	<10- 58
Ni	10	1	<10- 13	11	3	<10- 48	11	1	<10- 10
Pb	—	11	21	—	14	35	—	12	36
Rb	—	11	125	—	14	205	—	8	300
Sc	—	—	—	14	—	<10	7	4	<10- 14
Sr	—	11	245	—	14	190	—	12	290
V	2	9	<3- 34	—	14	31	2	10	<3- 24
Y	—	11	52	—	14	29	1	11	<10- 46
Zr	—	11	400	—	14	310	11	1	490->1000

389

TABLE 1(i)

Anorogenic intermediate rocks of Aden, Malawi and north-east Sudan

Element	ADEN, SOUTH ARABIAN FEDERATION				CHILWA ALKALINE PROVINCE, MALAWI				NORTH-EAST SUDAN			
	Number of samples			Arithmetic mean and indeterminate groups	Number of samples			Arithmetic mean and indeterminate groups	Number of samples			Arithmetic mean and indeterminate groups
	<	n	>	%	<	n	>	%	<	n	>	%
Si	—	6	—	28·77	—	22	—	28·91	—	6	—	29·23
Ti	—	6	—	0·56	—	22	—	0·43	—	6	—	0·28
Al	—	6	—	8·10	—	22	—	8·81	—	6	—	8·84
Fe^{3+}	—	6	—	3·46	—	22	—	1·58	—	6	—	2·99
Fe^{2+}	1	5	—	tr-2·05	—	22	—	2·32	—	6	—	0·78
Mn	—	6	—	0·14	—	22	—	0·17	2	4	—	0·15
Mg	—	6	—	0·65	—	22	—	0·40	—	6	—	tr-0·34
Ca	—	6	—	2·14	—	22	—	1·11	—	6	—	1·24
Na	—	6	—	4·14	—	22	—	4·84	—	6	—	5·12
K	—	6	—	3·04	—	22	—	4·57	—	6	—	3·33
H_2O+	—	6	—	0·34	—	22	—	0·53	—	6	—	0·82
H_2O-	—	6	—	0·77	—	22	—	0·19	—	6	—	0·22
P	—	6	—	0·15	—	22	—	0·11	—	6	—	0·07
F	—	6	—	0·21	—	—	—	—	—	—	—	—
				ppm				ppm				ppm
Ba	2	6	—	600	—	20	2	530->2000	—	6	—	480
Be	5	4	—	<3- 4	11	11	—	<3- 7	1	5	—	<3- 6
Co	6	1	—	<10- 10	22	—	—	<10	5	1	—	<10- 10
Cr	6	—	—	<10	9	13	—	<10- 11	—	6	—	46
Cu	—	6	—	8	—	10	—	16	—	—	—	—
Ga	—	5	—	33	—	22	—	34	—	6	—	<5
La	1	5	—	<100- 200	1	21	—	<100- 130	1	5	—	<100- 115
Li	—	6	—	17	2	20	—	<5- 29	2	4	—	<5- 11
Mn	—	5	1	1500->2000	—	17	5	1550->2000	—	5	1	1450->2000
Mo	—	5	—	<3- 4	7	15	—	<3- 6	3	3	—	<3- 8
Nb	1	6	—	180	—	22	—	110	—	6	—	120
Ni	6	—	—	<10	1	1	—	<10- 10	—	6	—	<10
Pb	6	—	—	<5	21	18	—	<5- 31->1000	6	—	—	11
Rb	—	6	—	105	3	22	1	170	1	5	—	<50- 59
Sc	1	5	—	<10- 15	14	8	—	<10- 13	4	2	—	<10- 10
Sr	—	5	—	235	1	21	—	<30- 185	2	4	—	<30- 140
V	1	5	—	<3- 40	4	18	—	<3- 17	5	1	—	<3- 90
Y	—	6	—	77	—	22	—	54	—	6	—	60

TABLE 1(*j*)

Orogenic intermediate rocks of Malawi and Somalia

Element	MALAWI Number of samples <	n	>	MALAWI Arithmetic mean and indeterminate groups	SOMALIA Number of samples <	n	>	SOMALIA Arithmetic mean and indeterminate groups
				%				%
Si	—	17	—	27·42	—	5	—	28·01
Ti	—	17	—	0·38	—	5	—	0·09
Al	—	17	—	9·93	—	5	—	11·20
Fe^{3+}	—	17	—	1·61	—	5	—	1·19
Fe^{2+}	—	17	—	1·87	—	5	—	1·14
Mn	—	17	—	0·08	—	5	—	0·06
Mg	—	17	—	0·71	1	4	—	tr-0·12
Ca	—	17	—	1·96	—	5	—	0·73
Na	—	17	—	4·19	—	5	—	5·44
K	—	17	—	5·15	—	5	—	4·98
H_2O+	—	17	—	0·37	—	5	—	0·56
H_2O-	1	16	—	tr-0·14	—	5	—	0·07
P	—	17	—	0·19	—	5	—	0·05
F	—	—	—	—	—	—	—	—
				ppm				ppm
Ba	—	8	9	410->2000	—	5	—	300
Be	7	10	—	<3- 8	1	4	—	<3- 5
Co	14	3	—	<10- 11	5	—	—	<10
Cr	7	10	—	<10- 18	4	1	—	<10- 30
Cu	5	3	—	<5- 12	2	—	—	<5
Ga	—	17	—	31	—	5	—	28
La	3	14	—	<100-240	3	2	—	<100-130
Li	2	15	—	<5- 16	3	2	—	<5- 6
Mn	—	17	—	880	—	5	—	780
Mo	9	8	—	<3- 6	4	1	—	<3- 3
Nb	—	17	—	75	1	4	—	<10- 73
Ni	12	5	—	<10- 21	4	1	—	<10- 10
Pb	1	16	—	<5- 35	2	3	—	<5- 12
Rb	—	15	—	215	—	5	—	58
Sc	15	2	—	<10- 13	5	—	—	<10
Sr	—	13	4	600->1000	—	5	—	115
V	—	17	—	42	4	1	—	<3- 8
Y	3	14	—	<10- 54	1	4	—	<10- 15
Zr	—	13	4	570->1000	—	4	1	250->1000

TABLE 2(a)-(s)

Tables of variance showing significant differences amongst the average analyses for selected groups of rocks. *Notation*: When no figure appears under the heading 'degrees of freedom' the data are insufficient for statistical testing, either through lack of analyses or because of the proportion of indeterminate results. The number of degrees of freedom is the number of values in a set which may be assigned arbitrarily. Thus, if the total number of determinations is n, and the number of sub-groups is m, then

$$\text{total number of degrees of freedom} = n-1$$
$$\text{degrees of freedom } between \text{ groups } = m-1$$
$$\text{degrees of freedom } within \text{ groups } = n-m$$

in contrast to thermodynamic usage in which 'degrees of freedom' represent the intensive and extensive variables of a system.

Significance has been tested by comparing the calculated variance ratio with tabulated percentage points of the F-distribution (Pearson and Hartley, 1958), and classified by the following scheme:

Calculated ratio exceeds tabulated value at 0·1% level—highly significant ★★★★★

Calculated ratio exceeds tabulated value at 1% level—significant ★★★

Calculated ratio exceeds tabulated value at 5% level—probably significant ★

Calculated ratio less than tabulated value at 5% level—not significant —

Visual assessment necessary because of indeterminate groups in analysis, but significant difference evident ?

TABLE 2(a)

Significant differences between anorogenic and orogenic acid rocks of Africa

	Degrees of freedom		NORMAL DATA		LOGARITHMIC DATA		
Element	Between groups	Within groups	Variance ratio F	Significance	Variance ratio F'	Significance	Critical F-value at % level
Si	1	147	1·58	—	4·20	★	3·9 at 5%
Ti	1	146	4·31	★	2·05	—	3·9 at 5%
Al	1	147	56·41	★★★★★	55·50	★★★★★	11·3 at 0·1%
Fe^{3+}	1	147	13·17	★★★★★	8·66	★★★	6·8 at 1%
Fe^{2+}	1	147	1·03	—	0·40	—	
Mn	1	146	22·67	★★★★★	14·89	★★★★★	11·3 at 0·1%
Mg	—	—	—	?	—		
Ca	1	142	22·96	★★★★★	20·39	★★★★★	11·3 at 0·1%
Na	1	147	14·64	★★★★★	13·72	★★★★★	11·3 at 0·1%
K	1	147	0·61	—	2·42	—	
H_2O+	1	144	0·00	—	0·46	—	
H_2O-	—	—	—		—		
P	—	—	—		—		
F	—	—	—		—		
Ba	—	—	—		—		
Be	—	—	—		—		
Co	—	—	—		—		
Cr	—	—	—		—		
Cu	—	—	—		—		
Ga	1	154	42·63	★★★★★	42·44	★★★★★	11·3 at 0·1%
La	—	—	—		—		
Li	—	—	—	?	—		
Mn	1	131	24·12	★★★★★	16·47	★★★★★	11·3 at 0·1%
Mo	—	—	—		—		
Nb	—	—	—	?	—		
Ni	—	—	—		—		
Pb	1	138	2·79	—	0·22	—	
Rb	1	127	3·50	—	2·07	—	
Sc	—	—	—		—		
Sr	—	—	—	?	—		
V	—	—	—		—		
Y	1	134	24·41	★★★★★	56·18	★★★★★	11·3 at 0·1%
Zr	1	141	30·75	★★★★★	22·84	★★★★★	11·3 at 0·1%

Table 2(b)

Significant differences between intrusive and extrusive anorogenic acid rocks

| | Degrees of freedom | | NORMAL DATA | | LOGARITHMIC DATA | | |
Element	Between groups	Within groups	Variance ratio F	Signifi-cance	Variance ratio F'	Signifi-cance	Critical F-value at % level
Si	1	68	5·00	★	5·12	★	3·99 at 5%
Ti	1	68	0·69	—	1·32	—	
Al	1	68	0·42	—	0·46	—	
Fe^{3+}	1	68	25·98	★★★★★	17·05	★★★★★	11·90 at 0·1%
Fe^{2+}	1	68	6·92	★	8·81	★★★	7·05 at 1%
Mn	1	67	10·67	★★★	9·12	★★★	7·05 at 1%
Mg	—	—	—		—		
Ca	1	64	0·54	—	1·61	—	
Na	1	68	4·38	★	2·94	—	3·99 at 5%
K	1	68	10·25	★★★	10·43	★★★	7·05 at 1%
H_2O+	1	68	3·14	—	1·72	—	
H_2O-	1	66	11·60	★★★	6·72	★	7·06 at 1%
P	—	—	—		—		
F	—	—	—		—		
Ba	1	65	1·89	—	2·19	—	
Be	—	—	—		—		
Co	—	—	—		—		
Cr	—	—	—		—		
Cu	—	—	—		—		
Ga	1	75	0·66	—	0·14	—	
La	—	—	—		—		
Li	—	—	—	?	—		
Mn	1	62	8·28	★★★	8·38	★★★	7·08 at 1%
Mo	—	—	—		—		
Nb	1	67	14·50	★★★★★	6·39	★	11·90 at 0·1%
Ni	—	—	—		—		
Pb	—	—	—		—		
Rb	1	62	8·99	★★★	14·77	★★★★★	11·95 at 0·1%
Sc	—	—	—		—		
Sr	—	—	—		—		
V	—	—	—		—		
Y	1	74	1·12	—	0·01	—	
Zr	1	63	4·52	★	2·59	—	4·00 at 5%

TABLE 2(c)

Significant differences amongst six localities of anorogenic acid rocks

	Degrees of freedom		NORMAL DATA		LOGARITHMIC DATA		
Element	Between groups	Within groups	Variance ratio F	Significance	Variance ratio F'	Significance	Critical F-value at % level
Si	5	64	12·58	★★★★★	11·82	★★★★★	4·74 at 0·1%
Ti	5	64	6·72	★★★★★	6·60	★★★★★	4·74 at 0·1%
Al	5	64	2·62	★	2·52	★	2·37 at 5%
Fe^{3+}	5	64	9·82	★★★★★	9·23	★★★★★	4·74 at 0·1%
Fe^{2+}	5	64	5·00	★★★★★	6·72	★★★★★	4·74 at 0·1%
Mn	5	63	15·25	★★★★★	11·67	★★★★★	4·74 at 0·1%
Mg	—	—	—	?	—		
Ca	5	60	4·67	★★★	4·46	★★★	3·34 at 1%
Na	5	64	16·67	★★★★★	16·36	★★★★★	4·74 at 0·1%
K	5	64	9·52	★★★★★	9·69	★★★★★	4·74 at 0·1%
H_2O+	5	64	1·34	—	1·24	—	
H_2O-	5	62	6·37	★★★★★	7·24	★★★★★	4·75 at 0·1%
P	—	—	—		—		
F	—	—	—		—		
Ba	5	61	5·16	★★★★★	8·54	★★★★★	4·76 at 0·1%
Be	—	—	—		—		
Co	—	—	—		—		
Cr	—	—	—	?	—		
Cu	—	—	—	?	—		
Ga	5	71	22·23	★★★★★	35·48	★★★★★	4·70 at 0·1%
La	—	—	—		—		
Li	—	—	—	?	—		
Mn	5	58	12·36	★★★★★	10·72	★★★★★	4·80 at 0·1%
Mo	—	—	—		—		
Nb	5	63	5·55	★★★★★	15·15	★★★★★	4·74 at 0·1%
Ni	—	—	—		—		
Pb	—	—	—		—		
Rb	5	58	2·87	★	5·64	★★★★★	4·80 at 0·1%
Sc	—	—	—		—		
Sr	—	—	—		—		
V	—	—	—		—		
Y	5	70	5·71	★★★★★	8·93	★★★★★	4·70 at 0·1%
Zr	5	59	3·94	★★★	5·85	★★★★★	3·35 at 1%

Table 2(d)

Significant differences amongst seven localities of orogenic acid rocks

	Degrees of freedom		NORMAL DATA		LOGARITHMIC DATA		
Element	Between groups	Within groups	Variance ratio F	Signifi-cance	Variance ratio F'	Signifi-cance	Critical F-value at % level
Si	6	72	2·49	★	2·33	★	2·23 at 5%
Ti	6	71	1·48	—	3·58	★★★	3·09 at 1%
Al	6	72	2·51	★	2·59	★	2·23 at 5%
Fe^{3+}	6	72	3·44	★★★	3·74	★★★	3·09 at 1%
Fe^{2+}	6	72	2·09	—	4·54	★★★★★	4·30 at 0·1%
Mn	6	72	1·72	—	2·23	★	2·23 at 5%
Mg	6	68	1·32	—	0·93	—	
Ca	6	71	1·42	—	1·93	—	
Na	6	72	4·54	★★★★★	5·48	★★★★★	4·30 at 0·1%
K	6	72	7·17	★★★★★	3·28	★★★	4·30 at 0.1%
H_2O+	6	69	4·63	★★★★★	4·41	★★★★★	4·32 at 0·1%
H_2O-	—	—	—		—		
P	—	—	—		—		
F	—	—	—		—		
Ba	—	—	—		—		
Be	—	—	—		—		
Co	—	—	—		—		
Cr	—	—	—	?	—		
Cu	—	—	—		—		
Ga	6	72	6·36	★★★★★	8·99	★★★★★	4·30 at 0·1%
La	—	—	—		—		
Li	—	—	—		—		
Mn	—	—	—		—		
Mo	—	—	—		—		
Nb	—	—	—		—		
Ni	—	—	—		—		
Pb	6	71	4·77	★★★★★	4·75	★★★★★	4·31 at 0·1%
Rb	—	—	—	?	—		
Sc	—	—	—		—		
Sr	6	69	2·10	—	1·42	—	
V	—	—	—		—		
Y	—	—	—		—		
Zr	6	71	6·76	★★★★★	6·08	★★★★★	4·31 at 0·1%

TABLE 2(e)
Significant differences between the acid rocks of Aden and Somalia

Element	Degrees of freedom		NORMAL DATA		LOGARITHMIC DATA		Critical F-value at % level
	Between groups	Within groups	Variance ratio F	Signifi-cance	Variance ratio F'	Signifi-cance	
Si	1	30	14·80	*****	14·68	*****	13·29 at 0·1%
Ti	1	30	0·62	—	3·50	—	
Al	1	30	0·06	—	0·13	—	
Fe^{3+}	1	30	38·27	*****	31·77	*****	13·29 at 0·1%
Fe^{2+}	1	30	0·61	—	2·82	—	
Mn	1	30	98·76	*****	49·93	*****	13·29 at 0·1%
Mg	—	—	—	?	—		
Ca	1	30	2·27	—	0·34	—	
Na	1	30	30·12	*****	33·89	*****	13·29 at 0·1%
K	1	30	0·40	—	0·88	—	
H_2O+	1	28	2·76	—	2·62	—	
H_2O-	1	30	21·61	*****	35·55	*****	13·29 at 0·1%
P	1	28	2·82	—	6·06	*	4·20 at 5%
F	—	—	—		—		
Ba	—	—	—		—		
Be	—	—	—		—		
Co	—	—	—		—		
Cr	—	—	—		—		
Cu	—	—	—		—		
Ga	1	30	16·32	*****	18·80	*****	13·29 at 0·1%
La	—	—	—		—		
Li	—	—	—		—		
Mn	1	30	91·45	*****	51·23	*****	13·29 at 0·1%
Mo	—	—	—		—		
Nb	—	—	—	?	—		
Ni	—	—	—		—		
Pb	—	—	—		—		
Rb	1	27	3·94	—	4·79	*	4·21 at 5%
Sc	—	—	—		—		
Sr	—	—	—		—		
V	—	—	—		—		
Y	—	—	—		—		
Zr	1	27	84·09	*****	26·26	*****	13·29 at 0·1%

TABLE 2(*f*)

Significant differences between the acid rocks of Aden and north-east Sudan

Element	Degrees of freedom		NORMAL DATA		LOGARITHMIC DATA		
	Between groups	Within groups	Variance ratio F	Signifi-cance	Variance ratio F'	Signifi-cance	Critical F-value at % level
Si	1	16	81·76	*****	73·90	*****	16·12 at 0·1%
Ti	1	16	11·64	***	12·60	***	8·53 at 1%
Al	1	16	8·87	***	9·18	***	8·53 at 1%
Fe^{3+}	1	16	20·27	*****	41·18	*****	16·12 at 0·1%
Fe^{2+}	1	16	5·18	*	9·40	***	4·49 at 5%
Mn	1	16	52·33	*****	40·37	*****	16·12 at 0·1%
Mg	—	—	—		—		
Ca	1	16	8·69	***	13·75	***	8·53 at 1%
Na	1	16	22·63	*****	19·17	*****	16·12 at 0·1%
K	1	16	6·06	*	6·00	*	4·49 at 5%
H_2O+	1	16	0·49	—	0·03	—	
H_2O-	1	16	5·74	*	12·56	***	8·53 at 1%
P	1	14	0·05	—	0·03	—	
F	—	—	—		—		
Ba	1	14	0·40	—	0·03	—	
Be	—	—	—		—		
Co	—	—	—		—		
Cr	—	—	—		—		
Cu	—	—	—		—		
Ga	1	16	38·84	*****	45·63	*****	16·12 at 0·1%
La	—	—	—		—		
Li	—	—	—		—		
Mn	1	16	55·83	*****	59·67	*****	16·12 at 0·1%
Mo	—	—	—		—		
Nb	—	—	—	?	—		
Ni	—	—	—		—		
Pb	—	—	—		—		
Rb	1	12	7·49	*	12·38	***	9·33 at 1%
Sc	—	—	—		—		
Sr	—	—	—		—		
V	—	—	—		—		
Y	1	16	24·04	*****	21·10	*****	16·12 at 0·1%
Zr	—	—	—	?	—		

TABLE 2(g)

Significant differences between the acid rocks of Aden and northern Nigeria

	Degrees of freedom		NORMAL DATA		LOGARITHMIC DATA		
Element	Between groups	Within groups	Variance ratio F	Signifi- cance	Variance ratio F'	Signifi- cance	Critical F-value at % level
Si	1	30	66·63	★★★★★	66·14	★★★★★	13·29 at 0·1%
Ti	1	30	11·15	★★★	8·01	★★★	7·56 at 1%
Al	1	30	24·83	★★★★★	23·81	★★★★★	13·29 at 0·1%
Fe^{3+}	1	30	52·10	★★★★★	41·78	★★★★★	13·29 at 0·1%
Fe^{2+}	1	30	2·43	—	4·28	★	4·17 at 5%
Mn	1	29	74·57	★★★★★	37·22	★★★★★	13·39 at 0·1%
Mg	—	—	—		—		
Ca	1	30	8·38	★★★	7·21	★	7·56 at 1%
Na	1	30	24·92	★★★★★	28·33	★★★★★	13·29 at 0·1%
K	1	30	11·01	★★★	10·92	★★★	7·56 at 1%
H_2O+	1	30	3·39	—	2·02	—	
H_2O-	1	29	13·28	★★★	12·19	★★★	7·60 at 1%
P	1	26	3·32	—	2·30	—	
F	1	27	0·27	—	1·00	—	
Ba	1	26	20·85	★★★★★	17·33	★★★★★	13·74 at 0·1%
Be	—	—	—		—		
Co	—	—	—		—		
Cr	—	—	—		—		
Cu	—	—	—		—		
Ga	1	30	10·55	★★★	13·40	★★★★★	13·29 at 0·1%
La	—	—	—		—		
Li	1	28	4·03	—	3·31	—	
Mn	1	30	42·54	★★★★★	32·16	★★★★★	13·29 at 0·1%
Mo	—	—	—		—		
Nb	1	28	12·40	★★★	15·57	★★★★★	13·50 at 0·1%
Ni	—	—	—		—		
Pb	—	—	—	?	—		
Rb	1	27	2·06	—	1·09	—	
Sc	—	—	—		—		
Sr	—	—	—		—		
V	—	—	—		—		
Y	1	30	7·55	★	9·02	★★★	7·56 at 1%
Zr	—	—	—		—		

TABLE 2(h)

Significant differences between the combined groups of Aswan (U.A.R.), north-east Sudan and Somalia compared with the acid rocks of Nigeria

	Degrees of freedom		NORMAL DATA		LOGARITHMIC DATA		
Element	Between groups	Within groups	Variance ratio F	Signifi-cance	Variance ratio F'	Signifi-cance	Critical F-value at % level
Si	1	48	0·75	—	0·00	—	
Ti	1	48	0·44	—	0·01	—	
Al	1	48	14·62	*****	14·61	*****	12·37 at 0·1%
Fe³⁺	1	48	0·96	—	0·47	—	
Fe²⁺	1	48	7·07	*	15·37	*****	12·37 at 0·1%
Mn	1	47	1·94	—	1·81	—	
Mg	—	—	—	?	—		
Ca	1	48	6·29	*	3·72	—	4·05 at 5%
Na	1	48	0·58	—	0·89	—	
K	1	48	3·60	—	2·70	—	
H₂O+	1	45	1·17	—	0·00	—	
H₂O−	—	—	—		—		
P	1	43	7·73	***	19·32	*****	7·28 at 1%
F	—	—	—		—		
Ba	—	—	—	?	—		
Be	—	—	—		—		
Co	—	—	—		—		
Cr	—	—	—		—		
Cu	—	—	—		—		
Ga	1	48	63·10	*****	76·98	*****	12·37 at 0·1%
La	—	—	—		—		
Li	—	—	—	?	—		
Mn	1	47	3·44	—	4·37	*	4·05 at 5%
Mo	—	—	—		—		
Nb	—	—	—	?	—		
Ni	—	—	—		—		
Pb	1	47	3·18	—	2·13	—	
Rb	1	47	7·13	*	6·33	*	4·05 at 5%
Sc	—	—	—		—		
Sr	—	—	—	?	—		
V	—	—	—		—		
Y	—	—	—	?	—		
Zr	—	—	—	?	—		

401

TABLE 2(i)

Significant differences between the acid rocks of south-east Uganda and Somalia

	Degrees of freedom		NORMAL DATA		LOGARITHMIC DATA		
Element	Between groups	Within groups	Variance ratio F	Signifi- cance	Variance ratio F'	Signifi- cance	Critical F-value at % level
Si	1	27	2·32	—	2·43	—	
Ti	1	26	0·51	—	1·72	—	
Al	1	27	3·47	—	3·39	—	
Fe^{3+}	1	27	1·99	—	1·17	—	
Fe^{2+}	1	27	4·79	★	9·83	★★★	7·68 at 1%
Mn	1	27	4·00	—	5·23	★	4·21 at 5%
Mg	1	26	0·93	—	0·09	—	
Ca	1	26	0·69	—	1·24	—	
Na	1	27	0·72	—	0·95	—	
K	1	27	0·22	—	0·10	—	
H_2O+	1	25	13·72	★★★	8·98	★★★	7·77 at 1%
H_2O-	1	27	0·54	—	1·52	—	
P	1	27	0·04	—	0·14	—	
F	—	—	—		—		
Ba	—	—	—		—		
Be	—	—	—		—		
Co	—	—	—		—		
Cr	1	27	4·28	★	1·80	—	4·21 at 5%
Cu	—	—	—		—		
Ga	1	27	0·15	—	0·63	—	
La	—	—	—		—		
Li	—	—	—		—		
Mn	1	26	4·91	★	6·45	★	4·23 at 5%
Mo	—	—	—		—		
Nb	—	—	—		—		
Ni	—	—	—	?	—		
Pb	1	26	1·33	—	1·87	—	
Rb	—	—	—		—		
Sc	—	—	—		—		
Sr	—	—	—		—		
V	—	—	—		—		
Y	—	—	—		—		
Zr	1	27	0·69	—	0·05	—	

TABLE 2(j)

Significant differences between the acid rocks of south-east Uganda and Zambia

Element	Degrees of freedom Between groups	Degrees of freedom Within groups	NORMAL DATA Variance ratio F	Signifi- cance	LOGARITHMIC DATA Variance ratio F'	Signifi- cance	Critical F-value at % level
Si	1	14	0·07	—	0·13	—	
Ti	1	13	0·96	—	1·42	—	
Al	1	14	0·03	—	0·00	—	
Fe^{3+}	1	14	2·56	—	2·89	—	
Fe^{2+}	1	14	0·57	—	1·19	—	
Mn	1	14	0·00	—	0·05	—	
Mg	1	14	0·50	—	0·92	—	
Ca	1	13	0·19	—	0·02	—	
Na	1	14	2·26	—	2·14	—	
K	1	14	0·01	—	0·05	—	
H_2O+	1	14	0·37	—	0·13	—	
H_2O-	1	14	0·70	—	1·23	—	
P	1	14	0·21	—	0·15	—	
F	—	—	—		—		
Ba	1	13	6·34	★	5·02	★	4·67 at 5%
Be	—	—	—		—		
Co	—	—	—		—		
Cr	1	14	0·49	—	0·02	—	
Cu	—	—	—		—		
Ga	1	14	34·41	★★★★★	48·75	★★★★★	17·14 at 0·1%
La	—	—	—		—		
Li	—	—	—		—		
Mn	—	—	—		—		
Mo	—	—	—		—		
Nb	—	—	—		—		
Ni	—	—	—		—		
Pb	1	14	0·07	—	0·33	—	
Rb	—	—	—		—		
Sc	—	—	—		—		
Sr	1	13	2·80	—	2·13	—	
V	—	—	—		—		
Y	—	—	—		—		
Zr	1	14	1·27	—	1·07	—	

TABLE 2(k)

Significant differences between the acid rocks of north-east Rhodesia and Zambia

| | Degrees of freedom | | NORMAL DATA | | LOGARITHMIC DATA | | |
| | Between groups | Within groups | Variance ratio F | Signifi-cance | Variance ratio F' | Signifi-cance | Critical F-value at % level |
Element							
Si	1	18	0·71	—	0·85	—	
Ti	1	18	0·24	—	0·02	—	
Al	1	18	0·58	—	0·71	—	
Fe³	1	18	0·13	—	0·37	—	
Fe²⁺	1	18	0·17	—	1·54	—	
Mn	1	18	3·80	—	3·07	—	
Mg	1	18	0·31	—	0·43	—	
Ca	1	18	0·15	—	0·00	—	
Na	1	18	10·22	***	9·24	***	8·29 at 1%
K	1	18	2·10	—	1·47	—	
H_2O+	1	18	8·76	***	5·67	*	4·41 at 5%
H_2O-	—	—	—		—		
P	—	—	—		—		
F	—	—	—		—		
Ba	—	—	—		—		
Be	—	—	—		—		
Co	—	—	—		—		
Cr	1	18	0·56	—	0·72	—	
Cu	—	—	—		—		
Ga	1	18	43·50	*****	48·90	*****	15·38 at 0·1%
La	—	—	—		—		
Li	1	17	2·45	—	1·55	—	
Mn	—	—	—		—		
Mo	—	—	—		—		
Nb	—	—	—		—		
Ni	—	—	—		—		
Pb	1	18	11·92	***	13·75	***	8·29 at 1%
Rb	—	—	—		—		
Sc	—	—	—		—		
Sr	1	18	4·73	*	3·77	—	4·41 at 5%
V	1	18	2·21	—	0·13	—	
Y	—	—	—		—		
Zr	1	18	10·69	***	14·67	***	8·29 at 1%

TABLE 2(*l*)

Significant differences between the acid rocks of South-West Africa and north-east Rhodesia

	Degrees of freedom		NORMAL DATA		LOGARITHMIC DATA		
Element	Between groups	Within groups	Variance ratio F	Signifi- cance	Variance ratio F'	Signifi- cance	Critical F-value at % level
Si	1	23	0·09	—	0·26	—	
Ti	1	23	0·24	—	0·05	—	
Al	1	23	0·40	—	0·53	—	
Fe^{3+}	1	23	1·49	—	2·12	—	
Fe^{2+}	1	23	6·73	★	8·94	★★★	7·88 at 1%
Mn	1	23	0·23	—	0·01	—	
Mg	—	—	—		—		
Ca	1	23	2·31	—	5·56	★	4·28 at 5%
Na	1	23	5·99	★	5·86	★	4·28 at 5%
K	1	23	1·33	—	1·49	—	
H_2O+	1	23	0·00	—	0·11	—	
H_2O-	—	—	—		—		
P	—	—	—		—		
F	—	—	—		—		
Ba	—	—	—		—		
Be	—	—	—		—		
Co	—	—	—		—		
Cr	1	23	6·55	★	13·53	★★★	7·88 at 1%
Cu	—	—	—		—		
Ga	1	23	0·03	—	0·09	—	
La	—	—	—		—		
Li	1	21	3·30	—	2·88	—	
Mn	1	23	6·47	★	3·37	—	4·28 at 5%
Mo	—	—	—		—		
Nb	—	—	—		—		
Ni	—	—	—		—		
Pb	1	23	14·93	★★★★★	15·45	★★★★★	14·19 at 0·1%
Rb	1	23	7·98	★★★	7·74	★	7·88 at 1%
Sc	—	—	—		—		
Sr	1	23	0·43	—	0·00	—	
V	1	21	0·11	—	0·38	—	
Y	1	23	5·24	★	5·72	★	4·28 at 5%
Zr	1	23	2·25	—	1·88	—	

Table 2(m)

Significant differences between the Nuanetsi Province of Rhodesia and the Gaberones Complex, Botswana

| Element | Degrees of freedom | | NORMAL DATA | | LOGARITHMIC DATA | | |
	Between groups	Within groups	Variance ratio F	Signifi- cance	Variance ratio F'	Signifi- cance	Critical F-value at % level
Si	1	18	0·71	—	0·55	—	
Ti	1	18	0·03	—	0·01	—	
Al	1	18	1·44	—	1·44	—	
Fe^{3+}	1	18	1·65	—	1·74	—	
Fe^{2+}	1	18	1·18	—	2·74	—	
Mn	1	18	0·44	—	0·37	—	
Mg	1	17	0·36	—	0·76	—	
Ca	1	18	0·25	—	0·12	—	
Na	1	18	1·56	—	1·72	—	
K	1	18	0·49	—	0·29	—	
H_2O+	1	18	0·83	—	0·89	—	
H_2O-	1	18	3·76	—	7·31	★	4·41 at 5%
P	1	18	0·04	—	0·01	—	
F	—	—	—		—		
Ba	—	—	—		—		
Be	—	—	—		—		
Co	—	—	—		—		
Cr	1	24	0·16	—	0·01	—	
Cu	—	—	—		—		
Ga	1	25	21·57	★★★★★	18·14	★★★★★	13·88 at 0·1%
La	—	—	—		—		
Li	—	—	—		—		
Mn	—	—	—		—		
Mo	—	—	—		—		
Nb	1	25	15·95	★★★★★	20·83	★★★★★	13·88 at 0·1%
Ni	—	—	—		—		
Pb	1	25	52·21	★★★★★	31·01	★★★★★	13·88 at 0·1%
Rb	—	—	—	?	—		
Sc	—	—	—		—		
Sr	1	23	1·33	—	2·67	—	
V	—	—	—		—		
Y	1	25	7·47	★	6·06	★	4·24 at 5%
Zr	1	25	2·14	—	3·75	—	

Table 2(n)

Significant differences amongst the acid rocks of South-West Africa, Gaberones Complex
of Botswana and Nuanetsi Province, Rhodesia

Element	Degrees of freedom Between groups	Within groups	NORMAL DATA Variance ratio F	Signifi- cance	LOGARITHMIC DATA Variance ratio F'	Signifi- cance	Critical F-value at % level
Si	2	28	0·31	—	0·24	—	
Ti	2	28	0·08	—	0·21	—	
Al	2	28	5·70	★★★	5·23	★	5·45 at 1%
Fe^{3+}	2	28	2·92	—	2·88	—	
Fe^{2+}	2	28	1·36	—	1·72	—	
Mn	2	28	0·42	—	0·27	—	
Mg	—	—	—	?	—		
Ca	2	28	0·11	—	0·32	—	
Na	2	28	1·16	—	1·32	—	
K	2	28	0·32	—	0·30	—	
H_2O+	2	28	0·51	—	0·53	—	
H_2O-	2	27	7·78	★★★	24·79	★★★★★	9·02 at 0·1%
P	2	28	0·02	—	0·06	—	
F	—	—	—		—		
Ba	—	—	—		—		
Be	—	—	—		—		
Co	—	—	—		—		
Cr	2	34	2·59	—	2·83	—	
Cu	—	—	—		—		
Ga	2	35	13·89	★★★★★	11·15	★★★★★	8·51 at 0·1%
La	—	—	—		—		
Li	—	—	—		—		
Mn	—	—	—		—		
Mo	—	—	—		—		
Nb	2	33	29·05	★★★★★	44·48	★★★★★	8·62 at 0·1%
Ni	—	—	—		—		
Pb	2	35	32·84	★★★★★	17·02	★★★★★	8·51 at 0·1%
Rb	—	—	—		—		
Sc	—	—	—		—		
Sr	2	33	4·55	★	3·69	★	3·29 at 5%
V	—	—	—		—		
Y	2	35	5·02	★	5·41	★★★	5·28 at 1%
Zr	2	35	1·38	—	2·14	—	

TABLE 2(o)

Significant differences amongst the acid rocks of South-West Africa, Gaberones (Botswana), Nuanetsi (Rhodesia) and north-east Rhodesia

Element	Degrees of freedom		NORMAL DATA		LOGARITHMIC DATA		Critical F-value at % level
	Between groups	Within groups	Variance ratio F	Signifi- cance	Variance ratio F'	Signifi- cance	
Si	3	41	0·26	—	0·09	—	
Ti	3	41	0·08	—	0·21	—	
Al	3	41	9·45	*****	9·08	*****	6·58 at 0·1%
Fe^{3+}	3	41	4·81	***	5·08	***	4·30 at 1%
Fe^{2+}	3	41	1·71	—	3·33	*	2·84 at 5%
Mn	3	41	0·65	—	0·27	—	
Mg	3	37	1·82	—	1·76	—	
Ca	3	41	1·46	—	2·13	—	
Na	3	41	2·43	—	2·43	—	
K	3	41	1·51	—	1·69	—	
H$_2$O+	3	41	0·46	—	0·49	—	
H$_2$O−	—	—	—	?	—		
P	—	—	—		—		
F	—	—	—		—		
Ba	—	—	—		—		
Be	—	—	—		—		
Co	—	—	—		—		
Cr	3	47	2·91	*	4·04	*	2·81 at 5%
Cu	—	—	—	?	—		
Ga	3	48	12·48	*****	9·57	*****	6·43 at 0·1%
La	—	—	—		—		
Li	—	—	—		—		
Mn	—	—	—		—		
Mo	—	—	—		—		
Nb	—	—	—	?	—		
Ni	—	—	—		—		
Pb	3	48	24·75	*****	16·09	*****	6·43 at 0·1%
Rb	3	41	9·44	*****	9·28	*****	6·58 at 0·1%
Sc	—	—	—		—		
Sr	3	46	3·20	*	3·87	*	2·82 at 5%
V	—	—	—		—		
Y	3	48	11·61	*****	13·59	*****	6·43 at 0·1%
Zr	3	48	2·24	—	2·37	—	

Table 2(p)

Significant differences amongst the acid rocks of Nuanetsi (Rhodesia), north-east Rhodesia and the Malawi basement

	Degrees of freedom		NORMAL DATA		LOGARITHMIC DATA		
Element	Between groups	Within groups	Variance ratio F	Signifi-cance	Variance ratio F'	Signifi-cance	Critical F-value at % level
Si	2	36	5·72	★★★	5·74	★★★	5·27 at 1%
Ti	2	36	0·01	—	0·61	—	
Al	2	36	21·20	★★★★★	21·35	★★★★★	8·47 at 0·1%
Fe^{3+}	2	36	6·50	★★★	9·06	★★★★★	5·27 at 1%
Fe^{2+}	2	36	1·45	—	4·67	★	3·26 at 5%
Mn	2	36	1·08	—	0·76	—	
Mg	2	35	3·11	—	2·79	—	
Ca	2	36	1·54	—	2·15	—	
Na	2	36	3·51	★	3·45	★	3·26 at 5%
K	2	36	11·22	★★★★★	8·25	★★★	8·47 at 0·1%
H_2O+	2	36	0·38	—	0·03	—	
H_2O-	—	—	—		—		
P	—	—	—		—		
F	—	—	—		—		
Ba	—	—	—		—		
Be	—	—	—		—		
Co	—	—	—		—		
Cr	2	41	1·98	—	2·08	—	
Cu	—	—	—	?	—		
Ga	2	43	15·25	★★★★★	13·04	★★★★★	8·17 at 0·1%
La	—	—	—		—		
Li	2	36	4·08	★	5·54	★★★	5·27 at 1%
Mn	—	—	—		—		
Mo	—	—	—		—		
Nb	2	36	30·93	★★★★★	64·06	★★★★★	8·47 at 0·1%
Ni	—	—	—		—		
Pb	2	43	8·29	★★★★★	9·16	★★★★★	8·17 at 0·1%
Rb	—	—	—	?	—		
Sc	—	—	—		—		
Sr	2	41	6·86	★★★	11·46	★★★★★	8·22 at 0·1%
V	—	—	—		—		
Y	2	42	11·22	★★★★★	11·11	★★★★★	8·20 at 0·1%
Zr	2	42	1·79	—	1·40	—	

TABLE 2(*q*)

Significant differences between the orogenic acid rocks of South–West Africa and Malawi

Element	Degrees of freedom		NORMAL DATA		LOGARITHMIC DATA		Critical F-value at % level
	Between groups	Within groups	Variance ratio F	Signifi- cance	Variance ratio F'	Signifi- cance	
Si	1	21	8·47	★★★	8·67	★★★	8·02 at 1%
Ti	1	21	0·02	—	1·08	—	
Al	1	21	2·52	—	2·53	—	
Fe³⁺	1	21	2·82	—	4·08	—	
Fe²⁺	1	21	5·07	★	6·18	★	4·32 at 5%
Mn	1	21	0·72	—	1·00	—	
Mg	—	—	—		—		
Ca	1	21	0·52	—	3·29	—	
Na	1	21	0·02	—	0·04	—	
K	1	21	14·28	★★★	13·09	★★★	8·02 at 1%
H₂O+	1	21	0·00	—	0·06	—	
H₂O−	1	20	8·89	★★★	3·86	—	8·10 at 1%
P	1	21	0·28	—	0·28	—	
F	—	—	—		—		
Ba	—	—	—		—		
Be	—	—	—		—		
Co	—	—	—		—		
Cr	1	20	2·47	—	2·17	—	
Cu	—	—	—		—		
Ga	1	21	24·38	★★★★★	22·73	★★★★★	14·59 at 0·1%
La	1	19	5·65	★	4·82	★	4·38 at 5%
Li	1	19	2·57	—	2·03	—	
Mn	1	19	0·01	—	0·16	—	
Mo	—	—	—		—		
Nb	—	—	—	?	—		
Ni	—	—	—		—		
Pb	1	21	6·97	★	8·58	★★★	8·02 at 1%
Rb	—	—	—	?	—		
Sc	—	—	—		—		
Sr	1	21	0·20	—	1·86	—	
V	—	—	—		—		
Y	1	20	0·21	—	0·16	—	
Zr	1	20	0·87	—	0·41	—	

TABLE 2(r)

Significant differences amongst the anorogenic acid rocks of Aden, Chilwa (Malawi) and northern Nigeria

Element	Degrees of freedom Between groups	Degrees of freedom Within groups	NORMAL DATA Variance ratio F	Signifi- cance	LOGARITHMIC DATA Variance ratio F'	Signifi- cance	Critical F-value at % level
Si	2	42	24·26	*****	22·40	*****	8·20 at 0·1%
Ti	2	42	6·43	***	4·75	*	5·16 at 1%
Al	2	42	4·73	*	4·59	*	3·22 at 5%
Fe^{3+}	2	42	21·91	*****	16·49	*****	8·20 at 0·1%
Fe^{2+}	2	42	7·24	***	5·40	***	5·16 at 1%
Mn	2	41	22·40	*****	24·54	*****	8·22 at 0·1%
Mg	—	—	—		—		
Ca	—	—	—	?	—		
Na	2	42	11·77	*****	11·74	*****	8·20 at 0·1%
K	2	42	8·50	*****	8·18	***	8·20 at 0·1%
H_2O+	2	42	2·30	—	2·08	—	
H_2O-	2	40	10·78	*****	8·78	*****	8·25 at 0·1%
P	—	—	—		—		
F	—	—	—		—		
Ba	—	—	—	?	—		
Be	—	—	—		—		
Co	—	—	—		—		
Cr	—	—	—		—		
Cu	—	—	—		—		
Ga	2	42	5·86	***	6·90	***	5·16 at 1%
La	—	—	—		—		
Li	2	39	1·29	—	1·73	—	
Mn	2	41	18·43	*****	20·95	*****	8·22 at 0·1%
Mo	—	—	—		—		
Nb	—	—	—		—		
Ni	—	—	—		—		
Pb	—	—	—	?	—		
Rb	2	37	1·41	—	1·74	—	
Sc	—	—	—		—		
Sr	—	—	—		—		
V	—	—	—		—		
Y	2	41	4·04	*	5·01	*	3·23 at 5%
Zr	—	—	—		—		

TABLE 2(s)

Significant differences between the acid rocks of Aden and Nuanetsi (Rhodesia)

Element	Degrees of freedom		Variance ratio F	Signifi-cance	Variance ratio F'	Signifi-cance	Critical F-value at % level
	Between groups	Within groups	NORMAL DATA		LOGARITHMIC DATA		
Si	1	24	9·70	***	9·14	***	7·82 at 1%
Ti	1	24	1·47	—	0·89	—	
Al	1	24	4·63	*	4·66	*	4·26 at 5%
Fe^{3+}	1	24	4·11	—	5·16	*	4·26 at 5%
Fe^{2+}	1	24	0·01	—	0·93	—	
Mn	1	24	44·17	*****	35·09	*****	14·03 at 0·1%
Mg	—	—	—	—	—	—	
Ca	1	24	1·27	—	0·45	—	
Na	1	24	140·93	*****	102·57	*****	14·03 at 0·1%
K	1	24	22·22	*****	20·42	*****	14·03 at 0·1%
H_2O+	1	24	0·64	—	0·09	—	
H_2O-	1	24	2·44	—	1·38	—	
P	1	22	4·04	—	3·94	—	
F	—	—	—		—		
Ba	—	—	—		—		
Be	—	—	—		—		
Co	—	—	—		—		
Cr	—	—	—		—		
Cu	—	—	—	?	—		
Ga	1	31	57·80	*****	48·22	*****	13·22 at 0·1%
La	—	—	—		—		
Li	—	—	—		—		
Mn	—	—	—	?	—		
Mo	—	—	—		—		
Nb	1	31	31·08	*****	31·73	*****	13·22 at 0·1%
Ni	—	—	—		—		
Pb	—	—	—	?	—		
Rb	—	—	—		—		
Sc	—	—	—		—		
Sr	—	—	—		—		
V	—	—	—		—		
Y	1	31	11·75	***	11·23	***	7·53 at 1%
Zr	1	28	13·24	***	13·83	*****	13·50 at 0·1%

412

Table 3

C.I.P.W. norms calculated from the average compositions of selected groups of samples, using a computer program written by Dr M. H. Hey and Dr R. W. Le Maitre of the British Museum, Department of Mineralogy, and Dr B. C. M. Butler, University of Oxford.

TABLE 3

C.I.P.W. norms calculated from average analyses

ACID ROCK GROUPS

Locality number	1 Nigeria	2 U.A.R., Aswan	3 N.E. Sudan	4 Aden, S. Arabia	5 Somalia	6 S.E. Uganda	7 Luapula, Zambia	8 S.W.A.	9 Gaberones, Botswana	10 Nuanetsi, Rhodesia	11 Chirwa, Rhodesia	12 Malawi	13 Chilwa, Malawi
q	29·25	23·69	37·09	18·16	30·10	24·72	26·91	28·05	29·24	29·11	29·42	19·75	21·73
c	—	0·68	0·60	—	—	0·11	0·14	0·33	—	—	0·29	—	—
or	28·13	28·90	22·93	25·77	24·35	22·93	22·64	30·20	30·26	31·33	27·37	37·24	30·50
ab	35·74	37·65	35·37	44·66	34·77	37·48	32·24	30·80	31·05	28·26	26·82	30·88	36·39
an	—	4·13	1·58	—	5·65	7·52	10·05	5·29	1·78	4·01	9·58	4·52	—
ne	1·24	—	—	0·75	0·66	—	—	—	—	—	—	—	1·53
di	—	—	—	—	—	—	—	—	2·51	1·42	—	1·41	—
ol	1·60	—	—	—	—	—	—	—	—	—	—	—	—
hy	—	1·24	0·08	1·17	1·12	4·05	3·64	1·37	0·75	0·19	3·36	1·18	3·12
wo	—	—	—	—	—	—	—	—	—	—	—	—	—
mt	1·18	3·26	1·04	3·34	2·06	1·44	2·03	1·91	2·81	2·65	1·84	3·22	1·15
il	0·38	0·42	0·25	0·63	0·51	0·57	0·85	0·82	0·80	0·76	0·80	0·80	0·44
hm	—	—	0·30	0·42	—	—	—	0·32	—	0·94	—	—	—
ap*	0·10	0·21	0·17	0·14	0·38	0·33	0·26	0·33	0·33	0·33	—	0·38	0·14
fr	0·40	—	—	0·30	—	—	—	—	—	—	—	—	—
ac	1·39	—	—	3·37	—	—	—	—	—	—	—	—	4·24
H₂O+	0·29	0·39	0·42	0·79	0·27	0·69	0·81	0·44	0·38	0·51	0·45	0·44	0·49
H₂O−	0·15	—	0·08	0·43	0·09	0·07	0·10	0·05	0·12	0·27	—	0·14	0·14
Total	99·85	100·57	99·91	99·93	99·96	99·91	99·67	99·91	100·03	99·78	99·93	99·96	99·87

INTERMEDIATE ROCK GROUPS

Locality number	3 N.E. Sudan	4 Aden, S. Arabia	5 Somalia	12 Malawi	13 Chilwa, Malawi
q	4·24	9·75	—	—	—
c	—	—	1·03	—	—
or	23·70	21·63	35·46	36·70	32·57
ab	58·38	47·21	44·65	40·11	54·44
an	2·79	5·92	4·34	7·52	—
ne	—	—	9·41	4·17	0·29
di	2·04	4·18	—	2·60	5·11
ol	—	—	1·20	2·43	1·46
hy	0·89	0·75	—	—	—
wo	0·89	—	—	—	—
mt	2·48	4·98	2·46	3·33	3·19
il	0·89	1·77	0·29	1·20	1·37
hm	2·57	1·51	—	—	—
ap*	0·38	0·81	0·26	1·04	0·59
fr	—	0·37	—	—	—
ac	—	—	—	—	0·17
H₂O+	0·81	0·34	0·56	0·35	0·52
H₂O−	0·22	0·77	0·07	0·13	0·19
Total	99·39	99·99	99·73	99·58	99·90

* Includes hydroxyapatite

REFERENCES

ABRAHAM, D. 1959. The stratigraphical and structural relationships of the Kundelungu System, Plateau Series and Basement rocks of the mid-Luapula Valley, Northern Rhodesia. Unpublished Ph.D. thesis, Univ. Leeds.

AHRENS, L. H. 1963. Lognormal-type distributions in igneous rocks—IV. *Geochim. et cosmochim. Acta*, **27**, 333.

—— 1966. Element distributions in specific igneous rocks—VIII. *Geochim. et cosmochim. Acta*, **30**, 109.

BAILEY, D. K., and SCHAIRER, J. F. 1966. The system $Na_2O-Al_2O_3-Fe_2O_3-SiO_2$ at 1 atmosphere, and the petrogenesis of alkaline rocks. *J. Petrology*, **7**, 114.

BLOOMFIELD, K. 1961. The age of the Chilwa Alkaline Province. *Rec. geol. Surv. Nyasaland*, **1**, 95.

—— 1965a. The geology of the Zomba area. *Bull. geol. Surv. Malawi*, **16**.

—— 1965b. A comparison between infracrustal syenites and granites of southern Malawi and plutonic rocks of the Chilwa Alkaline Province. *Rec. geol. Surv. Malawi*, **4**, 5.

—— 1968. The pre-Karroo geology of Malawi. *Mem. geol. Surv. Malawi*, **5**.

BROWN, G. M. 1963. Melting relations of Tertiary granitic rocks in Skye and Rhum. *Miner. Mag.*, **33**, 533.

BUTLER, J. R., BOWDEN, P., and SMITH, A. Z. 1962. K/Rb ratios in the evolution of the younger granites of northern Nigeria. *Geochim. et cosmochim. Acta*, **26**, 89.

CLIFFORD, T. N. 1968. Radiometric dating and the pre-Silurian geology of Africa. In *Radiometric dating for geologists*, p. 299. (Eds. E. I. Hamilton and R. M. Farquhar), London.

—— NICOLAYSEN, L. O., and BURGER, A. J. 1962. Petrology and age of the pre-Otavi basement granite at Franzfontein, northern South-West Africa. *J. Petrology*, **3**, 244.

—— ROOKE, J. M., and ALLSOPP, H. L. 1969. Petrochemistry and age of the Franzfontein granitic rocks of northern South-West Africa. *Geochim. et cosmochim. Acta*, **33**, 973.

COCHRAN, W. G. 1947. Some consequences when the assumptions for the analysis of variance are not satisfied. *Biometrics*, **3**, 22.

COX, K. G., GASS, I. G., and MALLICK, D. I. J. (*in preparation*). The petrology and petrochemistry of the volcanic rocks of Aden and Little Aden, South Arabia.

—— JOHNSON, R. L., MONKMAN, L. J., STILLMAN, C. J., VAIL, J. R., and WOOD, D. N. 1965. The geology of the Nuanetsi Igneous Province. *Phil. Trans. r. Soc. Lond.*, **257**, Ser. A, 71.

DELANY, F. M. 1955. Ring structures in the northern Sudan. *Eclog. geol. Helv.*, **48**, 133.

DICKINSON, D. R., DODSON, M. H., GASS, I. G., and REX, D. C. 1969. Correlation of initial $^{87}Sr/^{86}Sr$ with Rb/Sr in some late Tertiary volcanic rocks of south Arabia. *Earth Planetary Sci. Let.*, **6**, 84.

DIXEY, F., CAMPBELL SMITH, W., and BISSET, C. B. 1955. The Chilwa Series of southern Nyasaland. *Bull. geol. Surv. Nyasaland*, **5**.

GARSON, M. S. 1962. The Tundulu carbonatite ring-complex in southern Nyasaland. *Mem. geol. Surv. Nyasaland*, **2**.

GASS, I. G. 1955. The geology of the Dunganab area, Anglo-Egyptian Sudan. Unpublished M.Sc. thesis, Univ. Leeds.

—— MALLICK, D. I. J., and COX, K. G. 1965. Royal Society volcanological expedition to the South Arabian Federation and the Red Sea. *Nature, Lond.*, **205**, 952.

GELLATLY, D. C. 1960. Report on the geology of the Las Dureh area, Burao District. *Rep. geol. Surv. Somaliland Prot.*, **6**.

—— 1963. The geology of the Darkainle nepheline syenite complex, Borama District, Somali Republic. Unpublished Ph.D. thesis, Univ. Leeds.

GINDY, A. R. 1954. The plutonic history of the Aswan area, Egypt. *Geol. Mag.*, **91**, 484.

GREENWOOD, J. E. G. W. 1960. Report on the geology of the Las Khoreh-Elayu area, Erigavo District. *Rep. geol. Surv. Somaliland Prot.*, **3**.

HOLT, D. N. 1961. The geology of part of Fort Johnston District east of Lake Nyasa. *Rec. geol. Surv. Nyasaland*, **1**, 23.

HUNT, J. A. 1958. Report on the geology of the Adadleh area, Hargeisa and Berbera Districts. *Rep. geol. Surv. Somaliland Prot.*, **2**.

—— 1960. Report on the geology of the Berbera–Sheikh area, Berbera and Burao Districts. *Rep. geol. Surv. Somaliland Prot.*, **4**.

JACOBSON, R. R. E., MACLEOD, W. N., and BLACK, R. 1958. Ring-complexes in the younger granite province of Northern Nigeria. *Mem. geol. Soc. Lond.*, **1**.

—— SNELLING, N. J., and TRUSWELL, J. F. 1964. Age determinations in the geology of Nigeria, with special reference to the older and younger granites. *Overseas Geol. Mineral Resources*, **9**, 168.

JOHNSON, R. L. 1968. Structural history of the western front of the Mozambique Belt in northeast Southern Rhodesia. *Bull. geol. Soc. Am.*, **79**, 513.

KENNEDY, W. Q. 1964. The structural differentiation of Africa in the Pan-African (± 500 m.y.) tectonic episode. *8th Ann. Rep. Res. Inst. African Geol., Univ. Leeds*, 48.

LEGGO, P. J. 1968. Some recent isotope investigations. *12th Ann. Rep. Res. Inst. African Geol., Univ. Leeds*, 45.

LUTH, W. C., JAHNS, R. H., and TUTTLE, O. F. 1964. The granite system at pressures of 4 to 10 kilobars. *J. geophys. Res.*, **69**, 759.

MANTON, W. I. 1968. The origin of associated basic and acid rocks in the Lebombo-Nuanetsi Igneous Province, southern Africa, as implied by strontium isotopes. *J. Petrology*, **9**, 23.

MASON, J. E., and WARDEN, A. J. 1956. The geology of the Heis—Mait—Waqderia area, Erigavo District. *Rep. geol. Surv. Somaliland Prot.*, **1**.

McELHINNY, M. W. 1966. Rb–Sr and K–Ar measurements on the Modipe gabbro of Bechuanaland and South Africa. *Earth Planetary Sci. Let.*, **1**, 439.

MEHNERT, K. R. 1968. *Migmatites and the origin of granitic rocks*, Amsterdam.

MILLER, R. L., and KAHN, J. S. 1962. *Statistical analysis in the geological sciences*, New York.

OLD, R. A. 1968. The Masaba Granite Dome, S.E. Uganda. *12th Ann. Rep. Res. Inst. African Geol., Univ. Leeds*, 16.

PAGE, B. G. N. 1960. The stratigraphical and structural relationships of the Abercorn Sandstones, the Plateau Series and basement rocks of the Kawimbe area, Abercorn District, Northern Rhodesia. Unpublished Ph.D. thesis, Univ. Leeds.

PEARSON, E. S., and HARTLEY, H. O. 1958. *Biometrika tables for statisticians*. Vol. I, 2nd ed., p. 159, Cambridge.

QUENOUILLE, M. H. 1959. *Rapid statistical calculations*, London.

RODIONOV, D. A. 1962. Estimation of average content and dispersion of a lognormal distribution of components in rocks and ores. *Geochemistry*, **7**, 728.

ROOKE, J. M. 1964. Element distribution in some acid igneous rocks of Africa. *Geochim. et cosmochim. Acta*, **28**, 1187.

—— and FISHER, A. M. 1962. Validity of spectrographic determinations of trace elements in granite G–1 and diabase W–1. *Geochim. et cosmochim. Acta*, **26**, 335.

SHACKLETON, R. M. 1964. A preliminary study of orogenic belts in eastern Africa. *8th Ann. Rep. Res. Inst. African Geol., Univ. Leeds*, 49.

SNELLING, N. J. 1963. Age determination unit. In *Ann. Rep. Overseas geol. Survs.* (for 1961–62), p. 30.

—— 1966. Age determination unit. In *Ann. Rep. Overseas geol. Survs.* (for 1965), p. 44.

—— 1967. Age determination unit. In *Ann. Rep. Inst. geol. Sci.* (for 1966), p. 142.

—— HAMILTON, E. I., DRYSDALL, A. R., and STILLMAN, C. J. 1964a. A review of age determinations from Northern Rhodesia. *Econ. Geol.*, **59**, 961.

—— HAMILTON, E. I., REX, D., HORNUNG, G., JOHNSON, R. L., SLATER, D., and VAIL, J. R. 1964b. Age determinations from the Mozambique and Zambesi Orogenic Belts, central Africa. *Nature, Lond.*, **201**, 463.

STILLMAN, C. J., and COX, K. G. 1960. The Chikala Hill syenite-complex of southern Nyasaland. *Trans. geol. Soc. S. Afr.*, **63**, 99.

TURNER, F. J., and VERHOOGEN, J. 1960. *Igneous and metamorphic petrology*, 2nd ed., New York.

TUTTLE, O. F., and BOWEN, N. L. 1958. Origin of granite in the light of experimental studies in the system $NaAlSi_3O_8$-$KAlSi_3O_8$-SiO_2-H_2O. *Mem. geol. Soc. Am.*, **74.**

VAIL, J. R. 1967. The southern extension of the East African Rift System and related igneous activity. *Geol. Rdsch.*, **57,** 601.

—— and MALLICK, D. I. J. 1965. The Mongolowe Hills nepheline-syenite ring-complex, southern Nyasaland. *Rec. geol. Surv. Nyasaland*, **3,** 49.

—— and MONKMAN, L. J. 1960. A geological reconnaissance survey of the Chaone Hill ring-complex, southern Nyasaland. *Trans. geol. Soc. S. Afr.*, **63,** 119.

VON KNORRING, O. 1960. Some geochemical aspects of a columbite-bearing soda granite from south-east Uganda. *Nature, Lond.*, **188,** 204.

WALSHAW, R. D. 1966. Preliminary report on the geology of the Mlanje Mountains, southern Malawi. *10th Ann. Rep. Res. Inst. African Geol., Univ. Leeds*, 27.

WRIGHT, E. P. 1958. Geology of the Gaberones District. *Rec. geol. Surv. Bechuanaland Prot.* (for 1956), 12.

—— 1961. The geology of the Gaberones District. Unpublished Ph.D. thesis, Univ. Oxford.

WYLLIE, P. J., and TUTTLE, O. F. 1961. Hydrothermal melting of shales. *Geol. Mag.*, **98,** 56.

P. G. HARRIS

18   Convection and magmatism
      with reference to the African
      continent

ABSTRACT. *The main way in which the earth's radiogenic heat can be focussed sufficiently to cause local melting is through a terrestrial heat engine, in which the heat is converted into mechanical energy by means of convective movement, and then released in restricted areas. There are two environments of such energy release, one where a convective tongue rises along mind-ocean ridges and rift systems, the other at the region of downward ingestion of crust. The upward current consists entirely of mantle material, so magma formed there is mantle-derived (basaltic). The downward limb consists of various crustal as well as mantle materials, and resultant magmas are of several types, basaltic, andesitic and rhyolitic. Applied to present and past vulcanism in Africa this pattern of convection suggests that the rift valley systems are the site of penetrative upwell, though this has caused little crustal extension.*

Introduction

The major possible sources of energy within the earth at present are: solar energy stored in chemically metastable sediments; tidal and gravitational forces; and the radioactive decay of such naturally occurring radionuclides as ^{238}U, ^{235}U, ^{232}Th, and ^{40}K. Usually the first two sources are regarded as relatively unimportant within the solid earth beneath its superficial sedimentary cover. The third source, radioactive decay, is probably the only important continuous source of energy at present, especially if the gravitational segregation of the core was completed early in the earth's history (Birch, 1965). On this basis, the present processes of tectonism, magmatism and chemical segregation within the earth must be dependent for their energy on radioactive decay. However, there is no evidence that these processes are associated with regions of abnormally high radioactivity. Their restriction in time and place requires some method of transferring the energy from the decay of the dispersed radioactive elements, and concentrating it in specific regions. It has been suggested that the earth should be regarded as a heat engine, in which the radiogenic heat causes convective movement in the mantle. The geological processes of magmatism and tectonism and their locations are then controlled by this convective movement.

As a first step in understanding the past history of continents it is desirable to study the present distribution of magmatism and tectonism in relation to possible convective movement, to see if adequate explanations can be found for the major magmatic and tectonic features. Africa is a key continent for this.

419

This essay is particularly concerned with the pattern of convective movement in providing an explanation for the distribution and nature of modern magmatism in and around the African continent. Inevitably, much of it is speculative and descriptive and not quantitative. It should be emphasised that the ideas of others, particularly Dietz, Elsasser, Green and Ringwood, Hess, Oxburgh and Turcotte, and Wilson, have been drawn on extensively even though these may not always be referred to at each relevant point in the text. Also, I hope that this paper reflects the teaching of Professor Kennedy, in his insistence that magmatism and magma type must be considered not in isolation, but in relation to the tectonic environment. When this paper was nearly completed, a very comprehensive synthesis of convection and tectonism appeared by Isacks *et al.* (1968), in which some aspects of the subject are treated in a much more detailed and quantitative manner.

The world pattern of convection

The shapes and positions of continents and mid-ocean ridges, and the geophysical and geological evidence for ocean-floor spreading have led to a general acceptance of continental drift caused by convective movement in the mantle. The main rival explanation for such features as the fit of the two sides of the Atlantic Ocean and the mid-Atlantic Ridge, is the expanding earth hypothesis. However this cannot explain such compressive features as the circum-Pacific and Alpine-Himalayan mobile belts. Also the expanding earth hypothesis requires a very large rate of expansion. Although a small degree of expansion may have occurred during geological time (Birch, 1968), convective movement must be regarded as by far the dominant cause of the earth's crustal features.

The mechanism and causes of thermal convective processes in the mantle have been reviewed extensively in a number of papers (for example: Elsasser, 1968; Knopoff, 1967; Orowan, 1965; Tozer, 1965) and will not be discussed here. The form and geological results of thermal convection, in their effect on continental movement, are however directly relevant and are discussed briefly. There are two groups of rival hypotheses on the form of the convective movement:

(i) convective cells of more or less symmetrical cross-section within the mantle, the continental crust being pulled laterally by the viscous drag of the moving mantle; and

(ii) upward penetrating tongues or wedges forcing apart rigid plates of oceanic and continental crust.

Until very recently, most convectionists have considered that convection currents move symmetrically in paired, opposed cells. Some have thought these cells to extend down to the bottom of the mantle, and to be more or less equidimensional in cross-section. Runcorn (1965) has suggested five paired cells in a section through the earth, and Dearnley (1966) a four-paired system. However, more recent variants of the symmetrical convection cell hypothesis suggest that the viscosity of the lower mantle is much greater than that of the upper mantle, so that convection is possible only in the latter, possibly not extending much below the low velocity layer. Such convection cells must be very broad and shallow in cross-section.

Most hypotheses of cellular convection in the mantle assume that the mantle convection current has a greater speed than the lateral drift of the overlying continental mass, the continent being pulled along by the viscous drag of the upper surface of the cell until it reaches a rest position above the downcurrent of two opposing cells. In this position, a continent should develop margins with compressive or cordilleran type tectonic features, caused by the drag as the sub-crustal current and its coupled ocean floor are pulled beneath. If both the upcurrent and the downcurrent are bilaterally symmetrical, being formed by limbs of similar opposed cells, the tectonic features produced in the crust above should also be symmetrical. In the crust above upcurrents, the mid-ocean ridges, this bilateral symmetry is apparent.

Equally however, if the continents are above the downcurrents they should be expected to have a bilateral symmetry also, with the two opposed margins showing similar compressive tectonic features. In fact, as has been long recognised, continental margins differ markedly in their tectonic character. Those facing the Pacific Ocean have young fold mountain chains parallel to the coast, whereas the coasts of the Atlantic and Indian Oceans have no such young compressive features, and even have Precambrian shields truncated at the coast. The volcanic activity of the continental margins also show marked differences, the Pacific coasts being characterised by extensive Tertiary andesites, rhyolites and basalts, the Atlantic coasts by less frequent Tertiary and Mesozoic vulcanism, dominantly basaltic.

It is apparent that models with symmetrical convection cells, in which each continent lies above a downcurrent, are not satisfactory because of the dissimilarity of continental margins. In particular, there are no young tectonic structures attributable to downcurrents in the margins of the African continent, except in the Atlas region.

The second proposed type of convective movement abandons the idea of symmetrical convection cells in which each upcurrent has a corresponding downcurrent. Instead it assumes that sectors of the crust and upper mantle behave as rigid plates being wedged apart along mid-ocean rifts by the upward penetrating tongues of mantle material. These tongues or plumes have been termed penetrative convection by Elder (1965) and convective dykes by Orowan (1965). They are assisted in their upward movement, because as the pressure falls, some degree of melting occurs (Oxburgh and Turcotte, 1968a), and the viscosity decreases markedly. However, once the dykes penetrate to high levels, the newly-formed liquid is lost by eruption and the injected mantle material will cool. These processes cause an increase in the viscosity of the injected tongues so that in effect these become part of the rigid plates. As each tongue is continuously split longitudinally by successive injections of mantle, so the two halves will be added to the adjoining rigid plates. These rigid plates grow by addition along the mid-ocean rises. If this addition of mantle material occurs dominantly at a sub-surface level, then the zone of addition and forcing apart of the plates will be surmounted by a zone of tension, the rift along the crest of the rise. In this model the moving continents should have two types of margin. The outer, leading edge of the rigid plate pushing against the resistant material beyond will become crumpled, while the trailing edge, which is a margin between continent and ocean within the same plate, will be stable. The tectonic implications of such a model and recent ideas on this topic are discussed by Morgan (1968) and Isacks *et al.* (1968).

This second type of convective movement is best illustrated by recon-structing the continental movement, starting from the assumption that South America, Africa, India, Australia and Antarctica were once united as a single 'supercontinent', Gondwanaland, that subsequently became dis-rupted (Wegener, 1966). The disruption was initiated by tongues or diapir-like intrusions of mantle rising at sites corresponding to the present margin of the African continental shelf. In addition to the rifting developed along this major line, there were extending from it, radial rifts and lines of penetrative wedges corresponding to the boundaries between the peripheral fragments of the original supercontinent. Continued intrusion of the convective tongues forced the outer continental plates away from the central block of the African continent, by the addition of successive wedges of mantle around the conti-nental margins. In this way, the sites of convective upwelling and their rifts would have retreated progressively from Africa at half the speed of the peri-pheral continent blocks beyond them; as an example, the Mid-Atlantic Ridge would be moving away from Africa, at half the rate of South America's movement from Africa.

This mechanism of penetrative convection differs from the symmetri-cal cell hypothesis in that the spreading ocean floor is not thought of as being swept under the continent from both sides by convection currents, but added to the rigid crustal plate. Such a system of crustal plates moving away from a line of growth or dilation on the surface of a sphere of fixed size, requires that at some other places, crust is consumed and destroyed. There a rigid plate rides over a weaker or denser portion of crust, the overridden material being carried down to the region of counter-movement. The locations of the regions of upwelling and engulfment cannot be predicted from geometrical considerations but must be deduced from the geological evidence—regions of characteristic seismicity or vulcanicity, or where there is tectonic evidence of crustal ingestion or engulfment. In a section through the earth, for example at the Tropic of Capricorn, one would expect upward wedges at the Mid-Atlantic Ridge, Carlsberg Ridge and the East Pacific Rise, and perhaps in the Mozambique Channel between Madagascar and Africa; the downward engulfment would be at the east and west margins of the Pacific, along the 'andesite line'. In north-south sections, engulfment would occur along the Alpine–Himalayan–Indonesian Belt, and the Aleutian arc. Other zones of engulfment would be island arcs and regions of andesite vulcanism such as the South Sandwich and Caribbean arcs. These regions of upwell and engulfment correspond to those of extension and compression shown in the diagrams given by Isacks *et al.* (1968, pp. 5859 and 5861).

In the present convective cycle, the initial zone of engulfment presumably was at the original outer margin of Gondwanaland, where the rigid continental plates were pushed over the weaker sea floor. With the outward movement of the continents, and the formation and spreading of the new oceans within the original Gondwanaland, the zone of engulfment also moved outward. That is, the Pacific Ocean became smaller and its sea floor was consumed as the new oceans and their fringing continents spread outward across it. This spreading outward from the central keystone of Africa and the extension of the oceans around Africa, does not preclude convective movement and ocean-floor spreading within the Pacific Ocean itself. In fact the lack of pre-Mesozoic rocks in the Pacific as in the other oceans suggests a cleaning-off by convection of the Pacific floor. The only requirement is that the rim of the Pacific floor is overridden by the surrounding continents.

Once a zone of ocean-floor engulfment under the outer margin of the continent became established, it would persist at that point. Mechanically, the overriding and engulfment is self-lubricating. The shear zone between engulfed and overriding material is a zone of energy release, heating, magmatism and volatile release (Oxburgh and Turcotte, 1968b). The resultant decrease in viscosity at and above the shear ensures that this remains the weakest zone and the site of future overriding. This local heating at the shear zone will not materially increase the temperature of the engulfed material sliding beneath it, and so will not detract from the gravitational cause of engulfment. Also, the region of the ocean floor furthest from the point of ocean-floor spreading is the oldest and therefore the coldest. Sinking due to gravity will be more likely in this region than elsewhere.

Carried to its ultimate conclusion, this mechanism wherein the Pacific floor is overridden by continents propelled by the spreading of the floors of the other oceans, would result in the Pacific becoming smaller and smaller until finally it could become the site of a new super-continent composed of eastern Asia, Australia and America. In fact, if there have been distinct convective episodes in the past, each one might be expected to culminate in the sweeping together of continents into a new supercontinent, which in the next convective episode would be disrupted again.

The general pattern of convective wedging and asymmetrical convective movement is to some degree implicit in Wilson's (1965) scheme, and has been developed recently by Elsasser (1968), Morgan (1968), Isacks *et al.* (1968) and others.

Although this discussion has been of rigid plates overriding or being pushed over the sea floor as the result of sea-floor spreading elsewhere, this is an oversimplification. Ringwood and D. H. Green (1966) have indicated that the sinking downlimb is relatively cold, and that engulfed sea-floor material of basaltic composition readily undergoes a phase transformation to eclogite. Also the transformation from peridotite to garnet-peridotite in the ultramafic component occurs at shallower depths than elsewhere because of the lower temperatures. Because both eclogite and garnet-peridotite are denser than the neighbouring mantle, the down limb is as much sucked or dragged down by the sinking of the denser rocks as overridden or pushed down. In the 'convective engine' of Ringwood and Green (*ibid.*), this downward engulfment is as important as the upward limb in providing the motive force for the lateral movement; in effect, some of the rigid crustal plates are pulled as much as pushed.

Rifts and swells

Both the symmetrical type of convective cell and the wedging mechanism can explain some other features of earth structure. A mantle upcurrent is of hotter and therefore lighter material than neighbouring rocks, so regions of upcurrent will be 'structural highs' compared with neighbouring areas. This effect is accentuated by phase transformations in the mantle. In regions of high geothermal gradient, the boundaries between layers of different mineral facies are depressed; thus, at depth, increase in temperature will be accompanied by an expansion from dense low-temperature, high-pressure minerals, to less dense, high-temperature ones. Hence the elevation of ocean ridges above the general level of the ocean floor. The amount of this elevation will depend on the speed and size of the convective upwelling, and its temperature effects. The Mid-Atlantic Ridge is up to $2\frac{1}{2}$ km above the level of the neighbouring ocean basins, and the other mid-ocean ridges often have reliefs of 2-3 km. If one assumes that some Pacific guyots were along previous ridges or swells, then the present depth of the summits of these indicates an amount of subsidence of the swell of 1-2 km since mid-Cretaceous times (Hess, 1962). To this should be added any lowering of sea-level since then (see below).

If the regions of upcurrent and high geothermal gradient become structural highs, then regions of downcurrent and low geothermal gradient should become structural lows. It is significant that during and after the break-up of the Gondwanaland continent in the Mesozoic all the major continental masses became depressed areas, major sedimentary basins. This can be explained if one accepts symmetrical convection cells with upcurrents beneath oceans and downcurrents beneath continents. The regions of downcurrents should become depressed. In non-scientific terms, the lateral sector of the current cannot flow horizontally but only downhill; the region of upwelling must be a high, and that of downcurrent a low, for lateral movement to occur.

The apparent depression of continents in the Mesozoic is more difficult to explain by the penetrative convection hypothesis, because this does not require the continents to be the site of downcurrents or regions of low geothermal gradient. However, the period of maximum marine inundation, the Cretaceous, appears to have coincided with the greatest rate of continental drift, and is ascribed to the growth of the mid-ocean ridge system (Hallam, 1967). If continental drift was much more rapid than at present, not only would a much greater rate of convection and development of mid-ocean swells and ridges have occurred, but also the average sub-oceanic geothermal gradients would have been much steeper, and the consequent movement of phase boundaries would have made the oceans shallower relative to the continents. This relative elevation of the ocean floor and the increased volume of the ocean swells and mid-ocean ridges would have caused extensive overflow of the oceans onto the continental surface. With fall-off in convective activity and in the rate of continental drift at the end of the Cretaceous, the ocean floors would have subsided and the seas withdrawn from the inundated continents.

Thermal regime

If there was no convective movement within the mantle, the geothermal gradient should be controlled by thermal conductivity and by the distribution of the radioactive heat sources. The present world-average heat flow is about $1 \cdot 5 \times 10^{-6}$ cal/cm²/sec (Lee and Uyeda, 1965), but one cannot be sure if this is an equilibrium value or an abnormal value inflated by the effects of the Mesozoic-Tertiary convective movement. The mode of the histograms for heat flow measurements is about $1 \cdot 1 \times 10^{-6}$ cal/cm²/sec for both continental and oceanic areas, suggesting that this is the equilibrium value, the higher average value being due to the effects of convection. It is significant that the mean heat flow for mid-ocean ridges is $1 \cdot 82 \times 10^{-6}$ cal/cm²/sec, for ocean basins $1 \cdot 28$, and for ocean trenches $0 \cdot 99$ (*ibid.*). Traced from the most recently formed crust and uplifted mantle at the mid-ocean ridge towards the oldest ocean floor at the point of downward engulfment near or beneath the ocean trench, the heat flow falls progressively (McBirney and Gass, 1967), as would be expected if it is largely controlled by the convective movement.

Although the continents are well removed from the upcurrents of the mid-ocean ridges and may even be located over the downward limb, their mean heat flow values ($1 \cdot 42$) are not lower than those of ocean floors, as might be expected, because of the additional contribution of the radiogenic heat from within the continental crust itself. If surface heat flows are influenced by convective movement, the coincidence at the same value of about $1 \cdot 5 \times 10^{-6}$ cal/cm²/sec for the mean of both oceanic and continental areas must be accidental rather than meaningful.

To extrapolate from surface heat flows to geothermal gradients is difficult, and current estimates differ markedly from one another (MacDonald, 1965; Clark and Ringwood, 1964). As a further difficulty, these estimates are based on models assuming equilibrium heat flow, and ignoring the thermal effects of convective movement. Low heat flows indicate low geothermal gradients, and high heat flow high gradients, but this is true only for the crust and uppermost mantle. Below this there must be some depth at which the geotherms become uniform for all environments. This may be at the boundary between mantle and core, or more probably at the lower limit of convective movement. Below that depth, an equilibrium heat flow will have levelled out environmental differences. So if the greatest depth at which convective movement can occur is about 500–700 km then this is the possible depth below which geothermal gradients are similar.

The paucity of present vulcanism in areas of low and medium heat flow indicates that in general the average geothermal gradient is too low for melting to occur in the mantle and for magma to form. Where vulcanism does occur in these regions, for example in continental shields or in ocean basins, the rocks tend to be highly alkalic and fractionated, features attributable to a deep origin. Where vulcanism is extensive, it is in regions of high heat flow, indicating that the geothermal gradients are steep enough for partial melting to occur.

In a mantle undergoing convection, a very great increase in geothermal gradient can be attained in two types of region. For convection to be possible, the initial geothermal gradient must be steeper than the adiabatic gradient, so mantle material ascending in the convective plume will be hotter than static material on either side. This relationship could be made more complicated by the thermal effects of passing through zones of phase transformation (Verhoogen, 1965) but, even so, it can be assumed that a convective upcurrent is a region of markedly steepened geothermal gradient, sufficiently so for melting to occur at shallow depths in the mantle (Oxburgh and Turcotte, 1968a). The fact that both tholeiitic and alkalic basalts occur along mid-ocean ridges indicates that the thermal gradients there are variable, but locally are steep enough for melting to occur at perhaps less than 10 kb (30–40 km in depth) at local temperatures probably of about 1300°C. The dominance of tholeiitic over alkalic types, and their very low degree of fractionation as shown by their characteristically low potassium content (Engel et al., 1965; Melson et al., 1968), confirms the low pressure. In regions of convective intrusion at depth and of crustal spreading, the extension of the crustal surface may give rise to tension and release of pressure at shallower depths. This pressure release could itself assist melting and encourage both the formation of magma and its escape to the surface (Yoder, 1952). The flood basalts of fissure eruptions and dyke-swarms may owe their existence as much to tension as to high geothermal gradient.

The other regions of increased temperature are those where mechanical energy is converted into heat. At the downcurrents, rigid plates of crust and mantle are being forced over engulfed material. At the interface there will be a great deal of friction and movement, with the result that folded mountain chains and tectonic mobilities occur in these areas. It is difficult to estimate the energy release in these regions, but Oxburgh and Turcotte (1968b) have calculated that shearing along the zone of movement could permit the temperature within the shear zone to reach 1400°C, sufficient for magmatism. Another approach is to consider the energy release necessary to explain the volcanic phenomena of a typical area, such as the circum-Pacific volcanic region. In this region the energy release accompanying crustal engulfment is partly from sudden shear dislocations and partly from creep and plastic deformation, and is in the form of seismic waves and local heating. It has been estimated that the energy of seismicity in the circum-Pacific region is of the order of 5×10^{24} erg/year. The energy required for vulcanism is at least an order of magnitude greater. The annual volume of volcanic ejecta and intrusives in the circum-Pacific region is probably of the order of $\frac{1}{2}$ km³/year. If, conservatively, it is assumed that, for the production of magma by partial fusion, five parts by volume of solid are heated to produce one part of liquid, and that the solid has to be raised through 300°C to reach the temperature of initial melting, then the production of $\frac{1}{2}$ km³ of magma requires about $3 \cdot 5 \times 10^{25}$ ergs (heat capacity 0·3 cal/g/°C; heat of fusion 100 cal/g; S.G. of solid, 3·2 and of liquid, 2·5). A comparable amount of energy is lost from the hydrothermal fields of volcanic areas. These have high surface heat flows, as well as heat loss by the emission of steam and hot water. Elder (1965) estimates that a world figure for geothermal areas is about 5×10^{10} cal/sec ($6 \cdot 5 \times 10^{25}$ erg/year); and probably at least 4×10^{25} ergs of this is in the circum-Pacific area, so that there the total heat loss by vulcanism and hydrothermal activity is perhaps about $7 \cdot 5 \times 10^{25}$ ergs. This value is of course highly speculative, but even allowing for gross discrepancies, it seems that the energy released as heat in the circum-Pacific areas is much greater than the seismic energy, perhaps 10-20 times greater, and is about 1% of the total heat loss from the earth by conductivity (9×10^{27} erg/year).

However, whether or not one accepts estimates of this sort, most geologists will accept that the regions where the circum-Pacific continents and island arcs appear to ride over the Pacific floor are places of extensive vulcanism and hydrothermal activity, and that presumably the energy source for these is mechanical. Similarly the Indonesian and other island arcs, and the Alpine-Himalayan mountain belt are regions of energy release.

In summary, there are two types of environment where thermal gradients are steepened by the effects of convective movement, and where magmatism is particularly abundant: (*i*) along mid-ocean rises and rift systems where hot mantle material is injected towards the surface; and (*ii*) along the zone of movement where rigid crustal plates override other sectors of the crust. In the latter, heated material consists of oceanic crust and sediments, mantle material and perhaps also some continental crust.

The mantle and its melting products

Geothermal gradients indicate that, under most conditions, magmas must originate not in the crust but in the upper mantle. Isotopic and phase equilibria data also indicate that basic and ultrabasic magmas are of mantle origin. Accordingly it is useful to summarise current views on the chemical and mineralogical composition of the upper mantle. The best available samples of mantle material appear to be alpine peridotites, the lherzolitic xenoliths often found in lavas and pyroclastic deposits of nephelinite and strongly alkalic basalt, and the garnet-peridotite xenoliths of kimberlite. The exact significance of these groups is still uncertain; to what extent they represent undifferentiated mantle or the refractory residue left after removal of a fusible liquid fraction is not known. However, they do provide some guide to the chemical composition of the upper mantle (Harris et al., 1967).

The mineralogy of the upper mantle has been studied both in ultrabasic xenoliths and in experimental systems, particularly at the Geophysical Laboratory in Washington and at the Australian National University in Canberra, and has been reviewed by Ringwood (1966) and Clark and Ringwood (1964). In summary, the uppermost mantle is thought to be lherzolitic with olivine (about Fo_{90}), aluminous ortho- and clinopyroxene and spinel, the spinel disappearing with increasing temperatures. At still greater depth, pyropic garnet becomes stable and the orthopyroxene becomes low in Al_2O_3, so that the mineralogy corresponds to that of garnet lherzolite xenoliths from kimberlite pipes. At much greater depth in the region from 400 to 1000 km, there is probably a complete change in mineralogy to high density phases. These changes include polymorphic transformations, for example of olivine to a spinel-type and pyroxene to an ilmenite-type of structure, and mineral reactions such as the break-down of magnesium silicates to form periclase and stishovite, a high-pressure polymorph of silica. However, none of these products has been found as xenoliths in any volcanic debris, indicating that the *final* escape or eruption of magma is never from such deep regions of the mantle. Even the garnet-peridotite zone is represented as xenoliths only in kimberlite and a very few basaltic centres.

The next point of interest is the composition of magma that would be derived from different levels in the mantle. Here it is important to emphasise that the major element composition of a magma escaping at the surface is determined by the final environment of equilibrium. No matter where the magma formed or what its initial composition, if it subsequently resides for a long period at some particular level in the mantle, and, by reaction, approaches equilibrium with the local solid phases, then its final composition is determined by this last period of mantle reaction and equilibration. This means that the major element composition of the magma reflects the final environment in which it approaches equilibrium with its surroundings prior to eruption, and tells us nothing of its earlier history. On the other hand, the relative abundances of some other elements, those that are concentrated in the residual liquid because they cannot enter the crystal phases, do indicate the previous history of the magma by showing how much mantle has been processed by the liquid, or how much fractionation of the liquid has occurred (Harris, in press). These ideas are implicit in the mechanism of zone refining and mantle reaction (Harris, 1957) subsequently renamed as wall-rock reaction (Green, D. H. and Ringwood, 1967a).

The compositions of magmas derived under different pressure conditions, equivalent to different depths in the mantle, have been discussed in a number of recent Annual Reports of the Geophysical Laboratory, Washington, and by D. H. Green and Ringwood (*ibid.*). In summary, an ultramafic mantle, on partial melting or liquid-solid equilibration, would be expected to yield tholeiites and olivine tholeiites at pressures of up to 20 kb (Kushiro, 1964), alkalic basalts between 20 and 30-40 kb (Kushiro, 1965; D. H. Green and

Ringwood, 1967a) and ultrabasic or picritic liquids above 30-40 kb (Davis and Schairer, 1965) in the garnet-peridotite zone.

The crust and its melting products

Beneath the oceanic basins, the crust consists of about $\frac{1}{2}$ km of sediments underlain by 5 km or more of material, usually thought to be basaltic corresponding to the rock types at present forming at the mid-ocean ridges. However, Hess (1962) has suggested that this layer may be dominantly serpentine, hydrated mantle material. Both serpentinite and basalt have been dredged from scarps on the ocean floor and probably this layer is a composite one containing both rock types; and these materials will be largely hydrated with high water contents. It has been suggested that the engulfment of this oceanic floor by the convective downcurrent to a high pressure environment would result in its conversion to quartz eclogite (Ringwood and D. H. Green, 1966; D. H. Green and Ringwood, 1967b) or amphibolite (T. H. Green and Ringwood, 1967). The partial fusion of these would then yield andesite (T. H. Green and Ringwood, 1966, 1967).

Along the circum-Pacific margins, the sediments being deposited on the continental slope and in the ocean trenches appear to be largely derived from the andesite vulcanism and to have the composition of greywacke and argillite with a high water content. Some of this sediment will be accreted to the continental plate but part is likely to be engulfed along with the rest of the ocean-floor material. With their high water contents, this sediment and the hydrated igneous material are likely to provide a lubricating zone which will act as the shear surface against the underside of the overriding continental plate. The partial fusion of sediments such as greywacke and argillite yields granitic liquids (Winkler, 1967; Winkler and Von Platen, 1961; Ewart and Stipp, 1968).

The other material possibly subjected to fusion in a convective downcurrent is continental crust. The upper crust is assumed to correspond in average composition to granodiorite though, in fact, the composition will vary with the geological environment; for example, a Precambrian shield compared with a younger orogenic zone. The composition of the lower crust is uncertain. Formerly it was automatically thought to be gabbroic and of the same composition as the lower oceanic crust. However, Ringwood and D. H. Green (1966) have shown that the ease of the gabbro-eclogite transformation is such that a basic lower crust could not remain in a gabbroic mineral facies but would recrystallise as eclogite. Since seismic velocities confirm that the lower crust is not eclogite, it follows that the lower crust is not basic, but rather intermediate in composition. There is some evidence that intermediate rocks in the granulitic or eclogitic facies have suitable densities and radioactivities for the lower crust (D. H. Green and Ringwood, 1967b). Again it would be expected that the lower crust would differ geochemically in different tectonic environments and especially from a Precambrian shield to a marginal area of geosynclinal deposits. In general, partial fusion of acid or intermediate rocks in the lower continental crust would be expected to produce a granitic or granodioritic liquid, though this might have geo-chemical differences from the granitic liquid formed by the fusion of grey-wackes. Doe (1967) has found systematic differences in lead isotopes in granites from the two environments.

Oceanic vulcanism

Along the crest of the Mid-Atlantic Ridge, the submarine vulcanism is from fissure eruptions rather than central vents, and is dominantly tholeiitic or sub-alkalic, although mildly alkalic types do occur. The tholeiites, which would be classed by some as high-alumina tholeiites, indicate shallow depths of melting or equilibration, and are consistent with the high heat flow of the ridge zone (Oxburgh and Turcotte, 1968a). Away from the central ridge zone, in regions of lower heat flow, the volcanic rocks appear to be dominantly from central vent eruptions and are alkalic, often extremely so (McBirney and Gass, 1967). Where tholeiitic rocks occur away from the ridge, for example in the Tertiary vulcanism of the Thulean province of the Faroes and north-western Britain, these can be explained as the result of a short-lived convective episode and its consequent temperature effects. This is supported by the crustal extension shown in the dyke-swarms of the Scottish Tertiary volcanic province. This embryonic line of convective upwelling crossed the main Mid-Atlantic Ridge system at Iceland.

On the islands and floor of the Atlantic Ocean, the bulk of the rocks are basic and of mantle origin. The proportion of salic derivatives is very small (Baker, 1968; Ridley, 1969) and can be attributed to the fractionation of basic parental magmas or remelting of previously solidified basalts. Some of the acid rocks of the British Tertiary volcanic province are apparently of crustal origin, but these may be ascribed to local melting of continental material under the influence of intruded basaltic magma (Moorbath and Bell, 1965).

In other mid-oceanic areas, such as the Pacific and Indian Oceans, the vulcanism is typically basaltic. The non-basaltic volcanic rocks make up only a very small proportion and are the result of fractionation of the basaltic parent magmas. In the Pacific it is not always possible to attribute the more tholeiitic or sub-alkalic rocks to continuous ridge areas; the Hawaiian islands for example lie on a more restricted rise. However, it is possible to attribute the more tholeiitic members to a place in time and, in a single eruptive centre, the rocks appear to become more fractionated and more alkalic with time. In Hawaii, the volcanic activity has migrated from the north-west end of the group to the south-east. Within individual centres, the earliest phases have been the most voluminous, and the most tholeiitic or least alkalic. Here one can assume a migrating convective upwelling beginning in the north-west of the group. At any one point, the earliest vulcanism will be at the time of convective penetration and maximum temperature, the magmatism diminishing and coming from a deeper level of final equilibration as the temperature returns to normal.

In general, the vulcanism within oceanic regions can be attributed to the partial fusion of mantle in a convective upcurrent; there is nothing to indicate any parental continental material.

Circum-Pacific vulcanism

In the island arcs and mountain chains of the Pacific margins, and in the similar island arcs of Indonesia, the Antilles, and the South Sandwich Islands, the vulcanism is of a vastly different type from that of oceanic areas. It is characteristically andesitic, though basalts, rhyolites and ignimbrites are also important. The more basic lavas are from central vent volcanoes with few fissure eruptions, presumably because this is a compressional environment and not a tensional one. The basalts tend to be high-alumina types although, across an island arc or mountain chain, from the outer (oceanic) to the inner side, there is often a progressive change from tholeiites to high-alumina basalts and then to alkali basalts. This is found in Japan (Kuno, 1967) and Indonesia, to a lesser extent in New Zealand and elsewhere. The basalts and andesites often overlap, even coming from the same centre at different times.

The rhyolites and ignimbrites often have a geological distribution distinct from that of the andesites. Although the rhyolites and ignimbrites are particularly conspicuous in the continental areas, they occur also in the Melanesian islands, and even in island arcs such as the South Sandwich and Tonga Islands. A general feature of the andesites and ignimbrites is their strong pyroclastic or explosive tendencies, attributable to a high water content. The most recent discussion of the petrogenesis and geochemistry of a typical suite of andesites and rhyolites is that of Ewart and Stipp (1968).

The chemical features of the circum-Pacific volcanic rocks can be related to the chemical and physical environment of melting along the zone of energy release. The downward engulfment or ingestion of ocean floor beneath the overthrusting continental plate may carry with it four possible types of material: normal mantle; igneous ocean floor of basalt or serpentinite; sediments such as greywacke; and fragments of the continental crust from the overthrust plate (Coats, 1962; Isacks et al., 1968). Melting of these in the downcurrent will provide three dominant magma types, basalt, andesite and rhyolite. The andesite and rhyolite, being derived from hydrous parental material, will have high water contents which will influence their mode of eruption or intrusion (Harris et al., in press). Andesite will be strongly pyroclastic but not plutonic; the rhyolitic magma, however, will crystallise as plutons or escape to the surface as domes or plugs of highly viscous glass, or explosively as gas-fluidised systems in which the gas phase carries with it the quenched (glassy) rhyolitic liquid. Because each of the three primary magma types has a different liquidus temperature, and is formed from different materials in different parts of the downlimb, the magmas will remain as relatively discrete types with few intermediates. There will be a change in magma type with depth of melting of the ingested material, hence the change in basalt magma type across an island arc.

Where engulfment of the crust occurs in an intra-continental environment, as one continental plate overrides or is pushed against another, the engulfed material will contain mantle and continental crust. The latter is likely to be granitic or granulitic material rather than sedimentary and its engulfment and partial fusion should give rise to granitic liquids, presumably relatively dry. Granitic magmas of this type would be injected into the crust as batholithic bodies, or ring dykes with associated rhyolites but normally not ignimbrites. Ignimbrites require a high water pressure for eruption and are more likely to be formed in hydrous environments associated with the fusion of sediments.

Outside these principal regions of high heat flow—the mid-ocean ridges and the areas of overriding and engulfment—thermal gradients are relatively low. Any magmatism must be much more deep-seated in origin, and much more sporadic and less predictable in occurrence. Vulcanism does occur outside these two regions, as demonstrated by the several volcanic islands in the Atlantic well removed from the mid-ocean ridge. These are central vent volcanoes built of highly alkalic rock-types such as nephelinites. Volcanic rocks also have been injected in the past, in continental areas of low geothermal gradient, namely the Precambrian cratons. Here again, the typical rocks are highly alkalic, melilitites, lamprophyres and kimberlites.

All these highly alkalic rocks in both oceanic and continental areas are likely to have been derived from considerable depth and, in the case of kimberlite, perhaps from depths of 100-150 km, the depth of diamond stability at magmatic temperatures.

Applications to African vulcanicity

In the preferred convective model involving the wedging apart of rigid plates, the main feature, apart from the tectonic implications, is that magmatism of a dominantly basaltic nature is mantle-derived and therefore likely to be associated with a penetrative upwell, while magmatism of andesitic-rhyolitic type is due to the ingestion of oceanic crust. Granitic or rhyolitic bodies with no associated diagnostic rocks could be derived from the melting of continental crust in either environment. This may be true of past vulcanism as well as present, and the following discussion attempts to look at vulcanism in Africa in this context.

The most conspicuous volcanic zone in Africa is that of the Red Sea–East Africa Rift System (this volume: King, p. 263; Gass, p. 285). The change in volcanic type from north to south along the rift zone has been discussed recently (Harris, in press) but is reviewed briefly here. There is good evidence that the Red Sea is a region of continental separation, initiated in Tertiary times (Girdler, 1958; Gass and Gibson, 1969). Heat flow is high, and presumably the region overlies a convective upwell. Within the median trough of the Red Sea, the volcanic rocks are tholeiitic, but outside this trough and in the centres of the southern Arabian coast, the basalts are mildly alkalic (Gass, this volume, p. 288).

Southward through Ethiopia, the basalts again are mildly alkalic, but further south in Kenya, Uganda and Tanzania, are strongly alkalic, including such types as nephelinites, basanites, and are associated with carbonatites and carbonatitic lavas (King, this volume, p. 273). However, within the region less alkalic basalts also occur; for example at Rungwe where the Eastern and Western Rifts join at the northern end of Lake Malawi, the volcanic rocks include some with the chemical composition of olivine tholeiites. Also at the south end of Lake Kivu, at the intersection of the Albert and Tanganyika Rift Systems, some basalts have the chemical composition of silica-saturated tholeiites. In general, it seems that vulcanism is more likely at rift intersections and the volcanic products more abundant and less alkalic than elsewhere in the rift systems.

In the rift areas, the rocks are dominantly basic with relatively few salic derivatives, except in the Ethiopian Rift, and the phonolite sheets which are widespread in Kenya. Andesites of calcic or circum-Pacific type do not occur.

The high elevation of 1-2 km of the sides of the rift valleys above the general ground level gives a resemblance in cross-section to a mid-ocean ridge system. Presumably the upward swelling or distension of the surface is a thermal effect caused by an expansion from a high-density to a low-density mineral assemblage in the presence of abnormally high local temperatures at depth. In the view of Cloos (1939), rifting will occur along the crest of swells, because of the failure of the surface to stretch enough to provide for the distension. In Africa, this mechanism may have initiated the rifting, but the width of the rift valleys suggests some separation as well. This slight movement apart appears to have been arrested at an early stage, and only at the northern end of the rift system, in Ethiopia, are there signs of present crustal spreading (Gass and Gibson, 1969). Cloos suggested that at the regions of most extensive swelling or distension, more than one rift system may occur, the systems intersecting at the crests of the swells. If the degree of distension is related to the temperature effect at depth, the higher the temperature the greater would be the swelling, and the greater the possibility of intersecting rift systems; also the higher the temperature the more voluminous and the less alkalic the vulcanicity. Hence the possible relationship between the rift intersections and the nature of the vulcanism (Harris, in press).

Because the volcanic rocks are dominantly basaltic, derived from the mantle, and because of the structural analogies with mid-ocean rises and rifts, the African Rift System, like the Red Sea Rift, is deduced to be the site of a penetrative convective upwell. However, the absence of any large horizontal movement of the rift walls away from each other indicates that the pressure inwards towards Africa from the mid-Atlantic and mid-Indian Ocean convective upwells has been sufficient to prevent significant crustal extension in the African Rift Systems. The effects of this African convective upwell have been thermal but have not caused crustal spreading. The result could be considered as an arrested convective system. Only in the extreme north, where the mid-Indian Ocean Rift System joins the African Rift System, have the continental plates moved apart, to form the Red Sea (see Gass, this volume, p. 287).

P

Near the rift valleys of Kenya and Uganda there are extensive phonolite sheets. Their volume and lack of associated rocks of an obviously parental nature, suggest that the phonolitic magma is itself primary, formed by anatexis under the thermal influence of the same convective episode. The magma composition is not consistent with a direct origin from the mantle, so the anatexis must have been within the crust, perhaps the secondary melting of crustal material by ascending basalt. An analogy can be made with the granites and pitchstones of the Scottish Tertiary province which have also been attributed to secondary heating by ascending basalt magma. Normally, anatexis within the crust yields granitic not phonolitic liquids, especially in continental margins. The very extensive development of phonolite sheets indicates that this central continental region has an unusual crustal geochemistry. Presumably the anatexis has occurred in a lower crust, not silica-saturated as normal but silica-deficient and containing normative nepheline. Such a crustal composition could be developed in continental shields by fenitisation and alkali metasomatism, or by the accumulation of nepheline basalts at depth. Remelting of this material would yield phonolitic magma.

A further volcanic zone in which the magmatism is basaltic, and therefore mantle-derived, is the line Annobon, Saõ Thomé, Principe, Fernando Po and Mt. Cameroon (see p. 194). Is this also a convective upwell which has a strong thermal effect, but has not caused lateral extension of the crust?

Turning to older volcanic episodes, the tholeiitic lavas and intrusive rocks of the Stormberg and Karroo series suggest a period of shallow mantle melting and steepened geothermal gradient (this volume: Cox, p. 211; Woolley and Garson, p. 259). Like the Deccan Traps these rocks also indicate crustal extension by the presence of dyke-swarms, and mark the initial break-up of Gondwanaland. One can ascribe this period of vulcanicity to the increase in temperature around the continental margin following the beginning of the convective penetration, and to the effects of tension and pressure release on the crust and upper mantle.

Outside those regions of basalt vulcanism that can be attributed to the thermal effects of penetrative convection, one other rock-type is of interest. A very large number of kimberlite pipes and dykes has been described especially from southern, central and western Africa (see Dawson, this volume, p. 321). Nearly all occur in Precambrian cratons, the most diamondiferous tending to be in cratons older than 1500 m.y. (Clifford, 1966). This geological environment is characterised by its low heat flow and therefore low geothermal gradient. Here, magmas could be generated only at great depths in the mantle. Also some of the minerals in kimberlite, particularly diamond, could have formed only at depths of 150 km or more. It has been suggested that the low geothermal gradient and great depth of origin are the major factors in controlling the petrogenesis of kimberlite magma and its mode of intrusion (Harris, 1968). So, when magmatism does occur in such a geothermal environment, kimberlites or related rocks will be the expected products. Although many of the kimberlites of South Africa were intruded in the Cretaceous, the emplacement was in the solid state by gas fluidisation. In some cases the initial solidification appears to have been much earlier, and deep within the mantle below the depth required for diamond stability. This requires some mechanism for reactivation of the solidified kimberlite and its ejection to the surface. Since intrusion is by gas fluidisation, the only likely means of reactivation is by increasing the gas pressure in the solidified kimberlite (see Dawson, this volume, p. 332). This can be achieved by increasing the regional temperature.

It is suggested that the initial thermal effects of the convective movement causing the disruption of Gondwanaland were high-level, leading to the formation of the tholeiites of the Karroo vulcanism in the late Trias or early Jurassic. These might be followed in time or at a distance by increasingly fractionated basalts such as those of the Nuanetsi area (Cox et al., 1967) derived from deeper levels or colder environments. Finally, the thermal effects of the convective movement would reach even deeper levels within the craton, causing the reactivation of the solidified kimberlite, and its injection into the crust, sometimes into the older Karroo basalts and dolerites.

REFERENCES

BAKER, I. 1968. Intermediate oceanic volcanic rocks and the 'Daly Gap'. Earth Planetary Sci. Let., 4, 103.

BIRCH, F. 1965. Speculations on the Earth's thermal history. Bull. geol. Soc. Am., 76, 133.

—— 1968. On the possibility of large changes in the earth's volume. Phys. Earth Planetary Inter., 1, 141.

CLARK, S. P., and RINGWOOD, A. E. 1964. Density distribution and constitution of the mantle. Rev. Geophys., 2, 35.

CLIFFORD, T. N. 1966. Tectono-metallogenic units and metallogenic provinces of Africa. Earth Planetary Sci. Let., 1, 421.

CLOOS, H. 1939. Hebung–Spaltung–Vulkanismus. Geol. Rdsch., 30, 405.

COATS, R. R. 1962. Magma type and crustal structure in the Aleutian Arc. In The Crust of the Pacific Basin. (Eds. G. A. MacDonald and H. Kuno), Monog. Am. geophys. Un., 6, 92.

COX, K. G., MACDONALD, R., and HORNUNG, G. 1967. Geochemical and petrographic provinces in the Karroo basalts of southern Africa. Am. Miner., 52, 1451.

DAVIS, B. T. C., and SCHAIRER, J. F. 1965. Melting relations in the join diopside–forsterite–pyrope at 40 kilobars and one atmosphere. Yb. Carneg. Instn., 64, 123.

DEARNLEY, R. 1967. Orogenic fold belts and a hypothesis of earth evolution. Phys. Chem. Earth, 7, 3.

DIETZ, R. S. 1961. Continent and ocean basin evolution by spreading of the sea floor. Nature, Lond., 190, 854.

DOE, B. R. 1967. The bearing of lead isotopes on the source of granitic magma. J. Petrology, 8, 51.

ELDER, J. W. 1965. Physical processes in geothermal areas. In Terrestrial Heat Flow (Ed. W. H. K. Lee), Monog. Am. geophys. Un., 8, 211.

ELSASSER, W. M. 1968. Pattern of convective creep in the earth's mantle. Unpublished paper, N.A.T.O. Conf. Palaeogeophys., Univ. Newcastle.

ENGEL, A. E. J., ENGEL, C. G., and HAVENS, R. G. 1965. Chemical characteristics of oceanic basalts and the upper mantle. Bull. geol. Soc. Am., 76, 719.

EWART, A., and STIPP, J. J. 1968. Petrogenesis of the volcanic rocks of the central North

Island, New Zealand, as indicated by a study of Sr^{87}/Sr^{86} ratios, and Sr, Rb, K, U, and Th abundances. *Geochim. et cosmochim. Acta*, **32**, 699.

GASS, I. G., and GIBSON, I. L. 1969. The structural evolution of the rift zones in the Middle East. *Nature, Lond.*, **221**, 926.

GIRDLER, R. W. 1958. The relationship of the Red Sea to the East African Rift System. *Q.J. geol. Soc. Lond.*, **114**, 79.

GREEN, D. H., and RINGWOOD, A. E. 1967a. The genesis of basaltic magmas. *Contr. Miner. Pet.* **15**, 103.

—— and RINGWOOD, A. E. 1967b. An experimental investigation of the gabbro to eclogite transformation and its petrological applications. *Geochim. et cosmochim. Acta*, **31**, 767.

GREEN, T. H., and RINGWOOD, A. E. 1966. Origin of the calc-alkaline igneous rock suite. *Earth Planetary Sci. Let.*, **1**, 307.

—— and RINGWOOD, A. E. 1967. Crystallisation of basalt and andesite under high-pressure hydrous conditions. *Earth Planetary Sci. Let.*, **3**, 481.

HALLAM, A. 1967. The bearing of certain palaeozoogeographic data on continental drift. *Palaeogeog. Palaeoclim. Palaeoecol.*, **3**, 201.

HARRIS, P. G. 1957. Zone refining and the origin of potassic basalts. *Geochim. et cosmochim. Acta*, **12**, 195.

—— 1968. Genesis of kimberlite. *12th Ann. Rep. Res. Inst. African Geol., Univ. Leeds*, 26.

—— (in press). Basalt type and African Rift Valley tectonism. *Tectonophys.*

—— KENNEDY, W. Q., and SCARFE, C. M. (in press). Volcanism versus plutonism—the effect of chemical composition. In *Mechanism of igneous intrusion*, (Ed. N. Rast), Liverp. geol. Soc.

—— REAY, A., and WHITE, I. G. 1967. Chemical composition of the upper mantle. *J. geophys. Res.*, **72**, 6359.

HESS, H. H. 1962. History of ocean basins. In *Petrologic studies: a volume to honor A. F. Buddington*, p. 599. (Eds. A. E. J. Engel, H. L. James and B. F. Leonard), Geol. Soc. Am.

ISACKS, B., OLIVER, J., and SYKES, L. R. 1968. Seismology and the new global tectonics. *J. geophys. Res.*, **73**, 5855.

KNOPOFF, L. 1967. Thermal convection in the Earth's Mantle. In *The Earth's Mantle*, p. 171 (Ed. T. F. Gaskell), London.

KUNO, H. 1967. Volcanological and petrological evidences regarding the nature of the upper mantle. In *The Earth's Mantle*, p. 89. (Ed. T. F. Gaskell), London.

KUSHIRO, I. 1964. The system of diopside-forsterite-enstatite at 20 kilobars. *Yb. Carneg. Instn.*, **63**, 101.

—— 1965. The liquidus relations in the systems forsterite-$CaAl_2SiO_6$-silica and forsterite-nepheline-silica at high pressures. *Yb. Carneg. Instn.*, **64**, 103.

LEE, W. H. K., and UYEDA, S. 1965. Review of heat flow data. In *Terrestrial Heat Flow* (Ed. W. H. K. Lee), *Monog. Am. geophys. Un.*, **8**, 87.

MacDONALD, G. J. F. 1965. Geophysical deductions from observations of heat flow. In *Terrestrial Heat Flow* (Ed. W. H. K. Lee), *Monog. Am. geophys. Un.*, **8**, 191.

McBIRNEY, A. R., and GASS, I. G. 1967. Relations of oceanic volcanic rocks to mid-oceanic rises and heat flow. *Earth Planetary Sci. Let.*, **2**, 265.

MELSON, W. G., THOMPSON, G., and VAN ANDEL, T. H. 1968. Volcanism and metamorphism in the mid-Atlantic ridge, 22° N. latitude. *J. geophys. Res.*, **73**, 5925.

MOORBATH, S., and BELL, J. D. 1965. Strontium isotope abundance studies and rubidium-strontium age determinations on Tertiary igneous rocks from the Isle of Skye, north-west Scotland. *J. Petrology*, **6**, 37.

MORGAN, W. J. 1968. Rises, trenches, great faults, and crustal blocks. *J. geophys. Res.*, **73**, 1959.

OROWAN, E. 1965. Convection in a non-Newtonian mantle, continental drift, and mountain building. *Phil. Trans. r. Soc. Lond.*, **258**, Ser. A, 284.

OXBURGH, E. R., and TURCOTTE, D. L. 1968a. Mid-ocean ridges and geotherm distribution during mantle convection. *J. geophys. Res.*, **73**, 2643.

—— and TURCOTTE, D. L. 1968b. Problem of high heat flow and volcanism associated with zones of descending mantle convective flow. *Nature, Lond.*, **218**, 1041.

RIDLEY, I. 1969. The abundance of rock types on Tenerife and its petrogenetic significance. Abstract. *Trans. Am. geophys. Un.* **50**, 341.

RINGWOOD, A. E. 1966. Mineralogy of the mantle. In *Advances in Earth Science*, p. 357. (Ed. P. M. Hurley), Cambridge, U.S.A.

—— and GREEN, D. H. 1966. An experimental investigation of the gabbro-eclogite transformation and some geophysical implications. *Tectonophys.*, **3**, 383.

RUNCORN, S. K. 1965. Changes in the convection pattern in the Earth's mantle and continental drift: evidence for a cold origin of the Earth. *Phil. Trans. r. Soc. Lond.*, **258**, Ser. A, 228.

TOZER, D. C. 1965. Heat transfer and convection currents. *Phil. Trans. r. Soc. Lond.*, **258**, Ser. A, 252.

VERHOOGEN, J. 1965. Phase changes and convection in the Earth's mantle. *Phil. Trans. r. Soc. Lond.*, **258**, Ser. A, 276.

WEGENER, A. 1966. *The origin of continents and oceans*, 4th ed., London.

WILSON, J. T. 1965. Evidence from oceanic islands suggesting movement in the Earth. *Phil. Trans. r. Soc. Lond.*, **258**, Ser. A, 145.

WINKLER, H. G. F. 1967. *Petrogenesis of metamorphic rocks*, 2nd ed., Berlin.

—— and VON PLATEN, H. 1961. Experimentelle Gesteinsmetamorphose IV. Bildung anatektischer Schmelzen aus metamorphisierten Grauwacken. *Geochim. et cosmochim. Acta*, **24**, 48.

YODER, H. S. 1952. Change of melting point of diopside with pressure. *J. Geol.*, **60**, 364.

NAME INDEX

Page numbers in roman type indicate mention in the text; those in italic type indicate lists of references at the end of each chapter.

SUBJECT INDEX

449

Printed in Great Britain
by T. and A. CONSTABLE LTD., Hopetoun Street.
Printers to the University of Edinburgh